Communications in Computer and Information Science 2811

Series Editors

Rationale
The CCIS series is devoted to the publication of proceedings of computer science conferences. Its aim is to efficiently disseminate original research results in informatics in printed and electronic form. While the focus is on publication of peer-reviewed full papers presenting mature work, inclusion of reviewed short papers reporting on work in progress is welcome, too. Besides globally relevant meetings with internationally representative program committees guaranteeing a strict peer-reviewing and paper selection process, conferences run by societies or of high regional or national relevance are also considered for publication.

Topics
The topical scope of CCIS spans the entire spectrum of informatics ranging from foundational topics in the theory of computing to information and communications science and technology and a broad variety of interdisciplinary application fields.

Information for Volume Editors and Authors
Publication in CCIS is free of charge. No royalties are paid, however, we offer registered conference participants temporary free access to the online version of the conference proceedings on SpringerLink (http://link.springer.com) by means of an http referrer from the conference website and/or a number of complimentary printed copies, as specified in the official acceptance email of the event.

CCIS proceedings can be published in time for distribution at conferences or as post-proceedings, and delivered in the form of printed books and/or electronically as USBs and/or e-content licenses for accessing proceedings at SpringerLink. Furthermore, CCIS proceedings are included in the CCIS electronic book series hosted in the SpringerLink digital library at http://link.springer.com/bookseries/7899. Conferences publishing in CCIS are allowed to use our online conference service (Meteor) for managing the whole proceedings lifecycle (from submission and reviewing to preparing for publication) free of charge.

Publication process
The language of publication is exclusively English. Authors publishing in CCIS have to sign the Springer CCIS copyright transfer form, however, they are free to use their material published in CCIS for substantially changed, more elaborate subsequent publications elsewhere. For the preparation of the camera-ready papers/files, authors have to strictly adhere to the Springer CCIS Authors' Instructions and are strongly encouraged to use the CCIS LaTeX style files or templates.

Abstracting/Indexing
CCIS is abstracted/indexed in DBLP, Google Scholar, EI-Compendex, Mathematical Reviews, SCImago, Scopus. CCIS volumes are also submitted for the inclusion in ISI Proceedings.

How to start
To start the evaluation of your proposal for inclusion in the CCIS series, please send an e-mail to ccis@springer.com

Umapada Pal · Qiang Wu · Si Liu
Editors

Computing and Pattern Recognition

14th International Conference, ICCPR 2025
Beijing, China, October 24–26, 2025
Revised Selected Papers, Part I

Editors
Umapada Pal
Indian Statistical Institute
Kolkata, West Bengal, India

Qiang Wu
Beijing University of Technology
Beijing, China

Si Liu
Beihang University
Beijing, China

ISSN 1865-0929 ISSN 1865-0937 (electronic)
Communications in Computer and Information Science
ISBN 978-981-95-8314-0 ISBN 978-981-95-8315-7 (eBook)
https://doi.org/10.1007/978-981-95-8315-7

This Springer imprint is published by the registered company Springer Nature Singapore Pte Ltd.
The registered company address is: 152 Beach Road, #21-01/04 Gateway East, Singapore 189721, Singapore

Preface

The 14th International Conference on Computing and Pattern Recognition (ICCPR 2025), sponsored by Beijing University of Technology and endorsed by the International Association of Pattern Recognition (IAPR), was successfully held in Beijing, China, during October 24–26, 2025. As one of the leading forums in the field, ICCPR continues to serve as an international platform for researchers, academics, and practitioners to share recent advances, exchange innovative ideas, and explore emerging challenges in computing, pattern recognition, computer vision, and related domains.

Mohamed-Slim Alouini from King Abdullah University of Science and Technology, Saudi Arabia, Prof. Qingshan Liu from Nanjing University of Posts and Telecommunications, China and Zhaoxiang Zhang from Chinese Academy of Sciences, China, delivered wonderful keynote speeches. The valuable contributions of these distinguished scholars greatly enhanced the quality of the conference and encouraged further exploration and collaboration in the respective research fields.

These two volumes contain 82 high-quality papers selected for presentation at ICCPR 2025. The conference received 205 submissions, and all manuscripts underwent a strict double-blind evaluation process assessing originality, technical soundness, clarity, and thematic relevance, with at least two reviewers. The accepted papers represent both foundational studies and practical applications, showcasing current advances in methodologies, models, and systems across diverse research areas.

To provide better thematic coherence for readers, the papers have been organized into the following 12 chapters:

I. Pattern Recognition
II. Image Models and Image Reconstruction
III. Image Enhancement and Image Segmentation
IV. Intelligent Detection Models and Algorithms
V. Image-Based System Anomaly Detection Technology and Engineering Applications
VI. Key Technologies of Computer Vision and Image Processing Based on Multimodal Fusion
VII. Computer Models and Optimization Computation
VIII. Data Models and Data Computation
IX. Machine Learning and Deep Learning Models
X. Large-Scale Language Models and Multimodal Semantic Analysis
XI. Fault Detection and Security Analysis in Complex Systems
XII. Emerging Network Technologies and Signal Analysis

The wide spectrum of these topics illustrates the rapid progress and interdisciplinary nature of modern computing and pattern recognition research. The works included not only address classical challenges but also highlight novel approaches inspired by artificial intelligence, deep learning, and multimodal technologies.

We would like to express our sincere gratitude to all authors for their valuable contributions, and to the program committee members and reviewers for their dedication in ensuring the quality of the papers. We also thank the organizing committee, volunteers, and institutional partners for their tireless efforts, which made the conference and these proceedings volumes possible.

We hope that this collection of papers will serve as a useful reference for researchers, students, and practitioners, and will inspire further advancements in the field.

October 2025

Umapada Pal
Qiang Wu
Si Liu

Organization

Advisory Chair

David Zhang	Chinese University of Hong Kong (Shenzhen), China

General Chairs

Umapada Pal	Indian Statistical Institute, Kolkata, India
Qiang Wu	Beijing University of Technology, China

General Co-chairs

Zhitao Xiao	Tiangong University, China
Xiaojun Wu	Jiangnan University, China
Sushmita Mitra	Indian Statistical Institute, Kolkata, India

Program Committee Chairs

Yong Yang	Tiangong University, China
João Paulo Papa	São Paulo State University, Brazil
Kenji Suzuki	Tokyo Institute of Technology, Japan
Mounim A. El Yacoubi	Institut Polytechnique de Paris, France

Program Committee Co-chairs

Shengjin Wang	Tsinghua University, China
Sen Jia	Shenzhen University, China
Si Liu	Beihang University, China

Publication Chairs

Guangxu Li	Tiangong University, China
Huiyu Zhou	University of Leicester, UK

Publicity Chairs

Chaoyang Chen	Wayne State University, USA
Lin Zhang	Tongji University, China
Yanbei Liu	Tiangong University, China
Yatong Zhou	Hebei University of Technology, China

Technical Program Committee

Min-Ling Zhang	Southeast University, China
Mehmet Celenk	Ohio University, USA
Lihong Connie Li	City University of New York, USA
Wangmeng Zuo	Harbin Institute of Technology, China
Xu Chi	China University of Geosciences, Wuhan, China
Yongqiang Zhao	Northwestern Polytechnical University, China
Shaohui Liu	Harbin Institute of Technology, China
Mohamed Ismail Roushdy	Ain Shams University, Egypt
Huafeng Qin	Chongqing Technology and Business University, China
Ghazaleh Khodabandelou	University of Paris-Est Créteil, France
Olfa Jemai	University of Gabès, Tunisia
Mohamed Ibn Khedher	IRT SystemX, France
Chengjun Liu	New Jersey Institute of Technology, USA
Yun Tie	Zhengzhou University, China
Claudio Cusano	University of Pavia, Italy
Sai-Keung Wong	National Yang Ming Chiao Tung University, Taiwan
Weike Pan	Shenzhen University, China
Qiang Guo	Shandong University of Finance and Economics, China
Xiaofeng Zhang	Harbin Institute of Technology, China
Liang Ming	National Key Laboratory of Science and Technology on Information System Security, China

Ha Viet Uyen Synh	International University Ho Chi Minh City, Vietnam
Xiaoshuai Sun	Harbin Institute of Technology, China
Hsiang-Chieh Chen	National United University, Taiwan
Chun-Jen Tsai	National Yang Ming Chiao Tung University, Taiwan
Lim Seng Poh	Universiti Tunku Abdul Rahman, Malaysia
Lan Yao	Chengdu University of Information Technology, China
George Azzopardi	University of Groningen, The Netherlands
Yahui Liu	Shenzhen University, China
Weicheng Xie	Shenzhen University, China
Junmei Zhong	Marchex, Inc, USA
Boshi Huang	KLA-Tencor Corporation, USA
Lei Zhen	Chinese Academy of Sciences, China
Mu-Song Chen	Da-Yeh University, Taiwan
Luisito Lolong Lacatan	University of Cabuyao, Philippines
Lihong Zheng	Charles Sturt University, Australia
Du Huynh	University of Western Australia, Australia
Yanming Zhang	Chinese Academy of Sciences, China
Raja Kumar Murugesan	Taylor's University, Malaysia
Kai Yang	China Mobile Research Institute, China
Lei Geng	Tiangong University, China
Gaofeng Meng	Chinese Academy of Sciences, China
Xuyao Zhang	Chinese Academy of Sciences, China
Zhendong Guo	Nanyang Technological University, Singapore
Kunhong Liu	Xiamen University, China
Lizhong Yao	Chongqing University of Science and Technology, China
Wei Li	Xi'an University of Technology, China
Mohd Nazri Bin Ismail	Universiti Pertahanan Nasional Malaysia, Malaysia
Edwin Lughofer	Johannes Kepler University Linz, Austria
Hai-Ning Liang	Xi'an Jiaotong-Liverpool University, China
Rocio Gonzalez-Diaz	University of Seville, Spain
Guanglan Zhang	Boston University, USA
Minh-Tien Nguyen	Hung Yen University of Technology and Education, Vietnam
Guoqing Li	Chinese Academy of Sciences, China
Chuang Zhu	Beijing University of Posts and Telecommunications, China
Cai Chengtao	Harbin Engineering University, China

Dehong Qiu	Huazhong University of Science and Technology, China
Jaroslaw Krzywanski	Jan Długosz University in Częstochowa, Poland
Tse Guan Tan	Universiti Malaysia Kelantan, Malaysia
Zhang Wen	Icahn School of Medicine at Mount Sinai, USA
Liu Jianming	Jiangxi Normal University, China
Shurui Fan	Hebei University of Technology, China
Si Chen	Xiamen University of Technology, China
Kurban Ubul	Xinjiang University, China
Tan Chi Wee	Tunku Abdul Rahman University College, Malaysia
Danping Liu	Chongqing University, China
Ming Cheng	Xiamen University, China
Feng Chen	Xiamen University of Technology, China
Fangmei Chen	Dalian Minzu University, China
Chaoqun Hong	Xiamen University of Technology, China
Xujing Huang	Xiamen University of Technology, China
Yan Yan	Xiamen University, China
Guoqiang Zhong	Ocean University of China, China
Bing Zhang	Shanghai University, China
Yan Wang	East China Normal University, China
Chunjie Zhang	Beijing Jiaotong University, China
Shaohui Liu	Harbin Institute of Technology, China
Evgin Goceri	Akdeniz University, Turkey
Yu Li	Beijing University of Technology, China
Yafang Li	Beijing University of Technology, China
Zahid Akhtar	State University of New York Polytechnic Institute, USA
Jonghwa Kim	University of Augsburg, Germany
Xiwen Zhang	Beijing Language and Culture University, China
Sujing Wang	Lamar University, USA
Ji-hwan Woo	Shinhan Bank, South Korea
Shinan Lang	Beijing University of Technology, China
Ferran Torrent	Aimsun SLU, Spain
Gulustan Dogan	University of North Carolina Wilmington, USA
Francisco García-Sánchez	University of Murcia, Spain
Hong He	University of Shanghai for Science and Technology, China
Lingxia Lu	Zhejiang University, China
Li Fang	Nanyang Technological University, Singapore
Pratheepan Yogarajah	Ulster University, UK
Qijun Zhao	Sichuan University, China

Mei Wang	Xi'an University of Science and Technology, China
Wanli Xue	Tianjin University of Technology, China
Ali Reza Alaei	Southern Cross University, Australia
Bi-Ru Dai	National Taiwan University of Science and Technology, Taiwan
Srikanta Pal	Maynooth University, Ireland
Palainakote Shivakumara	University of Malaya, Malaysia
Xianhong Chen	Beijing University of Technology, China
Rajkumar Saini	Luleå University of Technology, Sweden
Pierrick Bruneau	Luxembourg Institute of Science and Technology, Luxembourg
Zhonghua Sun	Beijing University of Technology, China
Jian Chang	Bournemouth University, UK
Tianyang Xu	Jiangnan University, China
Yong Ma	Jiangsu Normal University, China
Tian Song	Tokushima University, Japan
Xin Song	National University of Defense Technology, China
Xiaoning Li	Sichuan Normal University, China
Hong Zhang	Beihang University, China
Qiuming Liu	Jiangxi University of Science and Technology, China
Foteini Liwicki	Luleå University of Technology, Sweden
Jiaxin Cai	Xiamen University of Technology, China
Tielin Zhang	Chinese Academy of Sciences, China
Huabao Qiang	Guilin University of Electronic Technology, China
Jia Guo	Tianjin Normal University, China
Han Yan	Beijing Institute of Control Engineering, China
Carlo Sansone	Università degli Studi di Napoli Federico II, Italy
Yuanyuan Gao	Qingdao Agricultural University, China
YangLi Jia	Liaocheng University, China
Raju Shrestha	Oslo Metropolitan University, Norway
Daniel P. Lopresti	Lehigh University, USA
Sukalpa Chanda	Østfold University College, Norway
Kenny Davila	DePaul University, USA
Palaiahnakote Shivakumara	University of Malaya, Malaysia
Robert Sablatnig	Technische Universität Wien, Austria
Mickaël Coustaty	La Rochelle University, France
Oriol Ramos Terrades	Universitat Autònoma de Barcelona, Spain
Ventzeslav Valev	Institute of Mathematics and Informatics, Bulgaria

Xiaoyi Jiang	University of Münster, Germany
K. V. Krishna Kishore	Vignan's Foundation for Science, Technology and Research, India
Haixin Wang	Fort Valley State University, USA
Xinxing Xia	Shanghai University, China
Tan Tse Guan	Universiti Malaysia Kelantan, Malaysia
Dongzhi Zhang	China University of Petroleum (East China), China
Xiulan Sun	Jiangnan University, China
Hui Li	Jiangnan University, China
Mingyue Niu	Yanshan University, China
Yew Kee Wong Eric	Hong Kong Chu Hai College, China
P.Velayutham	Aarupadai Veedu Institute of Technology, India
Nittaya Kerdprasop	Suranaree University of Technology, Thailand
Mukul Joshi	Nvidia, USA
Shahzad Ashraf	Gachon University, South Korea
Zejun Wang	China University of Petroleum, China

Contents

Intelligent Detection Models and Algorithms

Image-Based System Anomaly Detection Technology and Engineering Applications

Pattern Recognition

Modality-Specific and Modality-Agnostic Soft Contrastive Learning for Unsupervised Visible-Infrared Person Re-identification

Cong Zhang, Tao Wang(✉), Yanzhao Su, Nian Wang, Yunwei Lan, and Aihua Li

Rocket Force University of Engineering, Xi'an 710025, China
twang305@126.com

Abstract. Unsupervised visible-infrared person re-identification (USL-VI-ReID) presents significant challenges in cross-modality image matching without annotated training data. Recently, pseudo-label-based methods have emerged as a promising approach in USL-VI-ReID. However, existing methods ignore the discrepancy in the feature distribution caused by random data augmentation, resulting in noisy pseudo labels. Worse still, they usually explore the correspondence between the two modalities at modality-specific cluster-level, leading to insufficient cross-correspondences. In response, we propose a Modality-specific and Modality-agnostic Soft Contrastive Learning (MMSCL) method. Specifically, we design a modality-agnostic hybrid memory to initialize modality-shared prototypes to mitigate the impact of the cross-modality gap. In addition, we introduce soft contrastive learning with a mean-teacher framework to constrain the consistency of the feature distribution. Moreover, an Adaptive-weighted Memory Updating module is proposed to tackle the discrepancy distribution problem in the batch training process. Extensive experimental results on two benchmark datasets demonstrate that our proposed method achieves state-of-the-art performance.

Keywords: USL-VI-ReID · Contrastive learning · Cross modality

1 Introduction

The target of visible-infrared person re-identification (VI-ReID) [3,14,22,29,32,35] is to match person images captured by visible/infrared cameras corresponding to a given query from another modality. This task has attracted more attention recently due to its widespread application in intelligent surveillance systems. Existing VI-ReID methods [6,14,24] have achieved remarkable performance with deep learning methods. However, they heavily rely on well-annotated

P. Umapada et al. (Eds.): ICCPR 2025, CCIS 2811, pp. 3–17, 2026.
https://doi.org/10.1007/978-981-95-8315-7_1

cross-modality data, which is time-consuming and laborious in real-world scenarios. Recently, unsupervised visible-infrared person re-identification (USL-VI-ReID) methods [3,29,31,32] have been proposed to tackle the issue of expensive cross-modality annotations.

The primary challenge of USL-VI-ReID is to bridge two modalities by minimizing the cross-modality gap under unsupervised settings. Most methods follow a two stage paradigm: generating pseudo labels with the clustering algorithm [11,20,26,27] and performing contrastive learning based on cluster prototype memory banks. However, there are two key limitations to this paradigm. First, the cross-modality gap intensifies the generation of noise pseudo labels in the clustering process. PGM [29] alternately utilizes unidirectional metrics learning and MMM [22] designs two soft cluster-level intra- and inter-modality alignment losses to mitigate the impact of noisy pseudo labels. However, they train the model without considering the discrepancy in feature distribution caused by the random data augmentation strategy. Second, the cross-modality gap makes it difficult for the model to learn the common discriminative features of the two modalities. To this end, PGM [29] and MBCCM [3] perform graph matching to establish reliable cross-modality correspondences. OTLA [25] and DOTLA [5] employ the optimal transport strategy to assign pseudo labels from one modality to another modality. These methods are usually based on clustering-level exploration of cross-modality correspondence relationships. However, the centroid prototype are prone to losing the fine-grained features of instances, the learning of modality-agnostic features is insufficient.

In this paper, we propose a Modality-specific and Modality-agnostic Soft Contrastive Learning (MMSCL) framework for USL-VI-ReID. Specifically, we introduce a Soft Contrastive Learning (SCL) module, which utilizes the differences and complementarities between the two networks to alleviate the noise pseudo labels caused by inconsistent data augmentation. Moreover, we propose a modality-agnostic hybrid memory (MHM) module to fully exploit the fine-grained cross-modality information. In addition, to make full use of the information of each query feature in the mini-batch, we propose an Adaptive-weighted Memory Updating (AMU) module, which assigns a corresponding weight to query features and jointly updates the cluster representations. Extensive experimental results on SYSU-MM01 and RegDB datasets demonstrate the superiority of our method compared to existing USL-VI-ReID methods.

2 Related Work

2.1 Unsupervised Single-Modality Person ReID

The existing unsupervised single-modality person ReID methods can be summarized into unsupervised domain adaptation (UDA) methods and fully unsupervised learning (USL) methods depending on wether external labeled source domain data is used. Compared with UDA methods, the USL methods are more challenging. State-of-the-art methods [1,8,9,13,17,19,34] involve an iterative process alternating between generating the pseudo labels through clustering algorithm and optimizing network by representation learning. These methods try to design elaborate mechanism to improve the accuracy of clustered pseudo labels. BUC [19] employ a bottom-up clustering scheme to gradually group similar samples into one identity, and utilize a diversity regularization term to balance the number of samples in each cluster. HCT [34] design a hierarchical clustering strategy to make full use of similarity among samples, and a hard-batch triplet loss [15] to reduce the influence of hard samples. SpCL [13] adopt unified contrastive learning to progressively refine the clusters by evaluating the effectiveness of the clustering and alternatively updating. Cluster Contrast [9] initialize, update, and perform contrastive loss computation in cluster-level to solve the cluster updating inconsistency problem. ICE [1] introduce inter-instance pairwise similarity scores to boost the performance of contrastive learning. Although above single-modality ReID methods achieved impressive performance, but they are limited to solving the USL-VI-ReID task due to large cross-modality gaps.

2.2 Unsupervised Visible-Infrared Person ReID

Existing UDA methods use a labeled source domain to solve the USL-VI-ReID task, USL methods attract more attention due to the advantage of not relying on labeled data. Most of USL methods are following the Dual Contrastive Learning (DCL [32]) framework, which uses DBSCAN [11] to generate pseudo labels and learns intra-modality discriminance based on contrastive loss. To alleviate the effects of noisy pseudo labels, PGM [29] alternately utilizes unidirectional metrics learning to reduce the noisy pseudo labels. MMM [22] designs two soft cluster-level intra- and inter-modality alignment losses to mitigate the impact of noisy pseudo labels. However, they train the model without considering the discrepancy in feature distribution caused by the random augmentation strategy. To mine the cross-modality correspondences, ADCA [32] designs a cross-modality memory aggregation module to associate the reliable positive cross-modality cluster pairs. PGM [29] gathers the instance to its corresponding cross-modality proxy to enhance the learning of modality-invariant information. MMM [22] proposes a multi-memory learning and matching module to establish reliable cross-modality correspondences. However, due to the modality gap, the bidirectional contrastive learning with modality-specific memory banks would lead to unreliable cross-modality correspondences.

3 Method

3.1 Overview

Inspired by previous mean-teacher methods [7,12], we propose a Modality-specific and Modality-agnostic Soft Contrastive Learning framework for USL-VI-ReID, which is illustrated in Fig. 1. Let $\mathcal{X} = \{\mathcal{V},\ \mathcal{R}\}$ denote an unlabeled visible-infrared person ReID dataset, where $\mathcal{V} = \{x_i^v\}_{i=1}^{N_v}$ and $\mathcal{R} = \left\{x_j^r\right\}_{j=1}^{N_r}$ denote N_v visible images and N_r infrared images of two modalities, respectively. Following

DCL [32], we adopt a two-stream network to learn modality-specific features. Specifically, we use the student encoder to extract visible and infrared features, and generate pseudo labels with clustering algorithm. Meanwhile, we initialize two modality-specific memories with cluster centroids $\mathcal{M}^v = \{c_i^v\}_{i=1}^{K_v}$ and $\mathcal{M}^r = \{c_j^r\}_{j=1}^{K_r}$, where K_v and K_r are the numbers of clusters in these two modalities. Then, we perform contrastive learning to learn modality-specific information. In order to learn cross-modality information, we propose a modality-specific and modality-agnostic soft contrastive learning method, including a modality-agnostic hybrid memory, a soft contrastive learning module, and a adaptive-weighted memory updating module, which are detailed below.

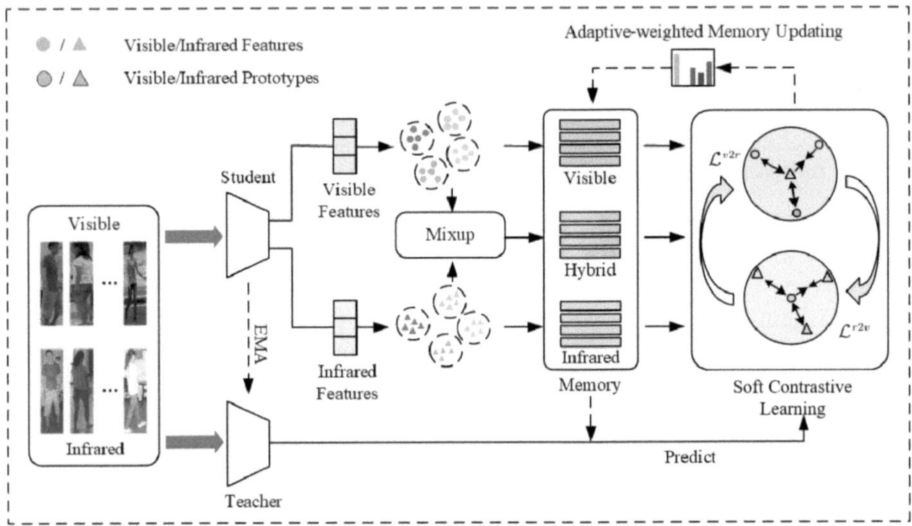

Fig. 1. The pipeline of our framework.

3.2 Modality-Agnostic Hybrid Memory

We propose a modality-agnostic hybrid memory module to reduce the cross-modality discrepancy and exploit modality-invariant information. As shown in Fig. 2, given visible features f^r and infrared features f^v, we first compute the similarity score between f^v with f^r as follows:

$$S_{i,k} = \frac{(f_{i,c}^v) \cdot \left(f_{j,k}^r\right)^T}{\|f_{i,c}^v\|_2 \cdot \|f_{j,k}^r\|_2}, \tag{1}$$

where $S_{i,k}$ denotes the cosine similarity between the extracted feature $f_{i,c}^v$ in c-th cluster and $f_{j,k}^r$ in k-th cluster. Lager $S_{i,k}$ indicates that the feature f_i^v extracted from instance x_i^v is more likely to belong to the k-th cluster of infrared modality. We regard the samples with top-K similarity score as reliable matched samples:

$$ind = \arg \max_{K} S_{i,k}, \tag{2}$$

where K denotes the order of similarity scores. ind denotes the indexes of top-K similarity counts. With the cross-modality correspondences derived from top-K similarity score, we initialize the hybrid memory $\mathcal{M}^h$ by mixing infrared cluster centroid vectors and matched visible features by

$$c_i^h \leftarrow \eta \times c_i^r + (1-\eta) \times \frac{1}{n_c^{ind}} \sum_{ind=1}^{n_c^{ind}} f_{i,ind}^v, \tag{3}$$

$$M_i^h \leftarrow c_i^h, \tag{4}$$

where c_i^r and c_i^h are infrared modality and hybrid modality cluster representations, $i \in \{1, 2, ..., n_c^r\}$. $f_{i,ind}^v$ denotes the top-K features of visible samples matched with the infrared samples, n_c^{ind} is the number of matched samples. η is the hyper-parameter to balance the fusion of cluster-level cross-modality information.

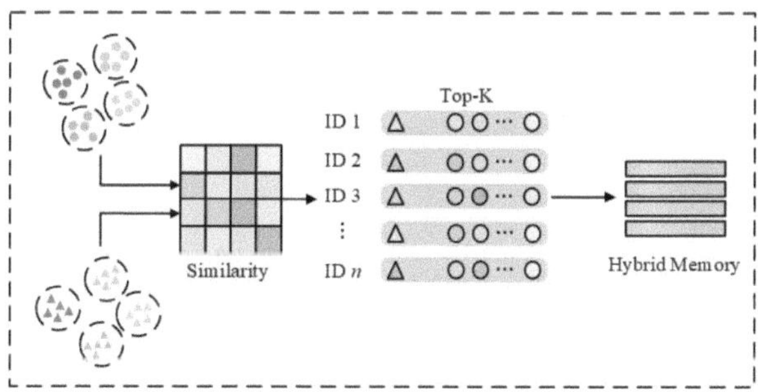

Fig. 2. The mixture of hybrid memory.

3.3 Soft Contrastive Learning

Existing contrastive learning methods [3,29,32] usually optimize models based on pseudo labels generated by clustering algorithms. However, due to the random data augmentation strategy, the intra-class feature distribution would deteriorate the quality of the pseudo labels. To address the discrepancy in feature distribution caused by data random augmentation, we introduce a mean-teacher framework to cast the USL-VI-ReID task as the gradual semi-supervised learning. To provide a more stable target, we adopt parameters from the student model to update the teacher model by Exponential Moving Average (EMA), which is proved to be effective in [12]. In addition, we introduce a soft contrastive loss to enhance the consistency of the same query feature distribution. As illustrate in Fig. 3, we denote the one-hot labels $y_i^{tv} [M^v]$ as the similarity probability of the query feature to the k-th cluster prototype, which can be formulated as:

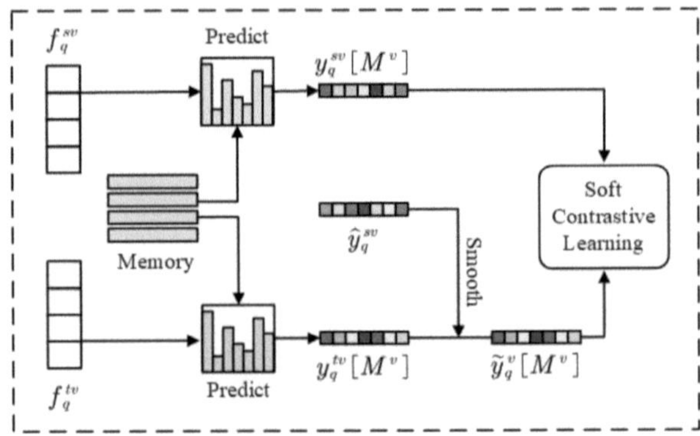

Fig. 3. The illustrate of soft contrastive learning.

$$y_i^{tv}[M^v] = \frac{\exp\left(f_q^{tv} \cdot m_i^v/\tau\right)}{\sum_{k=1}^{C^v} \exp\left(f_q^{tv} \cdot m_k^v/\tau\right)}, \tag{5}$$

where f_q^{tv} represents infrared query feature from teacher channel. c_k^v denotes the k-th cluster representation in M^v. n_c^v is the number of infrared modality clusters. τ denote the temperature hyper-parameter.

To constrain the consistency of the feature distribution of each instance, we adopt the clustering pseudo labels $\hat{y}_i^{sv}$ from the student channel to smooth the one-hot labels $y_i^{tv}[M^v]$, which is formulated as:

$$\tilde{y}_i^v[M^v] = (1-\mu) \times \hat{y}_i^{sv} + \mu \times y_i^{tv}[M^v], \tag{6}$$

where $\tilde{y}_i^v[M^v]$ denotes i-th visible soft pseudo label generated with memory bank M^v. μ is weight factor.

The clusterNCE [9] loss can be re-formulated to cross-entropy as:

$$\mathcal{L}^v = -\sum_{i=1}^{C^v} \hat{y}_i^{sv} \log y_i^{sv}, \tag{7}$$

where y_i^{sv} denote the similarity probability predict pseudo labels from student channel. C^v is the number of visible modality clusters. Replace the one-hot pseudo labels $\hat{y}_i^{sv}$ with the soft pseudo labels $\tilde{y}_i^v[M^v]$, our proposed soft contrastive loss can be formulated as:

$$\mathcal{L}_{scl}^v = -\sum_{i=1}^{C^v} \tilde{y}_i^v[M^v] \cdot \log y_i^{sv}, \tag{8}$$

Modality-specific Soft Contrastive Learning. To learn modality-specific information, we perform soft contrastive learning based on infrared modality memory M^r and visible modality memory M^v. The soft contrastive loss for the infrared modality is similar to Eq.(8):

$$\mathcal{L}_{scl}^{r} = -\sum_{j=1}^{C^r} \tilde{y}_j^r [M^r] \cdot \log y_j^{sr}, \tag{9}$$

where $\tilde{y}_j^r [M^r]$ denotes i-th infrared soft pseudo label generated with visible modality memory M^v following Eq. (6). y_j^{sr} is acquired from the student channel by clustering algorithm. C^r is the number of infrared modality clusters. The total modality-specific loss is:

$$L_{ms} = L_{scl}^{r} + L_{scl}^{v}. \tag{10}$$

Modality-Agnostic Soft Contrastive Learning. We exploit the modality-agnostic information from two aspects: soft contrastive learning with cross-modality alignment and with hybrid memory. In soft contrastive learning of cross-modality alignment, we employ graph matching [29] to build reliable cross-modality correspondences, e.g., y^{v2r} is the visible pseudo label y^v matched with the infrared pseudo label y^r, y^{r2v} is the infrared pseudo label y^r matched with visible pseudo label y^v. The soft contrastive learning with cross-modality alignment consists of two unidirectional learning, which can be expressed as:

$$\mathcal{L}_{acscl} = \begin{cases} -\sum\limits_{i=1}^{C^v} \tilde{y}_i^{r2v} [M^v] \cdot log y_i^{sr2v}, & epoch\%2 = 0 \\ -\sum\limits_{j=1}^{C^r} \tilde{y}_j^{v2r} [M^r] \cdot log y_j^{sv2r}, & epoch\%2 = 1 \end{cases}, \tag{11}$$

where $\tilde{y}_i^{v2r} [M^r]$ and $\tilde{y}_j^{r2v} [M^v]$ denote the two cross-modality soft pseudo label generated following Eq. (6). The soft contrastive learning with hybrid memory is formulated as in Eq. (11):

$$\mathcal{L}_{hacscl} = \begin{cases} -\sum\limits_{j=1}^{C^r} \tilde{y}_j^{v2r} \left[M^h\right] \cdot log y_j^{sv2r}, & epoch\%2 = 0 \\ -\sum\limits_{j=1}^{C^r} \tilde{y}_j^{r} \left[M^h\right] \cdot log y_j^{sr}, & epoch\%2 = 1 \end{cases}, \tag{12}$$

The total modality-agnostic soft contrastive loss is

$$\mathcal{L}_{mi} = \alpha \mathcal{L}_{ascl} + \beta \mathcal{L}_{ahscl}, \tag{13}$$

where α and β are balance coefficients.

Overall Loss Function. Considering the above descriptions, the overall loss function consists of the modality-specific and modality-agnostic soft contrastive loss as

$$L = L_{ms} + L_{mi}, \tag{14}$$

3.4 Adaptive-Weighted Memory Updating

To fully exploit the valid information in the global context, we propose an adaptive-weighted strategy for memory updating. Specifically, we assign a weight to each query instance corresponding to the similarities of query instance features with cluster centroids, and the instance with the lower similarity is assigned a larger weight. For the query instance feature $f_{j,q}$, the adaptive weight is illustrated as follows:

$$\omega_{ij} = \frac{\exp\left(-\left\langle f_{j,q} \cdot c_i \right\rangle / \tau_w\right)}{\sum_{k=1}^{N_i} \exp\left(-\left\langle f_{k,q} \cdot c_i \right\rangle / \tau_w\right)}, \tag{15}$$

where $\sum_{j=1}^{N_i} \omega_{ij} = 1$, τ_w is temperature factor, N_i is the instance number in i-th cluster. Our proposed adaptive-weighted memory updating strategy can be formulated as:

$$c_i \leftarrow \lambda c_i + (1 - \lambda) \sum_{j=1}^{N_i} \omega_{ij} f_{j,q}, \tag{16}$$

where λ is momentum updating factor.

4 Experiments

4.1 Experimental Settings

Datasets and Evaluation Metrics. We evaluate our proposed method on two widely used datasets SYSU-MM01 [28] and RegDB [10]. We evaluate the effectiveness of our method following the commonly used protocols: cumulative matching characteristic (CMC) and mean average precision (mAP).

Implementation Details. We utilize AGW [33] as the encoder to extract features. All images are resized to 288 $\times$ 144. We use Adam optimizer [16] to train the model with weight decay 5e−4. The initial learning rate is set to 3.5e−4, which is decreased to 1/10 per 20 epochs. The batch size is set to 128. The total number of training epochs is set to 100. In the first 50 training epochs, we follow the DCL [32] framework to learn modality-specific information. Then, our proposed network is trained for another 50 epochs. Following [32], the maximum distance for DBSCAN is set to 0.6 on SYSU-MM01 and 0.3 on RegDB, the momentum updating factor λ is 0.1 and the temperature factor τ is 0.05. The trade-off hyper-parameters α and β in Eq.(13) are set to 0.4/0.25 and 0.5/0.5 on SYSU-MM01 and RegDB, respectively.

4.2 Comparison With State-of-the-Art Methods

To demonstrate the efficiency of our proposed method, we compare it with two related VI-ReID settings, which are semi-supervised VI-ReID (SS-VI-ReID) and unsupervised VI-ReID (USL-VI-ReID). The results on SYSU-MM01 and RegDB are shown in Table 1. If not specified, we conduct an analysis on SYSY-MM01 under all search mode.

Table 1. Comparison with the state-of-the-art methods on SYSY-MM01 and RegDB. Bolded or underlined numbers indicate the best and second results, respectively.

Settings		SYSU-MM01				RegDB			
		All search		Indoor search		Visible2Thermal		Thermal2Visible	
Type	Method	Rank-1	mAP	Rank-1	mAP	Rank-1	mAP	Rank-1	mAP
SS-VI-ReID	OTLA [25]	48.2	43.9	47.4	56.8	49.9	41.8	49.6	42.8
	TAA [30]	48.8	42.3	50.1	56.0	62.2	56.0	63.8	56.5
	DPIS [23]	58.4	55.6	63.0	70.0	62.3	53.2	61.5	52.7
USL-VI-ReID	H2H [18]	30.2	29.4	–	–	23.8	18.9	–	–
	OTLA [25]	29.9	27.1	29.8	38.8	32.9	29.7	32.1	28.6
	ADCA [32]	45.5	42.7	50.6	59.1	67.2	64.1	68.5	63.8
	NGLR [4]	50.4	47.4	53.5	61.7	85.6	76.7	82.9	75.0
	MBCCM [3]	53.1	48.2	55.2	62.0	83.8	77.7	82.8	76.7
	CCLNet [2]	54.0	50.2	56.7	65.1	69.9	65.5	70.2	66.7
	PGM [29]	57.3	51.8	56.2	62.7	69.5	65.4	69.9	65.2
	GUR [31]	61.0	57.0	64.2	69.5	73.9	70.2	75.0	69.9
	MMM [22]	<u>61.6</u>	<u>57.9</u>	<u>64.4</u>	<u>70.4</u>	**89.7**	<u>80.5</u>	<u>85.8</u>	<u>77.0</u>
	MMSCL	**63.2**	**59.3**	**67.7**	**73.1**	<u>87.9</u>	**82.7**	**88.4**	**82.7**

Comparison with SS-VI-ReID Methods. SS-VI-ReID methods are proposed to alleviate the problem of labeling cost by using part of labeled data. Remarkably, our method outperforms all existing SS-VI-ReID methods and achieves a 3.7% improvement in mAP and a 4.8% improvement in Rank-1 on the SYSU-MM01 dataset compared to the DPIS method.

Comparison with USL-VI-ReID Methods. As reported in Table 1, our method achieves competitive performance compared with state-of-the-art USL-VI-ReID methods with 59.3%/73.1% in mAP and 63.2%/67.7% in Rank-1 on SYSU-MM01, 82.7%/82.7% in mAP and 87.9%/88.4% in Rank-1 on RegDB. This indicates that our proposed method achieves superior performance compared to existing USL-VI-ReID methods.

4.3 Ablation Study

In this subsection, we conduct ablation studies to validate the effectiveness of each component of our method in all search mode and indoor search mode on SYSU-MM01 dataset. We employ the DCL framework as the baseline. The results are reported in Table 2.

Effectiveness of the SCL Module. Comparing Order 1 and Order 2, the addition of soft contrastive learning (denote as "SCL") leads to an increase of

Table 2. Ablation studies on SYSU-MM01. Best results are bolded.

Order	Module				All Search		Indoor Search	
	DCL	SCL	MHM	AMU	mAP	Rank-1	mAP	Rank-1
1	✓				39.4	40.0	53.8	46.0
2	✓	✓			52.7	56.5	68.2	61.6
3	✓	✓	✓		57.6	61.7	71.0	65.4
4	✓	✓		✓	53.9	58.5	68.6	62.2
5	✓	✓	✓	✓	**59.3**	**63.2**	**73.1**	**67.7**

13.3% in mAP and 14.4% in Rank-1, respectively. It indicates that our proposed SCL module can effectively suppress the effect of inconsistent feature distribution caused by the data random augmentation strategy.

Effectiveness of the MHM Module. As shown in Order 3, the performance can reach 53.9% in mAP and 55.8% in Rank-1 when adding the modality-agnostic hybrid memory (denote as "MHM") module. It shows that our proposed MHM module effectively mitigates cross-modality discrepancies with joint learning modality-specific and modality-invariant information.

Effectiveness of the AMU Module. The performance of using adaptive-weighted memory updating (denote as "AMU") strategy is shown in Order 4. Compared to Order 2, performance improves by a large margin of 4.9% in mAP and 5.2% in Rank-1, respectively. This confirms that the proposed strategy can effectively tackle the distribution discrepancy in batch training and explore the relationship between all instances.

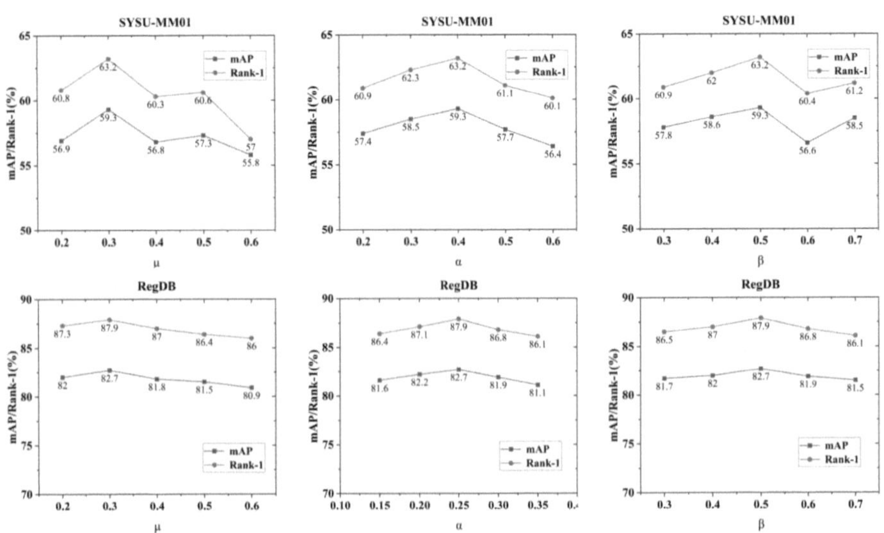

Fig. 4. The parameter analyses on SYSU-MM01 and RegDB datasets.

4.4 Further Analysis

Hyper-parameter Analysis. We conduct experiments to evaluate the three key hyper-parameters (e.g., μ, α and β) in our proposed method, and the quantitative results are shown in Fig. 4. It can be observed that the performance on SYSU-MM01 dataset is more sensitive to these hyper-parameters. Because images of SYSU-MM01 are captured through multiple cameras, the feature distribution caused by camera domain gaps is more discrete. As shown in Fig. 4, the optimal hyper-parameters are 0.3, 0.4 and 0.5 on SYSU-MM01, 0.3, 0.25, and 0.5 on RegDB, respectively.

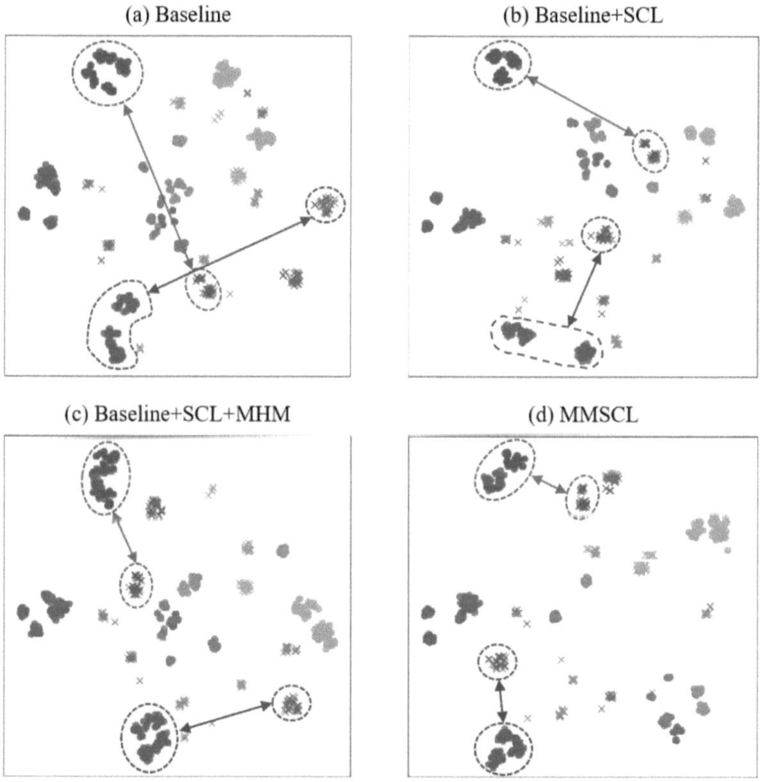

Fig. 5. Visualization of feature distribution on SYSU-MM01. Different shapes and colors represent different modalities and identities.

Visualization of Feature Distribution. To further illustrate the performance of our proposed MMSCL, we employ a set of data visualization experiments on SYSU-MM01 to visualize the visible and infrared feature distributions of person images with 7 randomly selected identities. As shown in Fig. 5, compared to baseline, the feature distributions of the same identities from different modalities are more compact (see red circles), and those overlapping feature distributions of different identities become distinguished (see blue circle).

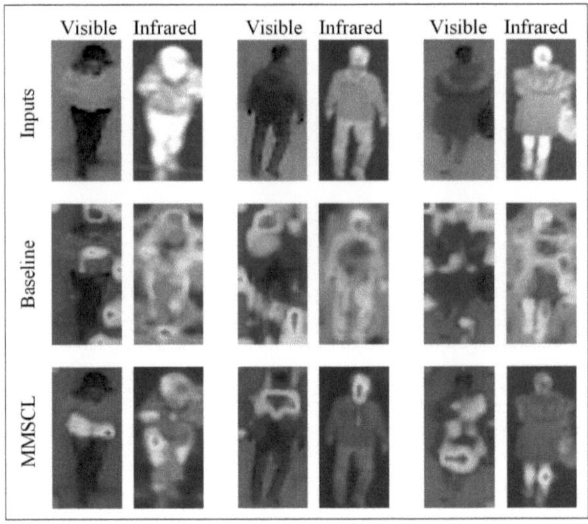

Fig. 6. Visualization of feature heatmaps on RegDB.

Visualization of Feature Heatmap. We randomly select three identities from the RegDB dataset and apply Grad-CAM [21] to visualize the heatmaps of the features learned from the baseline and our proposed, respectively. As shown in Fig. 6, we can observe that the baseline usually focuses on the semantic region misalignment of the two modality images and is affected by backgrounds. Compared to baseline, our MMSCL focuses on the cues of the person-semantic regions between the two modalities, effectively alleviating the modal differences.

5 Conclusion

In this paper, we propose a modality-specific and modality-agnostic soft contrastive learning approach to establish reliable cross-modality correspondences for USL-VI-ReID task. First, we propose a modality-agnostic hybrid memory to mitigate the modality gap. In addition, we introduce a soft contrastive learning module to constrain the inconsistency of the feature distribution. Moreover, we propose an adaptive-weighted memory updating strategy to learn global information efficiently. Comprehensive experiment results on two datasets demonstrate the effectiveness of our proposed method.

Acknowledgments. This work was supported in part by the Natural Science Foundation of Shaanxi Province under Grant 2023-JC-YB-501.

Disclosure of Interests. The authors declare that they have no known competing financial interests or personal relationships that could have appeared to influence the work reported in this paper.

References

1. Chen, H., Lagadec, B., Bremond, F.: ICE: inter-instance contrastive encoding for unsupervised person re-identification. In: 2021 IEEE/CVF International Conference on Computer Vision (ICCV), pp. 14940–14949 (2021). https://doi.org/10.1109/ICCV48922.2021.01469
2. Chen, Z., Zhang, Z., Tan, X., Qu, Y., Xie, Y.: Unveiling the power of clip in unsupervised visible-infrared person re-identification. In: Proceedings of the 31st ACM International Conference on Multimedia, MM 2023, pp. 3667–3675. Association for Computing Machinery, New York (2023). https://doi.org/10.1145/3581783.3612050
3. Cheng, D., He, L., Wang, N., Zhang, S., Wang, Z., Gao, X.: Efficient bilateral cross-modality cluster matching for unsupervised visible-infrared person reid. In: Proceedings of the 31st ACM International Conference on Multimedia, MM 2023, pp. 1325–1333. Association for Computing Machinery, New York (2023). https://doi.org/10.1145/3581783.3612073
4. Cheng, D., Huang, X., Wang, N., He, L., Li, Z., Gao, X.: Unsupervised visible-infrared person reid by collaborative learning with neighbor-guided label refinement. ArXiv abs/2305.12711 (2023)
5. Cheng, D., Huang, X., Wang, N., He, L., Li, Z., Gao, X.: Unsupervised visible-infrared person reid by collaborative learning with neighbor-guided label refinement. In: Proceedings of the 31st ACM International Conference on Multimedia, MM 2023, pp. 7085–7093. Association for Computing Machinery, New York (2023). https://doi.org/10.1145/3581783.3612077
6. Cheng, D., Wang, X., Wang, N., Wang, Z., Wang, X., Gao, X.: Cross-modality person re-identification with memory-based contrastive embedding. In: Proceedings of the AAAI Conference on Artificial Intelligence, vol. 37, pp. 425–432 (2023)
7. Cheng, D., Zhou, J., Wang, N., Gao, X.: Hybrid dynamic contrast and probability distillation for unsupervised person re-id. IEEE Trans. Image Process. **31**, 3334–3346 (2022). https://doi.org/10.1109/TIP.2022.3169693
8. Cho, Y., Kim, W.J., Hong, S., Yoon, S.E.: Part-based pseudo label refinement for unsupervised person re-identification. In: 2022 IEEE/CVF Conference on Computer Vision and Pattern Recognition (CVPR), pp. 7298–7308 (2022). https://doi.org/10.1109/CVPR52688.2022.00716
9. Dai, Z., Wang, G., Yuan, W., Zhu, S., Tan, P.: Cluster contrast for unsupervised person re-identification. In: Proceedings of the Asian Conference on Computer Vision, pp. 1142–1160 (2022)
10. Dat, N., Hyung, H., Ki, K., Kang, P.: Person recognition system based on a combination of body images from visible light and thermal cameras. Sensors **17**(3), 605 (2017)
11. Ester, M., Kriegel, H.P., Sander, J., Xu, X., et al.: A density-based algorithm for discovering clusters in large spatial databases with noise. In: kdd, vol. 96, pp. 226–231 (1996)
12. Ge, Y., Chen, D., Li, H.: Mutual mean-teaching: pseudo label refinery for unsupervised domain adaptation on person re-identification. arXiv preprint arXiv:2001.01526 (2020)

13. Ge, Y., Zhu, F., Chen, D., Zhao, R., et al.: Self-paced contrastive learning with hybrid memory for domain adaptive object re-id. Adv. Neural. Inf. Process. Syst. **33**, 11309–11321 (2020)
14. Gwon, S., Kim, S., Seo, K.: Balanced and essential modality-specific and modality-shared representations for visible-infrared person re-identification. IEEE Signal Process. Lett. **31**, 491–495 (2024). https://doi.org/10.1109/LSP.2024.3358725
15. Hermans, A., Beyer, L., Leibe, B.: In defense of the triplet loss for person re-identification. arXiv preprint arXiv:1703.07737 (2017)
16. Kingma, D.P., Ba, J.: Adam: a method for stochastic optimization. arXiv preprint arXiv:1412.6980 (2014)
17. Li, M., Li, C.G., Guo, J.: Cluster-guided asymmetric contrastive learning for unsupervised person re-identification. IEEE Trans. Image Process. **31**, 3606–3617 (2022). https://doi.org/10.1109/TIP.2022.3173163
18. Liang, W., Wang, G., Lai, J., Xie, X.: Homogeneous-to-heterogeneous: unsupervised learning for RGB-infrared person re-identification. IEEE Trans. Image Process. **30**, 6392–6407 (2021). https://doi.org/10.1109/TIP.2021.3092578
19. Lin, Y., Dong, X., Zheng, L., Yan, Y., Yang, Y.: A bottom-up clustering approach to unsupervised person re-identification. In: Proceedings of the AAAI Conference on Artificial Intelligence, vol. 33, pp. 8738–8745 (2019)
20. MacQueen, J.: Some methods for classification and analysis of multivariate observations, p. 281 (1965)
21. Selvaraju, R.R., Cogswell, M., Das, A., Vedantam, R., Parikh, D., Batra, D.: Grad-CAM: visual explanations from deep networks via gradient-based localization. In: 2017 IEEE International Conference on Computer Vision (ICCV), pp. 618–626 (2017). https://doi.org/10.1109/ICCV.2017.74
22. Shi, J., et al.: Multi-memory matching for unsupervised visible-infrared person re-identification. arXiv preprint arXiv:2401.06825 (2024)
23. Shi, J., et al.: Dual pseudo-labels interactive self-training for semi-supervised visible-infrared person re-identification. In: 2023 IEEE/CVF International Conference on Computer Vision (ICCV), pp. 11184–11194 (2023). https://doi.org/10.1109/ICCV51070.2023.01030
24. Tan, X., Chai, Y., Chen, F., Liu, H.: A Fourier-based semantic augmentation for visible-thermal person re-identification. IEEE Signal Process. Lett. **29**, 1684–1688 (2022). https://doi.org/10.1109/LSP.2022.3194841
25. Wang, J., et al.: Optimal transport for label-efficient visible-infrared person re-identification. In: European Conference on Computer Vision (2022)
26. Wang, N., Cui, Z., Li, A., Lu, Y., Wang, R., Nie, F.: Structured doubly stochastic graph-based clustering. IEEE Trans. Neural Netw. Learn. Syst. (2025)
27. Wang, N., et al.: Large-scale hyperspectral image-projected clustering via doubly stochastic graph learning. Remote Sens. **17**(9), 1526 (2025)
28. Wu, A., Zheng, W.S., Yu, H.X., Gong, S., Lai, J.: RGB-infrared cross-modality person re-identification. In: 2017 IEEE International Conference on Computer Vision (ICCV) (2017)
29. Wu, Z., Ye, M.: Unsupervised visible-infrared person re-identification via progressive graph matching and alternate learning. In: 2023 IEEE/CVF Conference on Computer Vision and Pattern Recognition (CVPR), pp. 9548–9558 (2023). https://doi.org/10.1109/CVPR52729.2023.00921
30. Yang, B., Chen, J., Ma, X., Ye, M.: Translation, association and augmentation: learning cross-modality re-identification from single-modality annotation. IEEE Trans. Image Process. **32**, 5099–5113 (2023). https://doi.org/10.1109/TIP.2023.3310338

31. Yang, B., Chen, J., Ye, M.: Towards grand unified representation learning for unsupervised visible-infrared person re-identification. In: 2023 IEEE/CVF International Conference on Computer Vision (ICCV), pp. 11035–11045 (2023). https://doi.org/10.1109/ICCV51070.2023.01016
32. Yang, B., Ye, M., Chen, J., Wu, Z.: Augmented dual-contrastive aggregation learning for unsupervised visible-infrared person re-identification. In: ACM MM, p. 2843–2851 (2022)
33. Ye, M., Shen, J., Lin, G., Xiang, T., Shao, L., Hoi, S.C.H.: Deep learning for person re-identification: a survey and outlook. IEEE Trans. Pattern Anal. Mach. Intell. **44**(6), 2872–2893 (2022). https://doi.org/10.1109/TPAMI.2021.3054775
34. Zeng, K., Ning, M., Wang, Y., Guo, Y.: Hierarchical clustering with hard-batch triplet loss for person re-identification. In: 2020 IEEE/CVF Conference on Computer Vision and Pattern Recognition (CVPR), pp. 13654–13662 (2020). https://doi.org/10.1109/CVPR42600.2020.01367
35. Zhang, Y., Kang, Y., Zhao, S., Shen, J.: Dual-semantic consistency learning for visible-infrared person re-identification. IEEE Trans. Inf. Forensics Secur. (2023)

The Design of a Time Watermark Recognition System for Forensic Video Images Based on PaddleOCR

Zefeng Zhang(✉), Chuan Guan, Weiwei Heng, and Hao Feng(✉)

Academy of Forensic Science, Shanghai Forensic Service Platform, Shanghai Key Laboratory of Forensic Medicine and Key Laboratory of Forensic Science, Ministry of Justice, Shanghai, China

{zhangzf,guanc,hengww,fengh}@ssfjd.cn

Abstract. This study proposes a video image time watermark detection method based on PaddleOCR, which addresses the issue of manual frame-by-frame reading of time watermarks in vehicle speed judicial appraisal practices. This study collects and labels 20 video materials containing time watermarks, which are classified into four types: pure black, pure white, black-and-white interlaced, and black edges filled with white, totaling over 15,000 frames. The images cover typical scene types in judicial appraisal practice, such as daytime, nighttime, rainy, foggy, sunny scenes, low-resolution CCTV, and compressd videos conditions. A four-stage detection model was designed, and the automatic detection of time watermarks in video images was achieved through a pipeline: PaddleOCR for single-image detection, image processing, PaddleOCR for video recognition and data cleaning with post-processing. Experiments demonstrate that the method proposed in this paper achieves an average accuracy rate of about 98% in the dataset and can be applied to identification practice.

Keywords: PaddleOCR · Video image · Time watermark · Judicial appraisal

1 Introduction

In handling road traffic accident cases in China, law enforcement authorities must investigate whether the target vehicle has violated the speed limit. The judicial appraisal of vehicle speed, while crucial for determining responsibility and making judgments, may have inaccuracies due to improper formula selection or parameter measurement, and must adhere to legal standards. Video-based vehicle speed assessment is the most widely used method in practice. Its principle involves selecting the reference distance of the target vehicle and the time elapsed to traverse the reference distance in the video and calculate the target vehicle's speed using the distance-to-time ratio. In this approach, time constitutes one of just two variables influencing the assessment outcome [1].

Video image time watermark serves as one of the data sources for calculating vehicle speed [2]. Figure 1 shows an example of a time watermark. According to national standards, if there are N frames of images per second, the time interval between frames

P. Umapada et al. (Eds.): ICCPR 2025, CCIS 2811, pp. 18–27, 2026.
https://doi.org/10.1007/978-981-95-8315-7_2

is 1/N. The time it takes for the target vehicle to pass the reference distance in n frames of images is n/N. Articles [2–5] conducted a verification study on video time using devices such as light boards capable of reflecting time frequency. In the context of judicial appraisal practice, the N value of the video time watermark may exhibit variations. For example, when the video frame rate is set at 25 FPS, the time watermark, measured in seconds, may show varying integer values between 21 and 29. In practice, technicians determine the N value by manually counting frames while playing the video sequentially. However, this approach is prone to high error rates and is not well-suited for analyzing lengthy videos.

Fig. 1. An example of a time watermark

In recent years, OCR technology has been widely applied across various sectors, including industrial automation [6–10]. For instance, in the industrial sector, OCR has been instrumental in enhancing production efficiency, reducing costs, and improving quality. PaddleOCR, in particular, has demonstrated significant success in projects such as bus route recognition [11] and license plate recognition [12], showcasing the technology's versatility and effectiveness. PaddleOCR, developed by Baidu's PaddlePaddle team, is an open-source OCR tool library known for its high performance, offering a comprehensive suite of OCR capabilities, including text recognition, text detection, table recognition, and direction classification. To tackle the practical issue where appraisers need to play videos frame by frame and manually count to obtain N values, this paper proposes developing a method for the automatic detection and recognition of time watermarks in video images based on PaddleOCR V3.2.0.

2 System Design

2.1 Design Concepts

This study proposes an automated video time watermark detection method based on PaddleOCR, addressing the manual effort traditionally required in forensic video analysis. The detection pipeline consists of six key steps in four stages: OCR for single-image detection, image precessing, OCR for video recognition, and data cleaning with post-processing. Initially, the system loads the first frame of the input video. Subsequently, PaddleOCR is employed to detect regions of interest (ROIs) containing temporal information. The third stage involves enhancing the ROI through binarization and pattern emphasis. The fourth stage applies PaddleOCR to recognize characters across all video frames within the ROI. The fifth stage outputs standardized recognition results, and finally, the system generates a temporal sequence for all frames incorporating the video frame rate. Should automatic ROI detection fail, the system allows manual selection of a reference frame or manual ROI specification. The overall technical workflow is illustrated in Fig. 2.

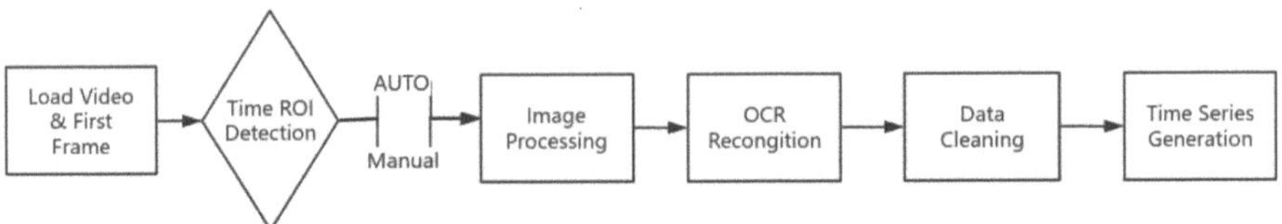

Fig. 2. The technical roadmap of the method proposed in this paper

2.2 ROI Localization for Time Watermark

A specific frame (defaulting to the first frame or user-selected) is used as the reference. PaddleOCR performs full-image text detection, after which regular expressions are applied to identify strings matching temporal patterns such as "xxxx-xx-xx", "xx:xx:xx", or "xxxx-xx-xx xx:xx:xx". Detected text regions are merged by computing the minimum bounding rectangle for all temporal text boxes, merging overlapping areas, and applying adaptive padding. Experimental results indicate that extending the width by 50% and height by 25% yields optimal performance. The final ROI is cropped ensuring it remains within image boundaries. If no ROI is detected automatically, the system either processes the entire image or prompts the user to manually delineate the ROI.

2.3 Image Processing

Conventional OCR preprocessing techniques, including grayscale conversion and binarization, are employed to improve recognition accuracy. Video time watermarks typically appear in pure black, pure white, or black-white interlaced patterns, with the white category further divided into pure white and white-filled with black edges. This paper utilizes dynamic thresholding for image enhancement, which effectively preserves critical temporal information while suppressing noise and irrelevant pixels. The processing modes are defined as follows:

Mode 0: No enhancement—outputs the original image.
Mode 1: Retains near-pure black and pure white regions, converting other areas to gray. This mode enhances image contrast and emphasizes light-dark boundaries.
Mode 2: Sets near-pure black to black, pure white to white, and all other regions to gray. This mode further strengthens contrast, reduces mid-tones, and mimics an extreme binarization process while preserving pure black and white areas.

Additionally, square expansion techniques standardize the image size, mitigating the impact of aspect ratio variations on OCR performance and establishing a foundation for high-precision time watermark recognition.

2.4 Data Cleaning and Time Prediction

OCR accuracy is influenced by factors such as image proportion, character size, text placement, and background noise. Although ROI may be successfully detected in the first frame, subsequent frames may suffer from recognition errors due to changing conditions. Thus, adaptive parameter tuning (e.g., ROI dimensions) is essential for robust detection across frames.

A four-stage text cleaning pipeline is implemented, comprising character correction, format normalization, invalid data filtering, and data imputation. Leveraging the temporal consistency of video watermarks, a time inference model is developed that integrates frame rate metadata and reference timestamps to correct misidentified values during the imputation stage. The system ultimately produces a comprehensive multi-dimensional report including raw OCR results, cleaned data, inferred time watermark, and time-frame statistics. An overview of the system's output data structure is provided in Table 1.

Table 1. An overview of the data generated by the system

Data layer	Content	Usage
Original Data	The Result OCR Recognized	Error Analysis
Cleaned Data	Standardized Time Format	Direct Use
Reasoning Data	Theoretical Calculation Of Time	Timeline Repair
Statistical Statement	Time-Frame Distribution	Performance Evaluation

3 Experiment

3.1 Dataset

Based on four distinct types of video time watermarks—namely, white-filled black edges, pure white, pure black, and black-and-white interlaced patterns—we collected five different scene samples for each type. These samples encompass typical judicial appraisal scenarios, including daytime, evening, rainy, foggy, sunny, low-resolution CCTV, and compressed videos conditions, resulting in a total of over 15,000 images. We manually marked the time watermark of each image for the subsequent evaluation of algorithm results. Figure 3 is four examples of the dataset.

Fig. 3. Four examples of the dataset

3.2 Automatic Recognition of Time Watermarks

The experiment employed PaddleOCR in succession to identify the ROI area of the original image and the processed ROI area. The detection results were evaluated to verify the applicability of the method proposed in this paper for vehicle speed identification in judicial appraisal. It has been experimentally verified that the time watermark positioning algorithm can automatically determine the time watermark position of each video in the dataset, eliminating the need for manual addition of ROI.

3.3 System Interface and Generated Results

The system interface is divided into two sections: the left section and the right section. The left section serves as the image preview area, while the upper part displays the preview of the original image. The following image displays the preview of the processed ROI area results. The right section functions as the operation area, where users can select files, adjust parameters, and view logs. Upon selecting the "Start Recognition" button, the system will initiate the OCR algorithm to identify the watermark time of the video image and generate information of concern in the identification practice such as "original data", "inference data", and "statistical reports". Figure 4 displays the system interface. Table 2 presents an example of a form for recording "raw_data" and "inferred_time", while Table 3 shows an example of a form for recording "statistical reports".

Fig. 4. The system interface

Table 2. An example of a form for recording "raw data" and "inference data"

filename	ocr_raw	inferred_time	time_sec
frame_0	2024年11月22日 星期五 01:08:23	2024-11-22 01:08:23	0
frame_1	2024年11月22日 星期五 01:08:23	2024-11-22 01:08:23	0.04
frame_2	2024年11月22日 星期五 01:08:23	2024-11-22 01:08:23	0.08
frame_3	2024年11月22日 星期五 01:08:23	2024-11-22 01:08:23	0.12
frame_4	2024年11月22日 星期五 01:08:23	2024-11-22 01:08:23	0.16
frame_5	2024年11月22日 星期五 01:08:23	2024-11-22 01:08:23	0.2
frame_6	2024年11月22日 星期五 01:08:23	2024-11-22 01:08:23	0.24
frame_7	2024年11月22日 星期五 01:08:23	2024-11-22 01:08:23	0.28
frame_8	2024年11月22日星期五01:08:23	2024-11-22 01:08:23	0.32
frame_9	2024年11月22日 星期五 01:08:23	2024-11-22 01:08:23	0.36
frame_10	2024年11月22日星期五01:08:23	2024-11-22 01:08:23	0.4

Table 3. An example of a form for recording "statistical reports"

Time	Frame_num
2024-11-22 01:08:23	23
2024-11-22 01:08:24	25
2024-11-22 01:08:25	24
2024-11-22 01:08:26	25

(continued)

Table 3. (*continued*)

Time	Frame_num
2024-11-22 01:08:27	25
2024-11-22 01:08:28	25

4 Evaluation

4.1 Testing Result

PaddleOCR recognition generates OCR_raw data, which contains date and time information for each frame. In vehicle speed identification, greater emphasis is placed on time information. For the standard time format "xx:xx:xx", if the OCR_raw data is not normalized, retrieving the standard time format solely from OCR_raw will result in very low recognition accuracy. Due to the presence of abnormal character interference in the ocr_raw of a large number of videos and the detection of non-standard format time information, such as "2024-11-2201:08:28" and "24-11-22 01:08:28", although regularization processing cannot completely cover all non-standard formats in ocr_raw. We apply optimal regularization processing to OCR_raw data, irrespective of processing status.

We use PaddleOCR to conduct ROI detection on the original image and the processed image, and evaluate the effectiveness of the algorithm by comparing the consistency of recognition time and calibration time. Figure 5 is an example of the original ROI image, and Fig. 6 is an example of the processed ROI image. Based on experience, whether it is the original image or the processed image, the one with less background interference will achieve a higher accuracy rate. The results show that the average ROI accuracy rate of the original image is approximately 93%, and that of the processed image is approximately 98%. For these four watermark characters, Type 1 is black-and-white interlaced, Type 2 is pure black, Type 3 is pure white, and Type 4 is black edges filled with white. The average value of the black threshold is 16, and the average value of the white threshold is 220. Table 4 presents the results of applying this method to the region of interest (ROI) in the original image and the ROI after the processing step.

Fig. 5. An example of the original ROI image

Fig. 6. An example of the image with the processed ROI

Table 4. The detection results of the method on both the original ROI and the preprocessed ROI

Type	original image	processed images	Type	original image	processed images
1–1	78.56%	98.80%	3–1	95.35%	100.00%
1–2	100.00%	100.00%	3–2	83.81%	99.78%
1–3	98.00%	97.34%	3–3	100.00%	100.00%
1–4	96.67%	100.00%	3–4	100.00%	100.00%
1–5	96.63%	97.76%	3–5	100.00%	100.00%
2–1	100.00%	100.00%	4–1	96.23%	99.56%
2–2	96.67%	100.00%	4–2	100.00%	100.00%
2–3	50.47%	86.68%	4–3	100.00%	100.00%
2–4	95.87%	100.00%	4–4	100.00%	100.00%
2–5	100.00%	99.60%	4–5	90.12%	99.41%

4.2 Limitations and Future Work

The GPU used in the experiment was one NVIDIA TITAN RTX, with 24 GB of video memory. After calculation, the average time required for each frame detection was approximately 0.076 s. For a 30-s video with a frame rate of 25 frames per second, the detection time for 750 frames was nearly 1 min. In judicial practice, the investigation authorities will download accident-related videos as evidence materials. The video duration is usually between 1 and 10 min. There are also videos with a duration longer than 1 h in practice. This method has relatively low efficiency in processing long videos.

From the experimental results, as long as the color of the time watermark is different from that of its background, and the image resolution can reach the level for human eye recognition, PaddleOCR can recognize most unprocessed images, and its success rate is independent of the font. If the resolution does not meet the conditions for human eye recognition, then OCR cannot recognize it either. Image preprocessing can enhance the difference between the time watermark and its background color, thereby improving the recognition accuracy. For the same ROI area of consecutive images, due to factors such as lighting and background changes, the detection results may be incorrect.

This method cannot effectively adapt to situations where the background changes significantly. Next, we will collect more datasets to conduct transfer learning in order to enhance the robustness of the method.

4.3 Application in Appraisal Practice

To address this issue, this paper integrates frame rate information extracted from video metadata with the periodic characteristics of time progression to generate inferred time stamps. This approach emphasizes the transformation of second-level positional information, which is critical for accurate identification in practical applications. Furthermore, by analyzing the organizational patterns of time and frame number data in statistical reports, the proposed method assists technicians in efficiently pinpointing abnormal video frames. Experimental results validate the effectiveness of this method in identification tasks.

5 Disclosure of Interests

The authors declare that they have no known competing financial intersts or personal relationships that could have appeared to influence the work reported in this paper.

Acknowledgments. This study was supported by grants from the Central Scientific Research Institutions of China(GY2023G-3), the National Key R&D Program of China(2023YFC3009703), and the Project of Shanghai Judicial Authentication Association(SHSFJD2023-015).

References

1. Feng, H., Fang, J., Chen, J., et al.: Current status of speed reconstruction technology in road traffic accidents. China Forensic Sci. **6**, 68–71 (2014)
2. Zhang, Z., Heng, W., Shi, S., et al.: A preliminary study on video time watermark analysis and application of vehicle speed appraisal. China Forensic Sci. **5**, 43–49 (2020)
3. Beauchamp, G., Pentecost, D., Koch, D., et al.: Speed Analysis from Video: a Method for Determining a Range in the Calculations. SAE Technical Paper 2021-01-0887 (2021). https://doi.org/10.4271/2021-01-0887
4. Wei, R., Li, P., Yu, Z.: Determination of Vehicle Speed and Trajectory Reconstruction for Discontinuous Time-Domain Video Using LED Ruler. SAE Technical Paper 2022-01-7116 (2022). https://doi.org/10.4271/2022-01-7116
5. Molnar, B., Terpstra, T., Voitel, T.: Accuracy of Timestamps in Digital and Network Video Recorders. SAE Technical Paper 2025-01-8690 (2025). https://doi.org/10.4271/2025-01-8690
6. Dong, K.: OCR technology in the application of the VAT invoice recognition analysis. J. Electron. Inf. Technol. **9**(6), 145–147+154 (2025). https://doi.org/10.19772/j.carolcarrollnki.2096-4455.2025.06.045
7. Kang, Z., Lin, S., Xu, J., et al.: Research on OCR recognition method for electrical cabinets oriented to intelligent inspection. Autom. Petrochem. Ind. **61**(04), 59–64 (2025)
8. Duan, Y., He, L.: Research on AI application in intelligent recognition and archiving of paper documents. Paper Inf. **07**, 150–151 (2025)

9. Meng, Q., Qin, L., Zong, C.: Research on optical character recognition technology and its application in customs supervision. China Port Sci. Technol. **7**(S1), 63–69 (2025)
10. Ding, S., He, J., Cheng, L., et al.: AI visual identification technology application in the liquor packaging printing detection. Autom. Appl. **66**(16), 51–56 (2025). https://doi.org/10.19769/j.zdhy.2025.16.015
11. Komal Adithya Reddy, B., Sohail, M., Kolla, M.: Bus route detection using YOLOv5 and PaddleOCR. In: Kumar, A., Ghinea, G., Merugu, S. (eds.) Proceedings of the Third International Conference on Cognitive and Intelligent Com-puting, Volume 1. ICCIC 2023. Cognitive Science and Technology. Springer, Singapore (2025). https://doi.org/10.1007/978-981-97-9262-7_1
12. Kaushal, A., Singh, A., Roy, A., Kumar, M.: Synergetic integration of YOLOv9, DeepSORT, and PaddleOCR: revolutionizing ANPR technology. In: Swaroop, A., Virdee, B., Correia, S.D., Polkowski, Z. (eds.) Proceedings of Data Analytics and Management. ICDAM 2024. Lecture Notes in Networks and Systems, vol. 1300. Springer, Singapore (2025). https://doi.org/10.1007/978-981-96-3361-6_15

Zero-ReID: Towards Modality-Unified Generalizable Person Re-identification with Inter-frame Identity Guidance

Yong Zhao[1,2], Yali Li[1,2], Zhaopeng Dou[1,2], and Shengjin Wang[1,2](✉)

[1] Beijing National Research Center for Information Science and Technology, Beijing, China

[2] Department of Electronic Engineering, Tsinghua University, Beijing, China

y-zhao23@mails.tsinghua.edu.cn, {liyali13,wgsgj}@tsinghua.edu.cn, dcp15@tsinghua.org.cn

Abstract. Generalizable person re-identification (ReID) focuses on learning identity-discriminative and domain-invariant features across multiple unseen scenarios, achieving significant progress on the canonical ReID. However, the generalization ability of visible-infrared person re-identification (VI-ReID) remains an open challenge, which has rarely been explored in the ReID community. To address this issue, we propose **Zero-Shot ReID** (Zero-ReID) framework, designed to tackle both generalizable canonical RGB and VI-ReID within a modality-unified model, which leverages large-scale unlabeled inter-frame data for contrastive learning. Zero-ReID intends to generalize to unseen real-world scenarios across both RGB and infrared modalities. Initially, we adopt the data synthesis techniques to translate RGB modality into missing infrared modalities. Then, Zero-ReID establishes identity correspondence priors between inter-frame visible images via bipartite graph matching. Finally, multiple priors-guided cross-modality correspondences are introduced to learn identity-discriminative and modality-shared features, while boosting performance on both modalities. Extensive experiments demonstrate that Zero-ReID achieves state-of-the-art zero-shot performance on 9 unseen public ReID datasets, **with 87.9% Rank-1 on Market-1501, and 21.7% Rank-1 on SYSU-MM01**.

Keywords: Person Re-identification · Self-Supervised Learning · Unified Modality · Generalization

1 Introduction

Person re-identification (ReID) [20,42] aims to identify a query person from a gallery of pedestrians captured by non-overlapping camera views, commonly used in surveillance. Most existing works concentrate on domain-specific ReID with an assumption of identically distributed training and test data. However,

P. Umapada et al. (Eds.): ICCPR 2025, CCIS 2811, pp. 28–42, 2026.
https://doi.org/10.1007/978-981-95-8315-7_3

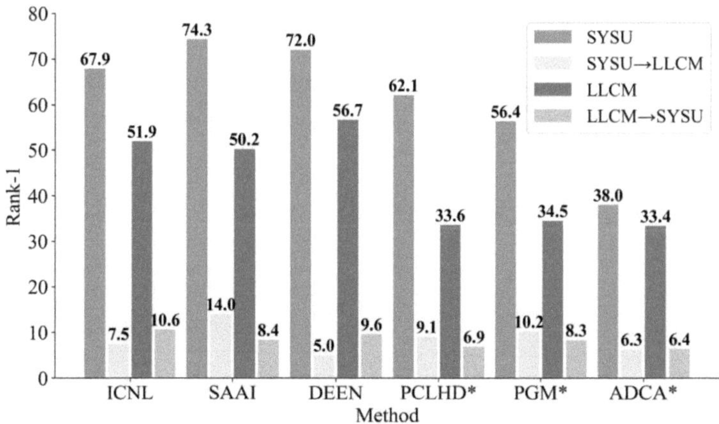

Fig. 1. The poor generalizability of existing VI-ReID methods. The methods marked with "*" are unsupervised. "A → B" denotes cross-domain retrieval on domain B after training on domain A.

such ReID models often suffer performance deterioration in real-world scenarios, due to significant variation of data distribution. Domain generalizable re-identification [3,4,43,46] (DG ReID) has emerged to address cross-domain challenge. Leveraging large-scale data for contrastive learning, recent works [6,32] have pioneered unsupervised domain generalizable ReID, surpassing their supervised counterparts. Visible-infrared person re-identification [36,41] (VI-ReID) is a variant of ReID designed for retrieving individuals captured in low-light environments. Figure 1 demonstrates existing VI-ReID models trained on small-scale annotated data exhibit poor generalization ability, primarily due to the modality gap and dramatic light condition change, making them difficult to deploy in real-world environments. To the best of our knowledge, domain generalizable VI-ReID (DG VI-ReID) remains rarely unexplored.

A generalizable ReID system should not only perform well across unseen scenarios but also tackle the challenge of modality uncertainty in real-world applications. However, existing approaches, which depend on limited data and are tailored to specific modalities, struggle with it. To this end, we propose a modality-unified representation learning framework Zero-ReID, which addresses both DG ReID and DG VI-ReID simultaneously. Within a unified model, Zero-ReID can generalize to unseen open-world scenarios across multiple modalities, including visible (RGB), near-infrared (abbreviated as infrared) and far-infrared (thermal) modalities.

To achieve modality-unified generalizable ReID, two key challenges must be addressed: large-scale data and identity discrimination. Zero-ReID leverages 47.8M unlabeled inter-frame images for contrastive learning, eliminating the need for labor-intensive human annotation while ensuring domain diversity. To overcome the lack of infrared modality data, we adopt data synthesis techniques to convert RGB pedestrians into near-infrared and thermal images. Zero-ReID establishes reliable identity association priors between visible inter-frame instances via bipartite matching. Then, the association priors serve as pseudo identity labels for cross-modality correspondence between visible and infrared person instances. Finally, multiple priors-weighted cross-modality correspondences are proposed to learn identity-discriminative and modality-shared features, while suppressing the noisy pseudo correspondence.

The main contributions can be summarized as follows:

- We propose Zero-ReID, a modality-unified ReID framework that tackles domain generalizable ReID across RGB and infrared modalities, via self-supervised learning.
- To learn identity-discriminative features for all modalities, we establish identity association priors between visible inter-frame instances and introduce multiple priors-guided cross-modality correspondences to learn modality-shared features.
- Through extensive experiments, Zero-ReID achieves 87.9% Rank-1 on Market-1501, 50.5% Rank-1 on MSMT17 and 21.7% Rank-1 on SYSU-MM01 simultaneously in the zero-shot scenario, outperforming state-of-the-art methods.

2 Related Work

2.1 Domain Generalizable Re-Identification

The goal of domain generalizable person re-identification (DG ReID) is to learn robust identity-discriminative features that generalize well across unseen domains without any fine-tuning. DG-ReID has gained much attention due to its promising application in real-world scenarios, which can be broadly categorized into two classes: supervised DG ReID and self-supervised DG ReID.

Supervised DG ReID methods typically utilize multi-source ReID datasets, trained with advanced network architectures [3,4,27,43] or sophisticated optimization techniques [46], to learn domain-invariant features. Nevertheless, these methods are trained with limited data, which hampers their generalization capabilities in real-world unseen scenarios. Recent DG ReID methods like [6,32] have focused on leveraging large-scale data for self-supervised learning, enabling further enhancement of ReID model's generalization ability. Both CycAs [32] and ISR [6] mine positive pairs from two adjacent video frames via sets matching algorithm and mitigate the impact of noisy pseudo labels, which have surprisingly outperformed supervised DG-ReID methods by a clear margin.

2.2 Visible-Infrared Person ReID

Existing VI-ReID works can be roughly divided into two classes: image-level alignment [28,45] and feature-level alignment [37,44]. The former focuses on generating cross-modality images to reduce modality discrepancy, while the latter

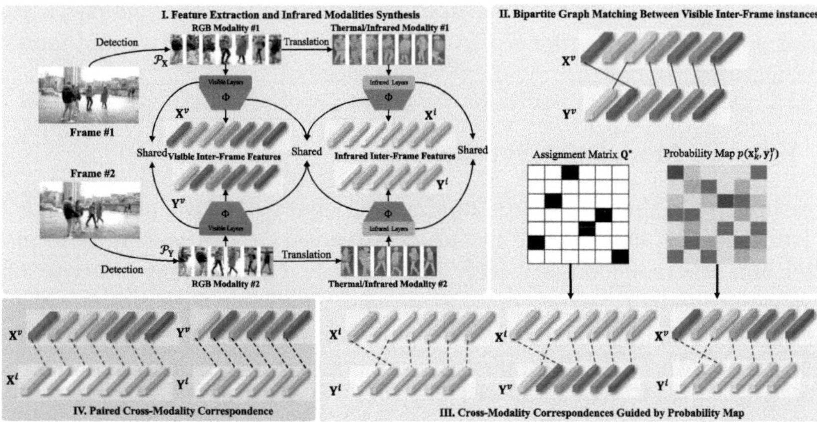

Fig. 2. The training framework of proposed Zero-ReID. The framework mainly consists of four parts: (I) Feature extraction and missing infrared modalities synthesis; (II) Reliable identity correspondence between **visible** inter-frame instances established by bipartite graph matching; (III) Multiple **inter-frame** cross modality correspondences guided by matching probability map to learn identity-discriminative and modality-shared features; (IV) Intrinsic **intra-frame** paired cross-modality correspondence.

aims to effectively learn a shared feature space across modalities. Unsupervised VI-ReID expands the concept of VI-ReID [36] by training in an unsupervised manner. The mainstream approaches [25,38] are largely inspired by Cluster-Contrast [5], which utilizes DBSCAN [7] to update cluster-based memory banks with pseudo-labels and then construct cross-modality correspondences.

As shown in Fig. 1, however, none of the above methods achieve satisfactory performance when testing on unseen domains. In contrast, our approach integrates VI-ReID with large-scale self-supervised learning to enhance the generalizability of VI-ReID while boosting zero-shot performance on the canonical ReID, a challenge that has rarely been explored in the ReID community.

3 Method

3.1 Overview

The framework of our proposed Zero-ReID is illustrated in Fig. 2. Two pedestrian sets $\mathcal{P}_\mathbf{X}$ and $\mathcal{P}_\mathbf{Y}$ are cropped from two inter-frame adjacent images with an appropriate time interval [33]. We begin by translating the RGB modality to missing infrared modalities. Two modality-specific shallow layers and one shared backbone are utilized to capture modality-specific information and modality-shared features. Firstly, visible and infrared person sets are mapped to embeddings in the feature space, which corresponds to two distinct modal features: visible inter-frame features $\mathbf{X}^v, \mathbf{Y}^v$ and infrared inter-frame features $\mathbf{X}^i, \mathbf{Y}^i$. Then, bipartite graph matching [15] is exploited to mine positive instances between visible inter-frame features, which establishes reliable identity association between inter-frame instances. Finally, multiple cross-modality correspondences guided by prior association are proposed to learn identity-discriminative and modality-shared features.

3.2 Infrared Modalities Synthesis

To synthesis the missing infrared modalities, Channel Augmentation [41] (CA) and TransUNet [1] network are introduced to generate the near-infrared modality [26] and thermal modality [35], respectively. The imaging principle of the near-infrared modality, which operates at shorter wavelength, closely resembles that of visible imaging, whereas thermal imaging, operating at longer far-infrared wavelength, captures the thermal radiation emitted by the human body.

CA is a channel-exchanging augmentation method used to synthesize near-infrared modality from RGB input, while TransUNet is trained on a large-scale paired visible-thermal dataset LLVIP [14] to produce high-fidelity thermal persons. The Zero-ReID training pipeline is identical for both near-infrared and thermal modalities, except for the translation operation. For simplicity, Fig. 2 illustrates only the TransUNet translation. For faster data access during training, thermal persons are generated through visible-thermal TransUNet translation before training, saved in the form of KVs. The generated samples, as shown in Fig. 3, demonstrate satisfactory style translation.

Fig. 3. The synthetic persons: RGB→infrared→thermal.

3.3 Bipartite Graph Matching Priors

The extracted visible inter-frame features are $\mathbf{X}^v = [\mathbf{x}_1^v, \mathbf{x}_2^v, \ldots, \mathbf{x}_m^v] \in \mathbb{R}^{m\times d}$ and $\mathbf{Y}^v = [\mathbf{y}_1^v, \mathbf{y}_2^v, \ldots, \mathbf{y}_n^v] \in \mathbb{R}^{n\times d}$, where d is the feature dimension, m and n are the number of person instances in $\mathcal{P}_\mathbf{X}$ and $\mathcal{P}_\mathbf{Y}$ sets respectively. All features are l_2-normalized. Bipartite matching is commonly used to tackle the one-to-one matching of two sets, and it can be seamlessly applied to establish correspondence between $\mathbf{X}^v$ and $\mathbf{Y}^v$. Such correspondence is formulated as a maximum bipartite matching problem, where the weight between $\mathbf{x}_k^v$ and $\mathbf{y}_j^v$ is defined as their cosine distance. The optimal assignment matrix $\mathbf{Q}^* \in \mathbb{R}^{m\times n}$ is solved by Hungarian algorithm [15] during training. $\mathbf{Q}^*$ is a binary matrix, where $\mathbf{Q}_{kj}^* = 1$ if $\mathbf{x}_k^v$ and $\mathbf{y}_j^v$ are matched (denoted as a positive pair), and $\mathbf{Q}_{kj}^* = 0$ otherwise.

Reliability-Guided InfoNCE Loss. The mined positive pairs may be incorrect, hindering the convergence of contrastive learning. This occurs when two pedestrian sets $\mathcal{P}_{\mathbf{X}}$ and $\mathcal{P}_{\mathbf{Y}}$ are disjoint due to large sampling interval. To address this issue, the reliability of positive pairs is used to suppress incorrect matching:

$$p(\mathbf{x}_k^v) = \frac{\exp(\mathbf{x}_k^v \cdot \mathbf{y}_{j^+}^v/\tau)}{\sum_{z=1}^{n} \exp(\mathbf{x}_k^v \cdot \mathbf{y}_z^v/\tau)}, \tag{1}$$

where $j^+ = \arg_j \mathbf{Q}_{kj}^* = 1$ and τ is the temperature hyper-parameter. Then, the reliability-guided InfoNCE [11] loss can be defined as:

$$\mathcal{L}_{\mathrm{RC}}(\mathbf{x}_k^v) = -sg(p^\gamma(\mathbf{x}_k^v))\log(p(\mathbf{x}_k^v)), \tag{2}$$

where $sg(\cdot)$ represents the stop-gradient and γ is the controlling hyper-parameter greater than 1. In order to stabilize the convergence during training, the average of $\mathcal{L}_{\mathrm{RC}}(\mathbf{x}_k^v)$ is scaled up to:

$$\mathcal{L}_{\mathrm{RC}}(\mathbf{X}^v) = sg(\overline{\mathcal{L}_{\mathrm{RC},\gamma=0}(\mathbf{x}_k^v)}/\overline{\mathcal{L}_{\mathrm{RC}}(\mathbf{x}_k^v)}) \cdot \overline{\mathcal{L}_{\mathrm{RC}}(\mathbf{x}_k^v)}, \tag{3}$$

where $\overline{(\cdot)}$ indicates the mean operation for all matched positive pairs. To improve contrastive learning, we adopt symmetric operation for $\mathcal{P}_{\mathbf{Y}}$. The final bidirectional reliability-guided InfoNCE loss is then computed as:

$$\mathcal{L}_{\mathrm{RC}} = \tfrac{1}{2}(\mathcal{L}_{\mathrm{RC}}(\mathbf{X}^v) + \mathcal{L}_{\mathrm{RC}}(\mathbf{Y}^v)). \tag{4}$$

3.4 Priors-Guided Cross-Modality Matching

The association obtained by bipartite graph matching between visible inter-frame instances, can serve as important priors for cross-modality correspondence, since the model struggle to learn identity-discriminative and modality-shared features during the initial training without the (pseudo) identity labels.

Priors-Guided InfoNCE Loss. Since $\mathbf{X}^i$ and $\mathbf{Y}^i$ are transformed from $\mathbf{X}^v$ and $\mathbf{Y}^v$ with shared identity for each instance, the inter-frame matching priors can be directly applied to cross-modality correspondence: $(\mathbf{X}^v, \mathbf{Y}^i)$ and $(\mathbf{X}^i, \mathbf{Y}^v)$. The intuition is that prior probability map $p(\mathbf{x}_k^v, \mathbf{y}_j^v)$ and assignment matrix $\mathbf{Q}^*$ can provide guidance for cross-modality correspondence naturally. Based on this, we propose bidirectional priors-guided InfoNCE loss:

$$\mathcal{L}_{\mathrm{GC}}(\mathbf{x}_k^v \to \mathbf{y}_{j^+}^i) = -sg(p(\mathbf{x}_k^v))\log(q(\mathbf{x}_k^v, \mathbf{y}_{j^+}^i)), \tag{5}$$

$$\mathcal{L}_{\mathrm{GC}}(\mathbf{y}_{j^+}^i \to \mathbf{x}_k^v) = -sg(p(\mathbf{y}_{j^+}^v))\log(q(\mathbf{y}_{j^+}^i, \mathbf{x}_k^v)), \tag{6}$$

$$q(\mathbf{x}_k^v, \mathbf{y}_j^i) = \frac{\exp(\mathbf{x}_k^v \cdot \mathbf{y}_j^i/\tau)}{\sum_{z=1}^{n} \exp(\mathbf{x}_k^v \cdot \mathbf{y}_z^i/\tau)}, \tag{7}$$

where $(\mathbf{x}_k^v, \mathbf{y}_{j^+}^i)$ is matched cross-modality positive pair. Finally, we can derive the symmetric priors-guided InfoNCE loss for $(\mathbf{X}^v, \mathbf{Y}^i)$:

$$\mathcal{L}_{\mathrm{GC}}(\mathbf{X}^v, \mathbf{Y}^i) = \tfrac{1}{2}(\overline{\mathcal{L}_{\mathrm{GC}}(\mathbf{x}_k^v \to \mathbf{y}_{j^+}^i)} + \overline{\mathcal{L}_{\mathrm{GC}}(\mathbf{y}_{j^+}^i \to \mathbf{x}_k^v)}). \tag{8}$$

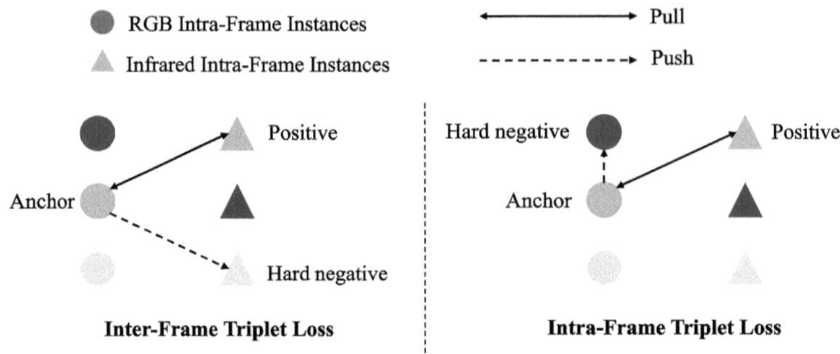

Fig. 4. Inter-Frame and Intra-Frame Triplet Losses.

Besides the InfoNCE loss, Triplet [24] loss is commonly used to boost ReID performance. We propose priors-guided cross-modality inter-frame and intra-frame Triplet losses to minimize the distance between inter-frame positive samples, while maximizing the distance between intra-frame negative samples, as shown in Fig. 4.

Inter-Frame Triplet Loss. For each anchor $\mathbf{x}_k^v$ in RGB frame, the cross-modality positive sample $\mathbf{y}_{j+}^i$ (via $\mathbf{Q}^*$) and hard negative sample $\mathbf{y}_{j-}^i$ can be selected with minimum Euclidean distance. The priors-guided inter-frame Triplet loss can be computed as:

$$\tilde{\mathcal{L}}_{\mathrm{TC}}^{\mathrm{inter}}(\mathbf{x}_k^v, \mathbf{Y}^i) = [d(\mathbf{x}_k^v, \mathbf{y}_{j+}^i) - d(\mathbf{x}_k^v, \mathbf{y}_{j-}^i) + m]_+, \tag{9}$$

$$q(\mathbf{x}_k^v, \mathbf{y}_{j+}^i, \mathbf{y}_{j-}^i) = sg([p(\mathbf{x}_k^v, \mathbf{y}_{j+}^v) - p(\mathbf{x}_k^v, \mathbf{y}_{j-}^v)]_+), \tag{10}$$

$$\mathcal{L}_{\mathrm{TC}}^{\mathrm{inter}}(\mathbf{x}_k^v, \mathbf{Y}^i) = q(\mathbf{x}_k^v, \mathbf{y}_{j+}^i, \mathbf{y}_{j-}^i) \cdot \tilde{\mathcal{L}}_{\mathrm{TC}}^{\mathrm{inter}}(\mathbf{x}_k^v, \mathbf{Y}^i), \tag{11}$$

where $[x]_+$ denotes $\max(x, 0)$, $d(\cdot)$ computes the Euclidean distance between two samples and m is the margin hyper-parameter. The reliability of the inter-triplet is determined by Eq. (10), which computes the difference between prior $p(\mathbf{x}_k^v, \mathbf{y}_{j+}^v)$ and $p(\mathbf{x}_k^v, \mathbf{y}_{j-}^v)$ and truncates it down to 0, to eliminate the effect of the noisy triplet.

Intra-frame Triplet Loss. However, the inter-frame triplet alone can't guarantee that intra-frame samples are separated from each other. Based on the inherent identity mutual exclusiveness of $\mathcal{P}_{\mathbf{X}}$, The priors-guided intra-frame Triplet loss is then introduced to distribute intra-modality's embeddings uniformly on the unit hypersphere, which is formulated as:

$$\tilde{\mathcal{L}}_{\mathrm{TC}}^{\mathrm{intra}}(\mathbf{x}_k^v, \mathbf{Y}^i) = [d(\mathbf{x}_k^v, \mathbf{y}_{j+}^i) - d(\mathbf{x}_k^v, \mathbf{x}_{l-}^v) + m]_+, \tag{12}$$

$$\mathcal{L}_{\mathrm{TC}}^{\mathrm{intra}}(\mathbf{x}_k^v, \mathbf{Y}^i) = sg(p(\mathbf{x}_k^v, \mathbf{y}_{j+}^i)) \cdot \tilde{\mathcal{L}}_{\mathrm{TC}}^{\mathrm{intra}}(\mathbf{x}_k^v, \mathbf{Y}^i), \tag{13}$$

where $\mathbf{x}_{l^-}^v$ denotes intra-frame hard negative sample. Then, we apply mean, symmetric and operation for Eqs. (11) and (13). The final priors-guided Triplet losses can be formulated as:

$$\mathcal{L}_{\mathrm{TC}}^{\mathrm{inter}}(\mathbf{X}^v, \mathbf{Y}^i) = \frac{1}{2}(\overline{\mathcal{L}_{\mathrm{TC}}^{\mathrm{inter}}(\mathbf{x}_k^v, \mathbf{Y}^i)} + \overline{\mathcal{L}_{\mathrm{TC}}^{\mathrm{inter}}(\mathbf{X}^v, \mathbf{y}_j^i)}), \tag{14}$$

$$\mathcal{L}_{\mathrm{TC}}^{\mathrm{intra}}(\mathbf{X}^v, \mathbf{Y}^i) = \frac{1}{2}(\overline{\mathcal{L}_{\mathrm{TC}}^{\mathrm{intra}}(\mathbf{x}_k^v, \mathbf{Y}^i)} + \overline{\mathcal{L}_{\mathrm{TC}}^{\mathrm{intra}}(\mathbf{X}^v, \mathbf{y}_j^i)}). \tag{15}$$

Finally, the overall priors-guided matching loss is computed by summing up Eqs. (8),(14) and (15):

$$\mathcal{L}_{\mathrm{gcm}}(\mathbf{X}^v, \mathbf{Y}^i) = \mathcal{L}_{\mathrm{GC}}(\mathbf{X}^v, \mathbf{Y}^i) + \mathcal{L}_{\mathrm{TC}}^{\mathrm{inter}}(\mathbf{X}^v, \mathbf{Y}^i) + \mathcal{L}_{\mathrm{TC}}^{\mathrm{intra}}(\mathbf{X}^v, \mathbf{Y}^i). \tag{16}$$

$\mathcal{L}_{\mathrm{gcm}}(\mathbf{X}^i, \mathbf{Y}^v)$ and $\mathcal{L}_{\mathrm{gcm}}(\mathbf{X}^i, \mathbf{Y}^i)$ can be derived in the same way.

Paired Cross-Modality Correspondence. As discussed in previous sections, $(\mathbf{X}^v, \mathbf{X}^i)$ and $(\mathbf{Y}^v, \mathbf{Y}^i)$ are one-to-one paired cross-modality samples with distinct identities. We directly apply standard InfoNCE loss and Triplet loss on these samples:

$$\mathcal{L}_{\mathrm{cm}} = \mathcal{L}_{\mathrm{nce}}(\mathbf{X}^v, \mathbf{X}^i) + \mathcal{L}_{\mathrm{tri}}(\mathbf{X}^v, \mathbf{X}^i) + \mathcal{L}_{\mathrm{nce}}(\mathbf{Y}^v, \mathbf{Y}^i) + \mathcal{L}_{\mathrm{tri}}(\mathbf{Y}^v, \mathbf{Y}^i). \tag{17}$$

3.5 Optimization and Training Objective

Modality-Shared Memory Banks. Large negative samples are essential for contrastive learning to uniformly distribute the feature space and improve generalization [11,31]. Therefore, modality-shared memory banks are established to store both RGB and infrared features alternately, along with their video IDs. During training, the memory banks are updated in a dequeue-enqueue manner, where samples from different video IDs are treated as negative samples. For each feature $\mathbf{f}^v$ in $[\mathbf{X}^v, \mathbf{Y}^v]$, M hardest negative samples are selected from the memory banks to minimize similarities:

$$\mathcal{L}_{\mathrm{XBM}}^v(\mathbf{f}^v) = \tfrac{1}{M}\textstyle\sum_{m=1}^{M} \log(1 + \exp(\mathbf{f}^v \cdot \mathbf{z}_m)), \tag{18}$$

where $\mathbf{z}_m$ represents the m-th hardest negative sample. Let $\mathcal{L}_{\mathrm{XBM}} = \frac{1}{2}(\overline{\mathcal{L}_{\mathrm{XBM}}^v(\mathbf{f}^v)} + \overline{\mathcal{L}_{\mathrm{XBM}}^i(\mathbf{f}^i)})$. Finally, the overall training loss can be:

$$\begin{aligned}\mathcal{L} = \lambda_1\mathcal{L}_{\mathrm{RC}} &+ \lambda_2(\mathcal{L}_{\mathrm{gcm}}(\mathbf{X}^v, \mathbf{Y}^i) + \mathcal{L}_{\mathrm{gcm}}(\mathbf{X}^i, \mathbf{Y}^v)) \\ &+ \lambda_3\mathcal{L}_{\mathrm{gcm}}(\mathbf{X}^i, \mathbf{Y}^i) + \lambda_4\mathcal{L}_{\mathrm{cm}} + \lambda_5\mathcal{L}_{\mathrm{XBM}},\end{aligned} \tag{19}$$

where $\lambda_1, \ldots, \lambda_5$ are hyper-parameters for trade-off.

4 Experiments

4.1 Datasets and Evaluation Protocols

Training Dataset. Inspired by [9,32], we collect large-scale inter-frame videos from YouTube for training. Specifically, we crawl 10K videos using queries like `cityname + streetview` to search videos of cities worldwide. The videos contains 100 major cities to ensure domain diversity. The data processing includes:

- Firstly, use the PySceneDetect[1] tool to split videos into non-overlapping clips, removing shot transitions in each clip.
- Next, a pedestrian detector JDE [33] is run on 5 uniformly sampled frames of each clip to crop the persons and discard the clip where the average number of persons is less than 5.
- Finally, apply the JDE detector to crop person images every 7 frames for the remaining video clips. The resulting dataset consists of 74K video clips, with about 73 frames per clip, 8.9 person instances per frame on average.

Evaluation Datasets and Protocols. Extensive experiments are conducted on RGB ReID datasets and VI-ReID datasets.

- **RGB ReID datasets.** Two common protocols are adopted. Protocol-1 uses a leave-one-out setting for Market-1501 [47], DukeMTMC [23] (has been withdrawn), CUHK03 [17] and MSMT17 [34], selecting one domain for evaluation and other domains for training. Protocol-2 combines all images in CUHK02 [16], CUHK03 [17], CUHK-SYSU [39], Market-1501 and DukeMTMC as training data, while evaluating on four small-scale datasets: PRID [13], GRID [19], VIPeR [10] and iLIDs [30]. Following [27], we report the average result of 10 random splits of the query and gallery sets.
- **VI-ReID datasets.** We evaluate our Zero-ReID on three benchmarks in VI-ReID: SYSU-MM01 [36], LLCM [44] and RegDB [21]. Both SYSU-MM01 and LLCM are infrared-visible ReID datasets, while RegDB is thermal-visible. Given the scarcity of works on DG VI-ReID, we establish a leave-one-out evaluation protocol for SYSU-MM01 and LLCM, referred to as Protocol-3. We only report zero-shot performance on RegDB, as no other visible-thermal datasets are known. For SYSU-MM01 and LLCM, we present the results with single-shot gallery setting. For RegDB, we report results from 10 repeated random splits of the training and testing sets.

Evaluation Metrics. Mean average precision (mAP) and cumulative matching characteristic (CMC).

4.2 Implementation Details

We use ResNet50 [12] with IBN [22] as backbone. The stride of `layer4` is set to 1. The visible and infrared layers in Fig. 2 correspond to the shallow layers before `layer1`, while the shared backbone consists of the remaining layers. All input images are resized to 288 $\times$ 144, augmented by random horizontal flipping and color jitter. Following [6,32], we combine multiple frames from different video clips into a super frame for large batch training. The number of person instances in a super frame is truncated to 65, with 3 adjacent frames sampled per clip. The total training batch size is $1560 = 65 \times 3 \times 4 \times 2$. Each video clip is sampled 16 times on average per epoch. The model is trained for 50 epochs and optimized using AdamW with a cosine learning rate decay, starting from 1×10^{-4}.

[1] http://scenedetect.com/.

Table 1. Comparison with state-of-the-art methods under Protocol-1 and Protocol-2 (RGB). "Sup." indicates whether the method is supervised or not. † indicates the results are re-implemented or re-produced based on the author's code on Github.

Method	Sup.	Protocol-1						Protocol-2							
		Market-1501		MSMT17		CUHK03		PRID		GRID		VIPeR		iLIDs	
		R-1	mAP	R-1	mAP	R-1	mAP	R-1	mAP	R-1	mAP	R-1	mAP	R-1	mAP
M^3L [46]	✔	75.9	50.2	36.9	14.7	33.1	32.1	55.0	65.3	40.0	50.5	60.8	68.2	65.0	74.3
RaMoE [4]	✔	82.0	56.5	34.1	13.5	36.6	35.5	57.7	67.3	46.8	54.2	56.6	64.6	**85.0**	**90.2**
MetaBIN [3]	✔	84.5	67.2	41.2	18.8	43.1	43.0	61.2	70.8	50.2	57.9	55.9	64.3	74.7	82.7
ACL [43]	✔	**90.6**	76.8	47.3	21.7	50.1	49.4	63.0	73.4	55.2	65.7	**66.4**	**75.1**	81.8	86.5
BAU† [2]	✔	90.5	**79.2**	**48.4**	**22.7**	**50.5**	**50.4**	**70.3**	**78.0**	**60.2**	**68.2**	64.1	73.3	**85.4**	89.6
MoCo [11]	✘	10.5	2.6	0.5	0.2	0.3	0.7	6.5	10.9	2.8	6.9	4.0	7.5	38.8	46.4
CycAs [32]	✘	80.3	57.5	43.9	20.2	25.8	26.5	58.8	67.7	52.5	62.3	57.3	66.0	85.2	90.4
ISR [6]	✘	85.1	65.1	45.7	21.2	26.1	27.4	59.7	70.8	55.8	65.2	58.0	66.6	87.6	91.7
OReID^{T}-R50 (Ours)	✘	81.3	54.4	48.7	18.4	17.4	19.8	55.3	66.6	48.9	56.6	48.9	60.4	**89.1**	**92.4**
OReID^{N}-R50 (Ours)	✘	**87.9**	**69.6**	**50.5**	**23.5**	**30.9**	**32.3**	**60.1**	**71.3**	**56.3**	**65.2**	**62.3**	**71.1**	88.4	91.8

We refer to Zero-ReID trained solely with near-infrared modality as OReID^{N}, while the model trained with both near-infrared and thermal modalities as OReID^{T}. Specifically, when training OReID^{T} network, RGB persons in Fig. 2 are augmented by CA with a low probability p, which denotes taht the RGB and near-infrared modalities share the same network.

The training pipeline follows a two-stage approach: In the first stage, the model is trained only using $\mathcal{L}_{\mathrm{RC}}$ loss on RGB inter-frame instances; then, it is fine-tuned using the proposed Zero-ReID with weights from the first stage. A large value of λ_1 helps maintain reliable bipartite graph matching priors during the second stage.

4.3 Zero-Shot Performance

We evaluate Zero-ReID on out-of-distribution datasets mentioned in Sect. 4.1 without any fine-tuning. All the compared methods adopt ResNet50 as backbone, unless otherwise specified.

Results on RGB ReID Datasets. As illustrated in Table 1, our OReID^{N}-R50 (ResNet50) achieves competitive results against supervised methods like MetaBIN [3] and ACL [43], outperforming BAU [2] by 2.1% Rank-1 and 0.8% mAP on the most challenging MSMT17 dataset. Compared to unsupervised methods, OReID^{N}-R50 demonstrates state-of-the-art performance, outperforming CycAs [32] and ISR [6] by a significant margin. MoCo [11] fails to differentiate between persons of different identities, because it is designed to learn instance-level discrimination instead of identity-level discrimination. OReID^{T}-R50 yields inferior performance relative to OReID^{N}-R50, which we attribute to the large modality gap between RGB and thermal modality. We argue that moderate multi-modal training can boost performance on both RGB and infrared modalities, while excessive multi-modal training may have negative effects. The impact

Table 2. Comparison with state-of-the-art methods under Protocol-3 (visible-infrared). † indicates the results are re-implemented based on the author's code. "A → B" denotes cross-domain retrieval on domain B after training on domain A.

Methods	Sup.	LLCM → SYSU-MM01						SYSU-MM01 → LLCM					
		All Search			Indoor Search			Infrared2Visible			Visible2Infrared		
		R-1	R-10	mAP	R-1	R-10	mAP	R-1	R-10	mAP	R-1	R-10	mAP
DEEN† [44]	✔	11.9	40.5	11.6	14.2	49.8	21.4	11.9	31.6	15.5	13.1	34.3	16.7
SAAI† [8]	✔	12.0	43.0	12.3	14.5	51.3	22.3	13.7	34.9	17.7	14.1	36.6	17.7
ICNL† [40]	✔	9.2	34.9	10.7	10.7	46.9	18.8	7.8	24.7	11.0	10.9	32.2	14.7
OReID^{N}-R50 (Ours)	✘	**21.7**	58.9	23.5	**29.6**	**71.5**	**39.1**	20.8	48.5	27.6	22.4	52.0	27.5
OReID^{T}-R50 (Ours)	✘	21.4	**63.6**	**23.6**	27.5	70.1	36.9	**22.2**	**51.2**	**28.9**	**24.5**	**54.7**	**29.5**

Table 3. Zero-shot performance on RegDB.

Method	Sup.	Seen.	RegDB			
			Thermo2Vis		Vis2Thermo	
			R-1	mAP	R-1	mAP
H2H [18]	✘	✔	–	–	23.8	18.9
OTLA [29]	✘	✔	32.1	28.6	32.9	29.7
OReID^{T}-R50	✘	✘	**41.6**	**44.7**	32.8	**33.4**

of multi-modal training will be discussed in Sect. 4.4. Despite the performance degradation caused by modality discrepancy, OReID^{T}-R50 still gains a 3.0% improvement over ISR on MSMT17.

Results on VI-ReID Datasets. Since few studies have explored the generalization of VI-ReID, we compare Zero-ReID with SOTA supervised VI-ReID methods under Protocol-3 in Table 2. The results present that Zero-ReID outperforms all the other methods substantially, confirming the potential generalization capability of Zero-ReID, while existing VI-ReID methods struggle with it. Table 3 demonstrates that OReID^{T}-R50 achieves over 40% Rank-1 and mAP on RegDB, the only visible-thermal ReID dataset, outperforming OTLA [29] by a clear margin.

4.4 Ablation Study

The Effectiveness of Loss Components. Table 4 demonstrates the impact of different loss functions. Experiments are conducted with OReID^{T}-R50 model. The baseline model is trained exclusively with $\mathcal{L}_{\mathrm{RC}}$ (with $\mathcal{L}_{\mathrm{GC}}$ replaced by $\mathcal{L}_{\mathrm{RC}}$). Applying the naive $\mathcal{L}_{\mathrm{RC}}$ for cross-modality correspondence results in a significant performance drop. However, this degradation can be alleviated by introducing $\mathcal{L}_{\mathrm{GC}}$, which weighs cross-modality correspondence via visible inter-frame matching priors, rather than relying on the probability predicted by itself. Incorporating $\mathcal{L}_{\mathrm{TC}}^{\mathrm{inter}}$ further enhances the cross-modality retrieval performance on RegDB,

Table 4. Ablation study of loss functions.

Index	Loss Component					Market-1501		MSMT17		SYSU-MM01		LLCM		RegDB		Average	
	$\mathcal{L}_{\text{RC}}$	$\mathcal{L}_{\text{GC}}$	$\mathcal{L}_{\text{TC}}^{\text{inter}}$	$\mathcal{L}_{\text{TC}}^{\text{intra}}$	$\mathcal{L}_{\text{XBM}}$	RGB		RGB		All search		IR2Vis		Vis2IR			
						R-1	mAP	R-1	mAP	R-1	mAP	R-1	mAP	R-1	mAP	R-1	mAP
Baseline	✓	✗	✗	✗	✓	56.6	24.6	29.1	7.8	17.5	18.9	17.3	22.6	40.1	43.2	32.1	23.4
1	✓	✓	✗	✗	✓	81.0	53.6	48.1	18.1	21.2	23.3	21.6	28.4	39.7	44.0	42.3	33.5
2	✓	✓	✓	✗	✓	80.7	53.4	48.4	18.1	20.9	23.0	22.1	28.8	**41.7**	**45.7**	42.8	33.8
3	✓	✓	✓	✓	✗	80.3	52.0	48.0	17.7	20.2	22.5	21.4	28.1	39.3	43.5	41.8	32.8
4	✓	✓	✓	✓	✓	**81.3**	**54.4**	**48.7**	**18.4**	**21.4**	**23.6**	**22.2**	**28.9**	41.6	44.7	**43.0**	**34.0**

Table 5. Ablation study of multi-modal training. "Thermo." indicates whether thermal modality is utilized, while "Infrared." means whether infrared modality branch network is trained. "50ep" and "100ep" refer to training for 50 and 100 epochs.

Method	CA p	Thermo.	Infrared.	Market-1501		MSMT17		SYSU-MM01		LLCM		RegDB	
				RGB		RGB		All search		Infrared2Visible		Thermal2Visible	
				R-1	mAP	R-1	mAP	R-1	mAP	R-1	mAP	R-1	mAP
ISR-50ep [6]	0.1	✗	✗	85.0	61.3	46.7	19.6	16.4	17.4	18.9	25.4	–	–
OReID^{T}-R50	**0.1**	✓	✗	81.3	54.4	48.7	18.4	**21.4**	**23.6**	**22.2**	**28.9**	**41.6**	**44.7**
ISR-100ep [6]	0.0	✗	✗	84.4	66.1	43.7	20.3	–	–	–	–	–	–
OReID^{N}-R50	0.0	✗	✓	**87.9**	**69.6**	**50.5**	**23.5**	21.7	23.5	20.8	27.6	–	–

but with little degradation on other datasets, which can be restored by additional $\mathcal{L}_{\text{TC}}^{\text{intra}}$. The synergy of $\mathcal{L}_{\text{TC}}^{\text{inter}}$ and $\mathcal{L}_{\text{TC}}^{\text{intra}}$ minimizes the distance between cross-modality positive samples, while ensuring that intra-modality's embeddings are distributed uniformly on the unit hypersphere, leading to the best average performance. $\mathcal{L}_{\text{XBM}}$ contributes to +1.2% improvements in both Rank-1 and mAP.

The Impact of Multi-modal Training. From previous experiments, OReID^{T} surprisingly exhibits slightly better visible-infrared retrieval performance than OReID^{N}, despite a low CA-augmented probability. One might wonder whether applying CA to ISR [6] in the same way as OReID^{T} could achieve comparable cross-modality retrieval performance to the proposed Zero-ReID. However, the results in Table 5 demonstrate its infeasibility. These can be attributed to the additional thermal modality regularization, which prevents the feature space of the infrared modality from collapsing into that of the RGB modality. CA-synthesized infrared modality acts as an intermediate modality between the thermal and RGB modality. The bottom rows of Table 5 also show that continual training of ISR alone doesn't work well, while OReID^{N} benefits from multi-modal training with the synthesized infrared modality, achieving better RGB retrieval performance. Moderate multi-modal training (OReID^{N}) can distribute the feature space and enhance generalization, while multi-modal training with large modality (OReID^{T}) discrepancy can harm the learning of RGB discriminative features. It should be mentioned that OReID^{T} is specifically tailored for VI-ReID task with additional CA to visible layers.

5 Conclusion

In this paper, we propose a modality-unified representation learning framework Zero-ReID that addresses both domain generalizable ReID and VI-ReID. Through large-scale self-supervised training, identity association priors mining, and multiple priors-guided cross-modality correspondences, Zero-ReID can generalize effectively to unseen open-world scenarios across RGB and infrared modalities without additional fine-tuning. Comprehensive experiments demonstrate the superiority of our proposed Zero-ReID in zero-shot scenario. Moreover, we analyze the regularization of multi-modal training on distributing the feature space, providing new insights for future research on domain-generalizable ReID.

Acknowledgment. This work is supported by the National Natural Science Foundation of China under Grant Nos. 61701277 and 61771288.

Disclosure of Interests. The authors have no competing interests to declare that are relevant to the content of this article.

References

1. Chen, J., et al.: TransUNet: transformers make strong encoders for medical image segmentation. arXiv preprint arXiv:2102.04306 (2021)
2. Cho, Y., Kim, J., Kim, W.J., Jung, J., Yoon, S.E.: Generalizable person re-identification via balancing alignment and uniformity. arXiv preprint arXiv:2411.11471 (2024)
3. Choi, S., Kim, T., Jeong, M., Park, H., Kim, C.: Meta batch-instance normalization for generalizable person re-identification. In: CVPR, pp. 3425–3435 (2021)
4. Dai, Y., Li, X., Liu, J., Tong, Z., Duan, L.Y.: Generalizable person re-identification with relevance-aware mixture of experts. In: CVPR, pp. 16145–16154 (2021)
5. Dai, Z., Wang, G., Yuan, W., Zhu, S., Tan, P.: Cluster contrast for unsupervised person re-identification. In: ACCV, pp. 1142–1160 (2022)
6. Dou, Z., Wang, Z., Li, Y., Wang, S.: Identity-seeking self-supervised representation learning for generalizable person re-identification. In: ICCV, pp. 15847–15858 (2023)
7. Ester, M., Kriegel, H.P., Sander, J., Xu, X., et al.: A density-based algorithm for discovering clusters in large spatial databases with noise. In: KDD, vol. 96, pp. 226–231 (1996)
8. Fang, X., Yang, Y., Fu, Y.: Visible-infrared person re-identification via semantic alignment and affinity inference. In: ICCV, pp. 11270–11279 (2023)
9. Fu, D., et al.: Unsupervised pre-training for person re-identification. In: CVPR, pp. 14750–14759 (2021)
10. Gray, D., Tao, H.: Viewpoint invariant pedestrian recognition with an ensemble of localized features. In: ECCV, pp. 262–275 (2008)
11. He, K., Fan, H., Wu, Y., Xie, S., Girshick, R.: Momentum contrast for unsupervised visual representation learning. In: CVPR, pp. 9729–9738 (2020)
12. He, K., Zhang, X., Ren, S., Sun, J.: Deep residual learning for image recognition. In: CVPR, pp. 770–778 (2016)

13. Hirzer, M., Beleznai, C., Roth, P.M., Bischof, H.: Person re-identification by descriptive and discriminative classification. In: Scandinavian Conference on Image Analysis, pp. 91–102 (2011)
14. Jia, X., Zhu, C., Li, M., Tang, W., Zhou, W.: LLVIP: a visible-infrared paired dataset for low-light vision. In: ICCV, pp. 3496–3504 (2021)
15. Kuhn, H.W.: The Hungarian method for the assignment problem. Naval Res. Logist. Q. 83–97 (1955)
16. Li, W., Wang, X.: Locally aligned feature transforms across views. In: CVPR, pp. 3594–3601 (2013)
17. Li, W., Zhao, R., Xiao, T., Wang, X.: DeepReid: deep filter pairing neural network for person re-identification. In: CVPR, pp. 152–159 (2014)
18. Liang, W., Wang, G., Lai, J., Xie, X.: Homogeneous-to-heterogeneous: unsupervised learning for RGB-infrared person re-identification. IEEE TIP **30**, 6392–6407 (2021)
19. Loy, C.C., Xiang, T., Gong, S.: Time-delayed correlation analysis for multi-camera activity understanding. IJCV 106–129 (2010)
20. Luo, H., et al.: A strong baseline and batch normalization neck for deep person re-identification. IEEE TMM 2597–2609 (2019)
21. Nguyen, D.T., Hong, H.G., Kim, K.W., Park, K.R.: Person recognition system based on a combination of body images from visible light and thermal cameras. Sensors **17**(3), 605 (2017)
22. Pan, X., Luo, P., Shi, J., Tang, X.: Two at once: enhancing learning and generalization capacities via ibn-net. In: ECCV, pp. 464–479 (2018)
23. Ristani, E., Solera, F., Zou, R., Cucchiara, R., Tomasi, C.: Performance measures and a data set for multi-target, multi-camera tracking. In: ECCV, pp. 17–35 (2016)
24. Schroff, F., Kalenichenko, D., Philbin, J.: FaceNet: A unified embedding for face recognition and clustering. In: CVPR, pp. 815–823 (2015)
25. Shi, J., Yin, X., Zhang, Y., Xie, Y., Qu, Y., et al.: Learning commonality, divergence and variety for unsupervised visible-infrared person re-identification. In: NeurIPS (2024)
26. Siesler, H.W., Ozaki, Y., Kawata, S., Heise, H.M.: Near-Infrared Spectroscopy: Principles, Instruments, Applications. Wiley (2008)
27. Song, J., Yang, Y., Song, Y.Z., Xiang, T., Hospedales, T.M.: Generalizable person re-identification by domain-invariant mapping network. In: CVPR, pp. 719–728 (2019)
28. Wang, G.A., et al.: Cross-modality paired-images generation for RGB-infrared person re-identification. In: AAAI, vol. 34, pp. 12144–12151 (2020)
29. Wang, J., et al.: Optimal transport for label-efficient visible-infrared person re-identification. In: ECCV, pp. 93–109. Springer (2022)
30. Wang, T., Gong, S., Zhu, X., Wang, S.: Person re-identification by video ranking. In: ECCV, pp. 688–703 (2014)
31. Wang, X., Zhang, H., Huang, W., Scott, M.R.: Cross-batch memory for embedding learning. In: CVPR, pp. 6388–6397 (2020)
32. Wang, Z., et al.: Generalizable re-identification from videos with cycle association. arXiv preprint arXiv:2211.03663 (2022)
33. Wang, Z., Zheng, L., Liu, Y., Li, Y., Wang, S.: Towards real-time multi-object tracking. In: ECCV, pp. 107–122 (2020)
34. Wei, L., Zhang, S., Gao, W., Tian, Q.: Person transfer GAN to bridge domain gap for person re-identification. In: CVPR, pp. 79–88 (2018)
35. Williams, T.: Thermal Imaging Cameras: Characteristics and Performance. CRC Press (2009)

36. Wu, A., Zheng, W.S., Yu, H.X., Gong, S., Lai, J.: RGB-infrared cross-modality person re-identification. In: ICCV, pp. 5380–5389 (2017)
37. Wu, Q., et al.: Discover cross-modality nuances for visible-infrared person re-identification. In: CVPR, pp. 4330–4339 (2021)
38. Wu, Z., Ye, M.: Unsupervised visible-infrared person re-identification via progressive graph matching and alternate learning. In: CVPR, pp. 9548–9558 (2023)
39. Xiao, T., Li, S., Wang, B., Lin, L., Wang, X.: End-to-end deep learning for person search. arXiv preprint arXiv:1604.01850 **2**(2), 4 (2016)
40. Yang, M., Huang, Z., Peng, X.: Robust object re-identification with coupled noisy labels. IJCV (2024)
41. Ye, M., Ruan, W., Du, B., Shou, M.Z.: Channel augmented joint learning for visible-infrared recognition. In: ICCV, pp. 13567–13576 (2021)
42. Zajdel, W., Zivkovic, Z., Krose, B.J.: Keeping track of humans: have i seen this person before? In: Proceedings of the 2005 IEEE International Conference on Robotics and Automation, pp. 2081–2086. IEEE (2005)
43. Zhang, P., Dou, H., Yu, Y., Li, X.: Adaptive cross-domain learning for generalizable person re-identification. In: ECCV, pp. 215–232. Springer (2022)
44. Zhang, Y., Wang, H.: Diverse embedding expansion network and low-light cross-modality benchmark for visible-infrared person re-identification. In: CVPR, pp. 2153–2162 (2023)
45. Zhang, Y., Yan, Y., Lu, Y., Wang, H.: Towards a unified middle modality learning for visible-infrared person re-identification. In: ACM MM, pp. 788–796 (2021)
46. Zhao, Y., et al.: Learning to generalize unseen domains via memory-based multi-source meta-learning for person re-identification. In: CVPR, pp. 6277–6286 (2021)
47. Zheng, L., Shen, L., Tian, L., Wang, S., Wang, J., Tian, Q.: Scalable person re-identification: a benchmark. In: ICCV, pp. 1116–1124 (2015)

SymphonyVision: Diffusion-Based System for Multimodal Emotion Recognition and Generation

Claire Guo(✉) and Ivy Zan

Lynbrook High School, San Jose, CA 95129, USA
{cguo696,izan030}@student.fuhsd.org

Abstract. Multimodal Emotion Recognition (MER) has made significant progress with the advancements in Transformer-based models in effectively processing data from various sources in recent years. However, these models still face challenges like computational cost, capturing long-range relationships over multimodal data inputs, and multimodal output generation in personalized environments. To address this, we present SymphonyVision, a novel multi-modal system that leverages different models to understand and generate personalized, emotion-driven multimedia content, including music, images, and dialogues, aimed at supporting stress management and emotional regulation. The system integrates Large Language Models (LLMs) for text-based emotion recognition, Speech Emotion Recognition (SER) with a Bidirectional Long Short-Term Memory (Bi-LSTM) network for speech analysis, and two independent Diffusion Models for generating emotionally aligned images and music. Semantic alignment between user inputs and generated outputs is achieved through contrastive learning with dynamic loss adjustment, enhancing the relevance and quality of the content. Experimental results validate the system's efficacy. For image generation, SymphonyVision achieves a Fréchet Inception Distance (FID) score of 15.432, with a recall of 0.970 and precision of 0.924. For music generation, it records a Fréchet Audio Distance (FAD) score of 18.765, with a MuLan similarity of 0.717 and a Cross-modal Retrieval Accuracy of 0.825. Additional metrics, including a Modality Gap of 0.085 and a Cross-modal Average Similarity Score of 0.782, highlight the robustness of our approach. To ensure practical utility, we developed an intuitive web interface, enabling high school users to seamlessly interact with the system.

Keywords: Generative AI · Multimodal Emotion Recognition system · Diffusion Model · Contrastive Learning

1 Introduction

Reliable emotion recognition is pivotal for a variety of applications, ranging from healthcare, to education, and customer service. Traditionally, the emotion recognition research has been primarily focused on single modalities [1–3], such as speech emotion detection, etc. However, the single modality approach often lacks the capability to capture the full

P. Umapada et al. (Eds.): ICCPR 2025, CCIS 2811, pp. 43–61, 2026.
https://doi.org/10.1007/978-981-95-8315-7_4

spectrum of an individual's emotion due to the complexity of the real-world situation. With the recent advancements of deep learning neural networks and Transformer-based language models, Multimodal Emotion Recognition (MER) can take in user data in text, audio, image, and video. Despite these significant progress in Transformer-based language models, these language models face several challenges:

(1) Emotions unfold over time. Even though Transformer-based language models are good at text data, modeling multimedia data such as audio and video using Transformer-based language models would not yield as compute/memory efficient as other models would do, especially in a real-time environment [4, 5].
(2) In recent years, one important application, Intelligent Personal Assistants (IPAs), demands personalized Multimodal Emotion Generation on top of MER [6]. In these applications, emotions are often mixed or ambiguous. Transformer-based language models are not well suited since these models usually output one deterministic classification prediction.

These challenges underscore the need for AI systems that integrate robust emotion recognition with ethically sound, personalized content generation.

To address these challenges, we propose SymphonyVision, a multimodal AI system designed to alleviate stress in high school students through personalized multimedia content, including music and images. SymphonyVision integrates speech-to-text (STT), speech emotion recognition (SER) using Bi-LSTM models, transformer-based natural language processing (NLP) for sentiment analysis, and diffusion models for generating emotionally resonant outputs [7–9]. The system processes text and voice inputs to create a unified emotional profile via a multimodal fusion layer, which informs the generation of tailored multimedia responses. Unlike existing tools, SymphonyVision dynamically adapts to users' emotional states, leveraging diffusion models to produce high-quality content that aligns with individual needs. Designed for deployment in schools and counseling centers, it aims to provide accessible, scalable mental health support with potential applications for broader populations.

This paper presents SymphonyVision as a novel approach to AI-driven mental health interventions, with the following contributions.

(1) Multimodal Framework: SymphonyVision combines STT, SER, NLP, and diffusion models to deliver a comprehensive system for emotion detection and personalized multimedia generation, addressing limitations in single-modality approaches.
(2) Robust Emotion Recognition: By integrating speech and text through multi-modal fusion, the system achieves high accuracy in detecting emotional states, enhancing the relevance of multimodal data outputs.
(3) Cross-Modal feature alignment and output generation: The system uses contrastive learning to align emotion states across multimodal data points to identify, track, and generate emotional and personalized interactions.

2 Related Work

2.1 Large Language Models

Large Language Models (LLMs) leverage vast text datasets to generate and comprehend human-like language, driving advancements in education, drug discovery, scientific research, and cybersecurity [10]. Despite their transformative potential, challenges such as bias, misinformation, and ethical concerns necessitate transparent evaluation and responsible deployment to ensure fairness and efficacy across applications [11].

In education, LLMs enable personalized learning, automated feedback, and adaptive tutoring, aligning with the goals of Education 4.0 to enhance student engagement [10]. Systems like AI-driven tutors tailor content to individual needs, but biases in training data can compromise fairness and accessibility. Adaptive techniques and transparent metrics are proposed to mitigate bias, promoting human-AI collaboration over fully automated instruction [11].

In drug discovery, LLMs streamline processes from target identification to clinical trials by optimizing molecular structures and predicting drug-drug interactions [12]. Tools like DrugAssist design lead compounds, while LLMs generate hypotheses for disease mechanisms, accelerating development [13]. In scientific research, LLMs facilitate bioinformatics and chemical innovation by translating complex 3D molecular structures into human-readable descriptions, aiding communication and hypothesis generation [14]. Their contributions to data-driven experimental design further enhance material discovery and chemical synthesis [15].

In cybersecurity, LLMs support threat detection, vulnerability assessments, and automated reporting, but face challenges from adversarial attacks, such as prompt injections [16]. Benchmarks like Purple Llama CyberSecEval evaluate LLMs' secure coding capabilities, improving AI-driven security solutions [17]. Addressing biases, misinformation, and security threats through rigorous evaluation is critical for ethical and effective LLM deployment across these domains.

2.2 Speech Emotion Recognition

Speech Emotion Recognition (SER) analyzes audio signals to classify emotions such as happiness, sadness, or anger based on acoustic features like tone, pitch, and speech rate. Its applications span healthcare, education, and forensics, leveraging advanced machine learning to provide objective emotional insights [18]. However, challenges such as acoustic variability, cultural differences, and the subjective nature of emotions often lead to inaccurate classifications, necessitating robust models and diverse datasets [19].

In healthcare, SER enables early detection of mental health issues by analyzing vocal cues, improving personalized treatment and patient-physician trust, particularly in telemedicine where visual cues are limited (e.g., during mask-wearing in COVID-19) [20]. SER also supports monitoring chronic conditions with high emotional variability, enhancing patient outcomes. However, models trained on limited datasets struggle to generalize to real-world scenarios, and the scarcity of high quality, annotated clinical speech data remains a significant barrier [21].

In education, SER analyzes students' vocal cues (e.g., mel-spectrograms) in real-time during online classes, enabling educators to adapt teaching strategies to enhance engagement without invasive monitoring [22]. However, background noise, accent variability, and high computational demands limit its applicability in resource-constrained settings, emphasizing the need for robust, accessible SER systems [23].

2.3 Diffusion Model

Diffusion models, a class of generative AI architectures, create novel data by progressively adding noise to existing data and learning to reverse this process, yielding high-fidelity outputs [24]. Their versatility enables applications in weather forecasting, materials science, healthcare, and financial modeling, though challenges like high computational costs and parameter tuning persist [25].

In weather forecasting, diffusion models assimilate historical data to generate realistic scenarios at lower computational costs than physics-based simulations. They correct biases in existing systems and improve predictions of extreme weather events using Continuous Ensemble Forecasting, which generates full trajectories in parallel, avoiding error accumulation from stepwise autoregressive methods [24]. However, limitations include diminishing influence of initial conditions over time and the need for frequent retraining [25].

In materials science, diffusion models design complex atomic structures, such as crystalline phases and defect configurations, using voxel-based grand canonical approaches like UniMat to handle variable particle numbers and scale to millions of structures [26]. Challenges include high computational demands and misalignment between generative metrics and material stability goals [27]. In healthcare, diffusion models enhance medical imaging by generating high-quality images, supporting anomaly detection, denoising, and reconstruction. Frameworks like MT-DDPM with Swim-transformer networks improve segmentation and diagnostic precision across modalities, but face issues with computational intensity and parameter optimization [28, 29].

In financial modeling, diffusion models manage noisy market data for stock predictions and portfolio management. Frameworks like FinDiff generate high-quality synthetic tabular data for economic scenario modeling, preserving privacy [30]. However, high computational costs and slow processing limit real-time applications [31].

3 Methodology

3.1 Dataset

The SymphonyVision initiative leverages diverse datasets to enable robust emotion recognition and multimedia generation tailored for adolescent mental health support. For text-based emotion recognition, the ChatGPT API, enhanced with custom prompt templates, is fine-tuned using the HOPE [32], MEMO [33], and Counsel-chats [34] datasets. These datasets facilitate feature engineering to extract salient emotional cues, improving classification accuracy and context understanding for high school students' linguistic expressions.

For speech emotion recognition (SER), audio data is sourced from the Kaggle Speech Emotion Recognition dataset [35], combining Crema, Ravdess, Savee, and Tess datasets in.wav format, enriched with emotion labels. Text-to-speech technology further augments this dataset, enhancing emotional speech sample diversity.

For multimedia generation, diffusion models utilize music data from MERP [36] and DEAM [37], which provide valence-arousal labels for emotionally resonant music generation and evaluation. Image data is drawn from RichHF-18K [38], incorporating extensive human feedback to ensure high-quality outputs aligned with emotional inputs from the LLM and SER models.

3.2 SymphonyVision System Design

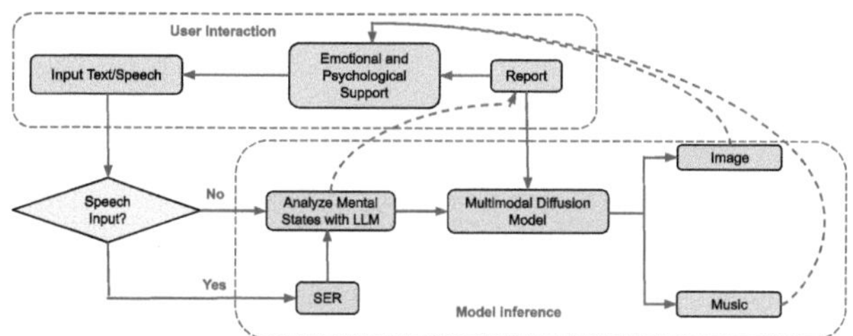

Fig. 1. Project Flow Chart

The emotionally intelligent AI system for psychological consultations takes input in either text or speech form. In the case of textual input, it would have to go through a process of cleaning, labeling, segmentation, augmentation, and normalization before passing it through ChatGPT 4o for emotion detection. For spoken input, it would first be standardized to WAV format and subjected to noise reduction followed by the elimination of filler words, segmentation, and normalization before being analyzed through a Bi-LSTM Speech Emotion Recognition model to extract emotional characteristics (Fig. 1).

From start to finish, the data transforms the simultaneous processes of preprocessing image and music data into a multimodal dataset that is fed into the Diffusion Model for the generation of images and music conditioned on the inferred emotional state. This guarantees that the generated content faithfully represents the emotional state of the user, hence creating a coherent, emotionally aware AI system for interventions in psychology.

The structure is in the form of a decision flow: speech input goes to Speech Emotion Recognition (SER), whereas text input is analyzed for emotional characteristics before passing into either the psychology consultation communication block or the SymphonyVision model for emotional analysis based on image and music inputs to generate a full report. If the system operates by the psychology consultation communication path, it directly generates a final psychological consultation report. This entire system is designed with embedded emotional detection and reaction mechanisms, making it a more emotionally intelligent and personalized AI-driven psychological consultation tool.

3.3 Data Processing

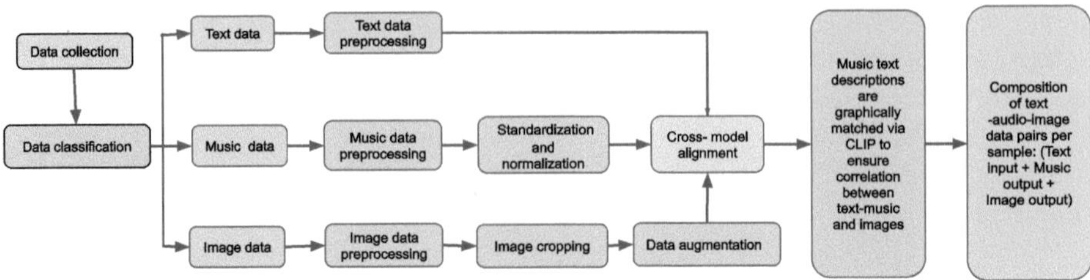

Fig. 2. Cross-modal Text-Music-Image Data Processing Flow

In the SymphonyVision initiative, data processing begins with collecting and classifying text, music, and image data from diverse datasets. Each modality undergoes specific preprocessing: text data are cleaned and tokenized, music data are processed for acoustic features, and image data are cropped and augmented. Subsequently, all modalities are standardized and normalized to ensure compatibility. A critical step involves cross-modal alignment using the CLIP model, which maps music text descriptions to ensure coherence among text, music, and image data. This process culminates in the creation of text-music-image pairs, forming a composite input-output structure to support robust model training and analysis for emotion-driven multimedia generation (Fig. 2).

3.4 SER

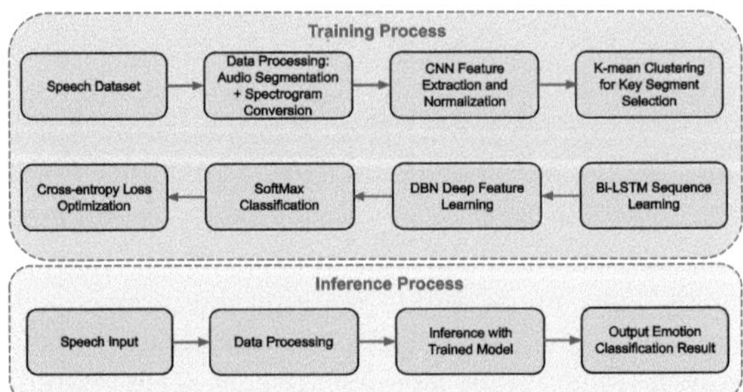

Fig. 3. Speech Emotion Classification Model Flow

Figure 3 reveals the training and inference phases of a model built for deep-learning-based speech emotion classification. During the training phase, a speech dataset is first prepared. Then, the data preprocessing consists of audio segmenting and spectrogram conversion for extracting important features. A convolutional neural network (CNN) will perform feature extraction and normalization and then follows that with K-means clustering for the selection of key speech segments. The model then goes on to learn features thoroughly by using the DBN (deep belief network) and combines that with Bi-LSTMs (bidirectional long short-term memory networks) to learn in a sequence format. Eventually, SoftMax scoring completes the emotion classification. Cross-entropy loss optimization is used as the continuous training process to improve the model's classification representation.

A key part of the inference process involves first processing the speech input, then doing inference with the trained model, and lastly outputting the result of the classification that indicates the emotion expressed in the speech input. Through integrating deep feature learning and temporal modeling, the system achieves more accurate recognition of the speech emotion. This is widely used in applications such as human-computer interaction, sentiment analysis, and the development of intelligent customer service systems.

The Speech Emotion Recognition (SER) block simply allows the system to recognize emotions from someone's voice. First, the given speech will be cleaned to remove the background noise. The system will then analyze the pitch, speed, and volume to recognize the emotion of a person. By using machine learning techniques, the system classifies the emotion (such as happy, sad, angry, etc.) on the basis of these voice characteristics. After emotion recognition, the emotion is used to form a reply or feedback.

3.5 Diffusion Model

The earlier paragraphs discussed diffusion models. They are a new way of generating images, but they can be heavy on computation. This is where Latent Diffusion Models (LDMs) come in. The main idea behind the LDM is to shift the diffusion from pixel space, where diffusion models work, to a compressed latent space. This change greatly decreases the training and inference costs.

LDMs consist of two stages. The encoder compresses the input image into a lower-dimensional latent form to perceptively encoding, in which perceivable details will be contained and unperceivable ones will be disregarded. This is achieved using a perceptual loss function and, combined with an adversarial objective, the reconstructed image looks realistic. The encoder E maps the image x to a latent variable E and then reduces the image by a f = H/h = W/w, where H and W are the height and width of the incoming image, while h and w are the height and width of the reconstructed image. The decoder D reconstructs the image.

Following this, semantic compression occurs. The model thus learns the conceptual and structural composition of the data rather than merely the details of the pixels. That is, the model looks for meaningful representations, which contributes further to efficiency and fidelity in synthesis.

The diffusion model is trained by denoising a normal variable that corresponds to the data distribution p(x), learning the reverse process of a fixed-length Markov Chain with length T. The formula it employs is:

$$L_{LDM} = E_{x_t \in \sim N(0,1)},\, t[\|\epsilon - \epsilon_0(x_t, t)\|]^2 \tag{1}$$

where ϵ_θ (x_t, t) is the denoising autoencoder at step t and x_t is the noisy version of the input x.

LDMs also have a cross-attention mechanism thrown into the mix. Cross-attention means that the multiple input sources that all contribute toward the output are enhanced with an ability to conditionally generate, allowing the LDM to bring in any text prompts or semantic layouts during the synthesis process. Cross attention is defined as:

$$Attention(Q, K, V) = softmax\left(\frac{QK}{\sqrt{V}}\right) \tag{2}$$

wherein Q, K, and V are calculated from the latent space representation along with any external condition input, resulting in a flexible multimodal synthesis.

LDMs also have a timeline-wise conditional U-Net backbone. This U-Net is applied primarily with data that has a spatial structure, such as that found in images. It follows the structure of an encoder decoder with skip connections. In addition, the model uses residual blocks that can enhance gradient flow and improve training stability. They will help maintain core information across layers, meaning it will reduce vanishing gradients.

LDMs achieve a fine balance of computation efficiency and perceived image quality through a variety of innovations, including perceptual and semantic compression, cross-attention mechanisms, and an optimally configured U-Net architecture with residual blocks. Thus, this innovative approach reduces the cost of training while simultaneously producing state-of-the-art images.

Diffusion models are very effective generators which can also be applied to music. In this entire pipeline for generating music, the first step is collecting 24 kHz raw audio file segments that would last for 30 s. Each of these audio clips goes to pseudo-labeling through MuLan, which serves as a pre-trained text and music audio model that provides captions for the unlabeled audio clips. To the above, a bigger language model, namely LaMDA, can also be prompted to come up with variations of musical descriptions that are factors in mood and genre, with all of these going into a text corpus named MuLaMCap. Now that audio-text pairing is established, the text can be sent to the CLIP encoder. This, in turn, would convert the user prompt or pseudo label text to that of a sequence of contextual embeddings. The diffusion process will then be guided by cross attention in the 1-dimensional Efficient U-Net, using these embeddings as a directional cue.

3.6 Text-Music-Image Generation Model

According to Fig. 4, the architecture of the model is built on CLIP Text Encoder for input texts and involves foreign cross-modal embedding and contrastive learning for associating and generating texts with music and images. In the first instance, the text is converted through embedding text-music, fed into the diffusion model for music. The music generated is post-processed with 1-D U-Net and music super-resolution before finally producing the music output. In parallel, the process occurs for the input text also; it is embedded as text-image and fed to the diffusion model for images before post-processed with 2-D U-Net in order to produce the corresponding image output.

To bridge the gap between different modalities, the model utilizes various forms of cross-modal contrastive learning. That includes text-music, text-image, and music-image contrastive learning as an integrated cross-modal technique that would help to ensure that text, music, and images correspond. Finally, all contrastive learning processes are

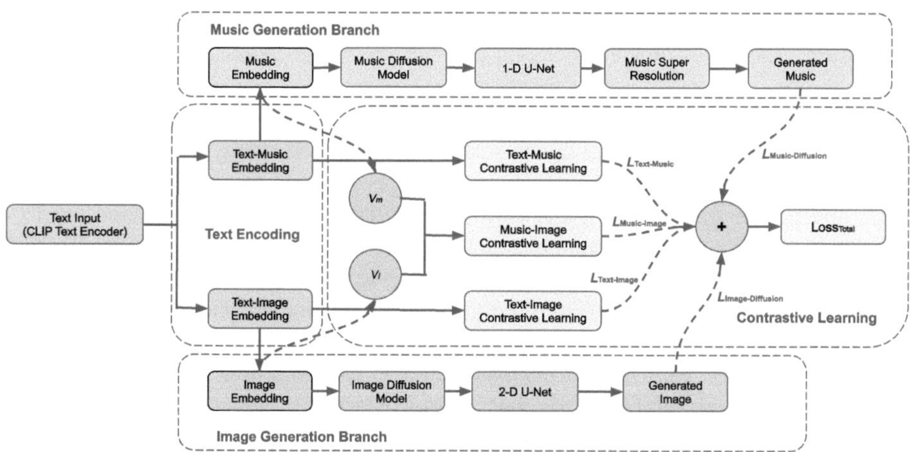

Fig. 4. Cross-modal Text-Music-Image Generation Model Architecture

optimized into the total loss function so that the generated music and images will be more semantically related with the input text, thereby increasing generation quality. This model leverages the simultaneous generation of music and images using text input, thus bridging the gap most pertinent for AI-driven multimodal creative content generation.

Another important characteristic of our model is contrastive learning, which encourages the system to distinguish between similar and different features. With Info-NCE (Information Noise Contrastive Estimation), the model focuses on distinguishing subtle emotional variations in the student data, including text and voice input, thus distinguishing stress from happiness and sadness. Such an improvement enhances the AI's ability to provide individualized emotional support.

In the design of our project, contrastive learning plays two very vital roles. First, it aligns the pairs of music-image embeddings within a common space in which music is able to make predictions about image and vice versa. This makes sure that the music and images that are generated reflect each other's properties reinforcing multi-modal consistency. Second, contrastive learning serves as the end-to-end joint optimization stage, fine-tuning different aspects, namely, text, music, and image embeddings, all for coherence through multiple contrastive loss functions: $L_{\text{text-music}}$, $L_{\text{text-image}}$, and $L_{\text{music-image}}$, which deal with the text-music, text-image, and music-image alignments, respectively. In addition to this, $L_{\text{image-diffusion}}$ and $L_{\text{music-diffusion}}$ ensure that the outputs from the diffusion models are optimized well. By the incorporation of these contrastive learning techniques, our model synchronizes generated music, images, and text in such a manner that all elements align meaningfully in an emotionally aware AI system.

3.7 Training Process

Generally speaking, there are four steps in the overall training process. In the first step, we extract embedding from pre-trained models. For instance, text descriptions are passed into the CLIP encoder to yield $V_{\text{text-music}}$, the text embedding. Input music processes through the audio MuLan encoder to yield V_{music}, the music embedding. Also, the text description is passed into the text encoder of CLIP to yield $V_{text\text{-}image}$, the text embedding. The input image processes through the image encoder of CLIP to yield V_{image}, the image embedding. At this stage, we have the separated embedding spaces, text-music and text-image, which are now available to use in later steps.

In the second step, we have music-image contrastive learning. The objective is to align the semantics between the music and images. Otherwise, in the next step with music and image generation by the diffusion model, the music and image could be completely unrelated. This alignment is achieved through contrastive learning. For this, the infoNCE loss is utilized to bring together the music and image embeddings into a joint embedding space. Preference is captured through a fine-tuned music-image embedding space, enabling V_{music} to potentially predict V_{image} and vice versa.

In the third step, we train the diffusion model. The goal here is to train diffusion models to create high-quality music and images using their embeddings. In the music diffusion model, the 1D U-Net diffusion model would use the text music embedding to generate music. The denoising loss value of this trained music diffusion model is $L_{music\text{-}diffusion}$. For the image diffusion model, the 2D U-Net diffusion model would leverage the text-image embedding to generate images. The diffusion loss for images would then be represented as $L_{image\text{-}diffusion}$. This yields two diffusion models which can be used to create high-quality music and images.

Finally, in the fourth step, we have end-to-end joint optimization. The goal is to tightly couple and adjust all the pieces so that the produced text, music, and images align together. We will join the different losses into one total cost function:

$$L_{total} = \lambda_1 L_{text-music} + \lambda_2 L_{text-image} + \lambda_3 L_{music-image} + \lambda_4 L_{music-diffusion} + \lambda_5 L_{image-diffusion} \tag{3}$$

This shows the beneficial effect of this approach. The loss $L_{text\text{-}music}$ operates contrastively for text-music alignment; the loss $L_{text\text{-}image}$ operates contrastively for text-image alignment; the loss $L_{music\text{-}image}$ operates contrastively for music-image alignment, while $L_{music\text{-}diffusion}$ captures the losses related to diffusion across music and $L_{image\text{-}diffusion}$ captures the losses related to diffusion and images. The defining measure for losses is reduced by using the weights and adjusted through training. It is set such that for 30% of the training period, firstly $L_{text\text{-}music}$, $L_{text\text{-}image}$, and $L_{music\text{-}image}$ will have the primary weight in order to assure well-optimized embeddings and their final alignment. The next 40% will use the reducing weight on the contrastive losses and increase weights on the diffusion model losses to reach a balance in the system. The remainder of the training phase involves teaching the diffusion losses to enable the system to generate high quality music and images that are aligned in a fully multi-modal system. In this regard, the entire system will be jointly optimized with respect to the generated music images that need to align well with each other and with the text description.

One of key innovations of our model is the learnable weights, λ_{1-5}, in formula 3. These loss weights are learnable tensors and registered as parameters. During training, they are updated by backdrop and optimizer, just like any other model weights. More specifically, we have five different channels here: image denoising, music denoising, text-image alignment, text-music alignment, and image-music alignment. Each dimension has its weight. If a loss is large or noisy during training, its weights will be turned down a little so it doesn't drown out the others. If a loss is small, its weight will be turned up a bit so that it keeps making progress. These balancing of weights are done automatically across two different objectives: contrastive vs diffusion. The end results then are achieved at the end:

- Stability: prevent one objective from wrecking image/music quality
- Adaptivity: as one dimension improves, its influence naturally rises.

4 Experiments

4.1 Evaluation Metrics

Fréchet Inception Distance (FID) compares real and generated image distributions using a pre-trained Inception V3 classifier to embed images and compute the Fréchet distance between their mean and covariance matrices. Lower FID indicates closer statistical similarity.

Fréchet Audio Distance (FAD) evaluates audio generation by comparing real and generated audio distributions, using a pre-trained model like VGGish to embed audio and calculate the Fréchet distance. Lower FAD signifies better statistical alignment.

Inception Score (IS) measures image quality and diversity via a pre-trained classifier, assessing class prediction certainty and variety. Higher IS reflects realistic and diverse generated images.

Precision gauges the realism of generated images by checking how many generated images align with the real image distribution. High precision indicates authentic looking images.

Recall assesses the diversity of generated images, measuring coverage of the real data distribution. High recall shows comprehensive representation of real data variety.

MuLan similarity score evaluates audio-text alignment by computing cosine similarity between embeddings of generated audio and its text prompt. Higher scores indicate better semantic and perceptual fidelity [40].

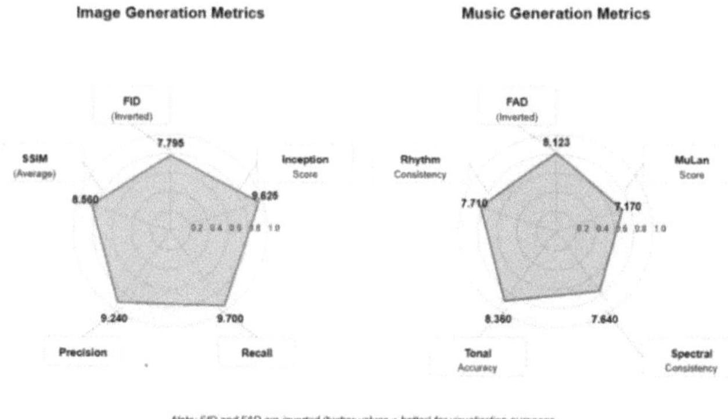

Fig. 5. Generation of task radar grams, the left figure (a) is the effect of image generation. The right figure, (b), is an evaluation of the effect of music generation.

4.2 Multimodal Generation Evaluation

In our experiments on the quality of multimodal generation, we apply a full-fledged evaluation framework that tests the performance of the system on three major dimensions: quality of the generated output, degree of semantic alignment across modalities, and effectiveness of specific types of comparative learning methods. This multi-dimensional evaluative design captures the extent to which the system understands, generates, and interrelates multimodal contents.

Table 1. Image Generation Quality Evaluation Metrics

Model	FID	Inception Score	Recall	Precision	SSIM (Average)
SymphonyVision	15.432	9.625	**0.970**	**0.924**	0.856
SDv1.5 (cfg = 3)	**8.78**	N/A	0.30	0.59	N/A
SDv2.1 (cfg = 3)	9.64	N/A	0.31	0.57	N/A
SDv2.1 (cfg = 4.5)	12.26	N/A	0.33	0.61	N/A

As depicted in Fig. 5(a) and Table 1, the performance of the system also stands exceptionally well in the quality assessments of image generation, with an FID (Fréchet Inception Distance) score of 15.432 higher than that of any other lately introduced dedicated image generation model but not quite as "good" as other multi-modal systems. In particular, the recall value is significantly high (0.970), implying that the system is capable of covering nearly all target distributions and thus producing diverse image content. Inception scores of 9.625 indicate further evidence of the quality and diversity of generated images. In contrast, the popular text-image generation model, Stable Diffusion (SD), has a better FID (lower is better) [41] while our model demonstrates better Precision and Recall values due to the unique contrastive learning with multimodal alignments.

Table 2. Music Generation Quality Evaluation Metrics

Model	FAD	MuLan Similarity	Spectral Consistency	Tonality Accuracy	Rhythm Consistency
SymphonyVision	18.765	**0.717**	0.764	0.836	0.771
Noise2Music Waveform	**2.134**	0.478	N/A	N/A	N/A
Noise2Music Spectrogram	3.840	0.434	N/A	N/A	N/A

In the music generation performance experiment, the results obtained from various experiments are shown in Table 2 and Fig. 5(b). The generated audio had a FAD (Fréchet Audio Distance) score of 18.765 for music generation performance evaluation, meaning that the spectral features of the generated audio have some reasonable degree of similarity to the actual audio with respect to music. The generated music was found to have a moderate MuLan similarity score of 0.717 which describes the extent to which generated music matches the meaning of the target description. It is important to know that tonal accuracy (0.836) is the best music generation metric as it shows the capability of this system in resonating the emotion in music.

As a comparison, model Noise2Music [40], is used to compare and contrast the characteristics of the model in single modality as in Table 2. Noise2Music trains a series of diffusion models to generate high quality music clips, which are reflected in the lower FAD scores, 2.134 and 3.840, for both Noise2Music in Table 2. FAD (lower is better) compares the distribution of embeddings from generated audio against the referenced dataset to measure how close they are. MuLan Similarity Score (higher is better), however, measures how well the generated audio semantically matches the text prompt. Our model performs better in this metric due to the unique contrastive learning with multimodality alignment.

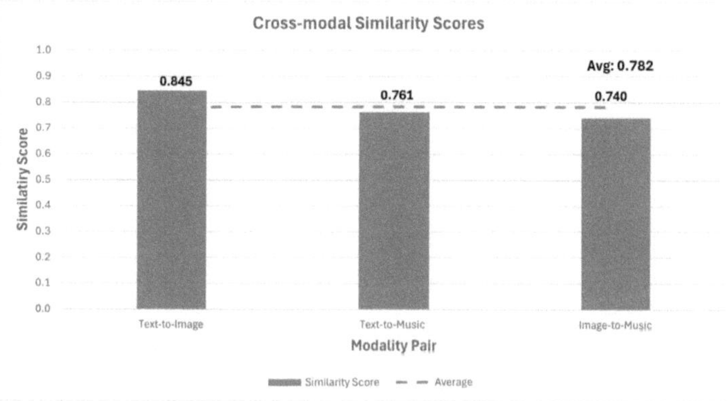

Fig. 6. Cross-modal Similarity Scores

In addition, cross-modal alignment analysis was performed, and the results are presented in Table 3 and Fig. 6. The experiments reported here indicate that this system establishes effective semantic connections between modalities. The highest scores were achieved for text-to-image alignment (0.845), proving the ability of the system to generate highly relevant image content regarding textual descriptions. Text-to-music alignment received the second highest score (0.761), consistent with the intrinsic complexity of music semantic representation. The image-to-music alignment was relatively low (0.740), which reveals the complicated nature of semantic mapping between these two modalities. Finally, the cross-modal average similarity reached 0.782, demonstrating the effectiveness of contrastive learning in building strong intermodal connections.

Table 3. Cross-Modal Alignment Evaluation Metrics

Alignment Type	Similarity Score
Text-to-Image Alignment	0.845
Text-to-Music Alignment	0.761
Image-to-Music Alignment	0.740
Cross-modal Average Similarity	0.782

To demonstrate the effectiveness of contrastive learning, experiments were conducted and the results are in Table 4, Figs. 7 and 8. The semantic gap between the different modalities was bridged to our satisfaction. Compared with the baseline approach, the text-image modality gap was reduced by 82.42%, the text-music gap by 82.66%, and the image-music gap by 81.21%. On average, our approach reduced the modal gap from 0.474 to 0.085, achieving an overall reduction of 82.10%. This improvement also directly translates into enhanced cross-modal retrieval performance, with a 46.54% increase in accuracy, improving from 56.3% in the baseline approach to 82.5% using our method.

Table 4. Contrastive Learning Performance Evaluation

Evaluation Metric	Baseline Method	Our Method	Improvement
Modal Gap(Text-Image)	0.421	0.074	82.42%
Modal Gap(Text-Music)	0.473	0.082	82.66%
Modal Gap(Image-Music)	0.527	0.099	81.21%
Average Modal Gap Reduction	0.474	0.085	82.10%
Cross-modal Retrieval Accuracy	56.30%	82.50%	46.54%

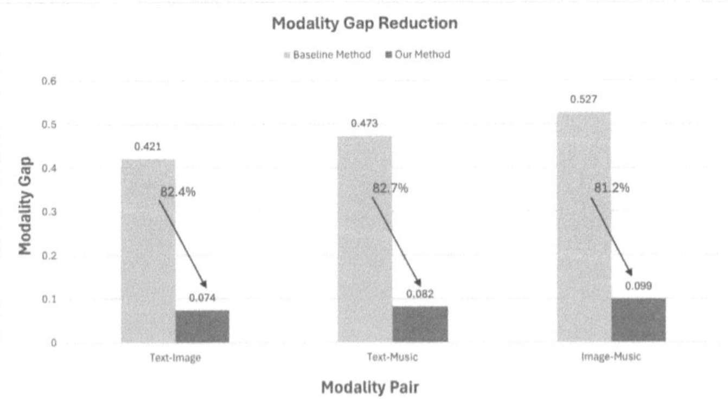

Fig. 7. Gap Reduction

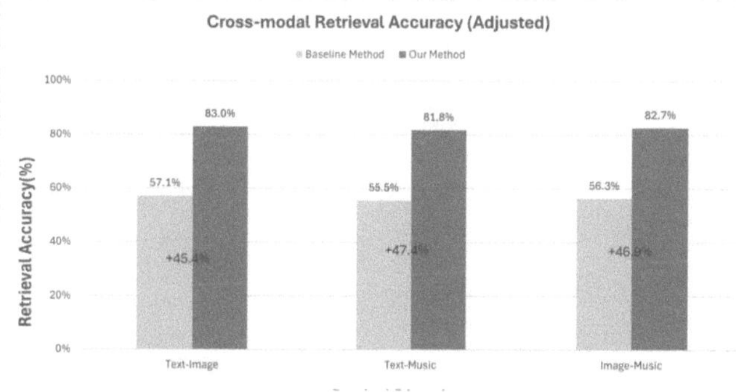

Fig. 8. Cross-modal Retrieval Accuracy

4.3 System Validation

This project also consists of an intuitive web interface that is accessible to users, as in Fig. 9. As depicted in the images, we created two major interfaces: the one for initiating emotional conversations, "Soul Dialogue," and "Text/Audio to Image and Music," which aims to generate multimedia content from emotional-based input. The first interface presents a welcoming environment and houses the tagline "Warmth through Technology, Hope through Understanding", highlighting the human-centered approach of our system. The second interface reflects the very nature of the system; that is, it allows users to put in the text or audio describing their state of emotions and receives personalized image and music outputs that fit the mood. This user-friendly design allows high schoolers to access and use the system without requiring proficiency in technical terminology.

SymphonyVision hopes to provide an accessible, private, and nonjudgmental platform for communication and emotion management. It will provide music, text, and images, which can help different people, as some may prefer to talk to others, some in visuals, and some in music.

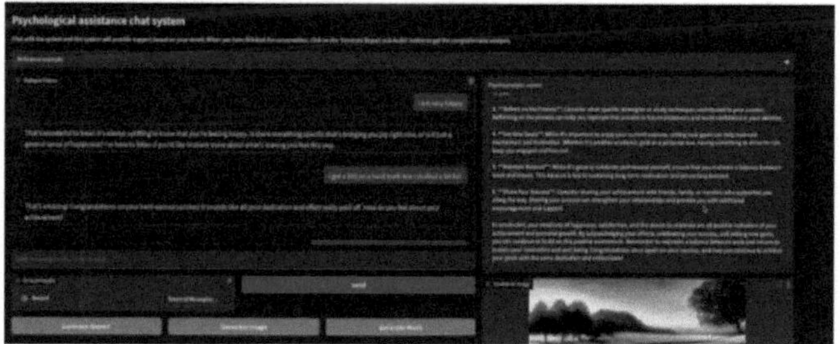

Fig. 9. Website Function Interface

4.4 User Feedback

We presented the system to a small group of high school student and here's the user feedbacks. With the positive feedbacks so far, the next phase is to expand to a bigger user community (Table 5).

Table 5. User Feedback

User ID	Strength	Improvement	Quantitative Ratings
Ruixuan Liu	This app feels groundbreaking, offering a truly multi-modal experience. I can type, speak, and click to see images or music generated that match my tone. It feels ahead of other AI tools, even ChatGPT	It would be great if the overall design were more visually appealing to create a more immersive experience	Usability: 5/5 Emotional Alignment: 5/5
Shannon	The ability to transform words or voice into music is incredibly creative and engaging	Music clips are too short—having longer tracks or curated song suggestions would improve immersion	Usability: 4/5 Emotional Alignment: 3/5
Isabelle	I genuinely love this app. The mood matching images and music feel unique and emotionally resonant	Music generation takes a bit of time—a faster response would make the experience smoother	Usability: 4/5 Emotional Alignment: 4/5

5 Discussion and Conclusion

We present SymphonyVision, a newly developed multimodal system designed to provide personalized emotional support for high school students through AI-generated multimedia content. Its approach integrates the use of large language models, speech emotion recognition, and diffusion methods for image and music generation into a coherent holistic approach of mental health support that draws upon all the different sensory modalities in their therapeutic potential.

As is evident from the results of experimental studies, any component of our system can perform optimally on its own. The chatbot feature therefore achieved an 83.5% adherence to guidance content due to its implied strength in applying the four phase counseling model. The speech emotion recognition model achieved 76.8% across the seven emotional categories, being particularly proficient in identifying neutral, happy and sad emotions. The multimodal generation system produced high-quality outputs, with an FID score of 15.432 for images and an FAD score of 18.765 for music, while achieving a cross-modal similarity score of 0.782 between different modalities.

Although we introduce Latent Diffusion Models (LDMs) to balance the computational costs of diffusion models for real-time deployment, we plan to further reduce latency through distillation, quantization, and accelerated sampling. The current system was designed and validated primarily with adolescent data drawn from high school populations, but future work will extend evaluation to adult and cross-cultural groups to test broader generalizability.

SymphonyVision has truly excelled in making artificial intelligence accessible to adolescents for recognizing their emotions and promoting emotion assimilated conversation. The fact that peace of mind can be achieved through emotional cognition and offering multimodal content using an easy-to-use web interface illustrates a new approach for emotional health in addition to traditional therapeutic means. Thus, as mental health continues to be an increasing concern for schools, initiatives such as SymphonyVision can provide intensive, personalized emotional assistance required for better well-being among high school students.

References

1. El Ayadi, M., Kamel, M.S., Karray, F.: Survey on speech emotion recognition: features, classification schemes, and databases. Pattern Recognit. **44**, 572–587 (2011). https://doi.org/10.1016/j.patcog.2010.09.020
2. Swain, M., Routray, A., Kabisatpathy, P.: Databases, features and classifiers for speech emotion recognition: a review. Int. J. Speech Technol. **21**, 93–120 (2018). https://doi.org/10.1007/s10772-018-9491-z
3. Zong, Y., Lian, H., Chang, H., Lu, C., Tang, C.: Adapting multiple distributions for bridging emotions from different speech corpora. Entropy **24**, 1250 (2022). https://doi.org/10.3390/e24091250
4. Poorna, S.S., Menon, V., Gopalan, S.: Hybrid CNN-BiLSTM Architecture with Multiple Attention Mechanisms to Enhance Speech Emotion Recognition. https://doi.org/10.1016/j.bspc.2024.106967
5. Redwan, U.G., Zaman, T., Mizan, H.B.: Spatio-Temporal CNN-BiLSTM Dynamic Approach to Emotion Recognition Based on EEG Signal. https://doi.org/10.1016/j.compbiomed.2025.110277
6. Li, Q., Huang, P., Xu, Y., Chen, J., Deng, Y., Yin, S.: Generating and Encouraging: an Effective Framework for Solving Class Imbalance in Multimodal Emotion Recognition Conversation. https://doi.org/10.1016/j.engappai.2024.108523
7. Khalil, R., Jones, E., Babar, M., Jan, T., Zafar, M., Alhussain, T.: Speech emotion recognition using deep learning techniques: a review. IEEE Access **7**, 117327–117345 (2019). https://doi.org/10.1109/ACCESS.2019.2936124
8. Gao, Z., Feng, A., Song, X., Wu, X.: Target-dependent sentiment classification with BERT. IEEE Access **7**, 154290–154299 (2019). https://doi.org/10.1109/ACCESS.2019.2946594
9. Karbout, K., Ghazouani, M., Lachgar, M., Hrimech, H.: Multimodal data fusion techniques in smart healthcare. In: 2024 International Conference on Global Aeronautical Engineering and Satellite Technology (GAST), pp. 1–6 (2024). https://doi.org/10.1109/GAST60528.2024.10520803
10. Lee, J., Hicke, Y., Yu, R., Brooks, C., Kizilcec, R.: The life cycle of large language models in education: a framework for understanding sources of bias. Br. J. Educ. Technol. **55**, 1982–2002 (2024). https://doi.org/10.1111/bjet.13505
11. Peláez-Sánchez, I.C., et al.: The impact of large language models on higher education: exploring the connection between AI and Education 4.0. Front. Educ. (2024). https://doi.org/10.3389/feduc.2024.1392091
12. Ye, G., et al.: DrugAssist: a Large Language Model for Molecule Optimization (2023). https://doi.org/10.48550/arXiv.2401.10334

13. Zheng, Y., et al.: Large Language Models in Drug Discovery and Development: from Disease Mechanisms to Clinical Trials (2024). https://doi.org/10.48550/arXiv.2409.04481
14. Li, J., et al.: Empowering molecule discovery for molecule-caption translation with large language models: a ChatGPT perspective. IEEE Trans. Knowl. Data Eng. **36**, 6071–6083 (2023). https://doi.org/10.1109/TKDE.2024.3393356
15. Zhang, Q., et al.: Scientific Large Language Models: a Survey on Biological & Chemical Domains (2024). https://doi.org/10.48550/arXiv.2401.14656
16. Zhang, J., et al.: When LLMs Meet Cybersecurity: a Systematic Literature Review (2024). https://doi.org/10.48550/arXiv.2405.03644
17. Bhatt, M., et al.: Purple Llama CyberSecEval: a Secure Coding Benchmark for Language Models (2023). https://doi.org/10.48550/arXiv.2312.04724
18. Chanchi Golondrino, G.E., Rodriguez Baca, L.S., Sierra Martinez, L.M.: Speech emotion recognition software system for forensic analysis. Ingeniería y Desarrollo **42**(01), 68–88 (2024). https://doi.org/10.14482/inde.42.01.519.019
19. Vásquez-Correa, J.C., Álvarez Muniain, A.: Novel speech recognition systems applied to forensics within child exploitation: Wav2vec2.0 vs. whisper. Sensors **23**(4), 1843 (2023). https://doi.org/10.3390/s23041843
20. Nigar, N.: Speech Emotion Recognition Using Convolutional Neural Network and Its Use Case in Digital Healthcare 11Dataset1. Peeref (2024)
21. Huang, C.-T., Huang, C.-W., Yang, H.-C., Li, Y.-C.: Speech emotion recognition applied to real-world medical consultation. In: Bichel-Findlay, J., Otero, P., Scott, P., Huesing, E. (eds.) Studies in Health Technology and Informatics. IOS Press (2024). https://doi.org/10.3233/SHTI231139
22. Abdelhamid, A.A.: Speech emotions recognition for online education. Fusion Pract. Appl. **10**(1), 78–87 (2023). https://doi.org/10.54216/FPA.100104
23. Liu, J., Wu, X.: Prototype of educational affective arousal evaluation system based on facial and speech emotion recognition. Int. J. Inf. Educ. Technol. **9**(9), 645–651 (2019). https://doi.org/10.18178/ijiet.2019.9.9.1282
24. Li, L., Carver, R., Lopez-Gomez, I., Sha, F., Anderson, J.: Generative emulation of weather forecast ensembles with diffusion models. Sci. Adv. (2024). https://doi.org/10.1126/sciadv.adk4489
25. Andrae, M., Landelius, T., Oskarsson, J., Lindsten, F.: Continuous Ensemble Weather Forecasting with Diffusion Models (2024). https://doi.org/10.48550/arXiv.2410.05431
26. Lei, B., et al.: Grand Canonical Generative Diffusion Model for Crystalline Phases and Grain Boundaries (2024). https://doi.org/10.48550/arXiv.2408.15601
27. Yang, S., et al.: Scalable Diffusion for Materials Generation (2024). https://doi.org/10.48550/arXiv.2311.09235
28. Pan, S., et al.: 2D medical image synthesis using transformer-based denoising diffusion probabilistic model. Phys. Med. Biol. (2023). https://doi.org/10.1088/1361-6560/acca5c
29. Kazerouni, A., Aghdam, E. K., Heidari, M., Azad, R., Fayyaz, M., Hacihaliloglu, I., Merhof, D.: Diffusion Models for Medical Image Analysis: A Comprehensive Survey (2023). https://doi.org/10.48550/arXiv.2211.07804
30. Sattarov, T., Schreyer, M., Borth, D.: FinDiff: Diffusion Models for Financial Tabular Data Generation. In: 4th ACM International Conference on AI in Finance, pp. 64–72 (2023). https://doi.org/10.1145/3604237.3626876
31. Daiya, D., Yadav, M., Rao, H.S.: DiffSTOCK: Probabilistic Relational Stock Market Predictions Using Diffusion Models (2024). https://doi.org/10.48550/arXiv.2403.14063
32. Malhotra, G., Waheed, A., Srivastava, A., Akhtar, S., Chakraborty, T.: Speaker and time-aware joint contextual learning for dialogue-act classification in counselling conversations. In: Proceedings of the Fifteenth ACM International Conference on Web Search and Data Mining (2022). https://doi.org/10.1145/3488560.3498509

33. Srivastava, A., Suresh, T., Lord, S.P., Akhtar, M.S., Chakraborty, T.: Counseling summarization using mental health knowledge guided utterance filtering. In: Proceedings of the 28th ACM SIGKDD Conference on Knowledge Discovery and Data Mining (2022). https://doi.org/10.1145/3534678.3539187
34. nbertagnolli. counsel-chat. GitHub (2023). https://github.com/nbertagnolli/coun-sel-chat
35. Speech Emotion Recognition (en). https://www.kaggle.com/datasets/dmitrybabko/speech-emotion-recognition-en
36. Koh, E.Y., Cheuk, K.W., Heung, K.Y., Agres, K., Herremans, D.: MERP: a music dataset with emotion ratings and raters' profile information. Sensors **23**(1), 382 (2022). https://doi.org/10.3390/s23010382
37. DEAM dataset - Database for Emotional Analysis of Music. https://cvm-l.unige.ch/databases/DEAM/
38. Rich Human Feedback for Text-to-Image Generation (2023). https://arxiv.org/html/2312.10240v1
39. Ho, J., Jain, A., Abbeel, P.: Denoising Diffusion Probabilistic Models (2020). https://doi.org/10.48550/arXiv.2006.11239
40. Huang, Q., et al.: Noise2Music: Text-Conditioned Music Generation with Diffusion Models (2023). https://doi.org/10.48550/arXiv.2302.03917
41. Dao, T., Nguyen, T.H., Le, T., Vu, D., Nguyen, K., Pham, C., Tran, A.: SwiftBrus v2: Make Your One-step Diffusion Model Better than its Teach. https://doi.org/10.48550/arXiv.2408.14176

Hybrid Attention Network for Online Handwritten Chinese Text Recognition

JingXing Ding, Feng Chen, Xuefeng Zhang, Zekai Cheng, and Xiwen Qu(✉)

School of Computer Science and Technology, Anhui University of Technology, Maanshan, China
{chenfeng,zxf_06,chengzk,qxw_ahut}@ahut.edu.cn

Abstract. In online handwritten Chinese text recognition (OHCTR), similar characters are one of the main causes of accuracy degradation. Exploring semantic relationships between characters and applying feature-specific weighting are effective approaches to distinguishing such similar characters. To this end, this paper proposes a hybrid attention network (HAN) to further enhance recognition accuracy. The HAN comprises a semantic attention module (SAM) and a channel attention module (CAM). The SAM captures multi-level semantic relationships among characters within a text line, while the CAM performs weighting across convolutional channels. Compared with the multi-head self attention, the SAM is nonlinear and achieves higher recognition accuracy with lower memory consumption and faster computation speed. Experiments on three publicly available online handwritten Chinese text datasets demonstrate that the proposed HAN outperforms the multi-head attention mechanism. When combined with an end-to-end convolutional recurrent network, the HAN achieves excellent recognition performance on OHCTR.

Keywords: Similar characters · Online handwritten Chinese text recognition · Semantic attention · Channel attention

1 Introduction

Online handwritten Chinese text recognition (OHCTR) has achieved a series of research results [2,12,13]. However, it remains highly challenging due to the large number of Chinese character categories, the existence of many structurally similar characters or characters with only subtle differences in strokes, the great diversity of writing styles and significant individual variations, as well as the impact of noise and irregular pen traces. Existing OHCTR algorithms can generally be divided into segmentation-based [16] and segmentation-free approaches [5,8–10]. Segmentation-based methods first divide a text line into components and then use a path optimization algorithm to find the optimal combination of components as the recognition result. These methods rely heavily on segmentation accuracy, the precision of single-character classifiers, and

P. Umapada et al. (Eds.): ICCPR 2025, CCIS 2811, pp. 62–73, 2026.
https://doi.org/10.1007/978-981-95-8315-7_5

the performance of path optimization algorithms, making the overall recognition process very complex. Currently, a new form of online handwriting–in-air handwriting–has emerged. As shown in Fig. 1, unlike traditional online handwritten Chinese text, in-air handwritten Chinese text is finished in a single continuous stroke, without pen-down or pen-up information, making segmentation extremely difficult. Segmentation-based algorithms thus become ineffective for recognizing in-air handwritten Chinese text.

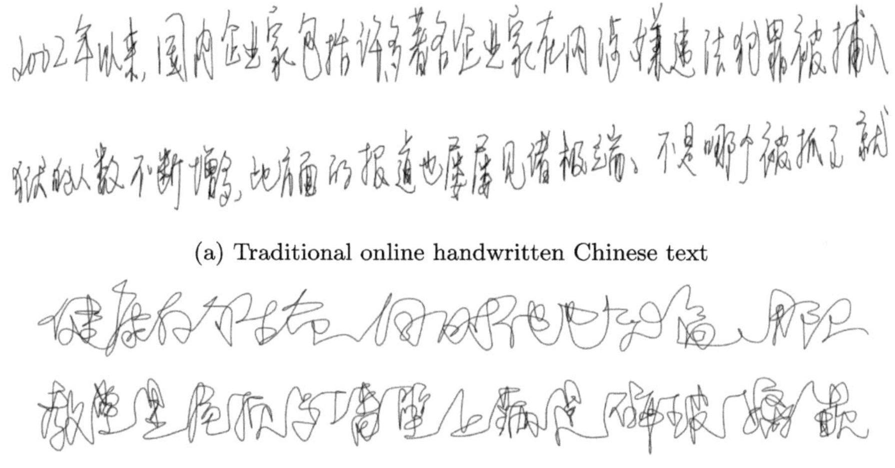

(a) Traditional online handwritten Chinese text

(b) In-air handwritten Chinese text

Fig. 1. Examples for traditional online handwritten Chinese text and in-air handwritten Chinese text.

Segmentation-free algorithms avoid the accuracy loss caused by segmentation and make the recognition process more streamlined. Recurrent neural network (RNN), which can process sequential data in an end-to-end manner [3,4], have been naturally applied to OHCTR. Sun et al. [11] proposed a recognition model that combines LSTM and connectionist temporal classification (CTC). This model encodes the raw coordinate sequence using LSTM and uses CTC to align input-output pairs, achieving higher recognition accuracy than traditional segmentation-based algorithms and Hidden Markov Models (HMMs) [17]. However, recognition algorithms based on RNN process the online handwritten text line point by point, resulting in slow recognition speed and poor real-time performance. In addition, RNN-based algorithms are sensitive to local noise, which can degrade recognition accuracy. To improve recognition speed and mitigate the impact of local noise, Xie et al. [14] combined convolutional neural network (CNN) with RNN and proposed a fully convolutional recurrent network model. In this model, path signature technology is first used to convert the coordinate sequence of the online handwritten text line into a signature feature map, which is then recognized using CNN and LSTM networks. CNN network can capture local semantic context through convolution and downsampling, effectively reducing the impact of local noise points, shortening the text line length, and improving recognition speed. LSTM networks can further capture the sequential information of local semantic context. Therefore, this model is well suited to OHCTR, achieving higher recognition accuracy and faster speed compared to RNN-based methods. However, the fully convolutional recurrent network model uses large convolution kernels, resulting in high computational cost and memory consumption. To reduce the convolution kernel size, Gan et al. [5] proposed a one-dimensional convolutional recurrent neural network model, which uses one-dimensional convolutions and downsampling to capture local semantic context, effectively reducing storage and computation time costs while maintaining recognition accuracy. CNN+RNN models can significantly improve recognition accuracy and speed, but these models usually require the sequential coordinate data to be converted into images or vectors first. This conversion process is complex, leads to information loss, and increases recognition time. To address this, Qu et al. [9] proposed an end-to-end multi-scale attention convolutional recurrent network (EMACRN) that completely avoids the data format conversion process. This model directly extracts multi-scale local features from the raw coordinate sequence, significantly improving recognition speed and freeing researchers from cumbersome data transformations.

Similar characters in text lines are one of the main reasons for the decline in recognition accuracy. The EMACRN [9] introduced multi-head self attention to OHCTR to explore multi-level semantic relationships between characters in text lines. Semantic dependency is an effective way to distinguish similar characters. For example, the Chinese characters "大" (big) and "太" (too) have minimal differences, and writing habits such as cursive strokes or omissions make these two characters extremely difficult to distinguish. However, in specific text lines, the dependency relationships between characters can help differentiate "大" and "太" such as in the texts "太阳很热" (The sun is very hot) and "大家好" (Hello everyone). However, the multi-head self attention is linear, and increasing the number of attention heads is required to capture rich semantic relationships, which leads to higher memory consumption and computational complexity. To address this, this paper proposes a hybrid attention network (HAN) to replace the multi-head attention mechanism. The HAN consists of a semantic attention module (SAM) and a channel attention module (CAM). The SAM is non-linear and is used to capture rich semantic relationships between characters in the text line. The channel attention module is used to weight different convolutional channels, thereby weakening or strengthening the influence of different features on the recognition results. Experiments were conducted on three public online handwritten Chinese text datasets, and the results demonstrate the effectiveness of the proposed HAN. The main contributions of this paper are summarized as follows:

(1) We propose HAN to replace the multi-head self attention in the EMACRN [9].

(2) The HAN consists of SAM and CAM. By exploring semantic relationships between characters in text lines and weighting character features, the model enhances the ability to distinguish similar characters.
(3) Experiments on publicly available benchmark datasets–the competition dataset ICDAR-2013, the traditional online handwritten Chinese text dataset CASIA-OLHWDB2, and the in-air handwritten Chinese text dataset IAHCT-UCA2018–show that compared to multi-head attention, the proposed HAN achieves higher recognition accuracy, faster recognition speed, and lower memory consumption.

The rest of this paper is organized as follows. Section 2 introduces the proposed method at length. The experimental results are reported in Sect. 3. Section 4 concludes the paper.

2 Proposed Method

In this section, we provide a detailed introduction to the proposed HAN. At present, Qu et al. [9] have introduced the multi-head self attention into online handwritten Chinese text recognition and have demonstrated that it can effectively improve recognition accuracy. However, multi-head self attention is linear, and obtaining richer semantic relationships requires increasing the number of heads, which leads to higher memory consumption and reduced recognition speed. To address this issue, this paper proposes HAN to replace the multi-head attention mechanism. The detailed structure of the HAN is shown in Fig. 2. In Fig. 2, multi-scale local contextual features $\boldsymbol{f}_i \in R^{256\times 1}$ is the output of the end-to-end convolutional recurrent network (See [9] for details), $\boldsymbol{F} = [\boldsymbol{f}_1, \boldsymbol{f}_2, \ldots, \boldsymbol{f}_S] \in R^{256\times S}$. S is the number of local semantic contexts, which varies depending on the input text lines to the end-to-end convolutional recurrent network. $\boldsymbol{p}_i \in R^{256\times 1}$ is the output of the HAN, $i = 1, \ldots, S$. In SAM,

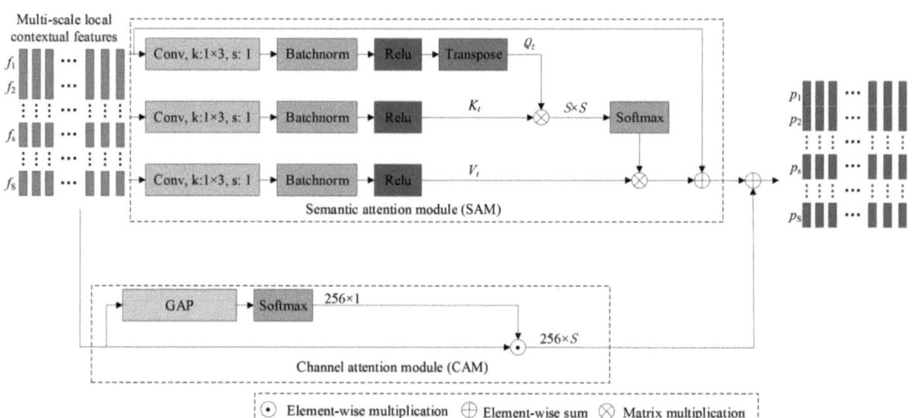

Fig. 2. Detailed structure of the proposed HAN.

$\boldsymbol{Q_t} \in R^{S\times d_k}$, $\boldsymbol{K_t} \in R^{d_k\times S}$, and $\boldsymbol{V_t} \in R^{256\times S}$ are the query matrix, key matrix, and value matrix, respectively. $\boldsymbol{d_k}$ is obtained through experiments. t represents the number of SAM, determined through experiments. The output $\boldsymbol{O}_t$ of the t-th SAM can be obtained by the following equation:

$$\boldsymbol{O}_t = (V_t \otimes softmax(\frac{Q_t \otimes K_t}{\sqrt{d_k}})) \oplus F. \tag{1}$$

As can be seen from Eq. 1, SAM is able to capture multi-level semantic relationships within the text line. When $t > 1$, the output O_{SAM} of SAM is

$$\boldsymbol{O}_{SAM} = concat(\boldsymbol{O}_1, \boldsymbol{O}_2, \ldots, \boldsymbol{O}_t) \otimes \boldsymbol{W}_{SAM}, \tag{2}$$

where $concat(\cdot)$ denotes matrix concatenation, $\boldsymbol{W}_{SAM}$ is weight matrix. In CAM, "GAP" denotes global average pooling, and the output of CAM can be obtained by the following equation:

$$\boldsymbol{O}_{CAM} = softmax(GAP(F)) \odot F. \tag{3}$$

As can be seen from Eq. 3, CAM can weight the features of $\boldsymbol{f}_i$, thereby enhancing or weakening their influence on the recognition results. The output $\boldsymbol{P}$ of HAN can be obtained from Eq. 2 and Eq. 3, that is:

$$\boldsymbol{P} = \boldsymbol{O}_{SAM} \oplus \boldsymbol{O}_{CAM}. \tag{4}$$

In the subsequent experiments, the EMACRN [9] is used for OHCTR, with only the proposed HAN replacing the multi-head self attention.

3 Experiments

In this section, we first introduce the experimental datasets in Sect. 3.1. Then, we give implementation details in Sect. 3.2. In Sect. 3.3, we conduct ablation experiments to evaluate the performance of the proposed HAN and its constituent modules. In Sect. 3.4, we conduct recognition experiments to evaluate the performance of EMACRN when the proposed HAN is used to replace the multi-head attention mechanism.

3.1 Datasets

To verify the effectiveness of our proposed method, we conduct experiments on three publicly available online handwritten Chinese text recognition datasets:

IAHCT-USA2018 [5]: is a publicly available in-air handwritten Chinese text dataset, divided into training and test sets. The training set comprises 11,807 text lines from 277 writers, containing 196,129 characters across 3,564 classes. The test set includes 3,864 text lines from 92 writers, with 64,095 characters spanning 2,397 classes.

CASIA-OLHWDB [6]: is a traditional online handwritten Chinese text dataset, split into training and test sets. The training set comprises 41,710 text lines from 815 writers, containing 1,082,220 characters across 2,650 classes. The test set includes 10,510 text lines from 204 writers, with 268,924 characters covering 2,631 classes.

ICDAR-2013 [15]: is a competition dataset for online handwritten Chinese text, containing 3,432 text lines with 91,375 characters across 1,375 classes. It is used exclusively for testing. Training and testing configurations of the three datasets as shown in Table 1.

Table 1. Training and testing configurations of the three datasets.

Datasets		Text Lines	Characters	Character classes	Writers
IAHCT-USA2018	Train	11,807	196,129	3,564	277
	Test	3,894	64,095	2,397	92
CASIA-OLHWDB2.0-2.2	Train	41,710	1,082,220	2,650	815
	Test	10,510	268,924	2,631	204
ICDAR-2013	Test	3,432	91,576	1,375	–

3.2 Implementation Details

Experiments are implemented on the PyTorch deep learning framework with Adam ($\alpha = 0.0001$, decay factor 0.1), batch size 32. Hardware: Intel i5-12490F, 32GB RAM, and RTX 4090.

To evaluate the performance of the proposed method, we adopt the evaluation metrics defined in ICDAR2013 [15]:

$$\mathrm{CR} = \frac{N - S_d - D_d}{N}. \tag{5}$$

$$\mathrm{AR} = \frac{N - S_d - D_d - I_d}{N}. \tag{6}$$

where AR measures correct recognitions, CR evaluates output similarity. S_d, D_d, and I_d are substitution, deletion, and insertion distances. These are computed via dynamic programming to optimize matching between predictions and ground truth.

3.3 Ablation Experiments

In this section, we conduct ablation experiments on the IAHCT-USA2018 dataset to assess the impact of each module and their combinations on accuracy, recognition speed, and storage cost. The effects of different values of

d_k in SAM and the impact of CAM on the recognition results are shown in Table 2. From Table 2, it can be observed that as d_k increases, the computational speed decreases, storage consumption increases, and recognition accuracy first improves before declining. When $d_k = 64$, the recognition accuracy is optimal, and compared to $d_k = 49$, it has a relatively minor impact on computational speed and memory consumption. Therefore, in the subsequent experiments in this paper, d_k is consistently set to 64.

Table 2. The effects of different values of d_k in SAM and the impact of CAM on the recognition results.

Modules	AR (%)	CR (%)	Speed (ms)	Storage (MB)
SAM (49)	83.89	85.40	**150**	**18.3**
SAM (64)	**84.04**	**85.53**	153	18.6
SAM (81)	83.92	85.43	158	18.9
SAM (256)	83.96	85.38	164	22.4
CAM	83.60	84.95	159	17.3

Similar to the multi-head self attention, the number of SAMs (t) also has a certain impact on recognition performance when it varies, as shown in the experimental results in Table 3. Table 3, it can be seen that when $t = 2$, the recognition accuracy is the best. Compared to $t = 1$, $t = 2$ has a relatively minor impact on memory consumption and computational speed. Therefore, in this paper, t is set to 2. It is worth noting that the multi-head attention mechanism achieves the highest recognition accuracy when the number of heads is 8 [10].

Table 3. The corresponding recognition performance when the number of SAMs (t) changes.

Number of Heads	AR(%)	CR(%)	Speed(ms)	Storage(MB)
1	84.53	85.85	**150**	**18.6**
2	**84.70**	**85.88**	153	19.1
4	84.39	85.54	164	20.1

3.4 Comparison Experiments

To compare the proposed HAN with the multi-head self attention, as well as the recognition performance of EMACRN when HAN replaces the multi-head attention mechanism, we evaluated the proposed method against state-of-the-art approaches on public datasets using four metrics: Accuracy Rate (AR), Character Recognition Rate (CR), Recognition Speed (ms), and Storage Cost (MB). The recognition algorithms involved in the comparison are as follows:

(1) **Semi-CRFs** [16]: Used over-segmentation to generate candidate sequences and a Semi-Markov CRF to estimate probabilities. Beam search was applied for final predictions.
(2) **2DCRNN** [13]: Combined 2D convolution and RNNs. Handwritten Chinese coordinates were converted into images, with LSTM used to capture context for final recognition.
(3) **2DFCRN** [13]: Similar to 2DCRNN, but incorporated domain knowledge from images (path signature) for feature modeling during recognition.
(4) **2DMCFCRN** [14]: An extended version of 2DFCRN that incorporates multi-scale information from images into the input data.
(5) **CPN** [2]: Proposed a segmentation-based approach for handwritten Chinese recognition using a convolutional prototype network.
(6) **VGG-DBLSTM and CharNet-DBLSTM** [1]: Recognition methods using improved VGG and CharNet with LSTM.
(7) **MLGRU** [7]: Used a multi-layer distilled GRU network with an information distillation mechanism for sequence modeling.
(8) **GLRNet** [8]: Extracted distance features from pen trajectories and replaced LSTM with Transformer for sequence modeling.
(9) **LSTM** [11]: Employed stacked multi-layer LSTM networks for recognition.
(10) **TCRN** [5]: Used 1D CNN for local features, LSTM for long-term dependencies, and CTC for final prediction.
(12) **EMACRN** [9]: Extracted multi-scale features with EMACNN, used BiLSTM and multi-head attention for dependencies and weighting, followed by Focal CTC for final prediction.
(13) **EMACRN(HAN)**: Proposed in this paper. Replaces Multi-head self attention with HAN.

The recognition performance of different methods on CASIA-OLHWDB2.0-2.2 and ICDAR2013 datasets is listed in Table 4. The experimental results demonstrate that the proposed method achieves superior recognition accuracy and faster processing speed compared to other methods. Particularly, it shows significant improvements over EMACRN in both computational efficiency (faster recognition speed) and model compactness (reduced storage cost). This enhanced performance can be attributed to the HAN which incorporates ReLU nonlinear activation functions, as opposed to multi-head self-attention. The nonlinear transformation enables HAN to capture richer semantic relationships between characters within text lines while using fewer SAM.

The experimental results on the IAHCT-USA2018 dataset are presented in Table 5. As can be observed, the proposed method outperforms all other comparative approaches. While RNN-based methods (i.e., LSTM in Table 5) exhibit certain advantages in terms of memory efficiency, they process online handwritten text coordinates point-by-point, resulting in significantly slower recognition speeds and poor real-time performance. Our proposed method leverages

Table 4. The recognition performance of different methods on CASIA-OLHWDB2.0-2.2 and ICDAR2013 datasets.

Methods	CASIA-OLHWDB2.0-2.2		ICDAR2013		Storage (MB)	Speed (ms)
	AR(%)	CR(%)	AR(%)	CR(%)		
Semi-CRFs	85.92	87.63	–	–	–	–
2DCRNN	91.86	92.59	89.24	90.18	90.0	–
2DFCRN	93.82	94.37	91.57	92.25	90.0	–
2DMCFCRN	94.73	95.50	92.86	93.53	192.1	–
CPN	89.97	91.98	87.71	89.97	–	–
VGG-DBLSTM	90.65	91.00	87.49	87.98	39.5	–
CharNet-DBLSTM	90.46	90.94	87.10	87.71	19.1	–
MLGRU	–	–	91.36	92.37	24.5	–
GLRNet	–	–	91.24	91.81	26.9	–
LSTM	95.30	95.82	89.12	90.18	**18.1**	1322
TCRN	96.28	96.91	93.95	94.82	22.8	582
EMACRN	97.09	97.75	94.39	94.95	21.6	318
EMACRN (HAN)	**98.02**	**98.40**	**94.98**	**95.47**	18.3	**119**

the strengths of a convolutional-recurrent network architecture: Convolutional layers capture local semantic context through convolution and downsampling, effectively reducing text line length while mitigating the impact of local noise on recognition accuracy. Recurrent layers model sequential dependencies to extract temporal information from local semantic contexts. The proposed HAN further enhances performance by learning multi-level semantic relationships between local contexts, adaptively weighting features to better distinguish similar characters, thereby significantly improving recognition accuracy. Compared to multi-head self-attention (i.e., EMACRN in Table 5), our method achieves higher recognition accuracy, lower memory consumption and over 2× faster recognition speed.

Table 5. The recognition performance of different methods on the IAHCT-USA2018 dataset.

Methods	AR (%)	CR (%)	Storage (MB)	Speed (ms)
LSTM	80.74	81.81	19.4	1507
TCRN	82.50	83.62	23.9	639
EMACRN	84.30	85.59	22.4	374
EMACRN (HAN)	**84.70**	**85.88**	**19.1**	**153**

To further clarify the roles of semantic and channel attention, Fig. 3 illustrates their patterns. In Fig. 3(a, c, d) (T = 64, Head = 2), semantic attention spreads across multiple positions rather than a single region, showing its capacity to capture long-range context. This explains confusions such as "主要思考虑" where continuity is prioritized over precise character shapes. Meanwhile, Fig. 3 (b) shows that channel attention selectively amplifies certain channels—most strongly near indices 64 and 256—rather than distributing attention evenly. Together, these mechanisms enhance robustness by combining global context modeling with channel-level feature refinement.

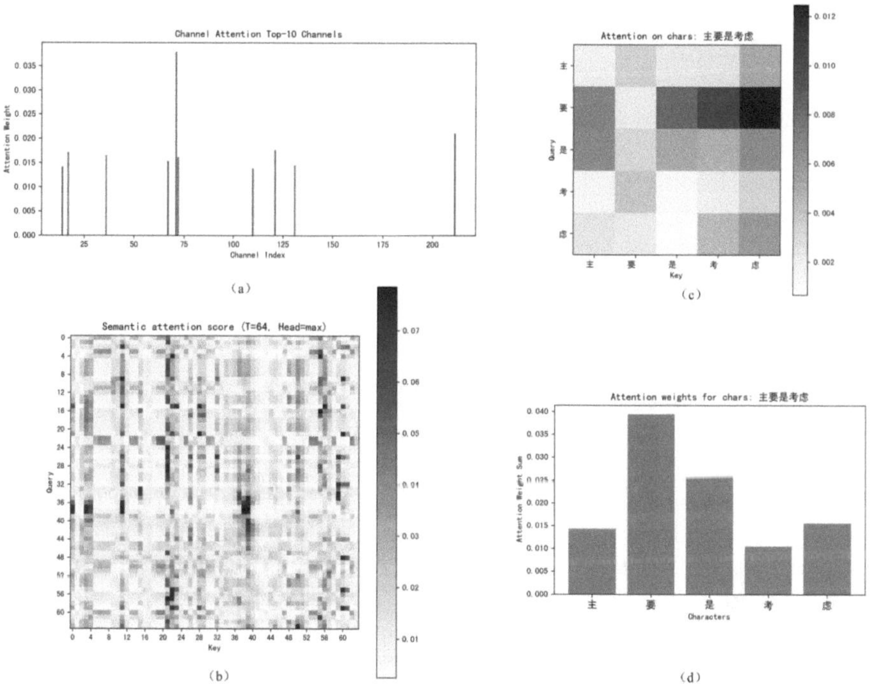

Fig. 3. Visualization of semantic and channel attention.

Some recognition examples are shown in Fig. 4. As observed, in correctly classified text line, the character "已" (ji)–despite its visual similarity to "乙" (yi)–was accurately identified, demonstrating the effectiveness of the proposed method in distinguishing similar characters. Among misclassifications, the character "是" (shi, meaning "is") was erroneously recognized as "思" (si, meaning "think"). In this text line, "是" functions as a verb with weak semantic dependencies on adjacent characters, highlighting a limitation in recognizing verb-type characters through semantic relationships. Addressing this challenge will be a key focus of future improvements.

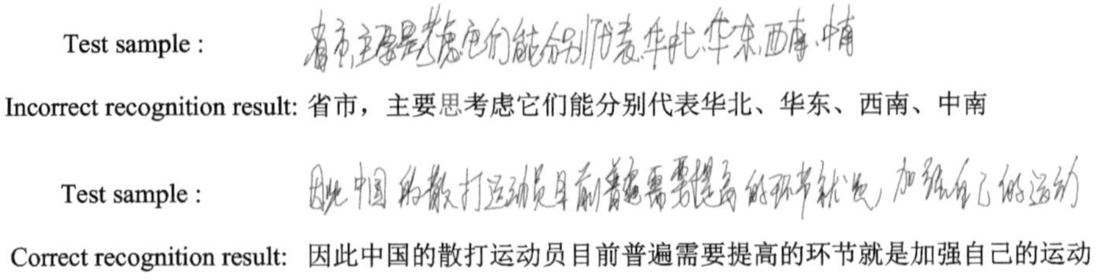

Test sample :

Incorrect recognition result: 省市，主要思考虑它们能分别代表华北、华东、西南、中南

Test sample :

Correct recognition result: 因此中国的散打运动员目前普遍需要提高的环节就是加强自己的运动

Fig. 4. Examples of incorrect (above) and correct (below) recognition results.

4 Conclusion

In OHCTR, there are a large number of similar characters. Due to differences in individual writing habits, these similar characters often have very poor discriminability, leading to decreased recognition accuracy. Leveraging semantic dependencies among characters within a text line and applying feature-specific weighting to distinguish similar characters is an effective way to further improve recognition accuracy. To this end, this paper proposed HAN. The HAN consists of SAM and CAM. The SAM captures multi-level semantic relationships among characters within a text line and is nonlinear, in contrast to the multi-head attention mechanism. The CAM performs convolutional channel weighting, which highlights or suppresses the influence of different features on the recognition result. Experimental results shown that, compared with the multi-head self attention, the HAN achieved higher recognition accuracy with lower memory consumption and faster computation speed. Combined with an end-to-end convolutional recurrent network, the proposed method achieved better recognition performance than state-of-the-art approaches in OHCTR. Our method also has potential for deployment in real-time or mobile applications, as its reduced computational cost and memory usage align well with the resource-constrained nature of mobile devices and real-time interaction requirements.

Acknowledgements. This work is supported by the Natural Science Research Project of Anhui Educational Committee under Grant No. 2024AH040028, the National Nature Science Foundation of China (NSFC) under Grant Nos. 62206006, 61906003, the Opening Foundation of Information Materials and Intelligent Sensing Laboratory of Anhui Province under Grant No. IMIS202215.

References

1. Chen, K., et al.: A compact CNN-DBLSTM based character model for online handwritten Chinese text recognition. In: Proceedings of the 14th IAPR International Conference on Document Analysis and Recognition (ICDAR), vol. 1, pp. 1068–1073. IEEE (2017)

2. Chen, Y., Zhang, H., Liu, C.L.: Improved learning for online handwritten Chinese text recognition with convolutional prototype network. In: Proceedings of the International Conference on Document Analysis and Recognition, pp. 38–53. Springer (2023)
3. Gan, J., Wang, W., Lu, K.: A unified CNN-RNN approach for in-air handwritten English word recognition. In: Proceedings of the 2018 International Conference on Multimedia and Expo, pp. 1–6. IEEE (2018)
4. Gan, J., Wang, W., Lu, K.: In-air handwritten English word recognition using attention recurrent translator. Neural Comput. Appl. **31**(7), 3155–3172 (2019)
5. Gan, J., Wang, W., Lu, K.: In-air handwritten Chinese text recognition with temporal convolutional recurrent network. Pattern Recogn. **97**, 107025 (2020)
6. Liu, C., Yin, F., Wang, D., Wang, Q.: CASIA online and offline Chinese handwriting databases. In: Proceedings of the 2011 International Conference on Document Analysis and Recognition, pp. 37–41. IEEE (2011)
7. Liu, M., Xie, Z., Huang, Y., Jin, L., Zhou, W.: Distilling GRU with data augmentation for unconstrained handwritten text recognition. In: Proceedings of the 16th International Conference on Frontiers in Handwriting Recognition (ICFHR), pp. 56–61. IEEE (2018)
8. Peng, D., et al.: Towards fast, accurate and compact online handwritten Chinese text recognition. In: Proceedings of the 16th International Conference on Document Analysis and Recognition, pp. 157–171. Springer (2021)
9. Qu, X., Wu, Z.: End-to-end multi-scale attention convolutional recurrent network for online handwritten Chinese text recognition. Expert Syst. Appl. 127626 (2025)
10. Qu, X., Wu, Z., Huang, J.: End-to-end attention convolutional recurrent network for online handwritten Chinese text recognition. Multimed. Tools Appl. **83**(23), 62541–62558 (2024)
11. Sun, L., Su, T., Liu, C., Wang, R.: Deep LSTM networks for online Chinese handwriting recognition. In: Proceedings of the 2016 International Conference on Frontiers in Handwriting Recognition, pp. 271–276. IEEE (2016)
12. Wu, Y.C., Yin, F., Liu, C.L.: Improving handwritten Chinese text recognition using neural network language models and convolutional neural network shape models. Pattern Recogn. **65**, 251–264 (2017)
13. Xie, Z., Sun, Z., Jin, L., Feng, Z., Zhang, S.: Fully convolutional recurrent network for handwritten Chinese text recognition. In: Proceedings of the 23rd International Conference on Pattern Recognition, pp. 4011–4016. IEEE (2016)
14. Xie, Z., Sun, Z., Jin, L., Ni, H., Lyons, T.: Learning spatial-semantic context with fully convolutional recurrent network for online handwritten Chinese text recognition. IEEE Trans. Pattern Anal. **40**(8), 1903–1917 (2017)
15. Yin, F., Wang, Q., Zhang, X., Liu, C.: ICDAR 2013 Chinese handwriting recognition competition. In: Proceedings of the 12th International Conference on Document Analysis and Recognition, pp. 1464–1470. IEEE (2013)
16. Zhou, X.D., Wang, D.H., Tian, F., Liu, C.L., Nakagawa, M.: Handwritten Chinese/Japanese text recognition using semi-Markov conditional random fields. IEEE Trans. Pattern Anal. **35**(10), 2413–2426 (2013)
17. Zhou, X., Zhang, Y., Tian, F., Wang, H., Liu, C.: Minimum-risk training for semi-Markov conditional random fields with application to handwritten Chinese/Japanese text recognition. Pattern Recogn. **47**(5), 1904–1916 (2014)

FusionNet for Multimodal Depression Recognition: Integrating Electroencephalography and Traditional Chinese Medicine Features

Junyu Liu[1], Banghua Yang[1,2(✉)], Junhua Chen[1], and Shuai Kuang[2]

[1] Mental Health Center Affiliated to Shanghai University School of Medicine, Shanghai, China
{yangbanghua,junhua_chen}@shu.edu.cn

[2] School of Mechatronic Engineering and Automation, Shanghai University, Shanghai, China
alex@shu.edu.cn

Abstract. Depression is a prevalent mental health disorder with significant impacts on individuals' quality of life and societal participation. Current diagnostic methods, primarily relying on self-reported scales and clinical interviews, are subjective, time-consuming, and susceptible to diagnostic inaccuracies. This study proposes FusionNet, a deep learning-based framework that integrates Traditional Chinese Medicine (TCM) consultation data with electroencephalography (EEG) signals obtained from an Oddball paradigm for depression state classification and recognition. The FusionNet model employs a feature-level fusion approach, utilizing EEGNet for spatial-temporal feature extraction and Long Short-Term Memory (LSTM) networks for temporal pattern recognition. Experimental results demonstrate that FusionNet achieves an 85% classification accuracy, highlighting its potential in improving diagnostic precision and efficiency for depression detection. The proposed method offers a robust, objective diagnostic tool that enhances early diagnosis and clinical decision-making, and provides a new approach for personalized treatment strategies.

Keywords: Depression Recognition · Electroencephalography (EEG) · Deep Learning · Traditional Chinese Medicine Consultation · Multimodal Fusion

1 Introduction

Depression is a common and debilitating mental health disorder that significantly affects individuals' emotional well-being, cognitive functions, and overall quality of life. The prevalence of depression has been increasing, partly due to rapid urbanization and growing societal pressures. In China, over 60 million individuals suffer from depression, with approximately 300,000 individuals dying from suicide each year due to the disorder [1]. Despite its high prevalence, nearly 90% of individuals in China fail to recognize their condition early, often missing the optimal treatment window [2]. The symptoms of depression are often persistent and include low mood, loss of interest in daily activities,

P. Umapada et al. (Eds.): ICCPR 2025, CCIS 2811, pp. 74–86, 2026.
https://doi.org/10.1007/978-981-95-8315-7_6

hopelessness, and significant impairment in social and occupational functions. However, depression often co-occurs with other mental health disorders, including anxiety and chronic pain, which complicates diagnosis and treatment.

Currently,the diagnosis of depression primarily relies on self-reported questionnaires and clinical interviews, which are subjective and prone to bias. Common diagnostic tools include the Self-Rating Depression Scale (SDS), Hamilton Depression Rating Scale (HAMD), and Patient Health Questionnaire-9 (PHQ-9). However, these diagnostic methods are subjective and vulnerable to factors such as patient concealment and physician bias. Therefore, there is an urgent need for more objective and reliable methods of diagnosing depression.

With the development and advancement of relevant theoretical research and hardware technology, the use of Electroencephalogram (EEG) technology to reflect the brain electrical activity of depression patients has been widely studied in clinical practice. EEG examinations can reflect changes in brain function at an early stage with high sensitivity [3]. It holds high value in the early diagnosis, disease assessment, and efficacy evaluation of depression, providing clinicians with an economical and practical diagnostic method [4]. Moreover, it can also be used to assess the effects of drug treatments, offering more reliable and objective evidence for clinical decision-making. It is a simple and practical examination method [5].

Traditional machine learning methods, such as support vector machines and random forests, rely on manually extracted features and relatively fixed model structures. Deep learning, on the other hand, improves upon these methods by automatically learning features and offering a more powerful model representation, which excels in handling complex and high-dimensional data. This approach has opened up new avenues for the diagnosis and research of brain disorders. For instance, Xiaowei Zhang et al. [6] developed a 1-D convolutional neural network (CNN) to obtain more effective EEG signal features. By integrating demographic factors such as age and gender using attention mechanisms, the 1-D CNN was able to explore the complex correlations between EEG signals and demographic factors, ultimately producing more effective high-level representations for depression detection. Furthermore, Hu Bin's team at Lanzhou University (2020) recorded EEG signals from 86 depression patients and 92 healthy controls under different stimuli. After extracting both linear and nonlinear features from these signals, they used a linear combination technique to fuse the features from different modalities, which improved the classification accuracy using the KNN classifier when combining positive and negative audio stimuli [7]. These studies have demonstrated that integrating different data sources can significantly improve the classification performance for depression, highlighting the potential of multi-modal fusion. However, current classification models still suffer from low robustness, and enhancing the model's performance and interpretability through deep learning requires further research.

Previous studies have successfully applied deep learning techniques in the field of Traditional Chinese Medicine (TCM), extending the understanding of TCM features such as constitution types and tongue diagnosis [8]. For example, Zhou Hao et al. [9] used a deep feature fusion method based on AlexNet to classify and identify constitution types such as Qi deficiency, phlegm-dampness, and damp-heat from tongue images, achieving an accuracy of 77.0%. These studies have verified the ability of deep learning models to assist in clinical diagnosis. However, in real-world clinical practice, patient data often exhibits significant individual differences, and high classification accuracy achieved in within-subject studies has limited practical guidance. One challenge is the cross-subject classification recognition, which remains a major issue in deep learning, especially in EEG/ERP signal processing. Additionally, a single diagnostic tool in clinical practice is often insufficient, and multiple evaluation methods are needed to improve diagnostic accuracy and reliability. This concept also applies to deep learning models, where appropriately combining different types of data features may improve cross-subject classification accuracy. Compared to imaging and resting-state EEG data, research using deep learning models combined with ERP data for disease diagnosis remains relatively limited. This study attempts to concatenate EEG data and TCM consultation data features to explore the classification accuracy of deep learning models in cross-subject depression recognition.

2 Materials and Methods

2.1 Participants

This study recruited 20 depression patients from Tongde Hospital in Zhejiang Province and 20 control subjects from Zhejiang Chinese Medical University. All participants signed written informed consent during the experiment, and the study was approved by the local research ethics committee.

The inclusion criteria for subjects were: Diagnosis of depression according to the International Classification of Mental Disorders, 4th Edition (DSM-IV);Age between 18 and 65, with no gender restrictions; Literacy and ability to cooperate with all assessments in the study; A clear diagnosis of depression by a psychiatrist or meeting the depression diagnostic criteria with a PHQ-9 score $\geq$ 10.

Exclusion criteria for subjects were: Severe cognitive impairment or hearing disability that prevented participation in the assessment; Severe head injury; Acute physical illness; Substance abuse (including drugs and alcohol) (Table 1).

Table 1. Basic information of depression and normal control groups

Attribute	Depression Group	Normal Control Group
Gender (Female: Male)	11:9	10:10
Age	27.30 ± 6.43	26.75 ± 5.55
Right-Handed	Yes	Yes
Native Language	Chinese	Chinese

2.2 EEG Data Acquisition and Preprocessing

Electroencephalogram (EEG) data were recorded using the Greentek Gelfree Net Cap device. Subjects were asked to sit comfortably on a chair, and based on the international

10–20 electrode placement system, the sampling frequency was set to 1000 Hz. To address the characteristic of delayed responses to external stimuli in depression patients, this study designed an auditory Oddball paradigm (Fig. 1), which induces the P300 component through auditory stimuli. This approach analyzes the neuromechanisms of cognitive impairment in depression from the perspective of differences in brain region activation levels when stimuli are presented. The stimuli consisted of two types: 30 high-frequency sound stimuli (1200 Hz) and 170 low-frequency sound stimuli (1000 Hz), presented in a random sequence. Before the experiment, the experimenter instructed the subjects to count mentally whenever a low-probability sound stimulus occurred, and to report the count at the end of the experiment.

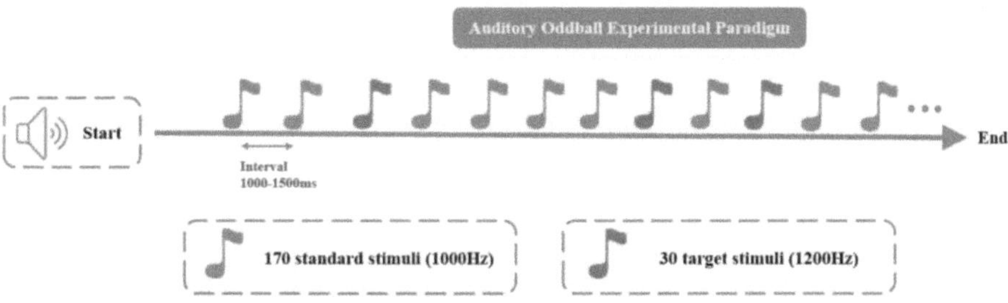

Fig. 1. Experimental paradigm

The experimental paradigm was presented using E-Prime software (E-Prime 3.0), and the data collected was preprocessed using the EEGLAB toolbox and MATLAB (R2019b) software [10]. First, a finite impulse response (FIR) filter with a passband of 0.5 to 45 Hz was applied (Fig. 2), followed by a notch filter (49–51 Hz) to remove 50 Hz interference. The data were then downsampled to 250 Hz, with the "Cz" electrode used as the reference. Independent Component Analysis (ICA) was performed to decompose the EEG signals into independent components. Artifactual components, such as eye movement and muscle artifacts, were identified and removed using visualization tools and statistical analysis (Fig. 3). For residual artifacts that could not be fully removed by ICA [11, 12], artifact correction tools were applied, such as using the pop_rmbad function in EEGLAB to remove bad channels or using artifact correction plugins for more precise removal. The data were segmented into 1-s epochs, from 200 ms before the stimulus to 800 ms after the stimulus, retaining only trials containing the target stimulus. After the initial preprocessing steps, the overall quality of the data improved significantly. However, some segments still exhibited abnormally high amplitudes, exceeding 100 μV. These anomalies could result from factors such as poor electrode contact, external electromagnetic interference, or involuntary movement by the subject. To ensure data accuracy and reliability, segments with abnormal amplitudes were carefully manually inspected and removed.

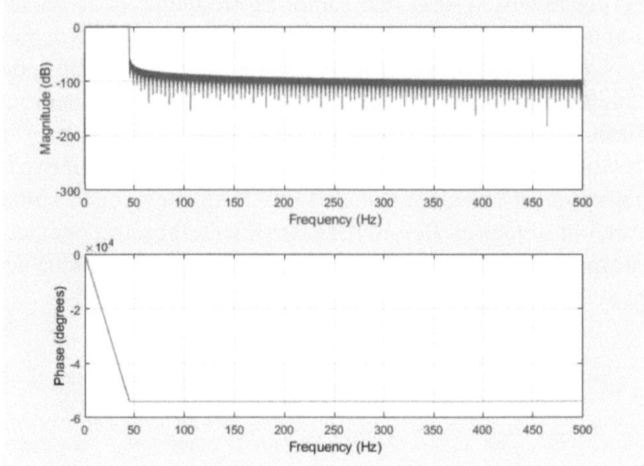

Fig. 2. Frequency response diagram of FIR filter in EEGLAB

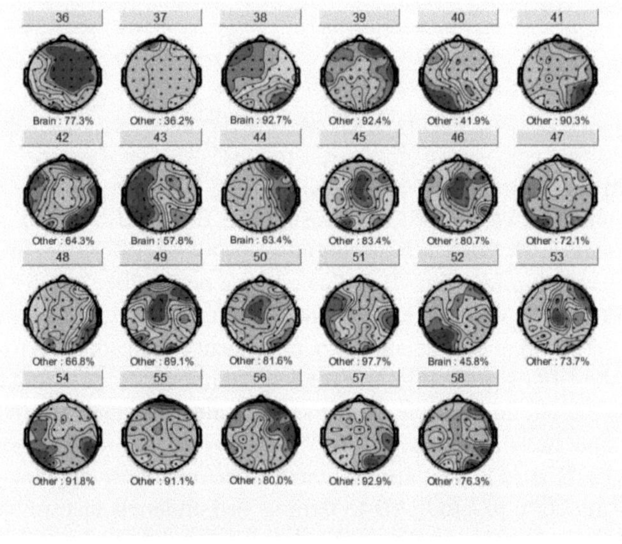

Fig. 3. Some of the ICA components for a subject

2.3 Feature Fusion of Multimodal Data

In this study, each option in the Traditional Chinese Medicine (TCM) diagnostic questionnaire was encoded using One-Hot encoding, transforming symptoms such as cold/heat, sweating, and sleep into numerical features that can be processed by machines (Fig. 4). The feature matrices extracted from EEG data and TCM diagnostic data are represented as: X_e $(x_1, x_2,\ldots, x_s)$ and $Y_e = (y_1, y_2,\ldots, y_t)$, where x_s is the s-th feature column vector of the EEG data, and y_t is the t-th feature column vector of the TCM diagnostic data. The feature matrix resulting from the linear combination operation (Fig. 5) is as follows:

$$H_e = k_1 \times X_e + k_2 \times Y_e \quad (1)$$

$$T_e = k_1 \times X_e + k_1 \times Y_e \quad (2)$$

where $k_1 = 1$, $k_2 = -1$. H_e is the linear subtraction of the feature matrix, and T_e is the linear addition of the feature matrix. The EEG feature vector has a length of 64 dimensions, the TCM diagnostic information feature vector has a length of 8 dimensions, and the fused data has 72 dimensions. In this way, the system can combine EEG signals with TCM diagnostic data for fusion analysis, improving the accuracy of depression recognition.

Sleep Condition
Normal
Difficulty falling asleep
Easily waking up
Frequent dreaming
Early waking
Somnolence

Normal	Difficulty falling asleep	Easily waking up	Frequent dreaming	Early waking	Somnolence
1	0	0	0	0	0
0	1	0	0	0	0
0	0	1	0	0	0
0	0	0	1	0	0
0	0	0	0	1	0
0	0	0	0	0	1

Fig. 4. One-Hot encoding of TCM diagnostic data

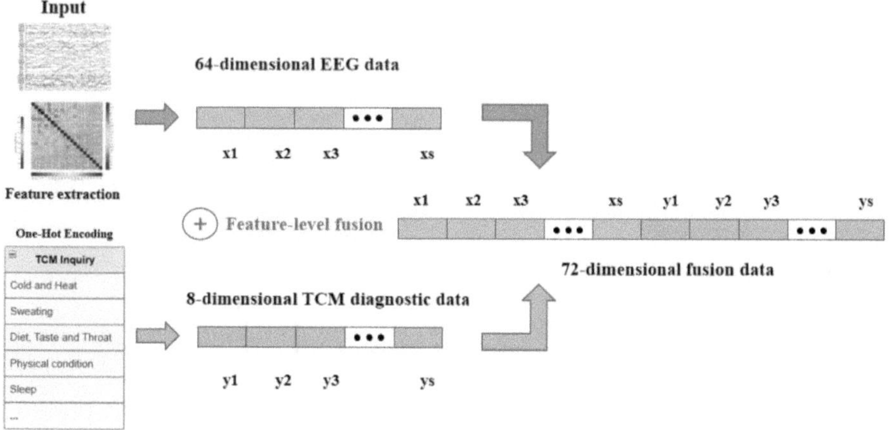

Fig. 5. Multimodal feature fusion method

3 Model Architecture

FusionNet is an innovative neural network model specifically designed for the efficient processing of electroencephalogram (EEG) signals. It combines the power of Convolutional Neural Networks (CNNs), specifically EEGNet, and Long Short-Term Memory (LSTM) networks, enabling it to learn long-term dependencies in time-series signals. The combination of both components contributes to effective multimodal data decoding.

Convolutional Neural Networks (CNNs) have a significant role in EEG-based cognitive state assessment, and EEGNet, designed specifically for processing EEG signals, reduces the number of model parameters by using deep separable convolutions, while retaining sufficient complexity to capture critical features in EEG data. The EEGNet model architecture consists of three core parts: 2D convolution, deep convolution, and separable convolution. Each part is optimized for different attributes of EEG signals, thereby improving feature extraction and information parsing (Fig. 6).

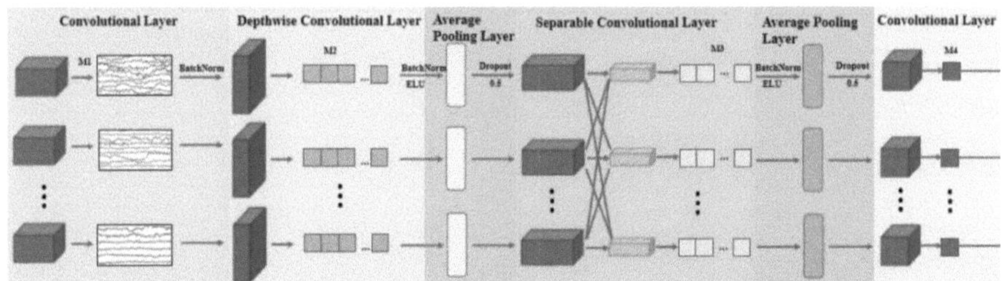

Fig. 6. EEGNet architecture with 2D, deep, and separable convolutions

After the essential features are extracted by EEGNet, the data is passed into the LSTM layer, which is used to capture the temporal dependencies in the EEG signal. LSTM networks are particularly well-suited for learning long-term patterns in time-series data, which allows them to recognize the dynamic features in EEG signals that change over time [13]. In LSTM architecture, there is a special hidden unit called the memory unit, which is used to remember the previous input for a long time. The LSTM architecture contains forgetting, learning, memory, and use gates that determine whether the input is important enough to be saved. Four different functions, sigmoid (σ), hyperbolic tangent (tanh), multiplicative ($\times$), and sum (+), are used in the LSTM cell, making it easier to update the weights during backpropagation (Fig. 7).

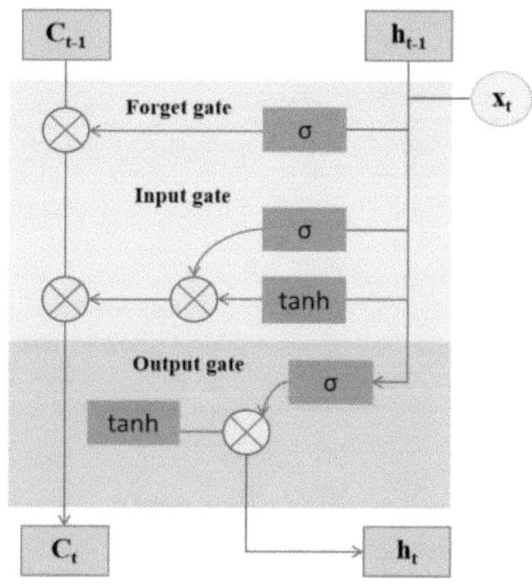

Fig. 7. Structure of LSTM with gates and memory units

The forgetting gate, input gate, and output gate of the LSTM cell can add and remove information from the cell state, as defined below:

$$f_t = \sigma\left(W_f\left[h_{t-1}, x_t\right] + b_f\right) \tag{3}$$

$$i_t = \sigma\left(W_i\left[h_{t-1}, x_t\right] + b_i\right) \tag{4}$$

$$o_t = \sigma\left(W_o\left[h_{t-1}, x_t\right] + b_o\right) \tag{5}$$

The current cell memory state is calculated as:

$$\widehat{C_t} = \tanh\left(W_C\left[h_{t-1}, x_t\right] + b_C\right) \tag{6}$$

The calculation of cell state is defined as:

$$C_t = f_t * C_{t-1} + i_t * \widehat{C_t} \tag{7}$$

The calculation of cell output is defined as:

$$h_t = o_t \tanh(C_t) \tag{8}$$

where σ (-) is the activation function, h_{t-1} is the output of the LSTM cell at the previous time step, C_{t-1} is the state of the LSTM cell at the previous time step, and b_i, b_o, and b_c are the biases.

Finally, the data is passed through the fully connected layer and classified using a sigmoid activation function. This classifies the processed data into two categories: Normal or Depressed. By combining the strengths of CNN for spatial feature extraction and LSTM for temporal pattern recognition, FusionNet effectively decodes multimodal data, improving depression detection accuracy based on EEG signals and TCM diagnostic information.

4 Classification Results

4.1 Dataset Division and Voting Principle

The data was divided into 1-s trials, each containing 250 sampling points. The sliding window method (with a 1-s interval) was applied to extract 30 high-frequency target auditory stimuli trials from each subject. The dataset includes data from 40 subjects, with a total of 1200 samples.

For the 10-Fold Cross-Validation(CV) method, the dataset was divided into ten folds. The training set contained 8 folds of data, corresponding to 16 depressed patients and 16 normal individuals, totaling 32 subjects. Each subject provided 30 samples, so the 8-fold training set had 960 samples in total. The validation set, used for model validation and parameter adjustment, included 1 fold of data (with data from 2 depressed patients and 2 normal individuals), totaling 120 samples; the test set, designed to evaluate the model's generalization ability, also contained 1 fold of data (120 samples). The entire dataset partitioning process was repeated 10 times to ensure each subject had the opportunity to be part of the test set. Before the training phase, data in the training, validation, and test sets was randomly shuffled to avoid the impact of data order on the results.

For the Leave-One-Subject-Out (LOSO) method, in each iteration, data from 1 subject was used as the test set, while data from the remaining 39 subjects was split into a training set (data from 35 subjects) and a validation set (data from 4 subjects, including 2 patients and 2 normal individuals to ensure balanced labels). The training set was used for model training, and the validation set for parameter optimization; after training, the test set was used for prediction to provide an unbiased performance assessment. Before training, data in all sets was randomly shuffled, and each subject's 30 samples were in only one fold to avoid cross-use of data.The dataset partitioning method for the LOSO approach is illustrated in (Fig. 8).

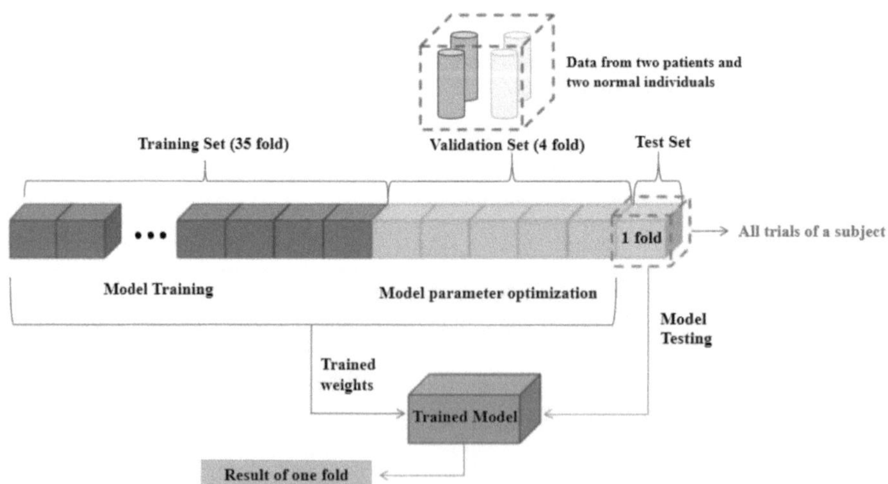

Fig. 8. Dataset partitioning diagram for the LOSO approach

In this study, the "Principle of Majority Voting" was applied to integrate the classification results of each sample predicted by trained model (Fig. 9). Specifically, for each subject, all 30 trials were classified, and the number of times each trial was assigned to the "positive" (depressed) or "negative" (normal) category was counted. The final classification for each subject was determined by the category that occurred the most frequently across the trials. This strategy combines the classification information from each trial, enhancing the accuracy of individual diagnosis and increasing the stability of the model. It provides a solid foundation for disease diagnosis, aligns with clinical standards for individual diagnosis, and ensures that all subsequent evaluation metrics are calculated based on individual subjects.

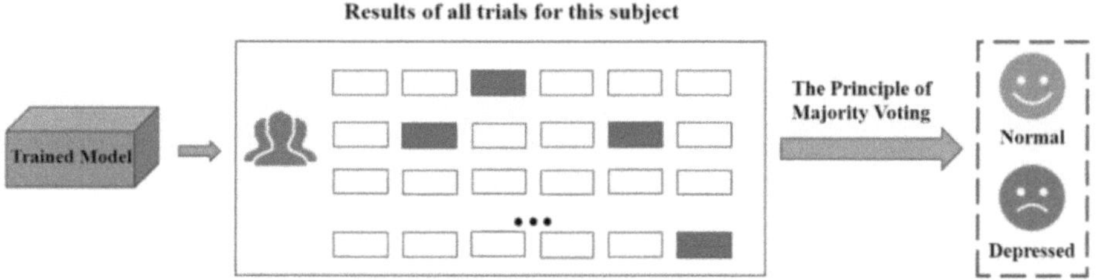

Fig. 9. Principle of majority voting result statistics diagram

4.2 Results

To validate the effectiveness of the proposed algorithm, the model was evaluated using accuracy as the primary metric, which represents the proportion of correctly predicted samples out of the total samples. The formula for accuracy is as follows:

$$\text{Accuracy} = \frac{TP + TN}{TP + TN + FP + FN} \tag{9}$$

where True Positives (TP) refers to the number of actual positives correctly identified as positive, True Negatives (TN) refers to the number of actual negatives correctly identified as negative, False Positives (FP) refers to the number of negative samples incorrectly classified as positive, and False Negatives (FN) refers to the number of positive samples incorrectly classified as negative.

For the LOSO method, the model correctly classified 34 out of the 40 subjects, resulting in an accuracy of 85%. This demonstrates the model's ability to effectively classify subjects into the two categories (Normal or Depressed) based on EEG and TCM diagnostic data. The confusion matrices are shown in (Fig. 10).

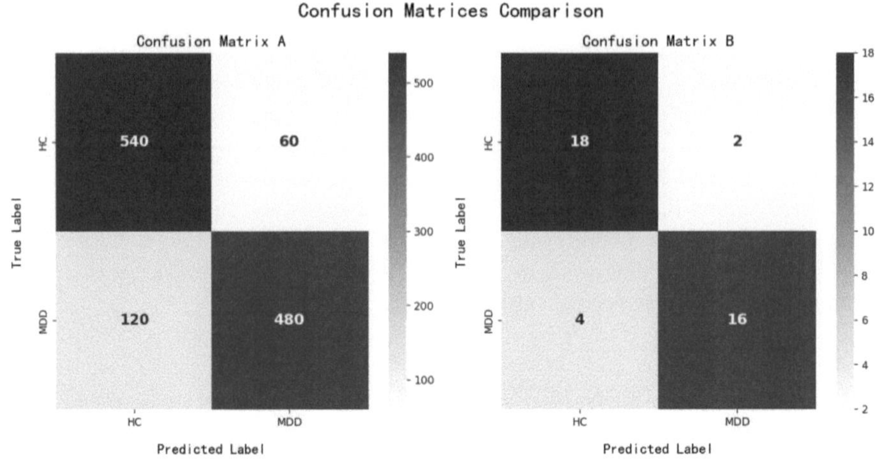

Fig. 10. Confusion matrix results under LOSO for two-class classification. Matrix A: Before adding the voting strategy in the two-class classification. Matrix B: After adding the voting strategy in the two-class classification.

4.3 Compared Methods

To evaluate the superiority of the FusionNet network architecture proposed in this study, three other algorithms were tested and compared on the same dataset, including EEGNet, FBCNet, and DeepConvNet.

FBCNet integrates the advantages of multi-band processing in FBCSP with the convolution-pooling characteristics of Convolutional Neural Networks (CNN). Its network architecture consists of spatial convolution layers, temporal convolution layers, a Softmax layer, and normalization layers [14]. After two convolutional operations, the network further processes the data through a variance layer, which calculates the temporal variance for each channel's time series using a specific formula, thereby enhancing the discriminative power of the features.

DeepConvNet is a deep convolutional neural network specifically designed for processing EEG data. The architecture of DeepConvNet includes four convolutional and max-pooling blocks [15]. Block1 utilizes a hierarchical convolution strategy, where the first layer performs temporal filtering, and the second layer applies spatial filtering. The network employs the ELU activation function, with batch normalization and normalization applied after each convolutional layer to ensure the input distributions between layers approximate a normal distribution. Except for the first convolutional layer, dropout with a probability of 0.5 is applied to the remaining layers to prevent overfitting and enhance the model's generalization ability (Table 2).

Table 2. Compare the accuracy and F1 Score with other algorithms between two validation methods

method	Accuracy (LOSO)	F1 Score (LOSO)	Accuracy (CV)	F1 Score (CV)
EEGNet	80.00%	79.17%	78.74%	79.42%
FBCNet	67.50%	68.30%	69.12%	67.95%
DeepConvNet	77.50%	76.92%	77.88%	77.54%
FusionNet	85.00%	84.10%	84.68%	84.22%

5 Discussion

In conclusion, the proposed FusionNet network structure offers a promising solution for the objective and quantitative identification of depression. By integrating the strengths of CNN, specifically EEGNet, with LSTM networks, FusionNet is capable of learning long-term dependencies in time series signals, making it ideal for decoding multimodal data. Notably, FusionNet effectively fuses EEG data with TCM information, leveraging the complementary advantages of both objective brain function data and subjective symptom data. The network demonstrates superior performance in comparison to other algorithms, achieving an accuracy of 85% and an F1 score of 84.1%, both of which are the highest among the tested models. This multimodal fusion significantly enhances the model's ability to recognize depression states.

Future research will focus on addressing the cultural specificity of TCM features, further optimizing the integration of TCM scales, and enhancing their applicability in different populations. In addition, exploring more advanced fusion strategies such as attention mechanisms and feature embedding will enable the model to better priori tize relevant features, thereby improving performance. Finally, further expanding the sample size of the dataset, introducing more diverse clinical data, and exploring data augmentation techniques will make the model more adaptable to a wide range of clinical applications.

6 Disclosure of Interests

All authors declare that the research was conducted in the absence of any commercial or financial relationships that could be construed as a potential conflict of interest.

Acknowledgments. This work is supported by the National Natural Science Foundation of China (No. 62376149).

References

1. Global, regional, and national incidence, prevalence, and years lived with disability for 301 acute and chronic diseases and injuries in 188 countries, 1990–2013: a systematic analysis for the Global Burden of Disease Study 2013. The Lancet **386**(9995), 743–800 (2015)

2. Lu, Y.: China's Rural Elderly Depression and Its Influencing Factors. PhD Thesis, Shandong University, Shandong (2014)
3. Wang, H.: Research status and prospects of EEG-based emotion recognition. China Med. Equip. **39**(1), 161–165 (2024)
4. Campbell, I.G., Kim, E.I., Darchia, N., et al.: Sleep restriction and age effects on waking alpha EEG activity in adolescents. Sleep Adv. **3**(1), zpac015 (2022)
5. Günes, S.S., Yesil, C., Gurdal, E.E., et al.: Primum non nocere: in silico prediction of adverse drug reactions of antidepressant drugs. Comput. Toxicol. **18**, 160–165 (2021)
6. Zhang, X., Li, J., Hou, K., et al.: EEG-based depression detection using convolutional neural network with demographic attention mechanism. In: 2020 42nd Annual International Conference of the IEEE Engineering in Medicine & Biology Society (EMBC), pp. 128–133 (2020)
7. Cai, H., Qu, Z., Li, Z., et al.: Feature-level fusion approaches based on multimodal EEG data for depression recognition. Inf. Fusion **59**, 127–138 (2020)
8. Wu, X., Xu, H., Lin, Z., et al.: A review of deep learning in tongue image classification. Comput. Sci. Explor. **17**(2), 303–323 (2023)
9. Zhou, H., Hu, G., Zhang, X.: Research on the method of Chinese medicine constitution classification based on tongue image deep feature fusion. Beijing Biomed. Eng. **39**(3), 221–226 (2020)
10. Delorme, A., Makeig, S.: EEGLAB: an open source toolbox for analysis of single-trial EEG dynamics including independent component analysis. J. Neurosci. Methods **134**, 9–21 (2004)
11. Jiang, X., Bian, G.B., Tian, Z.: Removal of artifacts from EEG signals: a review. Sensors **19**, 987 (2019)
12. Pion-Tonachini, L., Kreutz-Delgado, K., Makeig, S.: ICLabel:an automated electroencephalographic independent component classifier, dataset, and website. Neuroimage **198**, 181–197 (2019)
13. Bengio, Y., Simard, P., Frasconi, P.: Learning long-term dependencies with gradient descent is difficult. IEEE Trans. Neural Netw. **5**(2), 157–166 (1994)
14. Mane, R., Chew, E., Chua, K., et al.: FBCNet: a multi-view Convolutional Neural Network for Brain-Computer Interface (2021). arXiv:2104.01233
15. Schirrmeister, R.T., Springenberg, J.T., Fiederer, L.D.J., et al.: Deep learning with convolutional neural networks for EEG decoding and visualization. Hum. Brain Mapp. **38**(11), 5391–5420 (2017)

Vision Transformer for Robust Occluded Person Re-identification in Complex Surveillance Scenes

Bo Li[1], Duyuan Zheng[2], Xinyang Liu[2], Qingwen Li[1], Hong Li[1], Hongyan Cui[2,3](✉), Ge Gao[2], and Chen Liu[2]

[1] China Tower Corporation Limited, Beijing 100089, China
[2] China University of Petroleum-Beijing at Karamay, Karamay 834000, China
cuihy@bupt.edu.cn
[3] Beijing University of Posts and Telecommunications, Beijing 100876, China

Abstract. Person re-identification (ReID) in surveillance is challenged by occlusion, viewpoint distortion, and poor image quality. Most existing methods rely on complex modules or perform well only on clear frontal images. We propose Sh-ViT (Shuffling Vision Transformer), a lightweight and robust model for occluded person ReID. Built on ViT-Base, Sh-ViT introduces three components: First, a Shuffle module in the final Transformer layer to break spatial correlations and enhance robustness to occlusion and blur; Second, scenario-adapted augmentation (geometric transforms, erasing, blur, and color adjustment) to simulate surveillance conditions; Third, DeiT-based knowledge distillation to improve learning with limited labels.To support real-world evaluation, we construct the MyTT dataset, containing over 10,000 pedestrians and 30,000+ images from base station inspections, with frequent equipment occlusion and camera variations. Experiments show that Sh-ViT achieves 83.2% Rank-1 and 80.1% mAP on MyTT, outperforming CNN and ViT baselines, and 94.6% Rank-1 and 87.5% mAP on Market1501, surpassing state-of-the-art methods.In summary, Sh-ViT improves robustness to occlusion and blur without external modules, offering a practical solution for surveillance-based personnel monitoring.

Keywords: Person re-identification · Deep Learning · Vision Transformer

1 Introduction

Person re-identification (ReID) [1] is a fundamental problem in computer vision, aiming to identify the same pedestrian across different camera views. It has broad applications in public safety, intelligent surveillance, and operation and maintenance (O&M) management.

In base station O&M scenarios, the goal of ReID is to distinguish authorized from unauthorized personnel. As shown in Fig. 1, the base station scenario has unique imaging

Bo Li and Duyuan Zheng have contributed equally to this work and should be considered co-first authors.

P. Umapada et al. (Eds.): ICCPR 2025, CCIS 2811, pp. 87–98, 2026.
https://doi.org/10.1007/978-981-95-8315-7_7

characteristics and challenges: First, surveillance cameras are typically installed at elevated or oblique angles, leading to noticeable pose tilt and geometric distortion. Second, blur from poor lighting or hardware and frequent equipment-person occlusion further reduce fine-grained details. Third, in complex working environments, frequent interactions between personnel and equipment often result in partial or large-scale occlusion. Tilt, blur, and occlusion are the main bottlenecks of ReID in O&M applications, and form the focus of this study.

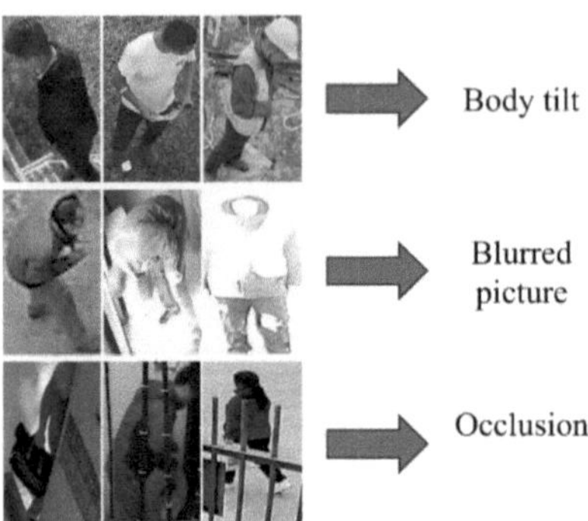

Fig. 1. Person re-identification problem, showing the limitations of convolutional kernel window sliding for feature extraction in complex scenes (including person tilt, illumination change, background interference), and the impact of strong data dependence and limited generalization ability of CNN on feature extraction robustness/accuracy and character recognition accuracy.

To address these issues, we propose Shuffled Vision Transformer (Sh-ViT), a lightweight yet robust framework for occluded person ReID. Sh-ViT incorporates three key innovations: First, a Shuffle module that randomly rearranges feature tokens to enhance robustness against occlusion and blurring; Second, scenario-adapted data augmentation that simulates top-down views, equipment occlusions, lighting variations, and blur degradations; Third, knowledge distillation from DeiT, improving stability and data efficiency under limited annotations.

In addition, we construct a new dataset, MyTT, containing over 10,000 pedestrians and 30,000+ images collected from real base station scenarios, which highlights challenges not well captured by public datasets.Extensive experiments demonstrate that Sh-ViT achieves 83.2% Rank-1 and 80.1% mAP on MyTT,94.6% Rank-1 and 87.5% mAP on Market1501, surpassing recent state-of-the-art methods. These results confirm that Sh-ViT offers a practical solution for person ReID in complex O&M surveillance scenarios.

The remainder of this paper is organized as follows: Sect. 2 reviews deep learning-based feature extraction methods. Section 3 introduces the proposed Sh-ViT model. Section 4 presents experimental results and comparisons with existing ViT models. Section 5 summarizes the work presented in this paper.

2 Related Work

2.1 Occluded Person Re-identification

Occluded person ReID is important in real surveillance. Traditional methods relied on part-based pooling or parsing (PCB-RPP [2], HOReID [3]). With Transformers, attention-based recovery has been studied, such as feature completion (FCFormer [4]) or dynamic patch selection (DPEFormer [5]). However, these methods depend on synthetic occlusion or complex modules. In contrast, our Shuffle mechanism directly enhances global robustness without explicit occlusion detection.

2.2 Vision Transformers in ReID

Transformers have been adapted for ReID, e.g., TransReID [6] with side information, self-supervised pre-training [7], and masked image modeling (PersonViT [8]). These works mainly address holistic recognition but lack designs for severe occlusion. Our Sh-ViT extends ViT with a shuffle module, improving tolerance to missing body regions.

2.3 Data Augmentation for Robust ReID

Conventional augmentations (e.g., random erasing [9]) and occlusion simulation [10] help robustness, but they overlook surveillance-specific challenges such as equipment occlusion and perspective distortion. We design scenario-adapted augmentation combining geometric transforms, realistic occlusion, blur, and lighting adjustments to better match real scenes.

2.4 Knowledge Distillation in ReID

Distillation improves ReID efficiency and performance. DeiT [11] introduced distillation tokens, while later works [12] adapted it for ReID tasks. We adopt DeiT distillation to enhance data efficiency, which is critical when labeled data are limited.

3 The Sh-ViT Method

This section introduces the proposed Shuffled Vision Transformer (Sh-ViT) for person Re-ID in complex surveillance scenarios. Sh-ViT is designed to address three key challenges—viewpoint distortion, image degradation, and occlusion—which frequently occur in base station inspection environments. As illustrated in Fig. 2, the framework is built upon a ViT-Base backbone and integrates three major components: a Shuffle module, scenario-adapted data augmentation, and knowledge distillation.

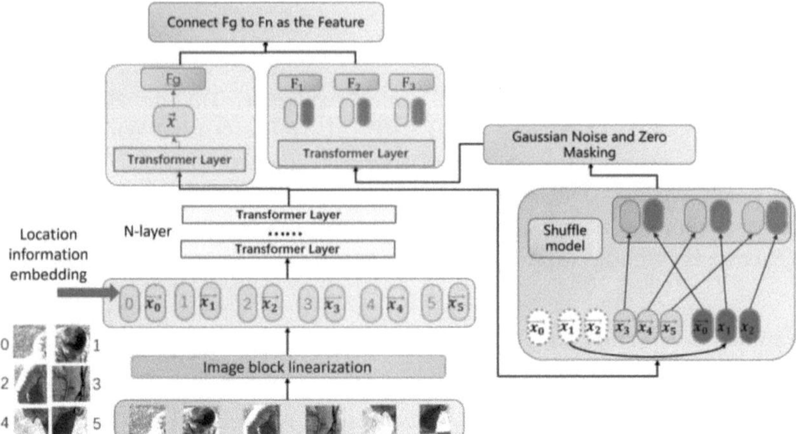

Fig. 2. Structure of Re-ID model. Structure of Shuffled Vision Transformer (Sh-ViT) for person re-identification, illustrating image patch linearization, location information embedding, multi-layer Transformer operations, shuffle module with Gaussian noise and zero masking, and feature concatenation of F_g and F_n.

3.1 Overall Framework

The overall pipeline of Sh-ViT follows the standard Transformer-based architecture for visual representation learning. Input images are divided into non-overlapping patches and mapped to high-dimensional embeddings, which are processed by a multi-layer Transformer encoder to capture long-range dependencies. To enhance robustness under practical conditions, Sh-ViT incorporates three innovations: First, a Shuffle module embedded in the last Transformer layer to improve resistance to occlusion and blur; Second, a scenario-adapted data augmentation strategy tailored to surveillance environments; Third, DeiT-based knowledge distillation to improve data efficiency. The final representation is normalized and fed into a fully connected classifier for identity prediction.

3.2 Shuffle Module

The Shuffle module distinguishes itself from existing patch-shuffling or occlusion-handling techniques in both conceptual design and practical implementation. Conceptually, unlike methods that rely on explicit occlusion reconstruction (e.g., FCFormer [4], which uses a complex decoder to recover missing occluded regions) or localized patch enhancement (e.g., PHA [19], which only augments high-frequency patches), the Shuffle module does not require prior occlusion detection or synthetic mask generation. Instead, it forces the model to learn spatially invariant global features by randomly permuting patch tokens, thus avoiding over-reliance on specific occlusion patterns (e.g., the fixed synthetic occlusions in [10]). Practically, in contrast to general patch-shuffling techniques applied in input preprocessing (e.g., the random erasing-based shuffling in [9]), our Shuffle module is embedded in the last layer of the Transformer encoder and integrates Gaussian noise injection and zero masking. This design ensures that the model retains high-level semantic information while enhancing robustness—existing input-level shuffling methods often disrupt spatial logic and degrade the discriminability of local features, leading to performance loss in identity recognition tasks.

The Shuffle module is the core contribution of Sh-ViT and aims to mitigate the adverse effects of partial occlusion and image blur.

- Token shuffling: Patch tokens are randomly grouped and permuted, breaking local spatial correlations and forcing the model to capture global dependencies across visible fragments.
- Reconstruction through attention: The shuffled tokens are reassembled via the self-attention mechanism, enabling the model to reconstruct robust identity features even when large portions of the body are missing.
- Robustness enhancement: Gaussian noise and zero masking are incorporated into the shuffled tokens, improving tolerance to image degradation and missing information.

Unlike input-level patch shuffling [9] or occlusion simulation [10], our shuffle module is embedded in the final Transformer layer. This design avoids semantic distortion and compels the model to learn spatially invariant global features.

This lightweight design introduces negligible computational overhead while substantially improving feature robustness under occlusion-heavy and blur-prone conditions.

3.3 Scenario-Adapted Data Augmentation

We design scenario-adapted augmentation (affine/perspective transforms, random erasing, Gaussian blur, and color adjustment) to simulate surveillance distortions.

- Affine and perspective transformations to simulate geometric distortion caused by elevated or tilted cameras;
- Random erasing to mimic occlusions from equipment or other personnel;
- Gaussian blur to handle motion blur and out-of-focus images;
- Color adjustment to address variations in illumination, saturation, and contrast (Figs. 3 and 4).

3.4 Self-Built MyTT Dataset

For base station inspection scenarios, we construct the MyTT dataset from real multi-camera surveillance videos, formatted to match Re-ID task standards (e.g., Market1501). Key details:

- Scale: 10,000+ pedestrians, 30,000+ images; split into training/test/validation sets at 6:3:1.
- Scenario Features: Top-down shooting angles, frequent equipment-person occlusion, unstable lighting, cross-camera color/clarity differences, and partial image blurriness—addressing gaps in public datasets (e.g., Market1501 [13]) that lack complex inspection-scene characteristics.

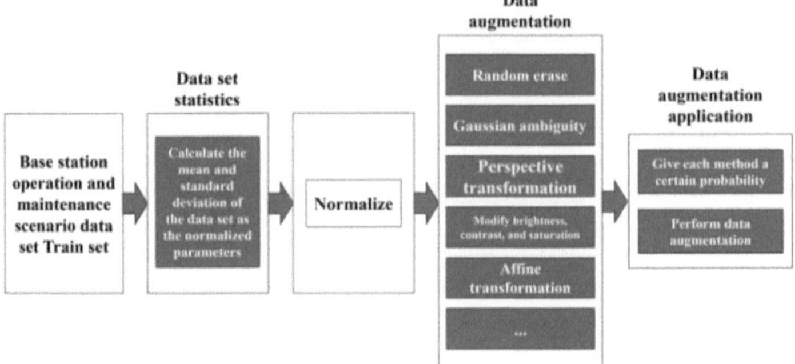

Fig. 3. Pipeline of Data Augmentation. Pipeline of data augmentation for base station operation and maintenance scenario dataset, showing steps of dataset statistics (calculating mean and standard deviation as normalization parameters), normalization, diverse augmentation methods (random erase, Gaussian ambiguity, perspective transformation, brightness/contrast/saturation modification, affine transformation, etc.), and probabilistic application of data augmentation.

Fig. 4. Examples of different data augmentation methods (Transform, Random Erase, Gaussian Blur, and Color transform) applied to personnel images in base station operation and maintenance scenarios.

4 Experiments

4.1 Experimental Setups

To verify the effectiveness of Sh-ViT, experiments are conducted on the self-built occluded person Re-ID dataset MyTT and the mainstream benchmark Market1501 [14].

MyTT: Constructed for security equipment room Re-ID, collected from multi-camera surveillance. It contains over 10,000 pedestrians and 30,000+ images, split into training (6,654 identities, 34,536 images), test (4,437 identities, 17,359 images), and validation (4,206 identities, 5,786 images) sets, formatted to match Market1501.

Market1501: A common holistic Re-ID benchmark captured by 6 cameras, with 751 training identities (12,936 images) and 750 test identities (19,732 query + 3,368 gallery images).

4.2 Comparative Experiments

We propose Sh-ViT by integrating a Shuffle module into the ViT-based Transformer, aiming to improve ViT's performance in occluded scenarios. Table 1 compares Sh-ViT with state-of-the-art (SOTA) methods on MyTT. Sh-ViT achieves 83.2% Rank-1 and 80.1% mAP, outperforming conventional CNNs (e.g., ResNet50: 82.1% Rank-1, 57.6% mAP; OSNet: 72.7% Rank-1, 54.2% mAP) and the ViT-based baseline TransReID (83.0% Rank-1, 79.6% mAP). This confirms Sh-ViT's superiority in handling occlusions and low-quality images in security inspection scenarios.

Table 1. Superiority of Sh-ViT on MyTT Dataset.

Method	Publication	Rank1 (%)	Rank5 (%)	Rank10 (%)	mAP (%)
OSNet [14]	ICCV2019	72.7	85.0	87.9	54.2
ABDNet [15]	ICCV2019	76.2	86.0	89.8	57.1
HOReID [3]	CVPR2020	69.6	85.7	88.6	53.9
TransReID [6]	ICCV2021	83.0	84.2	87.8	79.6
ResNet50 [16]	CVPR2016	82.1	86.5	90.3	57.6
ResNeSt50 [17]	CVPR2022	78.6	88.1	89.5	58.7
MSINet [18]	CVPR2023	72.5	89.7	92.7	59.7
PHA [19]	CVPR2023	76.9	89.6	91.5	58.8
UniHCP [20]	CVPR2023	76.5	92.5	94.6	57.6
DeepChange [21]	ICCV2023	70.8	82.2	86.0	46.6
Sh-ViT (Ours)	**This paper**	**83.2**	92.7	94.7	**80.1**

4.3 Cross-Domain Generalization Validation

To evaluate Sh-ViT's generalization, we test it on Market1501 (Table 2). Sh-ViT achieves 94.6% Rank-1, 98.2% Rank-5, 99.1% Rank-10, and 87.5% mAP, outperforming recent SOTA methods such as DCReID (94.2% Rank-1, 85.5% mAP) and ACFL (94.3% Rank-1, 85.3% mAP). This gain is attributed to the Shuffle module, which enhances the model's ability to mitigate occlusion interference and capture robust global features, enabling adaptation to different surveillance scenarios.

4.4 Performance on Occlusion-Specific Benchmark (DukeMTMC-ReID)

We further evaluate Sh-ViT on the occlusion-specific benchmark DukeMTMC-reID. As shown in Table 3, Sh-ViT achieves 89.6% Rank-1 and 80.3% mAP, surpassing recent methods such as ACFL [27] (85.2% Rank-1/76.8% mAP) and FCM [28] (84.7% Rank-1/75.9% mAP). These results confirm the strong generalization ability of Sh-ViT under real-world occlusion conditions.

Table 2. Performance of Sh-ViT on Market1501.

Method	Publication	Rank1 (%)	Rank5 (%)	Rank10 (%)	mAP (%)
D-MMD [22]	ECCV2020	70.6	87.0	91.5	48.8
MMCL [23]	CVPR2020	84.4	92.8	95.0	60.4
MLC [24]	PR2022	85.6	93.9	96.0	65.9
MDJL [25]	PR2023	80.3	87.4	89.9	59.8
AdaMG [26]	TCSVT2023	93.9	97.9	98.9	84.6
ACFL [27]	PR2024	94.3	98.0	98.8	85.3
FCM [28]	AAAI2024	93.2	96.7	97.6	83.5
DCReID [29]	CVDL2024	94.2	97.7	98.5	85.5
Sh-ViT (Ours)	This paper	94.6	98.2	99.1	87.5

Table 3. Performance comparison with state-of-the-art models on DukeMTMC-reID.

Method	Publication	Rank1 (%)	Rank5 (%)	Rank10 (%)	mAP (%)
D-MMD [22]	ECCV2020	63.5	78.8	83.9	46.0
MMCL [23]	CVPR2020	72.4	82.9	85.0	51.4
MLC [24]	PR2022	74.1	83.8	86.3	55.0
MDJL [25]	PR2023	78.6	86.6	88.7	62.8
ACFL [27]	PR2024	85.5	92.4	94.4	74.0
FCM [28]	AAAI2024	83.8	91.0	93.2	71.9
Sh-ViT(Ours)	**This papaer**	**89.6**	95.5	96.8	**80.3**

4.5 Algorithm Analysis

We analyze Sh-ViT's key components (data augmentation, backbone, optimizer) based on sufficient dataset images (about 30,000 images).

4.5.1 Effect of Data Augmentation

Table 4 compares ViT performance with and without the proposed scenario-adapted augmentation (affine transformation, random erasing, Gaussian blur, color adjustment). Augmentation improves Rank-1 by +8.2% with limited data, but has little effect with sufficient data.

4.5.2 Effect of Backbone and Shuffle Module

Table 5 compares ViT, DeiT, and Sh-ViT on MyTT. ViT (83.0% Rank-1, 79.6% mAP) outperforms DeiT (82.7% Rank-1, 78.7% mAP), so ViT is selected as the backbone.

Table 4. Results of Data Augmentation Experiment.

Method	Rank1(%)	Rank5(%)	Rank10(%)	mAP(%)
ViT (Insufficient data)	60.9	86.1	90.1	68.7
ViT + Data Augmentation (Insufficient data)	69.1	86.4	91.2	68.9
ViT (Sufficient data)	83.0	92.5	94.6	79.6
ViT + Data Augmentation (Sufficient data)	82.8	92.4	94.2	79.1

Adding the Shuffle module (Sh-ViT) further improves Rank-1 to 83.2% and stabilizes mAP, verifying the module's ability to enhance occlusion robustness.

Table 5. Results of Backbone and Shuffle module Experiment.

Method	Rank1(%)	Rank5(%)	Rank10(%)	mAP(%)
DeiT	82.7	92.3	94.3	78.7
ViT	83.0	92.5	94.6	79.6
Sh-ViT (Ours)	83.2	92.7	94.7	80.1

4.5.3 Effect of Optimizer

Table 6 compares SGD, Adan, and Adam on ViT. SGD achieves the best performance, while Adan and Adam underperform. SGD achieves the best performance and is adopted.

Table 6. Results of Optimizer Experiment.

Method	Rank1(%)	Rank5(%)	Rank10(%)	mAP(%)
ViT + SGD	82.8	92.5	94.4	79.4
ViT + Adan	77.3	88.4	91.0	68.9
ViT + Adam	72.8	83.7	86.7	65.6

4.6 Summary of Experiments

Our experiments across both public and self-built datasets confirm the robustness and effectiveness of Sh-ViT for occluded person re-identification under complex surveillance conditions.

Performance on Public Benchmarks. On Market1501, Sh-ViT achieves 94.6% Rank-1 and 87.5% mAP, outperforming recent state-of-the-art approaches such as DCReID and

ACFL. The consistent gains across Rank-5 and Rank-10 show that the Shuffle module enhances global feature robustness beyond the training domain. These improvements are particularly notable considering that Market1501 is less occlusion-heavy than our target scenarios, suggesting Sh-ViT generalizes well even in less constrained environments.

Evaluation on the MyTT Dataset. The MyTT dataset, designed to replicate real-world base station inspections, poses unique challenges: severe occlusion from equipment, top-down camera angles, unstable lighting, and cross-camera style shifts. On MyTT, Sh-ViT reaches 83.2% Rank-1 and 80.1% mAP, surpassing CNN-based baselines such as ResNet5 and OSNet. The results indicate that ViT's global attention combined with patch-level shuffling significantly improves resistance to partial occlusion and low image quality.

Effectiveness of Model Components. Ablation studies show that patch shuffling contributes directly to performance gains by breaking local correlations and forcing the Transformer to learn spatially invariant features. Although the performance increase from ViT to Sh-ViT on MyTT appears modest, the Shuffle module notably stabilizes mAP and improves robustness under severe occlusion (as observed in qualitative retrieval results). Scenario-specific augmentation—particularly random erasing and Gaussian blur—proves beneficial when labeled data are scarce, improving Rank-1 by 8.2% under limited-data conditions. Furthermore, the optimizer comparison demonstrates that SGD is more stable and effective than Adan or Adam in this setting, aligning with prior findings for ReID tasks.

Cross-Domain Generalization. The competitive performance on both Market1501 and MyTT validates Sh-ViT's ability to adapt across domains without relying on external parsing networks or pose estimators. Typical failure cases occur under heavy equipment occlusion (over 60% body covered) or extreme illumination, where performance drops significantly. These cases suggest directions for future refinement.

Limitations and Future Work. Despite its strengths, Sh-ViT still shows only incremental gains over TransReID on some metrics, indicating room for further refinement. Main failures are heavy occlusion and extreme lighting; future work may adopt adaptive token weighting and stronger illumination augmentation. The MyTT dataset, while diverse, focuses on base station environments; broader evaluation on additional occluded benchmarks could provide a more comprehensive assessment. Future work may explore adaptive shuffling strategies or integrating temporal cues from multi-frame sequences to enhance tracking stability in video-based ReID.

5 Conclusion

In this paper, we proposed Sh-ViT, a Vision Transformer framework tailored for occluded person re-identification in surveillance. Sh-ViT combines a shuffle module, scenario-adapted augmentation, and DeiT-based distillation, delivering strong robustness with lightweight design.

Experiments confirm its effectiveness: Sh-ViT reaches 83.2% Rank-1/80.1% mAP on MyTT, 94.6%/87.5% on Market1501, and 89.6%/80.3% on Occluded-Duke, showing consistent superiority and strong cross-dataset generalization. Nevertheless, gains over the TransReID baseline are incremental on some metrics, indicating room for further refinement.

Future work will explore adaptive shuffle strategies that dynamically adjust permutation intensity according to occlusion patterns, and temporal modeling for video-based ReID. Preliminary tests on MARS already suggest temporal extension can further improve performance, making Sh-ViT promising for real-time surveillance applications.

6 Disclosure of Interests

The authors have no competing interests to declare that are relevant to the content of this article.

Acknowledgments. This work was supported by National Natural Science Foundation of China (6217104), China University of Petroleum (Beijing) Karamay Campus introduction of talents and launch of scientific research projects(XQZX20240010), and China Tower Corporation Limited IT System 2023 Package Software Project - AI Algorithm and Services(23M01ZBZB011000017).

References

1. Zheng, L, Yang, Y., Hauptmann, A.G.: Person Re-identification: Past, Present and Future (2016). arXiv preprint arXiv:1610.02984
2. Sun, Y., Zheng, L., Yang, Y., Tian, Q., Wang, S.: Beyond part models: person retrieval with refined part pooling. In: Ferrari, V., Hebert, M., Sminchisescu, C., Weiss, Y. (eds.) ECCV 2018, LNCS, vol. 11205, pp. 480–496. Springer, Cham (2018)
3. H. Huang, D. Chen, X. Li, S. Wang, Z. Lei: HOReID: deep high-order mapping for occluded person re-identification. In: Proc. IEEE/CVF Conference on Computer Vision and Pattern Recognition (CVPR), pp. 3218–3227. IEEE Press, New York (2020)
4. Wang, T., Liu, H., Song, P., Sun, T., Jin, Y.: Feature completion transformer for occluded person re-identification. IEEE Trans. Multimed. **26**, 8529–8542 (2024)
5. Zhang, X., et al.: Dynamic Patch-Aware Enrichment Transformer for Occluded Person Re-identification (2024). arXiv preprint arXiv:2402.10435
6. He, S., Luo, H., Wang, P., Wang, F., Li, H., Jiang, W.: TransReID: transformer-based object re-identification. In: 2021 IEEE/CVF International Conference on Computer Vision (ICCV), pp. 14993–15002 (2021)
7. Luo, H., et al.: Self-Supervised Pre-training for Transformer-Based Person Re-identification (2021). arXiv preprint arXiv:2111.12084
8. Hu, B., et al.: PersonViT: large-scale self-supervised vision transformer for person re-identification. Mach. Vis. Appl. **36**(1), 32 (2024)
9. Zhong, Z., Zheng, L., Kang, G., Li, S., Yang, Y.: Random erasing data augmentation. In: Proceedings of the AAAI Conference on Artificial Intelligence, vol. 34, no. 07, pp. 13001–13008 (2020)
10. Wang, Y., et al.: Occlusion-aware R-CNN: detecting pedestrians in the wild with synthetic occlusion. in: 2018 IEEE/CVF Conference on Computer Vision and Pattern Recognition Workshops (CVPRW), pp. 639–647 (2018)

11. Touvron, H., et al.: Training data-efficient image transformers & distillation through attention. In: Proceedings of the 37th International Conference on Machine Learning. PMLR, vol. 119, pp. 10347–10357 (2020)
12. Yang, J., et al.: Knowledge Distillation Via Adaptive Instance Normalization (2020). arXiv preprint arXiv:2003.04289
13. Zheng, L., Shen, L., Tian, L., Wang, S., Wang, J., Tian, Q.: Scalable person re-identification: a benchmark. In: 2015 IEEE International Conference on Computer Vision (ICCV), pp. 1116–1124 (2015)
14. Zhou, K., Yang, Y., Cavallaro, A., Xiang, T.: Omni-scale feature learning for person re-identification. In: 2019 IEEE/CVF International Conference on Computer Vision (ICCV), pp. 3701–3711 (2019)
15. Chen, T., et al.: ABD-Net: attentive but diverse person re-identification. In: 2019 IEEE/CVF International Conference on Computer Vision (ICCV), pp. 8350–8360 (2019)
16. He, K., Zhang, X., Ren, S., Sun, J.: Deep residual learning for image recognition. In: 2016 IEEE Conference on Computer Vision and Pattern Recognition (CVPR), pp. 770–778 (2016)
17. Zhang, H., et al.: ResNeSt: split-attention networks. In: 2022 IEEE/CVF Conference on Computer Vision and Pattern Recognition Workshops (CVPRW), pp. 2735–2745 (2022)
18. Gu, J., et al.: MSINet: twins contrastive search of multi-scale interaction for object ReID. In: 2023 IEEE/CVF Conference on Computer Vision and Pattern Recognition (CVPR), pp. 19243–19253 (2023)
19. Zhang, G., Zhang, Y., Zhang, T., Li, B., Pu, S.: PHA: patch-wise high-frequency augmentation for transformer-based person re-identification. In: 2023 IEEE/CVF Conference on Computer Vision and Pattern Recognition (CVPR), pp. 14133–14142 (2023)
20. Ci, Y., et al.: UniHCP: a unified model for human-centric perceptions. In: 2023 IEEE/CVF Conference on Computer Vision and Pattern Recognition (CVPR), pp. 17840–17852 (2023)
21. Xu, P., Zhu, X.: DeepChange: a long-term person re-identification benchmark with clothes change. In: 2023 IEEE/CVF International Conference on Computer Vision (ICCV), pp. 11162–11171 (2023)
22. Mekhazni, D., Bhuiyan, A., Ekladious, G., Granger, E.: Unsupervised domain adaptation in the dissimilarity space for person re-identification. In: Vedaldi, A., Bischof, H., Brox, T., Frahm, J.-M. (eds.) ECCV 2020. LNCS, vol. 12370, pp. 159–174. Springer, Cham (2020)
23. Wang, D., Zhang, S.: Unsupervised person re-identification via multi-label classification. In: 2020 IEEE/CVF Conference on Computer Vision and Pattern Recognition (CVPR), pp. 10978–10987 (2020)
24. Li, Q., et al.: Unsupervised Person Re-Identification with Multi-Label Learning Guided Self-Paced Clustering (2021). arXiv preprint arXiv:2103.04580
25. Chen, F., Wang, N., Tang, J., Yan, P., Yu, J.: Unsupervised person re-identification via multi-domain joint learning. Pattern Recogn. **138**, 109369 (2023)
26. Peng, J., Jiang, G., Wang, H.: Adaptive memorization with group labels for unsupervised person re-identification. IEEE Trans. Circuits Syst. Video Technol. **33**(10), 5802–5813 (2023)
27. Ji, H., Wang, L., Zhou, S., Tang, W., Zheng, N., Hua, G.: Transfer easy to hard: adversarial contrastive feature learning for unsupervised person re-identification. Pattern Recogn. **145**, 109973 (2024)
28. Li, H., Hu, Q., Hu, Z.: Catalyst for clustering-based unsupervised object re-identification: feature calibration. In: Proceedings of the AAAI Conference on Artificial Intelligence, vol. 38, no. 4, pp. 3091–3099 (2024)
29. Liu, D., Fu, Y., Shi, W., Zhu, Z., Wang, D.: The double contrast for unsupervised person re-identification. In: Proceedings of the International Conference on Computer Vision and Deep Learning, pp. 1–8 (2024)

Image Models and Image Reconstruction

TAT-SatSR: Triple Attention Transformer for Super Resolution of Satellite Image

Shun Matsumoto(✉) and Tohru Kamiya

Department of Mechanical and Control Engineering, Kyushu Institute of Technology, 1-1, Sensui, Tobata, Kitakyushu, Fukuoka, Japan
matumoto.shun939@mail.kyutech.jp, kamiya@cntl.kyutech.ac.jp

Abstract. We propose a super-resolution method for satellite images that extends the SwinIR Transformer architecture. Our network introduces residual triple attention groups (RTAGs) and triple attention blocks (TABs), which combine three attention mechanisms to enhance the restoration of local textures and global structures. We perform knowledge distillation between models of similar scale to further improve performance. After being trained on 7,500 Google Earth images, our model outperformed RCAN, HAT, and SwinIR in terms of PSNR and SSIM.

Keywords: Satellite Image · Super Resolution · Channel Attention · Transformer · Knowledge Distillation · Attention Mechanism · State Space Model

1 Introduction

Image super resolution (SR) image processing aims to reconstruct high-resolution (HR) images from low resolution (LR) inputs, and recent deep learning methods such as SRCNN [1] and SwinIR [2] have greatly improved SR performance in natural images. However, applying SR to satellite imagery remains challenging due to differences in image structure, coarser ground sampling distances (GSD), atmospheric effects, and sensor variability. These factors limit the effectiveness of SR models trained on natural images, often resulting in artifacts and poor restoration of fine details in satellite data.

Satellite-based remote sensing provides essential data for applications such as agriculture, disaster management, and urban planning. This drives the need for higher-resolution imagery. Since acquiring natural high resolution (HR) satellite images are costly and physically limited, spatial resolution (SR) techniques offer a practical alternative for enhancing spatial detail.

Recent studies have developed specialized satellite remote sensing (SR) models to address the unique challenges of satellite imagery. These models often rely on external features, such as periodic patterns or multi-temporal data. In this study, we focus on making architectural improvements to the SR model itself. Our goal is to enhance performance with satellite data without requiring external information. Based on the SwinIR framework, we propose structural innovations tailored to satellite image characteristics to bridge the performance gap with conventional methods and achieve more accurate high resolution (HR) satellite image reconstruction.

P. Umapada et al. (Eds.): ICCPR 2025, CCIS 2811, pp. 101–110, 2026.
https://doi.org/10.1007/978-981-95-8315-7_8

2 Related Works

2.1 Super Resolution in Natural Images

Image super resolution techniques based on deep learning have evolved significantly over time. Pioneering methods such as SRCNN and EDSR laid the groundwork for these techniques. These early approaches used convolutional neural networks (CNNs) to transform low-resolution images into high-resolution images, improving performance significantly over traditional methods. More advanced architectures have since emerged, particularly those incorporating residual learning and attention mechanisms, which have further improved performance.

For example, the Residual Channel Attention Network (RCAN) [3] introduced a channel attention mechanism to dynamically adjust the importance of feature maps, significantly improving the accuracy of image restoration. Such methods improve the model's ability to focus on critical features, allowing for more accurate reconstruction. In addition, transformer-based methods have recently attracted attention in image restoration tasks. For example, the Hybrid Attention Transformer (HAT) [4] uses feature extraction and self-attention mechanisms like those used in RCAN, effectively combining local and global information to achieve further performance gains. HAT has demonstrated superior results in several benchmarks, establishing itself as a leading approach among transformer-based super resolution techniques.

2.2 Super Resolution for Remote Sensing

Despite their high performance, methods developed for natural images may not yield the expected results when applied directly to satellite imagery. Satellite images often contain multispectral or hyperspectral information and are affected by complex, remote sensing-specific factors, such as atmospheric scattering, sensor differences, and variations in orbital position during image acquisition. These characteristics necessitate specialized approaches to achieve super resolution in satellite imagery.

To address these challenges, super resolution methods designed specifically for remote sensing have been proposed, such as DSen2 [5] and PanNet [6]. These methods take advantage of the distinctive characteristics of satellite imagery, such as spectral correlations and temporal consistency, to reconstruct images with high precision.

However, high performance architectures developed for natural images, such as RCAN and HAT, have only been applied to satellite images in a few cases. Therefore, there is a recognized need to develop new hybrid architectures that adapt state-of-the-art methods to the specific characteristics of satellite imagery. The objective of this study is to enhance super resolution techniques for satellite imagery by improving the SwinIR-based feature extraction module.

3 Proposed Method

3.1 Overall Structure

This study proposes a novel architecture for super resolution processing based on the SwinIR framework and transformer structure. As illustrated in Fig. 1, the proposed network comprises three primary modules: (1) a shallow feature extraction module, (2) a deep feature extraction module, and (3) a high-resolution image reconstruction module. Given a low-resolution input image $I_{LR} \in mathbbR^{H \times W \times C_{in}}$, a 3×3 convolutional layer first extracts a shallow feature map $F_0 \in mathbbR^{H \times W \times C}$, where C_{in} is the number of input image channels and C represents the number of intermediate feature map channels. This layer primarily captures low-frequency components of the image.

Next, a deep feature extraction module, consisting of multiple Residual Triple Attention Groups (RTAGs) and a 3×3 convolutional layer $H_{conv}(\cdot)$, extracts high-frequency components and texture information, yielding a deep feature map $F_D \in mathbbR^{H \times W \times C}$. The shallow feature map F_0 and the deep feature map F_D are aligned in terms of channel dimensions and fused via a global residual connection to obtain the final feature representation:

$$F_{fusesd} = F_0 + F_D \tag{1}$$

Finally, the reconstruction module takes the fused feature map F_{fusesd} as input and employs pixel shuffle techniques to enhance the spatial resolution, generating a high-resolution output image $I_{SR} \in mathbbR^{rH \times rW \times C_{out}}$, where r is the upscaling factor and C_{out} is the number of output image channels.

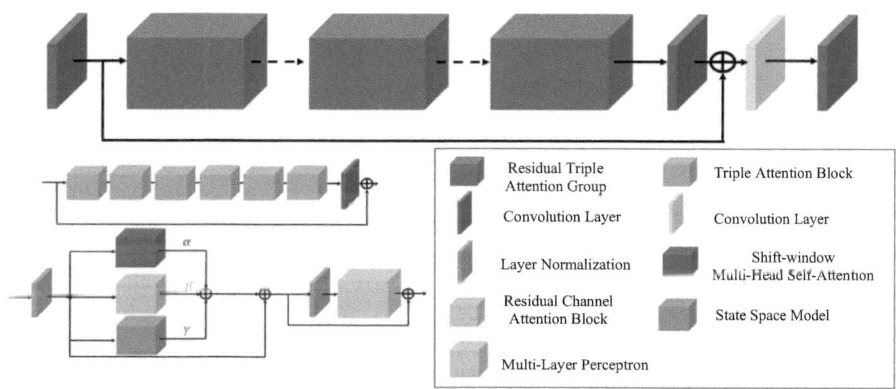

Fig. 1. Overall Architecture of the Network

3.2 Residual Triple Attention Group (RTAG)

As shown in Fig. 1, the RTAG consists of multiple Triple Attention Blocks (TABs), a 3×3 convolutional layer, and a shortcut connection. Compared to the Residual Swin Transformer Block (RSTB) of SwinIR, the RTAG integrates three different attention mechanisms to capture multi-scale features and model long-range spatial dependencies. The RTAG processing is formulated as:

$$F_{out}^{RTAG} = F_{in} + H_{Conv}(TAB_n(\ldots(TAB_1(F_{in})))) \tag{2}$$

where F_{in}, F_{out}^{RTAG} denote the input and output feature maps of the RTAG, respectively. Each RTAG applies TABs, followed by a convolutional layer and a shortcut connection with the input. This structure improves gradient flow and enables stable training of deep networks.

3.3 Triple Attention Block (TAB)

TAB is a novel structure that integrates three attention mechanisms: (1) Shift-window Multi-Head Self-Attention (Swin-MHSA), (2) Residual Channel Attention Block (RCAB), and (3) State Space Model (SSM). These mechanisms complement each other to extract local and global features. The processing flow of the TAB is defined as:

$$X_{fused} = X_{in} + \left(\begin{array}{c} \alpha \times Swin\text{-}MHSA \\ +\beta \times RCAB + \gamma \times SSM \end{array} \right) \tag{3}$$

$$X_{out} = X_{fused} + MLP\big(X_{fused}\big) \tag{4}$$

where $X_{in}, X_{fused}, X_{out}$ represent the input, fused, and output feature maps of the TAB, respectively. The parameters α, β, γ are learnable weights that dynamically adjust the contribution of each attention module. The MLP consists of two fully connected layers that nonlinearly transform the fused features.

3.3.1 Shift-Window Multi-Head Self-Attention (Swin-MHSA)

Swin-MHSA [2] alternates between window-based multi-head self-attention (W-MSA) and shifted window-based multi-head self-attention (SW-MSA). It reduces the computational complexity of standard MHSA from $O(N^2)$ to $O(M^2N)$, where M is the window size. W-MSA partitions the input feature map $(H \times W \times C)$ into non-overlapping $M \times M$ windows, generating queries (Q), keys (K), and values (V) within each window. Self-attention is computed with a relative position bias B:

$$Attention(Q, K, V) = Softmax\left(\frac{QK^T}{\sqrt{d_k}} + B\right)V \tag{5}$$

where d_k is the dimension of the keys. SW-MSA shifts the windows by $M/2$ pixels to indirectly capture dependencies between adjacent windows.

3.3.2 Residual Channel Attention Block (RCAB)

The RCAB [3] integrates two 3×3 convolutional layers with channel attention. For an input feature map $X \in mathbbR^{H \times W \times C}$, the first convolutional layer applies ReLU to extract local features. The second convolutional layer refines the features. This is followed by global average pooling and two fully connected layers that generate channel-wise scaling coefficients. These coefficients are then normalized by a sigmoid function.

The output is connected to the input via a residual connection, which improves the restoration accuracy of high-frequency components.

3.3.3 State Space Model (SSM)

State Space Model (SSM) [7, 8] is a mathematical framework that represents continuous-time dynamical systems by describing the evolution of internal states and their relationship to the output via state transition and output equations. Recently, SSMs have been generalized in the context of deep learning through the Mamba model, which enables their use as an efficient method for modeling sequences.

In this study, we extend the Selective Scan Module (SSM) to two-dimensional image data by introducing the 2D Selective Scan (SS2D) module. First, SS2D divides the input feature map into small patches. Then, it converts the patches into one-dimensional sequences by scanning in eight directions, including horizontal, vertical, and diagonal orientations. The selective scan mechanism dynamically adjusts weights based on the input, focusing on salient regions within the image. This approach effectively suppresses irrelevant information while maintaining a broad receptive field.

The SSM is constructed based on the following continuous-time formulation:

$$\frac{d}{dt}h(t) = Ah(t) + Bx(t) \tag{6}$$

$$y(t) = Ch(t) + Dx(t) \tag{7}$$

where $h(t)$ is the state vector, $x(t)$ is the input, $y(t)$ is the output, and A, B, C, D are learnable parameters. Upon discretization, this formulation allows the model to efficiently capture long-range dependencies within the input sequence. It achieves global context propagation with linear computational complexity $O(N)$, which is a key advantage.

After performing the SSM processing, the one-dimensional sequence reverts to its original two-dimensional spatial structure and is sent to the next layer. This sequential process enables the modeling of both local details and global context simultaneously, which contributes to the accurate reconstruction of fine-grained image components, such as edges and textures.

3.4 Optimization Objective

The proposed network is trained with a composite objective that combines an L1 reconstruction loss and a knowledge distillation (KD) loss [9]. The L1 term minimizes the pixel-wise difference between the student output and the ground-truth high-resolution (HR) image, ensuring visual fidelity. Unlike conventional KD that distills from a much larger teacher to a smaller student, we distill across architectures of comparable capacity: a CNN teacher to a Transformer student.

The loss function consists of three terms:

$$\begin{aligned} Loss = {} & \frac{\lambda_1}{m \times n} \sum_{i=0}^{m-1} \sum_{j=0}^{n-1} \left| I_{HR}(i,j) - I_{SR}^{S}(i,j) \right| \\ & + \frac{\lambda_2}{m \times n} \sum_{i=0}^{m-1} \sum_{j=0}^{n-1} \left| I_{SR}^{T}(i,j) - I_{SR}^{S}(i,j) \right| \\ & + \sum_{l=1}^{L} \left(\frac{\lambda_3}{a^{(l)} \times b^{(l)}} \sum_{i=0}^{a^{(l)}-1} \sum_{j=0}^{b^{(l)}-1} \left| F_T^{(l)}(i,j) - F_S^{(l)}(i,j) \right| \right) \end{aligned} \tag{8}$$

The first term is the L1 loss between the high-resolution target image I_{HR} and the student model's super resolution image I_{SR}^{S}, ensuring pixel-level accuracy. The second term is the L1 loss between the teacher model's super resolution image I_{SR}^{T} and the student model's I_{SR}^{S}, encouraging the student to mimic the teacher's output patterns. The third term is the L1 loss between the feature maps $F_T^{(l)}$ and $F_S^{(l)}$ of selected intermediate layers from the teacher and student models, weighted by layer-specific coefficients λ_3. Where, $m \times n$ denotes the image size, L is the number of layers used for feature map loss, and $a^{(l)}$, $b^{(l)}$ denote the height and width of the l-th feature map, respectively.

To avoid architectural mismatch between CNN and Transformer, we use the feature maps immediately before upsampling for the feature-level loss. This cross-architecture distillation helps the student inherit the teacher's ability to restore high-frequency details (edges and textures) while improving generalization and mitigating overfitting. Unless otherwise stated, we set the weights to $\lambda_1 = 0.8$, $\lambda_2 = 0.2$, and $\lambda_3 = 0.01$.

4 Experimental Results

4.1 Dataset Details

A total of 7,500 satellite images were collected using Google Earth Pro in the year 2021. The images were acquired across a wide range of locations throughout Japan, encompassing a diverse array of regions such as urban areas (approximately 50%), rural/farmland areas (approximately 30%), and coastal regions (approximately 20%). This ensures a comprehensive representation of various land cover types, structures, and textures encountered in real-world scenarios.

All images were captured using standard RGB optical sensors, with no additional spectral bands such as near-infrared (NIR). The raw imagery was exported at an approximate ground resolution of 0.5–1.0 m per pixel, corresponding to a viewing altitude of roughly 1 km, as determined by Google Earth Pro settings. Each image was then cropped into a patch of 224×224 pixels, which serves as the high-resolution (HR) ground truth.

The corresponding low-resolution (LR) input images were generated by applying bicubic downsampling to 56×56 pixels, simulating realistic degradation. To improve robustness and increase the diversity of the training data, we applied random rotations (0°, 90°, 180°, and 270°) and horizontal flipping as part of data augmentation.

For a fair and stable evaluation, fivefold cross-validation was adopted. The full dataset was randomly divided into five subsets, with four used for training (80%) and one for validation (20%) in each fold. This process was repeated across five iterations, and the performance was averaged.

4.2 Results and Discussion

As shown in Table 1, our proposed model outperforms existing methods, such as RCAN, HAT, and SwinIR, for satellite image super resolution. Building on SwinIR, our model incorporates structural enhancements to the feature extraction module and loss function.

The proposed method achieved a PSNR of 24.6620 and an SSIM of 0.7511, surpassing all baselines. Compared to RCAN, PSNR improved by 0.146 dB and SSIM by 0.0051; over SwinIR, improvements were 0.123 dB and 0.0041; and compared to HAT, gains were 0.067 dB and 0.0021, respectively. These improvements result from thoughtful module design and effective information fusion, rather than increased complexity.

Notably, although RCAN was used for knowledge distillation, the student model still outperformed it, indicating a synergistic effect with the structural modifications. Overall, the proposed approach demonstrates consistently superior quantitative performance and achieves accurate reconstruction and strong structural preservation, even in densely structured satellite imagery.

Table 1. Results of Experiment

Model	PSNR	SSIM
RCAN	24.5159	0.7460
HAT	24.5950	0.7490
SwinIR	24.5392	0.7470
Ours	**24.6620**	**0.7511**

Table 2. Contribution of Feature Extraction

RTAG	TAB	α	β	γ
1	1	0.0398	0.2547	0.0575
	2	0.0660	0.0426	0.1045
	3	0.0914	0.0275	0.0771
	4	0.0911	0.0268	0.0801
	5	0.1306	0.0299	0.1165
	6	0.1244	-0.0495	0.1125
2	1	0.5477	0.1016	0.4625
	2	0.5524	-0.2426	0.3766
	3	0.6296	-0.2791	0.3449
	4	0.5222	-0.2580	0.3377
	5	0.5304	-0.3139	0.3793
	6	0.5741	-0.3872	0.4888
3	1	0.7606	0.1763	0.6903
	2	0.7758	0.2304	0.7584
	3	0.9685	-0.3162	0.7250
	4	0.9923	-0.3326	0.8394
	5	1.000	-0.3902	0.9361
	6	1.030	-0.4767	1.026

The values shown in Table 2 represent the progression of weights associated with MHSA (α), Channel Attention (β), and SSM (γ) across layers TAB1 to TAB6 in each of the three RTAG blocks (RTAG1 to RTAG3). The variation in contribution of each architectural component with layer depth is revealed by these values.

In RTAG1, α gradually increases from 0.0398 to 0.1306, indicating that attentional mechanisms contribute more significantly to the mid and deeper layers than in shallow ones. β starts at 0.2547 and decreases to -0.0495 in the final layer, suggesting that channel attention is active in the early layers, but its influence diminishes in deeper layers. γ increases from 0.0575 to 0.1165, indicating that the contribution of SSM becomes more important in the later stages of the block.

In RTAG2, α remains consistently high, ranging from 0.5477 to 0.5741, which highlights the importance of attentional mechanisms across all layers. β transitions from a positive value of 0.1016 to a negative value of -0.3872, suggesting that overemphasis on channel features in deeper layers may lead to overfitting and is thus adaptively suppressed by the model. γ temporarily dips in the middle layers, but rises to 0.4888 in TAB6, indicating renewed emphasis on spatial processing toward the end of the block.

In RTAG3,α starts at 0.7606 and peaks at 1.030 in the final layer, emphasizing the dominant role of MHSA in the deepest layers. Although β begins with a positive value, it becomes negative from TAB3 onward and drops to -0.4767, again reflecting the model's decision to limit channel attention in deeper stages. γ steadily increases from 0.6903 to 1.026, showing that the influence of SSM is maximized in the deepest layers.

Overall, both α and γ increase with block depth, indicating the growing importance of MHSA and SSM in deeper layers. In contrast, β decreases and becomes negative in many layers, meaning that local emphasis via channel attention tends to become redundant and is actively suppressed in deeper layers to improve generalization. This suppressive behavior helps mitigate overfitting and increases the robustness of the model.

The model architecture prioritizes channel attention in the early layers to amplify local cues and progressively down-weights it in the middle and deeper layers, where MHSA and SSM take over to extract long-range context. This dynamic, layer-wise rebalancing yields a hierarchical representation that supports accurate and robust super-resolution.

Figure 2 illustrates both the strengths and the limitations of the proposed method using two representative scenes. In the top row (urban intersection), our method produces visually sharper crosswalk stripes and more continuous lane markings with fewer blocky artifacts than SwinIR, while the quantitative scores are comparable. In the bottom row (farmland with repetitive textures), our method preserves the overall structure but shows slight edge blurring and lower PSNR/SSIM (26.65/0.7957) than SwinIR (27.49/0.8408).

We attribute this gap in the farmland case to the texture characteristics: furrows are highly repetitive and exhibit low local contrast, so wide-context aggregation (MHSA + SSM) can average out high-frequency details and cause mild over-smoothing. By contrast, crosswalks benefit from a strong alternation between white stripes and gray asphalt, providing clear local boundaries that are easier to preserve.

In summary, the proposed model excels at reconstructing complex, man-made structures and sharp geometric boundaries, whereas highly homogeneous, high-frequency textures may be slightly over-smoothed. Refining the attention scheduling or incorporating frequency-aware loss terms is a promising direction to improve texture fidelity without sacrificing edge preservation.

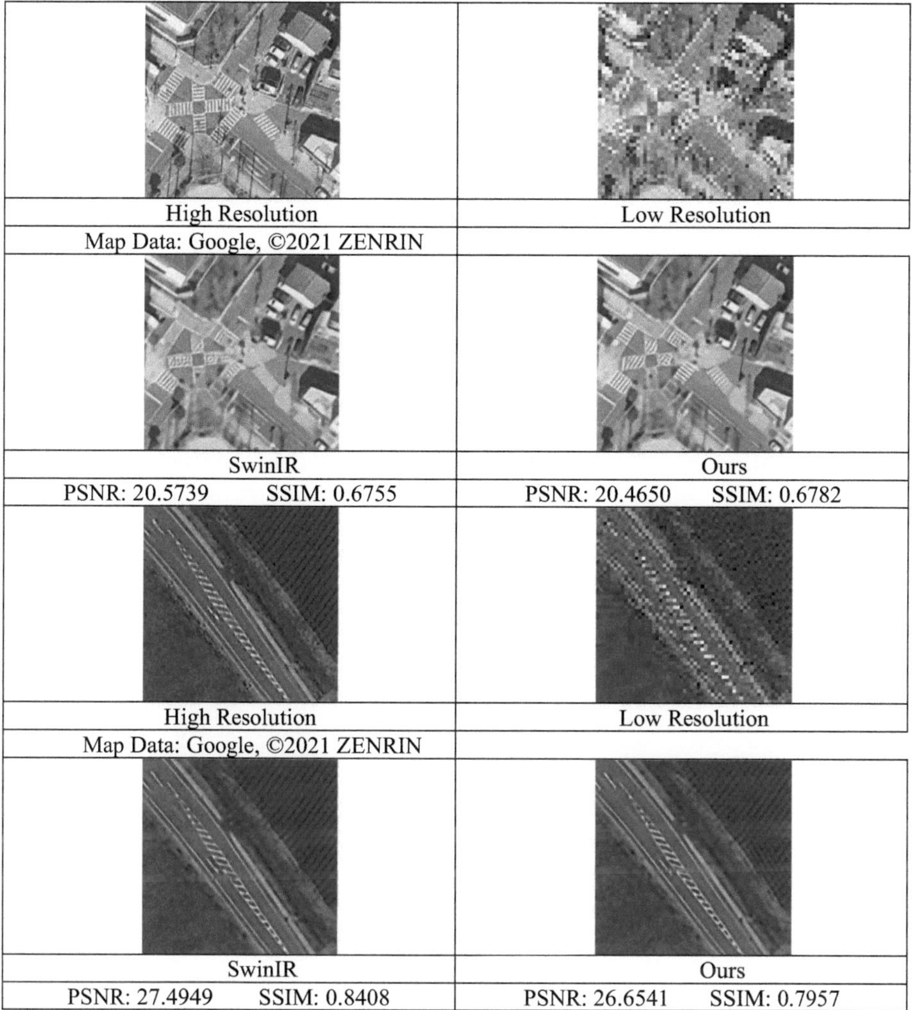

Fig. 2. Visual comparison on two representative scenes. Top: urban intersection. The proposed method maintains sharper crosswalk stripes and continuous lane markings with fewer blocky artifacts. Bottom: farmland with repetitive, low-contrast textures. In this failure case, the proposed method shows slight edge blurring and lower PSNR/SSIM than SwinIR.

5 Conclusion

This study proposes a super-resolution method for satellite imagery that integrates three attention mechanisms for high-fidelity reconstruction and is based on an enhanced SwinIR Transformer. We use knowledge distillation between models to further improve performance. Experiments demonstrate that our model outperforms RCAN and SwinIR, particularly regarding complex urban structures. However, some limitations remain regarding fine textures in rural areas. Future work will address these challenges and extend the approach to other satellite data.

Acknowledgments. In this paper, satellite images are used according to Google Earth guidelines [10].

References

1. Dong, C., Change Loy, C., He, K., Tang, X.: Image super-resolution using deep convolutional networks. IEEE Trans. Pattern Anal. Machine Intel. **38**, 295–307 (2015)
2. Liang, J., et al.: SwinIR: Image Restoration Using Swin Transformer. IEEE/CVF International Conference on Computer Vision, pp. 1833–1844 (2021)
3. Zhang, Y., Li, K., Li, K., Wang, L., Zhong, B., Yun, F.: Image super-resolution using very deep residual channel attention networks. European Conference on Computer Vision **7**(18), 294–310 (2018)
4. Chen, X., Wang, X., Zhou, J., Qiao, Y., Dong, C.: Activating More Pixels in Image Super-Resolution Transformer. IEEE/CVF Conference on Computer Vision and Pattern Recognition, pp. 22367–22377 (2023)
5. Lanaras, C., Bioucas-Dias, J., Galliani, S., Baltsavias, E., Schindler, K.: Super-resolution of Sentinel-2 images: learning a globally applicable deep neural network. ISPRS J. Photogramm. Remote. Sens. **146**, 305–319 (2018)
6. Yang, J., et al.: Pannet: A deep network architecture for pan-sharpening. IEEE International Conference on Computer Vision, pp. 1753–1761 (2017)
7. Zhu, L., et al.: Vision Mamba: Efficient Visual Representation Learning with Bidirectional State Space Model. arXiv preprint arXiv:2401.09417 (2024). Access 05 May 2025
8. Ma, H., Lei, S., Celik, T., Li, H.-C.: FER-YOLO-Mamba: Facial Expression Detection and Classification Based on Selective State Space, arXiv preprint arXiv:2405.01828 (2024). Access 05 May 2025
9. Matsumoto, S., Kamiya, T.: Super resolution of remote sensing image using improved SwinIR. Soft Computing and Intelligent Systems and International Symposium on Advanced Intelligent Systems, pp. 9–12 (2024)
10. Google: Brand Resource Center. https://about.google/brand-resource-center/products-and-services/geo-guidelines/. Access 04 May 2025

SSE-UVSR: Self-Supervised End-To-End Underwater Video Super-Resolution

Jingyi Wang[1], Huimin Lu[2], and Tohru Kamiya[1](✉)

[1] Kyushu Institute of Technology, Kitakyushu, Fukuoka, Japan
kamiya@cntl.kyutech.ac.jp
[2] Southeast University, Nanjing, Jiangsu, China

Abstract. Underwater videos are vital for marine science, yet often suffer from color distortion, blurring, and low resolution due to environmental complexity and hardware constraints. The scarcity of paired high-quality data further limits the applicability of supervised enhancement approaches. In this work, we present SSE-UVSR, a novel self-supervised end-to-end underwater video super-resolution framework which, unlike conventional blind VSR methods, explicitly integrates underwater physical degradation modeling into the reconstruction process. Our method first estimates clean video content, transmission maps, and atmospheric light via a Degradation-Aware Module (DAM). An Edge Enhancement and Detail Preservation Module (DRM) then restores fine textures and sharp structures by leveraging underwater-specific priors. The entire model is trained without high-quality references, employing a cycle consistency loss together with re-degradation validation, which ensures both perceptual quality and physical plausibility. Experiments on synthetic and real-world underwater datasets across diverse conditions demonstrate that SSE-UVSR consistently outperforms existing methods in terms of both resolution and visual quality.

Keywords: Underwater Video Super · Resolution · Self · Supervised Learning · End · To · End · Cycle Consistency Loss

1 Introduction

Underwater videos are critical for marine exploration, robotic navigation, and marine life monitoring. However, acquiring high-quality underwater videos faces significant challenges due to low contrast, blurring, and noise caused by insufficient lighting, water scattering, and absorption. Limited transmission bandwidth and storage capacity often necessitate compression and downsampling, further degrading video quality. Traditional enhancement methods typically depend on paired high and low quality datasets for supervised learning, which are costly and difficult to obtain in underwater environments. Existing research addresses underwater video enhancement through two main approaches: image enhancement methods [1, 2] that model degradation processes, and video super-resolution techniques [3, 4] based on deep learning. However, these approaches often rely heavily on extensive paired data, which is impractical for underwater scenarios. Recently emerged self-supervised image enhancement methods [5, 6] address limited paired data challenges but fail to effectively exploit temporal information in video sequences.

P. Umapada et al. (Eds.): ICCPR 2025, CCIS 2811, pp. 111–120, 2026.
https://doi.org/10.1007/978-981-95-8315-7_9

To overcome these limitations, we propose SSE-UVSR, a self-supervised end-to-end underwater video super-resolution framework. Our method employs convolutional neural networks to extract degradation information including latent clear videos, transmission maps, and atmospheric light parameters from low-resolution inputs. We introduce an attention-based module tailored for edge enhancement and detail preservation specific to underwater video characteristics. For self-supervised training, we design a comprehensive cycle consistency loss incorporating consistency verification, re-degradation validation, and edge enhancement regularization. We develop the UVD dataset containing 180 diverse underwater videos based on the MVK dataset [7]. Extensive experiments demonstrate that our method significantly improves visual quality and resolution, surpassing existing state-of-the-art methods in both subjective and objective evaluations. The main contributions include:

- A novel self-supervised underwater video enhancement framework using a degradation-aware module to accurately model and mitigate underwater-specific degradations without requiring high-quality references.
- A re-degradation validation mechanism ensuring enhancements adhere closely to underwater physical optics principles.
- The UVD dataset establishing a new benchmark, with demonstrated superior performance in detail preservation and visual quality enhancement.

2 Related Work

2.1 Underwater Video Enhancement

Underwater video enhancement is crucial for marine research, autonomous navigation, and seafloor exploration. Early methods based on image fusion [1] and wavelet transform [2] suffer from limited modeling capacity and poor handling of complex degradations. With the rise of deep learning, Li et al. [8] introduced a physics-based approach using underwater scene priors. Recent end-to-end models further advance this field by learning direct mappings from degraded to enhanced videos. Yang et al. [9] proposed an effective architecture with a dedicated dataset, and Wang et al. [3] introduced the UVEB dataset to support practical use. However, most existing methods rely on paired training data, which is difficult to obtain in real underwater scenarios.

2.2 Video Super-Resolution

Video super-resolution (VSR) has experienced remarkable progress driven by deep learning advancement. Early works [10] established the foundation by introducing CNN architectures for VSR tasks. Subsequent research concentrated on exploiting temporal information through recurrent neural networks [11] and deformable alignment modules [12], achieving substantial quality improvements by leveraging temporal dynamics and inter-frame correlations. Parallel research focused on developing efficient real-time models [13] balancing performance and computational efficiency. Contemporary state-of-the-art architectures, exemplified by BasicVSR [14] and BasicVSR + + [15], incorporate advanced innovations including detail enhancement mechanisms, transformer-based architectures [16], and adaptive processing approaches [17]. Despite significant achievements, existing VSR methodologies exhibit fundamental limitations: they predominantly rely on paired training datasets and adopt oversimplified degradation models such as bicubic downsampling, which fail to capture complex real-world degradation patterns, limiting practical applicability in challenging environments like underwater conditions.

2.3 Self-Supervised Video Restoration

Self-supervised video restoration has emerged as a promising paradigm to address blind video super-resolution challenges where degradation models are unknown or complex. Recent advances leverage self-supervision principles [a5], dynamic adaptation strategies [a6], and expanded synthetic degradation modeling [18] to enhance model performance and generalization. The self-supervised approach eliminates paired training dataset requirements and enables learning from intrinsic structural patterns and temporal consistencies in video sequences. Accurate modeling of complex real-world degradation phenomena is critical for effective restoration, particularly for challenging scenarios such as satellite imagery [19, 20] and underwater environments.

3 Methodology

3.1 Overview

SSE-UVSR addresses underwater video super-resolution by estimating degradation parameters, reconstructing high-resolution frames, and validating results via re-degradation. It first extracts key parameters—atmospheric light (A), clean content (J), and transmission map (T). Then, the Detail Reconstruction Module (DRM) fuses multi-scale features to produce high-resolution outputs. Finally, a re-degradation step with cycle consistency and edge-aware losses ensures structural and detail fidelity. The overall architecture of SSE-UVSR is illustrated in Fig. 1. Unlike conventional blind VSR pipelines, our design combines a Degradation-Aware Module (DAM) for explicit estimation of underwater-specific physical parameters, a Detail Reconstruction Module (DRM) for fine-grained texture recovery, and a re-degradation verification stage guided by cycle consistency. This coordinated design ensures both high perceptual quality and physical plausibility in reconstructed underwater videos.

Underwater video enhancement is crucial for marine research, autonomous navigation, and seafloor exploration. Early methods based on image fusion [1] and wavelet transform [2] suffer from limited modeling capacity and poor handling of complex degradations. With the rise of deep learning, Li et al. [8] introduced a physics-based approach using underwater scene priors. Recent end-to-end models further advance this field by learning direct mappings from degraded to enhanced videos. Yang et al. [9] proposed an effective architecture with a dedicated dataset, and Wang et al. [3] introduced the UVEB dataset to support practical use. However, most existing methods rely on paired training data, which is difficult to obtain in real underwater scenarios.

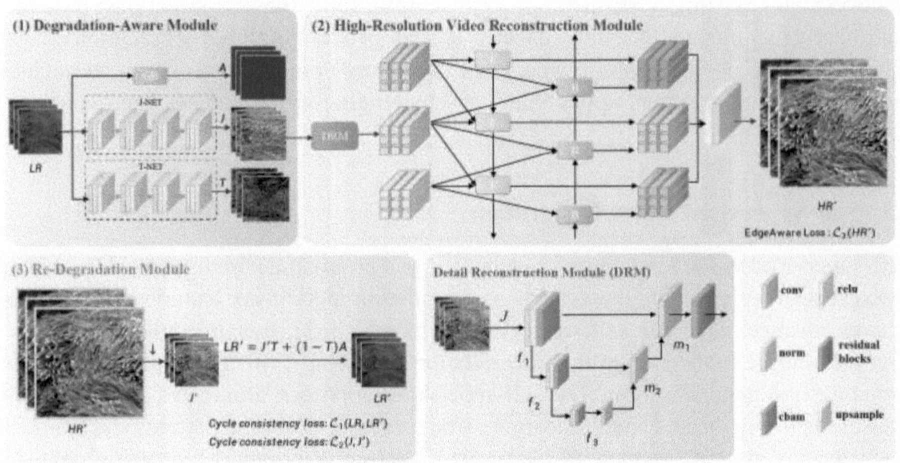

Fig. 1. Overview of our proposed SSE-UVSR method. It consists of three modules: (1) Degradation-Aware Module} employing deep convolutional networks to estimate key degradation parameters; (2) High-Resolution Video Reconstruction Module, using a multi-branch structure to generate the high-resolution (HR') video; (3) Re-Degradation Module, verifying HR' by re-degrading it into LR'. The entire framework is trained end-to-end in a self-supervised manner using cycle consistency constraints.

3.2 Underwater Degradation Estimation

Based on the Jaffe-McGlamery model [22], the degradation process of underwater images is described as

$$I(x) = J(x)e^{-\beta d(x)} + A\left(1 - e^{-\beta d(x)}\right)$$

where $J(x)$ represents object radiance, β is the light attenuation coefficient, $d(x)$ is the object-to-camera distance, and A represents ambient light.

For video sequences with N frames, we extend this model assuming consistent degradation parameters across adjacent frames during short capture durations. We simplify the model as

$$I = J \cdot T + A(1 - T)$$

where I represents the degraded video, J the clean video, T the transmission map, and A the atmospheric light.

Degradation parameters are estimated using dedicated networks

$$J = J - \text{NET}(LR), T = T - \text{NET}(LR)$$

while atmospheric light A is estimated using Gaussian blur operations.

3.3 High-Resolution Video Reconstruction

Based on the Jaffe-McGlamery model [22], the degradation process of underwater images is described as.

Detail Reconstruction Module (DRM). The DRM addresses underwater-specific issues through a spatial attention pyramid structure. Input video J undergoes progressive feature extraction:

$$f_i = \text{CBAM}(\text{ReLU}(\text{Conv}(f_{i-1}))), i = 1, 2, 3$$

where $f_0 = J$. Multi-scale features are fused through concatenation:

$$m_1 = \text{concat}(f_3, f_2, f_1),$$

providing comprehensive feature representation for subsequent super-resolution.

Bidirectional Video Super-Resolution. We employ a BasicVSRbased bidirectional recurrent structure operating at feature level. For frame i with features m_i, forward and backward propagation branches utilize optical flow estimation and feature alignment:

$$s_i^{f,b} = S(m_i, m_{i\pm1}), h_i^{f,b} = W\left(h_{i\pm1}^{f,b}, s_i^{f,b}\right),$$

where S represents Spynet optical flow estimation and W denotes the warping module. Features are refined through residual modules and combined for final reconstruction:

$$HR' = U\left(h_i^f, h_i^b\right)$$

where U represents the upsampling operation using pixel shuffle.

Cycle Consistency Constraints. To enable self-supervised learning, we reconstruct the low-resolution video using estimated high-resolution output and degradation parameters. The cycle consistency constraint minimizes:

$$\min LR - D\left(HR'\right)^2,$$

where $D(\cdot)$ represents the degradation operator. Combining underwater imaging principles:

$$HR_{LR'} = HR' \downarrow, HR_{LR\_recon'} = HR_{LR'} \cdot T + (1 - T) \cdot A$$

where $\downarrow$ denotes downsampling operation.

4 Experiments and Analysis

4.1 Implementation Details

We built the Underwater Video Dataset (UVD) by collecting 180 diverse underwater video sequences from public, copyright-free sources such as YouTube and the MVK dataset [7]. The selection process ensured scene diversity while avoiding duplicates, focusing on relatively clear and stable footage. Each clip was standardized to 1920

× 1080 resolution, and trimmed to 100 consecutive frames to form one sample. The dataset covers a broad range of environments, including clear and turbid waters, shallow to mid-depth scenes, and varied lighting conditions from bright sunlight to low-light situations. Experiments were conducted on an A6000 server (Ubuntu 20.04) using the Adam optimizer with a learning rate of 2×10^{-4}, with 256 × 256 random crops and batch size 8. We compared our method against state-of-the-art approaches, including Bicubic [28], RBPN [29], BasicVSR [14], BasicVSR + + [15], SSL [30], SAVSR [31], IART [32], and self-supervised models like DeepVSR [13] and BlindVSR [12].

Evaluation was performed using six no-reference image quality metrics: four general-purpose natural image quality measures (BRISQUE [23], NIQE [24], PIQE [25], AG [33]) and two underwater-specific quality measures (UIQM [26], UCIQE [27]). The general metrics assess perceptual naturalness, sharpness, and distortion without requiring ground truth, while the underwater metrics focus on color balance, contrast, and clarity in marine scenes. This choice reflects the self-supervised nature of our framework, where ground-truth high-resolution references are unavailable.

Table 1. Quantitative performance comparison of different super-resolution methods on the UVD dataset. The best results are marked in **red**, and the second-best results are marked in **blue**.

Method	BRISQUE↓[23]	NIQE↓[24]	PIQE↓[25]	AG↓[33]	UIQM↑[26]	UCIQE↑[27]
Bicubic [28]	55.513	7.420	3.661	6.831	2.689	0.511
RBPN [29]	51.188	4.826	6.233	4.301	2.895	**0.514**
BasicVSR [14]	48.373	4.557	6.436	4.064	2.973	**0.513**
BasicVSR ++ [15]	42.936	**4.088**	**6.683**	**3.732**	2.827	**0.513**
DeepVSR [13]	46.881	5.758	5.513	5.148	2.921	0.510
BlindVSR [12]	**42.708**	5.199	5.685	4.761	2.988	0.511
SSL [30]	48.853	4.837	6.506	4.179	**3.095**	0.506
SAVSR [31]	44.483	5.738	4.859	5.515	2.852	0.505
IART [32]	48.203	4.571	6.614	3.989	**3.127**	0.506
Ours	**20.205**	**3.395**	**6.714**	**3.388**	3.089	0.510

4.2 Comparisons with State-Of-The-Art Methods

Quantitative Analysis. As shown in Table 1, our method achieves the best overall performance across all evaluation metrics on the UVD dataset. Specifically, it outperforms all competing methods in BRISQUE (20.21), NIQE (3.40), NRQM (6.71), and PI (3.39), indicating superior perceptual quality and naturalness. It also ranks among the top in UIQM (3.09) and UCIQE (0.51), demonstrating its effectiveness in preserving underwater-specific visual characteristics. Compared to the second-best method, BasicVSR + +, our approach reduces the NIQE by 16.9% and improves NRQM by 0.5 points, showing significant gains in both no-reference image quality and structural fidelity.

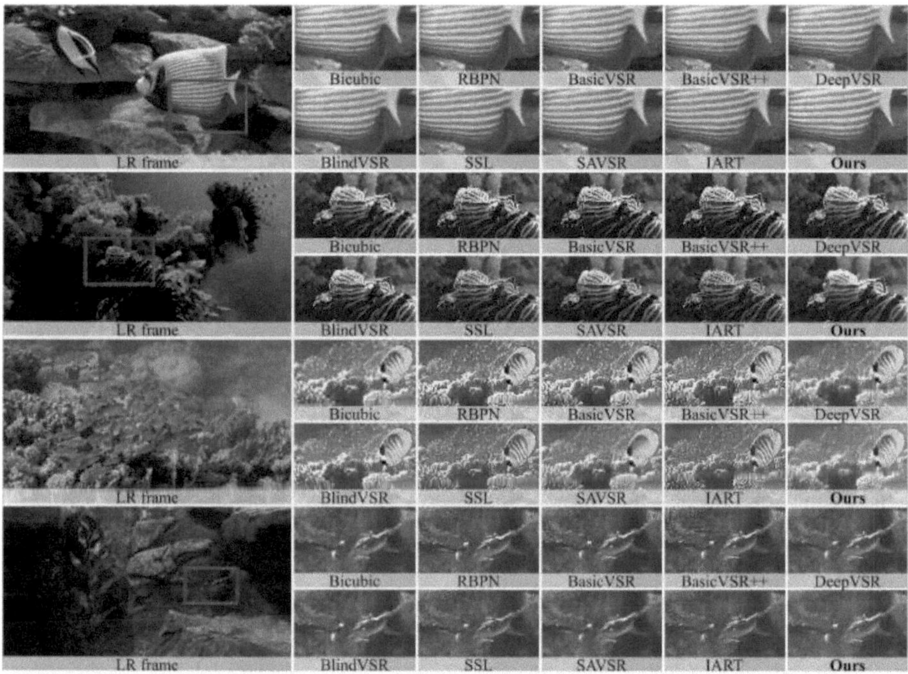

Fig. 2. Qualitative comparisons on the UVD dataset for 4 × scaling. Zoom in for the best view.

Qualitative Analysis. Figure 2 presents visual comparisons on representative frames from the UVD dataset. Our method consistently restores finer textures, smoother edges, and more natural colors across diverse scenes. It eliminates artifacts in small fish, sharpens contours of coral reefs, and preserves structural details of marine animals. Benefiting from self-supervised learning and underwater-specific priors, SSE-UVSR effectively reduces noise and blur, producing visually pleasing results that align with underwater characteristics.

In terms of computational efficiency, SSE-UVSR achieves inference speed comparable to mainstream bidirectional VSR models such as BasicVSR + + under the same resolution and hardware conditions. While our framework is primarily designed for offline enhancement, its runtime characteristics also make it suitable for near real-time processing in practical underwater applications.

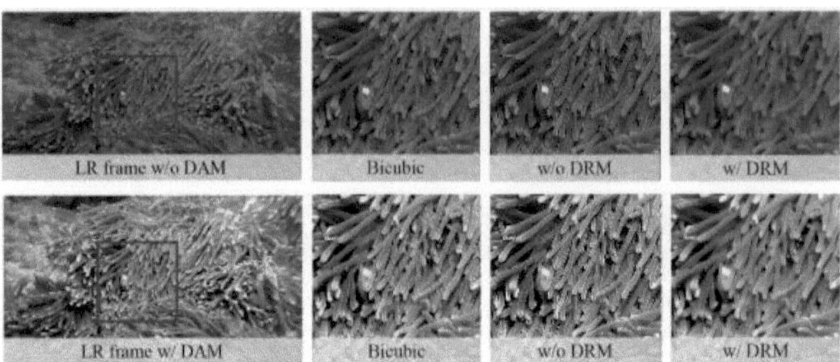

Fig. 3. Visual comparison of the ablation study for SSE-UVSR.

4.3 Ablation Study

To evaluate the contribution of the Degradation-Aware Module (DAM) and Detail Reconstruction Module (DRM), we present qualitative comparisons in Fig. 3. The top row shows results without DAM, where color distortion and low contrast persist. In contrast, the bottom row with DAM exhibits improved color and clarity. Bicubic upsampling fails to recover fine structures, while DRM significantly enhances texture details and edge sharpness. The combination of DAM and DRM delivers the best visual quality, effectively restoring intricate coral textures and producing clear, natural results. This confirms the complementary roles of DAM in degradation correction and DRM in fine-detail reconstruction.

5 Conclusion

We present a novel self-supervised end-to-end underwater video super-resolution method that effectively addresses the challenges faced by existing techniques in practical applications. By leveraging deep convolutional neural networks, an edge enhancement and detail preservation module, and a cycle consistency loss function, our approach enables the recovery of high-definition videos from degraded low-resolution underwater footage without the need for high-quality reference videos. Extensive experiments on both a constructed dataset and real-world data demonstrate the superiority of our method over state-of-the-art techniques in terms of visual quality and resolution enhancement. The ablation study further validates the necessity and complementary roles of the key components in our method. Our research contributes to the advancement of underwater video processing technology and has the potential to greatly benefit marine science research and related applications. Future work may focus on further optimizing the network architecture, training strategies, and incorporating domain-specific knowledge to improve the robustness and adaptability of the method in various underwater scenarios.

Acknowledgments. This work was supported by JST SPRING, Japan Grant Number JPMJSP2154.

References

1. Ancuti, C., Ancuti, C.O., Haber, T., Bekaert, P.: Enhancing underwater images and videos by fusion. In: Proceedings of the IEEE Conference on Computer Vision and Pattern Recognition, pp. 81–88 (2012)
2. Singh, G., et al.: Underwater image/video enhancement using wavelet based color correction (WBCC) method. In: Proceedings of the IEEE Underwater Technology (UT), pp. 1–5 (2015)
3. Xie, Y., et al.: UVEB: A large-scale benchmark and baseline towards real-world underwater video enhancement. In: Proceedings of the IEEE/CVF Conference on Computer Vision and Pattern Recognition, pp. 22358–22367 (2024)
4. Sun, H., Yue, J., Li, H.: An image enhancement approach for coral reef fish detection in underwater videos. Eco. Inform. **72**, 101862 (2022)
5. Bai, H., Pan, J.: Self-supervised deep blind video super-resolution. IEEE Trans. Pattern Anal. Mach. Intell. **46**(7), 4641–4653 (2024)
6. Lee, S., Choi, M., Lee, K.M.: Dynavsr: Dynamic adaptive blind video super-resolution. In: Proceedings of the IEEE/CVF Winter Conference on Applications of Computer Vision, pp. 2093–2102 (2021)
7. Truong, Q.-T., et al.: Marine video kit: a new marine video dataset for content-based analysis and retrieval. In: International Conference on Multimedia Modeling, pp. 539–550 (2023)
8. Li, C., Anwar, S., Porikli, F.: Underwater scene prior inspired deep underwater image and video enhancement. Pattern Recogn. **98**, 107038 (2020)
9. Du, D., Li, E., Si, L., Xu, F., Niu, J.: End-to-end underwater video enhancement: Dataset and model. arXiv preprint arXiv:2403.11506 (2024)
10. Liu, H., et al.: Video super-resolution based on deep learning: a comprehensive survey. Artif. Intell. Rev. **55**(8), 5981–6035 (2022)
11. Sajjadi, M.S.M., Vemulapalli, R., Brown, M.: Frame-recurrent video super-resolution. In: Proceedings of the IEEE Conference on Computer Vision and Pattern Recognition, pp. 6626–6634 (2018)
12. Tian, Y., Zhang, Y., Fu, Y., Xu, C.: TDAN: Temporally-deformable alignment network for video super-resolution. In: Proceedings of the IEEE/CVF Conference on Computer Vision and Pattern Recognition, pp. 3360–3369 (2020)
13. Shi, W., et al.: Real-time single image and video super-resolution using an efficient sub-pixel convolutional neural network. In: Proceedings of the IEEE Conference on Computer Vision and Pattern Recognition, pp. 1874–1883 (2016)
14. Chan, K.C.K., Wang, X., Yu, K., Dong, C., Loy, C.C.: BasicVSR: The search for essential components in video super-resolution and beyond. In: Proceedings of the IEEE/CVF Conference on Computer Vision and Pattern Recognition, pp. 4947–4956 (2021)
15. Chan, K.C.K., Zhou, S., Xu, X., Loy, C.C.: BasicVSR++: Improving video super-resolution with enhanced propagation and alignment. In: Proceedings of the IEEE/CVF Conference on Computer Vision and Pattern Recognition, pp. 5972–5981 (2022)

16. Cao, J., Li, Y., Zhang, K., Van Gool, L.: Video super-resolution transformer. arXiv preprint arXiv:2106.06847 (2021)
17. Liu, C., Sun, D.: A Bayesian approach to adaptive video super resolution. In: Proceedings of the IEEE Conference on Computer Vision and Pattern Recognition, pp. 209–216 (2011)
18. Jeelani, M., Cheema, N., Illgner-Fehns, K., Slusallek, P., Jaiswal, S., et al.: Expanding synthetic real-world degradations for blind video super resolution. In: Proceedings of the IEEE/CVF Conference on Computer Vision and Pattern Recognition, pp. 1199–1208 (2023)
19. Xiao, Y., Yuan, Q., Zhang, Q., Zhang, L.: Deep blind super-resolution for satellite video. IEEE Transactions on Geoscience and Remote Sensing (2023)
20. Liu, H., Gu, Y.: Deep joint estimation network for satellite video super-resolution with multiple degradations. IEEE Trans. Geosci. Remote Sens. **60**, 1–15 (2022)
21. Zhang, K., Liang, J., Van Gool, L., Timofte, R.: Designing a practical degradation model for deep blind image super-resolution. In: Proceedings of the IEEE/CVF International Conference on Computer Vision, pp. 4791–4800 (2021)
22. McGlamery, B.L.: A computer model for underwater camera systems. In: Ocean Optics VI, vol. 208, pp. 221–231 (1980)
23. Gao, S., Gruev, V.: Bilinear and bicubic interpolation methods for division of focal plane polarimeters. Opt. Express **19**(27), 26161–26173 (2011)
24. Haris, M., Shakhnarovich, G., Ukita, N.: Recurrent back-projection network for video super-resolution. In: Proceedings of the IEEE/CVF Conference on Computer Vision and Pattern Recognition, pp. 3897–3906 (2019)
25. Xia, B., et al.: Structured sparsity learning for efficient video super-resolution. In: Proceedings of the IEEE/CVF Conference on Computer Vision and Pattern Recognition, pp. 22638–22647 (2023)
26. Li, Z., et al.: SAVSR: Arbitrary-scale video super-resolution via a learned scale-adaptive network. In: Proceedings of the AAAI Conference on Artificial Intelligence, vol. 38, no. 4, pp. 3288–3296 (2024)
27. Xu, K., Yu, Z., Wang, X., Mi, M.B., Yao, A.: Enhancing video super-resolution via implicit resampling-based alignment. In: Proceedings of the IEEE/CVF Conference on Computer Vision and Pattern Recognition, pp. 2546–2555 (2024)
28. Mittal, A., Moorthy, A.K., Bovik, A.C.: No-reference image quality assessment in the spatial domain. IEEE Trans. Image Process. **21**(12), 4695–4708 (2012)
29. Zhang, L., Zhang, L., Bovik, A.C.: A feature-enriched completely blind image quality evaluator. IEEE Trans. Image Process. **24**(8), 2579–2591 (2015)
30. Pandey, A., et al.: Evaluation of perception based image quality evaluator (PIQE) no-reference image quality score for 99mTc-MDP bone scan images. Society of Nuclear Medicine (2020)
31. Yang, M., Sowmya, A.: An underwater color image quality evaluation metric. IEEE Trans. Image Process. **24**(12), 6062–6071 (2015)
32. Panetta, K., Gao, C., Agaian, S.: Human-visual-system-inspired underwater image quality measures. IEEE J. Oceanic Eng. **41**(3), 541–551 (2015)
33. Mustafa, Z., Sims, B.: A new approach to generalized metric spaces. J. Nonlin. Conv. Anal. **7**(2), 289 (2006)

Depth-Constrained Kernel Estimation-Based Full-Focus Blind Super-Resolution for Light Field Images

Kong Deqian[1(✉)], Guan Ling[1], and Su Lijuan[2]

[1] National Key Laboratory of Scattering and Radiation, Beijing, China
kongdq08@163.com

[2] Key Laboratory of Precision Opto-Mechatronics Technology, Ministry of Education, Beihang University, Beijing, China

Abstract. In light field(LF) imaging technology, both the direction and intensity of light rays are recorded by a special four-dimensional(4D) imaging structure, thereby enabling depth information to be obtained and the relative positions of captured objects to be acquired. Consequently, unique advantages are demonstrated across multiple three-dimensional(3D) application domains. Microlens array-based LF cameras are confronted with the challenge of low spatial resolution because the device - a single detector - is limited in size and simultaneous spatial and angular sampling is required. Currently, numerous non-blind super-resolution(SR) algorithms have been proposed to enhance spatial resolution, where the blur kernel is provided as a known prior input to the network. However, these blur kernels are configured with a single type, and thus remain limited when applied to real world LF images suffering from complex degradations. Therefore, to address the limitation of non-blind algorithms in handling complex degradation blur, a full-focused blind light field super-resolution(LFSR) network based on deep kernel estimation, named the Depth-Constrained Kernel Estimation Network (DCKE-Net), is constructed in this paper. An iterative optimization architecture is adopted. A depth-constrained spatially variant (SV) kernel estimation module is designed. A feature interaction module is introduced to extract contextual blur information. A parallel window transformer structure is proposed to process spatial and angular features separately for attention calculation. Finally, the network's capability in detail restoration and multi-defocus blur removal is validated through simulation and ablation experiments, providing a blind solution for the full-focus SR task.

Keywords: Light Field · Blind Super-resolution · Full-Focus · Kernel Estimation · Swin Transformer

1 Introduction

As a novel snapshot imaging technique, LF imaging has gained significant attention for its unique four-dimensional structure. During a single exposure, it [1] captures multi-view images of a scene from different perspectives, enabling comprehensive visual information acquisition. With advancements in computational imaging, LF technology has

P. Umapada et al. (Eds.): ICCPR 2025, CCIS 2811, pp. 121–133, 2026.
https://doi.org/10.1007/978-981-95-8315-7_10

evolved rapidly over the past two decades. Its inherent advantages offer substantial potential in applications such as digital refocusing [2], 3D display [3], and depth estimation [4]. However, because LF imaging simultaneously records both spatial and angular information within the physical limits of the sensor, its spatial resolution is typically lower than that of conventional imaging systems, which capture only spatial data. Currently, LFSR reconstruction has become the most widely adopted resolution enhancement approach [5, 6] due to its lower cost, easier implementation, and superior efficacy.

Nevertheless, most existing SR networks do not fully utilize the depth cues unique to LF imagery for targeted deblurring, and often rely on a predefined non-blind blur kernel for degradation modeling. To overcome these limitations, we propose a depth-constrained spatially variant kernel matrix estimation module that resolves complex degradations by estimating 3D SV kernel matrices. Based on this, we construct an end-to-end full-focus blind super-resolution network—termed Depth-Constrained Kernel Estimation Network (DCKE-Net)—through iterative optimization, achieving full-focus SR reconstruction in a single step. Additionally, a weighted loss function is designed in accordance with the kernel estimation mechanism. Simulated experiments and ablation studies convincingly demonstrate the effectiveness of the proposed method.

2 Related Work

To enable full-focus blind SR for light field images with depth-variant deblurring, this section reviews relevant LFSR and blind SR methods. After identifying their key limitations, we present the corresponding improvements that are integrated into our proposed algorithm.

2.1 LFSR Techniques

An increasing number of neural network architectures have recently been deployed for LFSR [7–14].The Light Field Convolutional Neural Network (LFCNN) [7] pioneered the integration of CNNs with LFSR. A macro-pixel-based method [8] employs two independently trained CNNs for spatial and angular SR via supervised learning, demonstrating enhanced visual performance in de-ringing. A two-stage CNN [9] decouples LF structures by jointly leveraging intrinsic correlations (among sub-aperture images) and extrinsic correlations (across image instances), thereby improving reconstruction quality. A self-supervised learning-based autoencoding method with a hybrid loss function is employed for spatial-domain LFSR imaging [10].

However, existing methods reconstruct images with fixed focal depths and universally adopt predefined non-blind blur kernels during degradation modeling. These approaches typically disregard depth information, consequently failing to generate high-resolution (HR) full-focus images directly. Thus, deep learning-based techniques for full-focus LFSR require further development and refinement.

2.2 Blind SR Techniques

Blind image reconstruction achieves restoration without prior knowledge of the specific degradation process. Significant progress has recently been made in neural network-based blind SR approaches. Contemporary blind SR networks are categorized into two paradigms: explicitly modeled [15–18] and implicitly modeled [19–21] architectures.

Explicitly modeled networks must capture degradation characteristics (e.g., blur kernel k and additive noise n) inherent in the high-to-low-resolution transformation. In 2018, Zhang *et al.* proposed SRMD [15], which first estimates degradation parameters then performs reconstruction. Departing from isolated estimation, Gu et al. developed Iterative Kernel Correction (IKC) [17] for optimized kernel refinement. This method utilizes characteristic error patterns under kernel mismatch as empirical cues to rectify inaccurate blur kernels during reconstruction. The Deep Alternating Network (DAN) [18] extends IKC's iterative paradigm by unifying kernel estimation and reconstruction within an end-to-end architecture.

For scenarios where degraded low-resolution(LR) counterparts are unavailable, implicitly modeled networks employ unpaired image collections. For instance, Cycle-in-Cycle Generative Adversarial Networks (CinCGAN) [19] utilize dual-cycle learning for real-world SR. Bulat et al. proposed a stepwise framework [20] integrating degradation simulation and SR networks. Liu et al. systematically categorized such implicit modeling architectures [21].

In summary, while blind SR has advanced rapidly for single-image restoration, its application to LF imagery remains underexplored. LFSR must concurrently preserve spatial resolution and angular information. This necessitates handling both pixel-intensity variations and ray-direction alterations, thereby compounding the problem complexity. Consequently, we design an end-to end blind SR network for light fields by integrating their intrinsic properties with blind SR architectures, aiming to streamline processing and enhance practical utility.

3 Dataset Construction

Existing public LF datasets typically provide only high-resolution LF images and their ground truth disparity maps. For training, a common practice is to either apply a fixed image filter to blur the HR images or to bypass blurring entirely by directly downsampling them. They fail to simulate the depth-dependent defocus blur variations that occur in real-world imaging.

In practical multi-depth scenes, objects at different depths exhibit distinct defocus blur characteristics. As illustrated in Fig. 1, which depicts the imaging process of object points at varying depths through a single sub-aperture, an in-focus point (shown in blue) converges through the optical system onto a single microlens. Its projection onto the sensor approximates a point spread function (PSF) close to the size of a single pixel. In contrast, an out-of-focus point (green rays) converges either behind or in front of the microlens array, resulting in a blurred projection that manifests as defocus in the captured image.

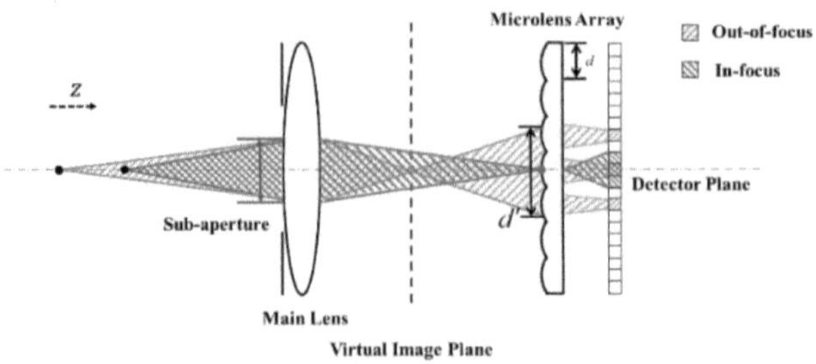

Fig. 1. Light field sub-aperture imaging (SAI) under different focusing conditions

We construct the multi-defocus light field dataset using the following procedure, with the mathematical formulation provided in Eq. (1):

$$\mathbf{I}_i^{\mathrm{lr}} = (\mathbf{I}_i^{\mathrm{hr}} \otimes \mathbf{K}_i) \downarrow_\alpha + \mathbf{N}_i \tag{1}$$

where $\mathbf{I}_\mathrm{i}^{\mathrm{hr}} \in \mathbb{R}^{H \times W}, i = 1, \ldots, N$ represents the HR SAI at the i depth; $\mathbf{K}_i \in \mathbb{R}^{k \times k}$ is the Gaussian blur kernel at the i-th depth with standard deviation σ_i and kernel size k; $\mathbf{N}_i$ denotes the additive noise; $\mathbf{I}_i^{\mathrm{lr}} \in \mathbb{R}^{(H/\alpha) \times (W/\alpha)}$ is the degraded sub-aperture image at the i-th depth, where α is the downsampling factor, reducing the resolution to $1/\alpha$ of the ground truth image.

We select three publicly available LF depth datasets—HCInew [22], HCIold [23], and DUTLF [24]—as the ground truth HR images. A set of N masks is generated from the ground truth disparity maps. Each HR SAI is pixel-wise multiplied with the mask corresponding to its depth, producing separate SAIs for each depth level. To emulate depth-variant defocus, we convolve the depth-separated SAIs with Gaussian blur kernels of different sizes, where the kernel variance is a function of depth. The blurred SAIs are subsequently downsampled by a factor of k, and additive noise is introduced to obtain the final low-resolution (LR) multi-defocus SAIs.

4 The Proposed Framework

This section presents the construction of an end-to-end blind SR convolutional neural network, termed DCKE-Net, whose overall framework is illustrated in Fig. 2.

The architecture of the DCKE-Net consists primarily of two components: a blind kernel estimation(BKE) module and a full-focused SR module. The process can be expressed by the following two equations:

$$\mathbf{I}^{\mathrm{sr}} = \Gamma_1(\mathbf{I}^{\mathrm{lr}}, \mathrm{K}) \tag{2}$$

Where $Gamma_1$ denotes the full-focus SR process, $\mathrm{K} \in \mathbb{R}^{k \times k \times n}$ represents the SV kernel matrix, k indicates the size of the blur kernel, n denotes the number of SV blur kernels, and $\mathbf{I}^{\mathrm{lr}}$ and $\mathbf{I}^{\mathrm{sr}}$ refer to the LR input LF image and the super-resolved reconstructed LF image, respectively. The network first receives a LR LF image, which is then processed through a full-focus SR module. Meanwhile, the SV kernel matrix is embedded as critical

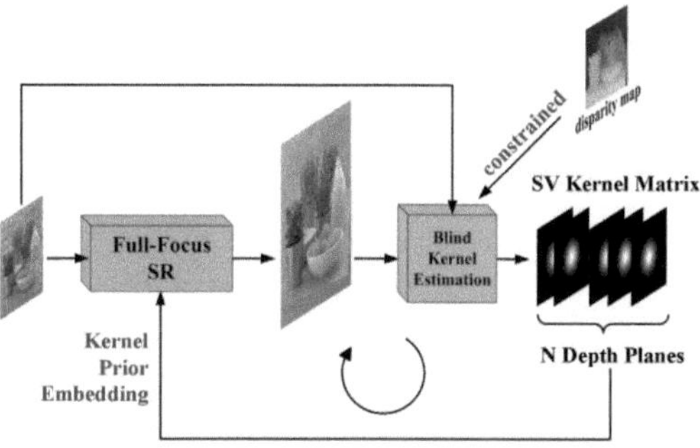

Fig. 2. Framework of DCKE-Net

prior information into the full-focus SR module, ultimately outputting a super-resolved image.

$$\mathbf{K} = Gamma_2(\mathbf{I}^{sr}, \mathbf{I}^{lr}) \tag{3}$$

where $Gamma_2$ denotes the blind kernel estimation process. After the super-resolved image $\mathbf{I}^{sr}$ is obtained, both $\mathbf{I}^{sr}$ and the $\mathbf{I}^{lr}$ are fed into the disparity-map-constrained blind kernel estimation module to generate a SV kernel matrix. This process is then repeated iteratively to alternately refine the results, ultimately converging to the optimal solution.

4.1 Feature Interaction Based Full-Focus SR Module

The input to the blind SR network DCKE-Net is organized in an image pyramid structure. The overall architecture comprises two main modules: a U-net-based feature interaction module and a window transformer-based full-focus module (Fig. 3).

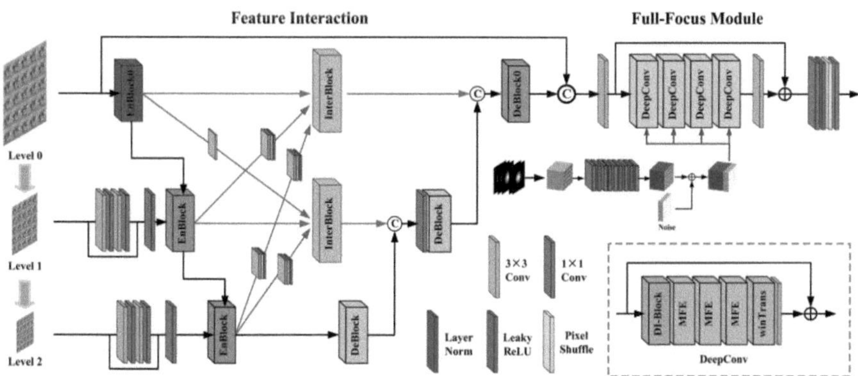

Fig. 3. Framework of Full-Focus SR Module

U-Net-Based Feature Interaction Module. Since blind sr networks require blur kernel estimation, accurate extraction of blur features is particularly critical. To capture more comprehensive blur information, a U-net architecture [25] with blur feature interaction is employed on the image pyramid.

The U-net structure consists of an encoder and a decoder, which are symmetrically organized. The encoder extracts and reduces the dimensionality of image features. The decoder then gradually restores spatial features and details through upsampling and convolutional layers, enabling precise pixel-level prediction. Based on the U-net architecture, DCKE-Net incorporates feature interaction blocks (InterBlocks) to enable interaction among blur features extracted from the three-level encoder. The InterBlocks are applied at Level 0 and Level 1 of the pyramid. Their main function is to integrate features from three pyramid levels and perform feature interaction via fully connected layers and standard convolutional layers. When integrating lower-level features into an InterBlock, the PixelShuffle upsampling layer is used for resolution alignment. Conversely, when incorporating higher-level features, a convolutional layer with specific parameters is applied for downsampling.

Window Transformer-Based Full-Focus Module. The full-focus module consists of four DeepConv modules, which utilize depth masks to separate input features into features at different depth layers, forming a feature matrix. This matrix is then convolved and embedded with the SV kernel matrix obtained from kernel estimation. Each DeepConv comprises a degeneration-integration Block (DI-block), three Multiple Feature Extraction (MFE) modules, a window transformer (winTrans) and a 3×3 convolutional layer. The DI-block and MFE modules follow the design established in our previous work [26].

The Swin Transformer is an innovative model based on the classic Transformer architecture, designed to extend the powerful capabilities of Transformers from natural language processing to computer vision [27]. DCKE-net uses a parallel winTrans structure improved from the Swin Transformer specifically for LF features. This structure can separately process spatial features and angular features to compute attention, with its detailed architecture shown in Fig. 4.

Compared to traditional convolution operations, the self-attention mechanism incorporated in the winTrans is not constrained by a fixed receptive field size. It can adaptively adjust the region of focus based on image content, thereby enabling more flexible processing of characteristics across different spatial areas—particularly beneficial in this study for handling images with blended blur regions. Furthermore, by capturing global dependencies, the window transformer can effectively identify and address spatial complexities.

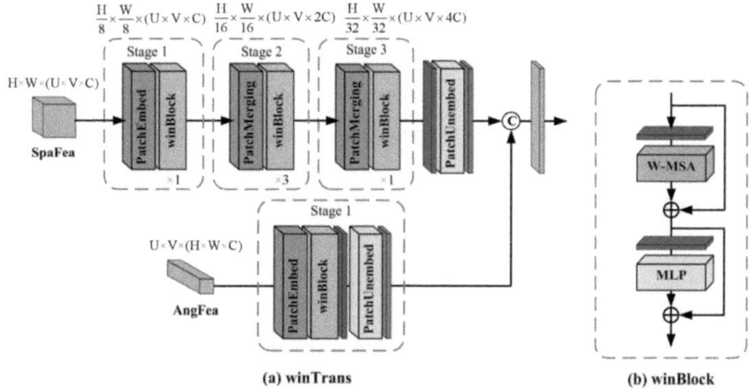

Fig. 4. Structure of winTrans

4.2 SV Kernel Matrix Estimation Module

The objective of the SV kernel matrix estimation module is to estimate the SV kernel matrix required by the full-focus SR module, using the input LR LF image. To achieve this, the module accepts the LR LF image as input. In addition, to enable iterative optimization, the output from the full-focus SR module is fed back into the kernel estimation module, which requires the latter to also accept the HR image generated by the former as input. Finally, following the principle of depth constraint, masks corresponding to different depths are also fed into the module to guide the estimation process of the SV kernels.

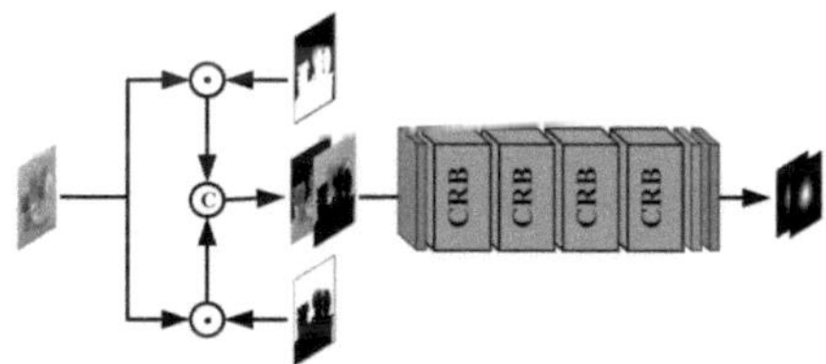

Fig. 5. Structure of SV kernel matrix estimation

Figure 5 illustrates the detailed architecture of the blind kernel estimation module and the process of incorporating depth constraints, using two depth levels as an example. The super-resolved image is first downsampled via a convolution layer with specific parameters and then concatenated with the LR image along the channel dimension. The combined result is subsequently multiplied pixel-wise by masks corresponding to different depths to extract image features specific to each depth level. All depth-specific features are then concatenated along the channel dimension and fed into a Conditional Residual Block (CRB) [18], which integrates conditional inputs with primary features to enhance structural representation.

5 Experiments

5.1 Implementation Details

In this section, the implementation details of the proposed method are introduced. The training data consist of light field RGB images with an angular resolution of 5 × 5. To conserve GPU memory, each SAI in the ground truth is cropped into patches of size 64 × 64. Furthermore, each pair is randomly augmented through flipping, transposition, and color channel shuffling to enhance robustness. The network is implemented under the PyTorch framework using an NVIDIA Titan XP GPU and is optimized with the Adam optimizer with an initial learning rate of 2.5×10^{-4}. A StepLR scheduler is employed to update the learning rate every 10 epochs, where the gamma value is set to 0.5, resulting in the learning rate being reduced to half of its previous value at each update.

Due to the incorporation of the SV kernel matrix estimation module, errors originating from kernel estimation are incorporated into the loss function, forming a weighted loss function as shown in Eq. (4).

$$Loss = \alpha_{sr} \cdot L_1(\mathbf{I}^{sr}, \mathbf{I}^{hr}) + \alpha_{lr} \cdot L_1(\mathbf{I}^{hr} \otimes \widehat{\mathbf{K}}, \mathbf{I}^{lr}) \tag{4}$$

where $\mathbf{I}^{hr}$ denotes the HR ground truth image, $\mathbf{I}^{lr}$ represents the LR input image, $\mathbf{I}^{sr}$ is the super-resolved image obtained from the network, and $\widehat{\mathbf{K}}$ refers to the estimated kernel. The loss function consists of two components: the super-resolution loss and the kernel estimation loss. The first part computes the loss between the super-resolved image and the ground truth image while another part computes the loss between the original LR image and the pseudo LR image obtained by convolving the HR light field image with $\widehat{\mathbf{K}}$. Here, α_{sr} and α_{lr} represent the weight coefficients for the two loss terms, both set to 0.5 as hyperparameters during training.

5.2 Comparison with the State-Of-The-Art Networks

This section presents the training results of several existing super-resolution networks on the multi-defocus light field dataset and compares them with DCKE-Net. The compared algorithms include four spatial super-resolution networks for light field images, namely: the DKnet [28], the LF-DMnet [29], the DistgSSR [14], and the Detail-Preserving Transformer (DPT) [30]. The quantitative results in Table 1 demonstrate the state-of-the-art performance of the proposed method on the test dataset.

Table 1. Quantitative Results of Comparison

Methods	PSNR(dB)			SSIM		
	Test1	Test2	Test3	Test1	Test2	Test3
DKNet	34.90	32.74	35.96	0.959	0.940	0.956
LF-DMnet	33.03	31.71	34.56	0.945	0.930	0.949
DistgSSR	34.91	32.87	35.18	0.965	0.943	0.965

(continued)

Table 1. (*continued*)

Methods	PSNR(dB)			SSIM		
	Test1	Test2	Test3	Test1	Test2	Test3
DPT	34.73	32.86	35.73	0.959	0.944	0.962
DCKE-Net	**34.99**	**33.17**	**36.53**	**0.966**	**0.945**	**0.965**

Owing to the adoption of a depth-guided strategy, the proposed method achieves comparable or even superior PSNR and SSIM performance relative to state-of-the-art light field super-resolution approaches. The results represent the average value over all data in each test set, with bolded values indicating that the DCKE-Net algorithm achieves the best performance across all test sets.

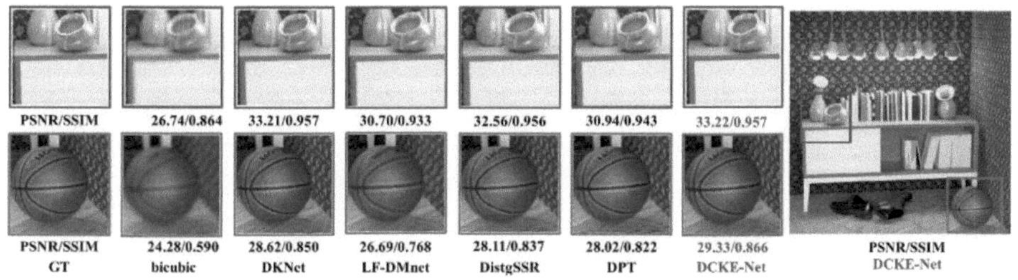

Fig. 6. Qualitative Results of Comparison

Figure 6 provides a detailed visual comparison of the central SAIs reconstructed by various algorithms through full-focus super-resolution. Results from the proposed method are **highlighted in red**. In the multi-defocus scenario, the foreground is configured to exhibit stronger blur than the background. As illustrated, while other methods can reconstruct the background satisfactorily, they often introduce ringing artifacts in the foreground, leading to visible distortions. In contrast, the proposed method delivers enhanced super-resolution performance for both foreground and background regions.

Compared to other algorithms, our network effectively performs SR on objects at different depths and yields superior outcomes. Other methods, by comparison, are only capable of enhancing objects at a single depth level and fail to adequately handle images with multiple depth layers. This demonstrates the superiority of the proposed network in processing light field images and its effective full-focus super-resolution capability.

5.3 Ablation Experiment

This section conducts a series of ablation studies on specific components of DCKE-Net to evaluate the impact of each structural module on the network's performance. Both quantitative analysis and visual comparisons are presented using 2 × super-resolution as an example. Table 2 summarizes the quantitative results, where the symbols "✓" and "✗" indicate whether a module is included or removed, respectively. Three distinct ablation experiments are subsequently introduced, and the corresponding results in the table are analyzed.

Table 2. Quantitative Results of Ablation Experiments

InterBlock	winTrans		SV Kernel Matrix Estimation	PSNR (dB)			SSIM		
				Test1	Test2	Test3	Test1	Test2	Test3
①				33.05	31.80	35.01	0.952	0.929	0.951
②				31.83	31.24	34.37	0.942	0.920	0.945
③				32.29	31.77	34.94	0.945	0.929	0.951
				34.99	33.17	36.53	0.966	0.944	0.965

Figure 7 presents a visual quality assessment of the ablation studies conducted on DCKE-Net. The first column on the left displays the super-resolution reconstruction results of the full DCKE-Net, while the columns on the right show locally magnified reconstruction results at different depths, along with comparisons to corresponding results from various ablated configurations. Visually, the local reconstructions produced by DCKE-Net exhibit strong consistency across different depth levels, confirming the network's capability to achieve effective full-focus super-resolution reconstruction when processing multi-depth light field images. These results not only demonstrate the superiority of DCKE-Net in super-resolution reconstruction but also validate its ability to maintain consistency and accuracy in complex multi-depth light field scenarios.

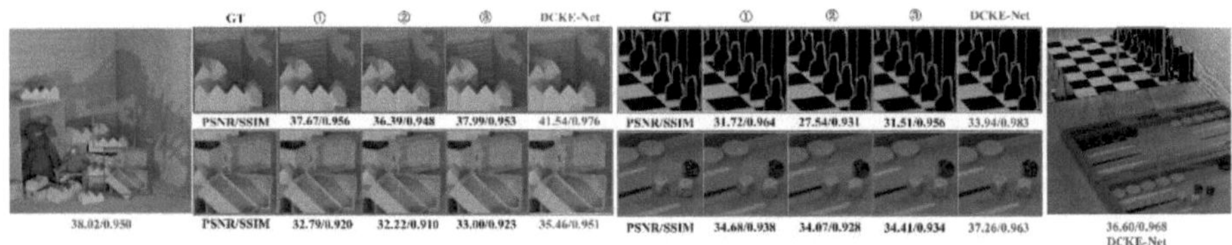

Fig. 7. Qualitative Results of Ablation Experiments

Impact of the Feature Extraction-Based U-net Structure. The InterBlock was removed from the network, reverting to a conventional U-net structure for the ablation experiment. The InterBlock is designed to facilitate information flow between different network layers and enhance inter-layer feature fusion, thereby improving super-resolution performance. In the absence of this block, a decline in network performance was observed. As shown in the first row of Table 2, the average PSNR decreased by 1.61 dB, and the average SSIM dropped by 0.014 compared to the results of the full network (third row).

Impact of the Window Transformer. To evaluate the effect of the improved winTrans, it was replaced with a sliding window transformer (Swin Transformer). The winTrans, based on a Transformer architecture and employing a parallel structure, is designed to effectively capture both spatial and angular information in light field images, which is essential for performing efficient super-resolution transformations. After replacing winTrans with the Swin Transformer, the network lost its ability to extract angular features, resulting in a noticeable reduction in the accuracy of detail reconstruction during HR image recovery.

Impact of the Spatially Variant Kernel Matrix Estimation Module. The spatially variant kernel matrix estimation module utilizes depth masks as constraints to handle multi-depth light field images. A third ablation study was conducted by removing the depth constraint from the kernel estimation module and replacing the multi-depth kernel matrix with a uniform blur kernel during training, in order to verify the influence of depth-aware constraints on network performance. Both the quantitative results in the third row of Table 2 and the visual comparisons in Fig. 7 demonstrate degraded performance when depth constraints are omitted, confirming the necessity of incorporating depth constraints in the full-focus super-resolution network.

6 Conclusion

Based on depth-guided spatially variant kernel estimation, we propose an end-to-end full-focus blind light field super-resolution network named DCKE-Net. Unlike traditional non-blind super-resolution methods, DCKE-Net incorporates a depth-constrained kernel estimation module that adaptively estimates blur kernels at different depths, enabling effective handling of complex blur types and supporting end-to-end training for multi-depth scenarios. Additionally, the network integrates a window-based transformer and an enhanced U-Net feature interaction module, which collectively improve the capture of both local details and global context, further boosting reconstruction accuracy.

Through comprehensive simulation and ablation studies, DCKE-Net demonstrates superior performance in detail recovery and deblurring. The ablation experiments confirm that both the feature interaction block and the window transformer are essential to the network's performance—their removal leads to significant degradation. In summary, by combining depth-aware kernel estimation with an advanced network architecture, DCKE-Net offers an efficient and accurate solution for full-focus light field super-resolution, particularly in multi-depth environments, thereby facilitating broader application of light field imaging technology.

Disclosure of Interests.. The authors have no competing interests to declare that are relevant to the content of this article.

References

1. Adelson, E.H., Wang, J.Y.A.: Single lens stereo with a plenoptic camera. IEEE Trans. Pattern Anal. Mach. Intell. **14**(2), 99–106 (1992)

2. Dayan, S.B., Mendlovic, D., Giryes, R.: Deep sparse light field refocusing. arXiv:2009.02582 [cs.CV] (2020)
3. Lee, B., Park, J.-H., Min, S.-W.: Three-dimensional display and information processing based on integral imaging. In: Javidi, B., Okano, F. (eds.) Digital Holography and Three-Dimensional Display: Principles and Applications, pp. 333–378. Springer, Boston (2006)
4. Jeon, H.-G., et al.: Accurate depth map estimation from a lenslet light field camera. In: Proc. IEEE Conference on Computer Vision and Pattern Recognition (CVPR), pp. 1547–1555 (2015)
5. Yu, L., Ma, Y., Hong, S., et al.: Review of light field image super-resolution. Electronics **11**(12), 1904 (2022)
6. Xiong, Y., Wang, A., Zhang, K.: Review of light field super-resolution algorithm based on deep learning. Laser Optoelectron. Prog. **61**(18), 1800003 (2024)
7. Yoon, Y., et al.: Learning a deep convolutional network for light-field image super-resolution. In: Proceedings of the IEEE International Conference on Computer Vision Workshops (ICCVW), pp. 24–32 (2015)
8. Gul, M.S.K., Gunturk, B.K.: Spatial and angular resolution enhancement of light fields using convolutional neural networks. IEEE Trans. Image Process. **27**(5), 2146–2159 (2018)
9. Fan, H., et al.: Two-stage convolutional neural network for light field super-resolution. In: Proceedings of the IEEE International Conference on Image Processing (ICIP), pp. 1167–1171 (2017)
10. Liang, D., Zhang, H., Qiu, J.: Self-supervised learning for spatial-domain light-field super-resolution imaging. Laser Optoelectron. Prog. **61**(4), 0411007 (2024)
11. Wang, Y., et al.: Spatial-angular interaction for light-field image super-resolution. In: Vedaldi, A., Bischof, H., Brox, T., Frahm, J.-M. (eds.) ECCV 2020. LNCS, vol. 12357, pp. 290–308. Springer, Cham (2020)
12. Ko, K., et al.: Light-field super-resolution via adaptive feature remixing. IEEE Trans. Image Process. **30**, 4114–4128 (2021)
13. Jin, J., et al.: Light-field spatial super-resolution via deep combinatorial geometry embedding and structural consistency regularization. In: Proceedings of the IEEE/CVF Conference on Computer Vision and Pattern Recognition (CVPR), pp. 2260–2269 (2020)
14. Wang, Y., et al.: Disentangling light fields for super-resolution and disparity estimation. IEEE Trans. Pattern Anal. Mach. Intell. **45**(1), 425–443 (2022)
15. Zhang, K., Zuo, W., Zhang, L.: Learning a single convolutional super-resolution network for multiple degradations. In: Proceedings of the IEEE Conference on Computer Vision and Pattern Recognition (CVPR), pp. 3262–3271 (2018)
16. Xu, Y.-S., et al.: Unified dynamic convolutional network for super-resolution with variational degradations. In: Proceedings of the IEEE/CVF Conference on Computer Vision and Pattern Recognition (CVPR), pp. 12496–12505 (2020)
17. Gu, J., et al.: Blind super-resolution with iterative kernel correction. In: Proceedings of the IEEE/CVF Conference on Computer Vision and Pattern Recognition (CVPR), pp. 1604–1613 (2019)
18. Huang, Y., et al.: Unfolding the alternating optimization for blind super-resolution. In: Advances in Neural Information Processing Systems (NeurIPS), vol. 33, pp. 5632–5643 (2020)
19. Yuan, Y., et al.: Unsupervised image super-resolution using cycle-in-cycle generative adversarial networks. In: Proceedings of the IEEE Conference on Computer Vision and Pattern Recognition Workshops (CVPRW), pp. 701–710 (2018)
20. Bulat, A., Yang, J., Tzimiropoulos, G.: To learn image super-resolution, use a GAN to learn how to do image degradation first. In: Ferrari, V., Hebert, M., Sminchisescu, C., Weiss, Y. (eds.) ECCV 2018. LNCS, vol. 11209, pp. 185–200. Springer, Cham (2018)

21. Liu, A., et al.: Blind image super-resolution: a survey and beyond. IEEE Trans. Pattern Anal. Mach. Intell. **45**(5), 5461–5480 (2022)
22. Honauer, K., et al.: A dataset and evaluation methodology for depth estimation on 4D light fields. In: Lai, S.-H., Lepetit, V., Nishino, K., Sato, Y. (eds.) ACCV 2016. LNCS, vol. 10112, pp. 19–34. Springer, Cham (2017)
23. Wanner, S., Meister, S., Goldluecke, B.: Datasets and benchmarks for densely sampled 4D light fields. In: Proceedings of Vision, Modeling and Visualization (VMV), pp. 225–226 (2013)
24. Piao, Y., et al.: DUT-LFSaliency: Versatile dataset and light field-to-RGB saliency detection. arXiv preprint arXiv:2012.15124 (2020)
25. Ronneberger, O., Fischer, P., Brox, T.: U-net: Convolutional networks for biomedical image segmentation. In: Navab, N., Hornegger, J., Wells, W.M., Frangi, A.F. (eds.) MICCAI 2015. LNCS, vol. 9351, pp. 234–241. Springer, Cham (2015)
26. Kong, D., et al.: Depth-guided full-focus super-resolution network for light field images. In: Proceedings of the 7th International Conference on Image and Graphics Processing (ICIGP 2024), pp. 260–266. ACM, New York, NY, USA (2024)
27. Liu, Z., et al.: Swin transformer: hierarchical vision transformer using shifted windows. In: Proceedings of the IEEE/CVF International Conference on Computer Vision (ICCV), pp. 10012–10022 (2021)
28. Hu, Z., et al.: Texture-enhanced light-field super-resolution with spatio-angular decomposition kernels. IEEE Trans. Instrum. Meas. **71**, 1–16 (2022)
29. Wang, Y., et al.: Real-world light-field image super-resolution via degradation modulation. arXiv preprint arXiv:2206.06214 (2023)
30. Wang, S., et al.: Detail-preserving transformer for light-field image super-resolution. In: Proceedings of the AAAI Conference on Artificial Intelligence, vol. 36(3), pp. 2522–2530 (2022)

Image Enhancement and Image Segmentation

A Novel Deep Learning-Based Approach for Facial Part Segmentation from 3D Point Cloud Data

Narumi Kihara[1], Namiko Kimura-Nomoto[2], Takako Okawachi[2,3], Guangxu Li[4], Norifumi Nakamura[2,5], and Tohru Kamiya[1](✉)

[1] Kyushu Institute of Technology, 1-1, Sensui, Tobata, Kitakyushu, Japan
kihara.narumi101@mail.kyutech.jp, kamiya@cntl.kyutech.ac.jp
[2] Kagoshima University, 8-35-1, Sakuragaoka, Kagoshima, Japan
[3] National Hospital Organization Kagoshima Medical Center, 8-1, Shiroyama, Kagoshima, Japan
[4] Tiangong University, 399, BinShuiXi Road, Tianjin, XiQing District, China
[5] Universitas Indonesia, Jawa Barat 16424, Indonesia

Abstract. Cleft lips are one of the most common birth defects worldwide, and a quantitative method for analyzing facial symmetry is required to enable more effective surgical planning. To realize a detailed analysis of the 3D facial data, facial part segmentation from the 3D point cloud plays an important role, however this area has not been explored yet. In this paper, we introduce a new deep learning-based approach for facial part segmentation from 3D point cloud data. Our main contributions are as follows: (1) we present a new approach for creating a large-scale synthetic dataset of facial 3D point clouds, and (2) we test the performance of a deep learning model using the synthetic dataset and our private dataset. The experimental results demonstrate the great potential of deep learning for understanding 3D point cloud data of the face.

Keywords: Cleft lip · 3D point cloud · Facial part segmentation · Synthetic dataset · Deep learning

1 Introduction

Cleft lips are one of the most common birth defects worldwide [1]. It causes a lot of problems such as aesthetic, speech, and psychological disorders. In particular, the deformity of cleft lip has a profound psychological impact on cleft lip patients [2]. Furthermore, according to M.S. Yusof et al. [3], many studies that met their criteria of the review reported that the cleft lip condition has a negative impact on patients' QOL(Quality Of Life). Therefore, forming the symmetrical and natural lip and nose is one of the key factors to improve their QOL. However, it is concerned that the surgical policy depends on the subjective opinion of the surgeon because the criterion of the facial symmetry is not clear. To address this issue, a quantitative indicator of facial symmetry is required.

P. Umapada et al. (Eds.): ICCPR 2025, CCIS 2811, pp. 137–145, 2026.
https://doi.org/10.1007/978-981-95-8315-7_11

In previous studies, it is common to use 3D data to analyze facial symmetry. For example, D. Hosoki et al. [4] and D. Al-Rudainy et al. [5] proposed methods to analyze the left-right difference of 3D facial data using ICP (Iterative Closest Point) algorithm. Although these methods are useful to quantify the facial symmetry, they cannot evaluate the asymmetrical parts that remain after the first surgery. We hypothesize that this is due to the lack of information used for the analysis; a single 3D image is just the moment of human expressions. This observations motivates the use of 4D data, 3D point clouds that contain temporal changes. The 4D data enables us to analyze faces under more natural conditions such as speaking and smiling. To take the advantage of the 4D data, facial part segmentation for each 3D point cloud plays an important role. Therefore, in this paper, we propose a new approach for facial part segmentation from 3D point cloud data.

The rest of this paper is organized as follows. Section 2 presents related work from two perspectives: facial part segmentation and point cloud segmentation. Section 3 describes our new approach for facial part segmentation. This framework consists of three steps: generating a large-scale synthetic dataset, pre-training on this dataset, and fine-tuning on our small-scale private dataset. Section 4 presents the experimental results for each dataset, and finally Sect. 5 concludes the paper with a summary.

2 Related Work

2.1 Facial Part Segmentation and Dataset

Facial part segmentation is a key process in many face analysis tasks such as facial expression recognition, head pose estimation, and facial landmark detection [6]. To the best of our knowledge, all existing research works on 2D images for facial part segmentation, and no method has been proposed using the 3D data. The reason for this is simple; annotating in 3D space is very challenging, time-consuming, and labor-intensive.

Regarding the 2D facial image, some datasets provide not only large number of images but also detailed annotations of facial parts. For example, HELEN [7] is one of the most commonly used datasets. It contains 2,330 images and is manually annotated with 11 classes (e.g., “Face”, “Nose”, “Upper lip”, and “Lower lip”). State-of-the-art methods for facial part segmentation working on HELEN dataset can be seen in [8, 9]. In terms of 3D space, numerous 3D facial datasets have been made publicly available in recent years, including FaceWarehouse [10], KinectFaceDB [11], and FaceScape [12], to name a few. These datasets provide 3D facial data with various facial expressions and additional information such as face landmark position. However, unlike HELEN, they do not contain the ground truth of facial parts and thus cannot be used to train a deep learning model in a supervised manner.

Although we do not have any raw 3D datasets for facial part segmentation, methods have been proposed to reconstruct the facial 3D structure from a single 2D image. Furthermore, some of these methods also generate pseudo-labels of facial parts [13, 14], which can be used to train a model for 3D segmentation. In this paper, we use the more recent one named 3DDFA-V3 (3D Dence Face Alignment V3) [15] because it provides more detailed pseudo-labels than existing methods [13, 14].

2.2 Point Cloud Segmentation and Dataset

Point cloud data is commonly used to describe the surfaces of objects and environments in a scene, and it has many applications, including object identification, classification, and obstacle detection in robotics and autonomous vehicles [16]. The most commonly used datasets for point cloud segmentation are ScanNet [17], S3DIS [18] and SemanticKITTI [19]. ScanNet and S3DIS consist of indoor scenes, while SemanticKITTI is a large-scale outdoor dataset. Judging from the aforementioned fields of application and public datasets for 3D point cloud segmentation, existing methods focus on indoor and outdoor scenarios, and their applicability to facial 3D point clouds has not been explored yet. In this paper, we use PTv3 (Point Transformer V3) [20] because of its memory efficiency and high performance. We then test its segmentation accuracy on datasets of facial 3D point clouds.

3 Facial Part Segmentation from 3D Point Cloud

3.1 Synthetic Dataset

As mentioned in Sect. 2, although there is no dataset of raw 3D faces with facial part annotations, methods for 3D face reconstruction [13–15] can help to train a 3D segmentation model because these methods generate a 3D face and corresponding pseudo-label from a single 2D image. In this paper, we use 3DDFA-V3 (3D Dense Face Alignment V3) [15] to generate a large-scale synthetic dataset. It is a state-of-the-art method in the field of 3D face reconstruction, and most importantly, it provides more detailed pseudo-labels of facial parts than previous methods [13, 14]. Each point in a generated 3D face data is labeled with one of eight classes: "Face", "Nose", "Left eyebrow", "Right eyebrow", "Left eye", "Right eye", "Upper lip" and "Lower lip".

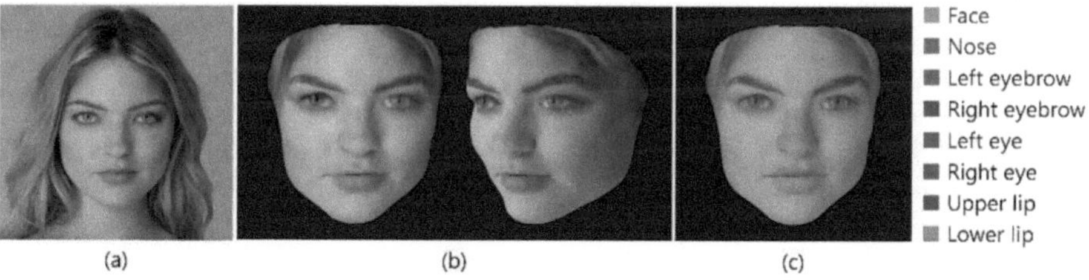

Fig. 1. 3D reconstruction. Best viewed in color. (a) is a 2D image from CelebA dataset. (b) is a 3D point cloud made by 3DDFA-V3. (c) shows pseudo-labels of facial parts.

Following 3DDFA-V3, we collected CelebA-HQ dataset [21, 22] and used the first 24,000 images for this study. We then applied 3DDFA-V3 to the images to generate 3D faces and pseudo-labels (Fig. 1). The generated 3D data is in the ".obj" format, which includes the colors, vertices, and connections between vertices. As our private dataset is in the ".pcd" format, we use the object format as point cloud data by omitting the connections. As a result, we obtain facial 3D point clouds with detailed pseudo-labels. In this paper, we refer to this synthetic dataset as CelebA-3D.

It is important to note that we have skipped image number 7863 of CelebA-HQ. We are considering using additional facial expressions provided by Z. Wang et al. [15] in our future work, however we have found that the synthesized expression of the image number 7863 is currently problematic. This may be updated in the future, so please refer to their latest information.

3.2 Pre-Training

We randomly shuffled the order of data in CelebA-3D, and used the first 4,800 point clouds for testing and other 19,200 point clouds for training.

To compute the segmentation of facial parts, we use PTv3 (Point Transformer V3) [20]. This model requires less memory during training and outperforms previous methods in both indoor and outdoor scenarios. Most of the hyperparameters follow the setting of PTv3 for S3DIS dataset. We set the "batch size" to 16, "num worker" to 2, and "epoch" to 5. Regarding the number of epochs, we select this small value for two main reasons. First, we have found that 5 epochs are sufficient for PTv3 to learn "what is facial part segmentation" from the dataset. Second, we aim to minimize the risk of overfitting pseudo-labels.

3.3 Fine-Tuning

Labelling. Since PTv3 has learnt 8-class facial segmentation, we also need the pseudo-labels for fine-tuning and use 3DDFA-V3 again. However, this workflow is more complicated than making the CelebA-3D because our data is in the point cloud format.

Algorithm 1 shows the pseudocode to obtain pseudo-labels for our dataset. Firstly, a 3D point cloud of the face is converted to a 2D image by making color and real coordinate arrays. In detail, each point is mapped to the nearest pixel in the xy plane, and the RGB values and coordinates in the point cloud are retained in these arrays. If more than two points are assigned to the same pixel, the point with the largest z-coordinate is selected because the other points are not visible due to the occlusion.

Secondly, 3DDFA-V3 is applied to the image to get pseudo-labels as a "2D array". Thirdly, a sparse point cloud is reconstructed using 2D arrays of real coordinates and pseudo-labels. Fourthly, the pseudo-labels of the sparse point cloud are assigned to the original point cloud data. Finally, and most importantly, the pseudo-labels of "Upper lip" and "Lower lip" are replaced with the manually created ground truth under the guidance of doctors. As a result, we obtain our small-scale private dataset with manual annotations ("Upper lip" and "Lower lip") and detailed pseudo-labels (the other classes).

Algorithm 1. Python-like pseudocode for labelling a facial 3D point cloud using 3DDFA-V3

```
cloud = load_from_file(filepath)
label = empty array with the same size as cloud
image, real_coordinate = convert_cloud_to_image(cloud)
pseudo_label_2d = 3DDFA-V3(image)
cloud_reconstruct = empty array # "sparse point cloud"
for y in range(image height):
  for x in range(image width):
    if real_coordinate[y, x] is empty:
      continue
    coordinate = real_coordinate[y, x]
    color = pseudo_label_2d[y, x] # save label as color
    data = concatenate(coordinate, color)
    cloud_reconstruct.append(data)
radius = 0.01 * (diagonal of cloud)
tree = kd-tree made from cloud_reconstruct
for index, point in enumerate(cloud):
  nn_indices = tree.search_neighbor(point, radius)
  label_counter = neighbors' labels using
cloud_reconstruct and nn_indices
  label[index] = most frequent label in label_counter
  if the same frequency exists:
    label[index] is randomly selected
 label = replace_label(label) # apply ground truth
```

Implementation Detail. We fine-tune the pre-trained model on our dataset. The point clouds were captured at Kagoshima University using VECTRA®H1 manufactured by Canfield Scientific. Participants were asked to sit on a chair and relax their facial expression (Table 1(a)). To analyze the temporal change of facial expression, some participants were also asked to open their mouth and keep it while the point cloud was captured (Table 1(b) and (c)). In our dataset, all participants with cleft lips have already undergone the first operation.

Our private dataset totally consists of 40 point clouds. Using this dataset, we test the performance of PTv3 by irregular 7-fold cross-validation. In the normal n-fold cross-validation, a dataset is divided equally into n sets, one of which is used for testing and the others for training. The overall result is obtained by averaging the results of n iterative experiments, ensuring that each set is once used for testing. On the other hand, in our irregular 7-fold cross-validation, we select one person from the participants of Table 1(b) and (c) to be used as test data, while the remaining participants are used for training. Each participant in Table 1(b) and (c) is selected for testing in the iterative experiments, ensuring that those already selected are not selected again. We then get the overall score by averaging the results of 7 experiments. We take this strategy not to use the same person for training and testing at the same time. As shown in Table 1, we have 3 point clouds per participant in case of Table 1(b), and 4 point clouds in case of Table 1(c). If the normal cross-validation is applied to our experiment, the same person can be assigned to both the training and test sets, which is not ideal to evaluate the model performance. Therefore, we divide the dataset into training and test sets at person level. In addition, we note that the participants in Table 1(a) are only used for training because testing relaxed expression alone is less meaningful than testing multiple expressions.

In the fine-tuning stage, most of the hyperparameters follow the pre-training setting. We set the "batch size" to 4 and "epoch" to 200. Both the pre-training and fine-tuning are carried out on a single GPU (NVIDIA GeForce RTX 3060 Ti).

Table 1. Detail of our private dataset

	Cleft lip	Participants	Point clouds per person
(a)	w	15	1
(b)	w	3	3
(c)	w/o	4	4

4 Experimental Result and Discussion

We used IoU (Intersection over Union) to evaluate the segmentation performance. Let *GT* be the region of ground truth and *Pred* be the region of model's prediction, IoU is calculated using the following formula. The value of IoU ranges from 0 to 1, and the larger it is, the more accurate the segmentation.

$$IoU = \frac{GT \cap Pred}{GT \cup Pred} \tag{1}$$

The experimental results on the CelebA-3D and our private dataset are shown in Table 2. PTv3 was significantly accurate at segmenting the facial parts on the synthetic CelebA-3D dataset, and also obtained favorable results on our small-scale dataset. Figure 2 shows the segmentation results, and we can see that PTv3 had great potential to learn how to segment the facial parts, even in the challenging data (Fig. 2 bottom row).

Furthermore, we can see its potential from another perspective. According to X. Wu et al. [20], in terms of the IoU metric, PTv3 achieved 78.6% on ScanNet [17], 80.8% on S3DIS 6-fold [18], and 71.3% on SemanticKITTI [19]. When compared with these results obtained in indoor and outdoor scenarios, PTv3 demonstrated competitive performance on facial datasets. These results encourage further research to comprehend 3D point cloud data of the face using PTv3.

Table 2. Results on CelebA-3D and our private datasets (Metric: IoU)

	CelebA-3D	Private dataset
Face	0.9833	0.9650
Nose	0.9489	0.8792
Left eyebrow	0.8479	0.6373
Right eyebrow	0.8355	0.6617
Left eye	0.8788	0.7769
Right eye	0.8929	0.7494
Upper lip	0.9144	0.6712
Lower lip	0.8652	0.6575
Average	0.8965	0.7498

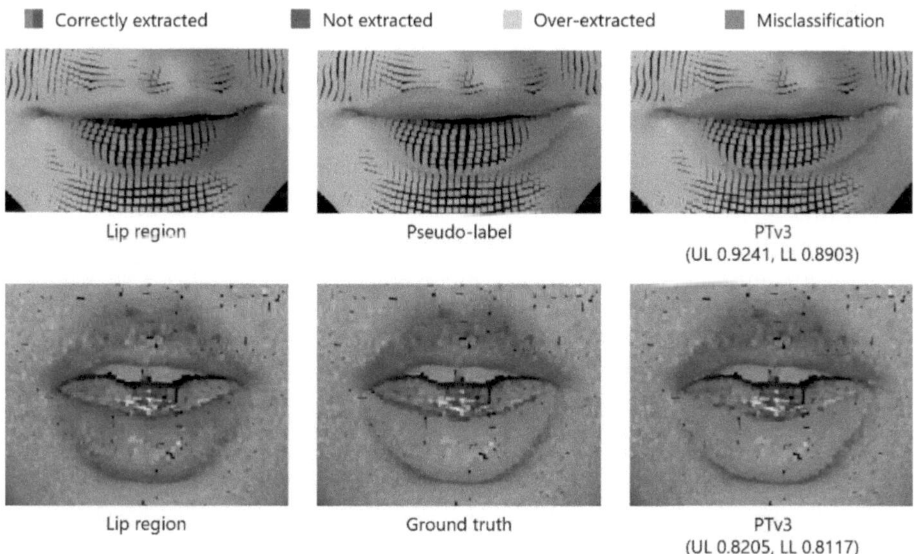

Fig. 2. Segmentation examples of the lip region. Best viewed in color. The data of the top and bottom rows comes from CelebA-3D and our own dataset, respectively. UL represents Upper Lip, and LL represents Lower Lip.

5 Conclusion

In this paper, we presented a new approach for deep learning-based segmentation of facial parts from 3D point cloud data, an area that has not been explored yet. To overcome the limitation of facial 3D point clouds with detailed annotations, we employed a novel method of 3D face reconstruction named 3DDFA-V3 (3D Dense Face Alignment V3) and created a large-scale synthetic dataset. We then trained an efficient, high performance model named PTv3 (Point Transformer V3) in two steps: pre-training on the synthetic dataset and fine-tuning on our small-scale private dataset. The experimental results showed the great potential of PTv3 to comprehend 3D point cloud data of the face.

In the future, we will test the performance of other deep learning methods for facial part segmentation from 3D point cloud data. In addition, we need to improve the segmentation accuracy to make the face analysis more accurate and reliable. Possible solutions include modifying the model architecture and designing a loss function.

Acknowledgement. This work was supported by JST SPRING, Japan Grant Number JPMJSP2154.

References

1. National Clearinghouse for Birth Defects Surveillance and Research, http://www.icbdsr.org/wp-8content/annual_report/Report2014.pdf. Last accessed 2 May 2025
2. Shaw, W.C.: Folklore surrounding facial deformity and the origins of facial prejudice. Br. J. Plast. Surg. **34**, 237–246 (1981)
3. Yusof, M.S., Ibrahim, H.M.: The impact of cleft lip and palate on the quality of life of young children: a scoping review. Med. J. Malaysia **78**(2), 250–258 (2023)
4. Hosoki, D., et al.: Symmetric plane detection and symmetry analysis from a 3D point cloud data of face. In: International Conference on Control, Automation and Systems, pp. 402–406 (2020)
5. Al-Rudainy, D., Ju, X., Stanton, S., Mehendale, F.V., Ayoub, A.: Assessment of regional asymmetry of the face before and after surgical correction of unilateral cleft lip. Journal of Cranio-Maxillofacial Surgery **46**, 974–978 (2018)
6. Khan, K., Khan, R.U., Ahmad, K., Ali, F., Kwak, K.: Face segmentation: a journey from classical to deep learning paradigm, approaches, trends, and directions. IEEE Access **8**, 58683–58699 (2020)
7. Smith, B.M., Zhang, L., Brandt, J., Lin, Z., Yang, J.: Exemplar-based Face Parsing. In: IEEE Conference on Computer Vision and Pattern Recognition, pp. 3484–3491 (2013)
8. Khan, K., Attique, M., Khan, R.U., Syed, I., Chung, T.: A multi-task framework for facial attributes classification through end-to-end face parsing and deep convolutional neural networks. Sensors **20**(2), 328 (2020)
9. Lin, Y., Shen, J., Wang, Y., Pantic, M.: RoI tanh-polar transformer network for face parsing in the wild. Image Vis. Comput. **112**, 104190 (2021)
10. Chen, C., Weng, Y., Zhou, S., Tong, Y., Zhou, K.: FaceWarehouse: A 3D facial expression database for visual computing. IEEE Trans. Visual Comput. Graphics **20**(3), 413–425 (2014)
11. Min, R., Kose, N., Dugelay, J.: KinectFaceDB: A Kinect Database for Face Recognition. IEEE Transactions on Systems, Man, and Cybernetics: Systems **44**(11), 1534–1548 (2014)
12. Zhu, H., et al.: FaceScape: 3D Facial Dataset and Benchmark for Single-view 3D Face Reconstruction. IEEE Conference on Computer Vision and Pattern Recognition **45**(12), 14528–14545 (2020)
13. Li, T., Bolkart, T., Black, M.J., Li, H., Romero, J.: Learning a model of facial shape and expression from 4D scans. ACM Transactions on Graphics **36**(6), 194:1–194:17 (2017)

14. Graphics University of Basel and Vision Research, https://github.com/unibas-gravis/parametric-face-image-generator. Last accessed 14 May 2025
15. Wang, Z., Zhu, X., Zhang, T., Wang, B., Lei, Z.: 3D face reconstruction with the geometric guidance of facial part segmentation. In: Computer Vision and Pattern Recognition, pp. 1672–1682 (2024)
16. He, Y., et al.: Deep learning based 3D segmentation in computer vision: a survey. Information Fusion **115**, 102722 (2025)
17. Dai, A., et al.: ScanNet: Richly-annotated 3D Reconstructions of Indoor Scenes. In: Computer Vision and Pattern Recognition, pp. 2432–2443 (2017)
18. Armeni, I., et al.: 3D semantic parsing of large-scale indoor spaces. In: Computer Vision and Pattern Recognition, pp. 1534–1543 (2016)
19. Behley, J., et al.: SemanticKITTI: A Dataset for Semantic Scene Understanding of LiDAR Sequences. In: International Conference on Computer Vision, pp. 9297–9307 (2019)
20. Wu, X., et al.: Point Transformer V3: Simpler, Faster, Stronger. In: Computer Vision and Pattern Recognition, pp. 4840–4851 (2024)
21. Liu, Z., Luo, P., Wang, X., Tang, X.: Deep Learning Face Attributes in the Wild. In: International Conference on Computer Vision, pp. 3730–3738 (2015)
22. Karras, T., Aila, T., Laine, S., Lehtinen, J.: Progressive Growing of GANs for Improved Quality, Stability, and Variation. arXiv preprint arXiv:1710.10196 (2018). Last accessed 29 July 2025

A Dual-Encoder Hybrid CNN-Mamba Network for Breast Tumor Segmentation

Pengju Shao[1], Jie De[2], Ting Hu[1], Yingnan Dang[1], Mengfan Geng[1], Zhe Li[1], Zhonghua Sun[1], Kebin Jia[1], and Jinchao Feng[1(✉)]

[1] School of Information Science and Technology, Beijing University of Technology, Beijing, China
fengjc@bjut.edu.cn

[2] Beijing Geriatric Hospital, Beijing, China

Abstract. Breast tumor segmentation using dynamic contrast-enhanced magnetic resonance imaging (DCE-MRI) holds great potential for clinical application. In particular, pre-/post-contrast images and subtraction images emphasize complementary aspects of tumor characteristics, with the former capturing global information such as location and size, and the latter highlighting local details such as boundary delineation. Motivated by this observation, we propose a novel Dual-Encoder Hybrid CNN-Mamba Network (DEHN), which leverages the complementary features of pre- and post-contrast images and their associated subtraction counterparts. The encoder consists of two parallel branches: a Local-Focused Encoder and a Global-Focused Encoder. The Local-Focused Encoder, constructed with convolutional blocks, extracts detailed local features from both pre- and post-contrast images. In parallel, the Global-Focused Encoder integrates convolutional layers with Mamba blocks to capture high-level global contextual features from the subtraction images. The decoder processes the fused local and global features to generate the final tumor segmentation. We evaluated the proposed DEHN on 1,606 cases from two public DCE-MRI datasets (Yunnan and MAMA-MIA). The proposed DEHN consistently outperformed baseline and seven state-of-the-art models across key metrics, including Dice coefficient, precision, and recall. Specifically, DEHN achieved Dice scores of 81.52% on the Yunnan dataset and 75.93% on the MAMA-MIA dataset.

Keywords: Breast tumor segmentation · DCE-MRI · Convolutional neural networks · Mamba · Hybrid network

1 Introduction

Breast cancer remains a major cause of cancer-related mortality among women, where early detection and timely treatment significantly improve patient outcomes [1]. Among imaging modalities, dynamic contrast-enhanced MRI (DCE-MRI) offers superior sensitivity for tumor detection [2, 3], but accurate interpretation is highly dependent on radiologist expertise and is prone to subjectivity and delays. This has motivated the development of automatic segmentation algorithms to enhance diagnostic efficiency.

P. Shao and J. De—These authors contributed equally.

P. Umapada et al. (Eds.): ICCPR 2025, CCIS 2811, pp. 146–158, 2026.
https://doi.org/10.1007/978-981-95-8315-7_12

Despite the success of convolutional neural networks (CNNs), particularly U-Net [4], in biomedical image segmentation [5, 6], their limited receptive fields restrict modeling of long-range dependencies. Vision Transformers (ViTs) [7] address this limitation through self-attention mechanisms, enabling global context modeling, though they often fail to capture fine-grained local features [8, 9]. Recent work combining CNN-based (e.g., nnU-Net [10]) and Transformer-based models (e.g., nnFormer [11]) highlights their complementary strengths. Hybrid architectures [12] aim to integrate both types of features but often suffer from high computational cost. Mamba [13], a recent state-space model, improves efficiency with linear complexity, and its application in U-Mamba [14] demonstrates strong performance in long-sequence modeling. For 3D medical image segmentation, Li et al. [15] introduced the MISM module, which combines CNNs with Mamba by decomposing 3D volumes into three orthogonal 2D views, allowing for comprehensive multi-view spatial correlation analysis.

In breast tumor segmentation, DCE-MRI provides spatiotemporal priors, with subtraction images—derived from pre- and post-contrast differences—offering enhanced tumor-specific contrast while suppressing background noise. Due to the distinct feature distributions of pre-contrast, post-contrast, and subtraction MRI sequences, simply aggregating them within a single encoder is insufficient for capturing their complementary information. To address this, we propose a hybrid encoder-decoder network with two dedicated branches: a Local-Focused Encoder that extracts spatial features from pre- and post-contrast images, and a Global-Focused Encoder that leverages a Multi-View Mamba (MVM) module to capture global contextual features from subtraction images. This design is motivated by the distinct feature representations observed among the different image types. It enables the network to extract both explicit visual cues (e.g., lesion size, shape) and implicit hemodynamic information associated with tumor vascularization. Guided by a lightweight design philosophy, our model takes pre-contrast images, post-contrast images, and their corresponding subtraction images as input, ensuring efficient yet comprehensive feature learning for breast tumor segmentation.

The main contributions of our work are as follows:

- We propose a dual-branch encoder architecture to independently process pre-/post-contrast images and subtraction images, accounting for their distinct feature characteristics.
- We introduce a Multi-View Mamba (MVM) module that combines the local feature extraction capability of CNNs with the global modeling strength and computational efficiency of Mamba.
- We evaluate our method on two public breast DCE-MRI datasets, and demonstrate that our approach outperforms existing state-of-the-art segmentation methods in terms of accuracy and robustness.

2 Methods

Let $Ds = \{(X_L n, Yn)\}^{N}_{n=1}$ denote the training dataset consisting of N pairs of input patches $\{X_L n\}$ and their corresponding annotations $\{Yn\}$, where $n = \{1, \cdots, N\}$. Each input patch $X_L n$ has a size of $2 \times D \times H \times W$, and each annotation Yn has a sizef D

× H × W, where D, H, and W represent the depth, height, and width of the input patch, respectively.

$$X_G n = X_{post} n - X_{pre} n \in R^{D\times H\times W} \tag{1}$$

The architecture of the proposed DEHN is illustrated in Fig. 1. The proposed model adopts a dual-encoder architecture that explicitly incorporates DCE-MRI priors by extracting complementary features from distinct input modalities. The Local-Focused Encoder processes pre- and post-contrast images, $X_L n = \{(X_{pre} n, X_{post} n)\}$, to capture fine structural and boundary details essential for accurate tumor delineation. In parallel, the Global-Focused Encoder receives the subtraction image $X_G n$, as defined in Eq. (1), which enhances tumor-specific contrast patterns while suppressing background information. Although the subtraction may partially obscure fine details, it provides critical global context regarding tumor location and extent. To efficiently model this information, the Global-Focused Encoder is implemented as a lightweight hybrid module.

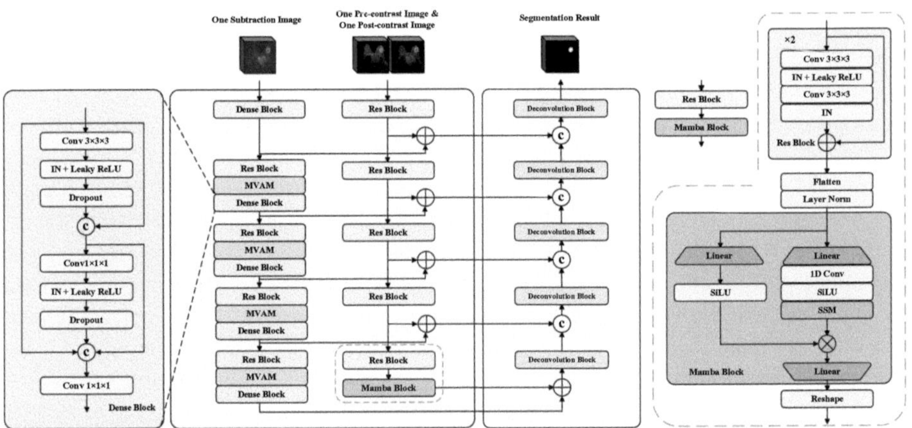

Fig. 1. Architecture of our proposed DEHN model. Both pre-contrast images and post-contrast images are fed into the Local-Focused Encoder, while subtraction images are input to the Global-Focused Encoder.

2.1 Local-Focused Encoder

To extract fine-grained structural and boundary features from pre- and post-contrast images, we design a residual convolution-based Local-Focused Encoder. These sequences contain rich local textures and ambiguous tumor boundaries with high variability in intensity and morphology. Stacked 3D residual convolutional blocks are employed to capture spatial details, with residual connections enabling deeper network construction and stable training.

To overcome the limited receptive field of CNNs, a lightweight Mamba module is introduced at the bottleneck to model long-range dependencies. As an enhanced state-space model, Mamba incorporates an input-dependent selection mechanism for efficient global context extraction in linear time. The integration of residual convolutions and Mamba allows the encoder to effectively capture both local and global features essential for accurate tumor segmentation.

As shown in Fig. 1, the Local-Focused Encoder is primarily composed of stacked convolution optimized for local feature extraction. Each convolutional module consists of two consecutive Res Blocks. Each Res Block employs standard $3 \times 3 \times 3$ convolutions, followed by Instance Normalization (IN) and a Leaky ReLU activation. The resulting bottleneck feature map is then passed through the Mamba Block, with implementation details are shown in Fig. 1.

2.2 Global-Focused Encoder

As shown in Fig. 1, Global-Focused Encoder begins with a Dense Block to expand the input channel dimension, followed by a stack of hybrid modules. Each hybrid module consists of three blocks in sequence: a Res Block, an MVM Module, and a Dense Block. Given that the input to the Global-Focused Encoder has fewer channels than the Local-Focused Encoder, this branch is deliberately designed to be lightweight to avoid redundancy while still effectively modeling long-range dependencies and preserving local feature details.

Res Block.
The Res Block is placed before the MVM module, and takes as input the intermediate feature map generated by the preceding Dense Block. While this feature map contains certain local spatial information, it lacks the semantic richness required for effective global modeling. To this end, we designed a Res Block aiming at enhancing features expressiveness and providing a stronger foundation for the subsequent long-range dependency modeling by the MVM module. To improve computational efficiency, the Res Block integrates point convolution ($1 \times 1 \times 1$) and group convolution ($3 \times 3 \times 3$). The residual connection alleviates vanishing gradient problem, facilitates deep feature propagation, and improves the stability and effectiveness of the subsequent MVM module's global modeling.

MVM Module.
To comprehensively model spatial information in subtraction images, we introduce the MVM Module with Vmamba as its core. VMamba employs a selective scanning mechanism (SS2D)[16]: a 2D image is first divided into patches, which are then unfolded along multiple directions to form different sequences. Each sequence is processed with Mamba to model spatial dependencies, enabling each pixel to aggregate long-range information from various directions. However, VMamba is inherently designed for 2D images. Directly flattening a 3D volume along a single axis can disrupt spatial coherence—voxels that are adjacent in 3D may become distant in the sequence, weakening neighborhood integrity. Considering that tumor voxels typically form compact 3D masses, we address this by designing a Multi-View Mamba Block (MVMB).

As illustrated in Fig. 2, the 3D volume is decomposed into 2D slices along the sagittal, coronal, and axial planes. Each slice set is independently processed by the MVMB to analyze spatial correlations within its respective view. Importantly, voxel pairs that are distant in one view may appear adjacent in another, thereby preserving neighborhood structures and enhancing spatial context modeling. Each MVMB applies the SS2D mechanism to its 2D slices, scanning along multiple directions to generate patch sequences and then uses SSMs to extract global tumor features in each view, as shown in Fig. 3.

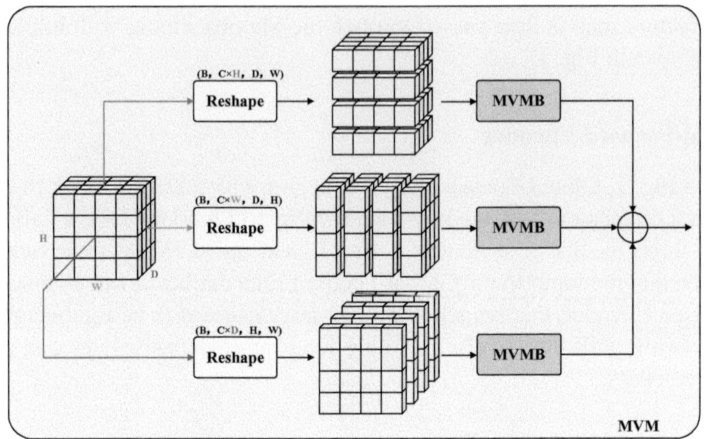

Fig. 2. Overview of the MVM module

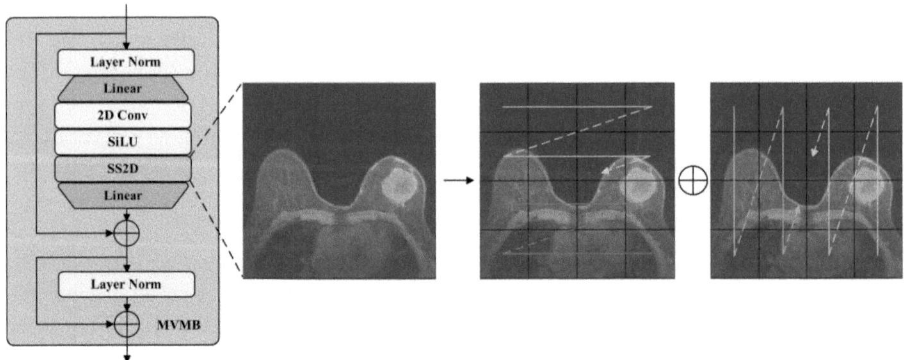

Fig. 3. Schematic diagram of the MVMB module.

Dense Block.

The feature maps generated by the MVM encapsulate comprehensive global contextual information across multiple anatomical planes. However, due to the inherent global nature of MVM, these features often lack the fine-grained local structural details necessary for precise tumor boundary delineation.

To address this limitation, we introduce a dense block as an improved module to strengthen the network's ability to represent local features. To maintain computational efficiency, the Dense Block is implemented using a lightweight design based on depthwise separable convolutions. Furthermore, the module employs dense connections to facilitate effective feature reuse and encourage deeper feature propagation. Additional components, including IN, Leaky ReLU activation, and dropout, are integrated to improve training stability and generalization capability. This design allows the network to progressively recover and integrate discriminative local features while preserving the learned global context from MVM. As a result, it enables more accurate depiction of tumor boundaries and contributes to improved segmentation performance.

2.3 Decoder

The decoder adopts a fully convolutional residual architecture that progressively restores spatial resolution while preserving semantic consistency. Each decoding stage begins with upsampling, followed by pointwise convolution to reduce channel dimensions. The resulting features are concatenated with skip connections from the corresponding encoder layers.

In the dual-encoder framework, skip connections are obtained by fusing features from both encoder branches. This strategy exploits the complementary strengths of the two encoders: the Local-Focused Encoder captures fine-grained structural details from pre- and post-contrast images, while the Global-Focused Encoder extracts semantic context from subtraction images. Element-wise addition enhances shared activations and suppresses irrelevant features, improving robustness and reducing false positives. Residual blocks within the decoder enhance feature propagation and mitigate gradient vanishing, enabling deeper refinement of integrated features. This design ensures effective integration of local and global information while maintaining computational efficiency for accurate breast tumor segmentation.

3 Experiments and Results

3.1 Dataset

Two public datasets were used to evaluate the performance of the proposed DEHN. The first public dataset is Yunnan dataset [21], which contains 100 cases collected from Yunnan Cancer Hospital. Each case also includes image in the pre-contrast phase and five post-contrast phases, along with annotations for both the breast tumor and the entire breast. For each case, the images from the pre-contrast phase and the first post-contrast phase were used. The dataset was randomly divided into a training set of 75 cases and a testing set of 25 cases.

The second dataset is the MAMA-MIA dataset [22], which comprises 1,506 multicenter DCE-MRI cases. Each case includes images in the pre-contrast phase and 2–5 post-contrast phases. The pixel spacing of dataset ranges from 0.3 (mm) to 1.4 (mm), and the slice thickness ranges from 0.8 (mm) to 4.0 (mm). Tumors were annotated by breast cancer experts with an average of nine years of experience, and all annotations were reviewed and corrected by a panel of 16 experts. In our experiments, each case used the images in the pre-contrast phase and the first post-contrast phase. The dataset was randomly divided into a training set with 1,200 cases and a test set with 306 cases.

3.2 Implementation Details

We compared our method with seven state-of-the-art (SOTA) 3D medical image segmentation models, including Attention Unet [17], nnU-Net [10], Swin UNETR [18], UNETR ++ [19], U-Mamba [14], PLHN [20], and HCMA [15]. All networks were implemented using Pytorch and trained on NVIDIA RTX 3090 GPUs. Data preprocessing and augmentation followed the nnU-Net [10]. The voxel spacing distribution of the entire dataset is then analyzed, and a task-specific target spacing is determined, typically by taking the median voxel spacing along each dimension. All images are subsequently resampled to this target spacing using third-order spline interpolation for images and nearest-neighbor interpolation for labels. To reduce computational overhead, each image is cropped to the tight bounding box of non-zero voxels, effectively removing irrelevant background regions. Intensity normalization is applied on a per-case basis, where the voxel intensities are z-score standardized to zero mean and unit variance. During training and inference stages, image patches of size $64 \times 192 \times 192$ were randomly cropped and used as input to the network. The one-cycle learning rate policy was adopted, and models were trained using the AdamW optimizer for 500 epochs. To comprehensively evaluate the proposed DEHN, four metrics including Dice, Intersection of Union (IoU), Precision, and Recall were used to measure the agreement between manually and automatic segmentations. Additionally, two distance-based metrics—95th percentile Hausdorff distance (HD95) and average surface distance (ASD) were used to quantify the surface-level discrepancies between the segmented label maps, where the unit of HD95 is voxel.

3.3 Results

Yunnan Dataset. Table 1 summarizes the segmentation performance of all models on the 25 Yunnan test cases. The proposed DEHN model achieves the highest scores across key metrics, including a Dice of 81.52% and IoU of 71.63%, outperforming all seven SOTA methods. Compared with two recently developed models PLHN and HCMA, DEHN improves the Dice score by 2.32% and 0.81%, respectively. It also achieves the highest precision (86.67%) and recall (83.78%), with recall exceeding HCMA by 2.75%, indicating enhanced tumor coverage. These results highlight the effectiveness of DEHN's dual-encoder design and input allocation strategy over existing approaches.

When compared with UNETR + +, the strongest baseline on the Yunnan dataset, DEHN achieves a notable Dice improvement of approximately 0.57%. Furthermore, DEHN reduces HD95 to 8.96, outperforming UNETR + + (10.45), and achieves the lowest ASD of 2.03, indicating superior spatial accuracy. These improvements of 1.49 in HD95 and 1.31 in ASD over UNETR + + further validate the structural advantages and segmentation precision of the DEHN model. These results confirm that the proposed DEHN architecture offers a more efficient and effective network design for the task of breast tumor segmentation.

The experimental results indicate that the proposed hybrid network, DEHN, is particularly well-suited for breast tumor segmentation, benefiting from a more effective architectural design. To rigorously assess the statistical significance of its performance improvements, we conducted Wilcoxon signed-rank tests comparing DEHN with the leading baseline, UNETR + +, using both Dice and HD95 metrics. The results show that DEHN achieves average p-values consistently below the 0.05 threshold (Dice: 0.037, HD95: 0.041), thereby confirming the statistical significance of the observed differences. Collectively, these findings substantiate that DEHN delivers a meaningful advancement over the previous state-of-the-art method in breast tumor segmentation.

Table 1. Segmentation metrics of DEHN and comparison model on the Yunnan dataset

Models	Dice↑	IoU↑	Precision↑	Recall↑	HD95↓	ASD↓
Attention Unet [17]	79.97	69.84	84.75	80.19	12.71	2.41
nnU-Net [10]	78.40	68.00	84.92	81.06	22.38	2.56
Swin UNETR [18]	79.53	69.37	84.44	82.25	11.42	2.11
UNETR + + [19]	80.95	69.94	83.77	82.79	10.45	3.34
U-Mamba [14]	80.45	69.96	82.81	83.61	10.76	2.50
PLHN [20]	79.20	68.26	85.91	79.00	9.37	3.93
HCMA [15]	80.71	70.12	85.75	81.03	9.97	2.43
DEHN	**81.52**	**71.63**	**86.67**	**83.78**	**8.96**	**2.03**

MAMA-MIA Dataset.
To further illustrate the robustness of our DEHN model, we evaluated its performance on the MAMA-MIA dataset. The segmentation results on 306 test cases are summarized in Table 2. The DEHN model achieved an average Dice score of 75.93%, outperforming the current SOTA HCMA model, which achieved 75.55%. In addition to achieving the highest Dice score, also outperforms all compared methods in terms of IoU (64.78%), Precision (76.16%), and Recall (75.25%). Although a slight decrease in performance was observed on the MAMA-MIA dataset compared to the Yunnan dataset, DEHN consistently outperformed all other models across multiple evaluation metrics. Notably, in terms of boundary accuracy, DEHN achieved the lowest HD95 (73.53) and ASD (10.31), representing reductions of 3.25 and 1.12 respectively, compared to best-performing transformer-based method, PLHN. These results further validate the robustness and generalization capability of DEHN for breast tumor segmentation.

We employed the Wilcoxon signed-rank test, as in the previous analysis, to compare the average Dice results of the proposed DEHN model with those of the lightest and strongest baseline method, HCMA. Owing to the substantially larger sample size of the MAMA-MIA dataset compared with the Yunnan dataset, the DEHN model yielded average p-values for Dice that were far below the conventional 0.05 threshold. This outcome suggests that the performance differences between DEHN and the baseline are statistically significant and consistently observed across the majority of cases.

Table 2. Segmentation performance of DEHN and comparison models on the MAMA-MIA dataset.

Models	Dice↑	IoU↑	Precision↑	Recall↑	HD95↓	ASD↓
Attention Unet [17]	74.63	63.30	75.14	73.87	79.67	11.75
nnU-Net [10]	73.82	62.92	74.86	72.98	80.48	13.61
Swin UNETR [18]	72.66	60.88	73.99	75.80	89.82	14.08
UNETR + + [19]	74.50	63.76	75.08	73.76	78.93	11.83
U-Mamba [14]	73.28	61.97	73.98	74.41	78.79	12.82
PLHN [20]	73.24	61.83	74.87	74.85	76.78	11.43
HCMA [15]	75.55	64.36	76.00	71.56	74.02	10.47
DEHN	**75.93**	**64.78**	**76.16**	**75.25**	**73.53**	**10.31**

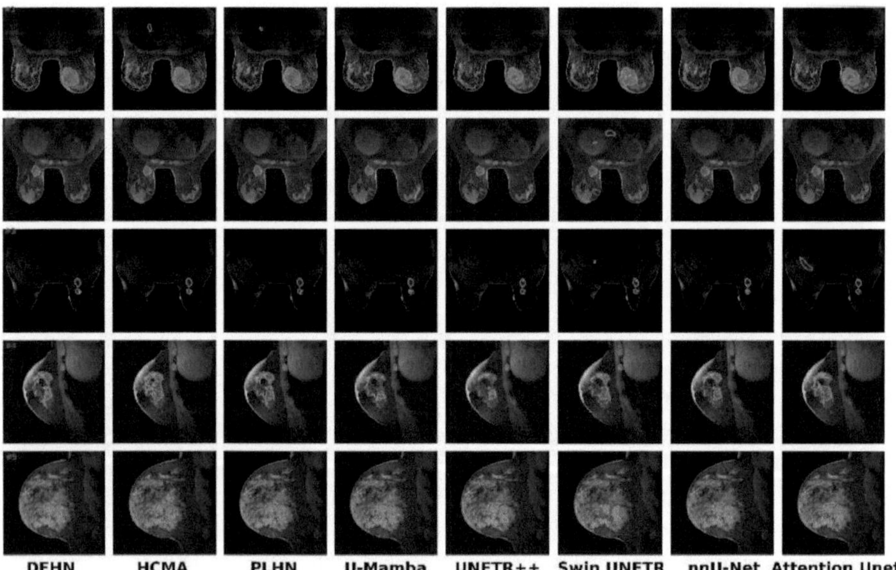

Fig. 4. Visual comparison of segmentation results produced by each model on the MAMA-MIA test set. Each row corresponds to a different subject. Post-contrast axial-plane images are shown, with ground truth contours overlaid in red and the automatic segmentation results from various methods in green.

Figure 4 presents a visual comparison on five subjects from the MAMA-MIA test set. DEHN consistently outperforms competing methods across diverse tumor scenarios, including large, small, and complex-margin lesions. For well-defined tumors (#1, #2) and those with ambiguous boundaries (#4, #5), DEHN achieves superior boundary accuracy. CNN-based models (e.g., nnU-Net) are prone to contralateral false positives, while Transformer-based models (e.g., UNETR + +) often miss small tumors or struggle with boundary precision. In contrast, DEHN's hybrid architecture effectively integrates local and global features, reducing false detections and capturing tumor-related voxels more accurately—especially in challenging cases (#3–#5). These results highlight the robustness of DEHN's feature fusion strategy and its advantage in complex breast tumor segmentation tasks.

3.4 Ablation Studies on the Dual-Encoder Architecture

Ablation Studies on the Dual-Encoder Architecture. To investigate the contribution of the dual-encoder design in DEHN, we conducted an ablation study by systematically modifying the architecture and comparing the performance of different variants on the yunnan dataset. The results are analyzed from two perspectives: architectural effectiveness and network complexity.

Analysis of Structure. To assess the effectiveness of the dual-encoder design, we conducted ablation studies using two single-encoder variants: the local-focused Encoder (L) and the global-focused Encoder (G), with the full model denoted as L + G. All variants received identical input (pre-contrast, post-contrast, and subtraction images) to ensure fair comparison. As shown in Table 3, the L and G models achieved Dice scores of 79.33% and 78.65%, respectively. The combined L + G model outperformed both, confirming the complementary advantages of the dual-encoder architecture. Additionally, the G model yielded lower HD95 and ASD values than the L model, indicating better boundary delineation due to enhanced global context modeling.

Table 3. Ablation study results evaluating the effectiveness of the dual-encoder structure in the DEHN model.

	Encoder	Dice	IoU	Precision	Recall	HD95↓	ASD↓
Single Encoder	L	79.33	70.51	83.62	81.34	14.57	3.87
	G	78.65	69.24	82.45	82.68	11.13	3.49
Dual Encoder	L + G	**81.52**	**71.63**	**86.67**	**83.78**	**8.96**	**2.03**

Analysis of Network Complexity. With an input size of $64 \times 192 \times 192$, we evaluated and compared the computational complexity of various network architectures. The computational costs and corresponding Dice scores of different methods are summarized in Table 4. In general, more complex models tend to yield higher segmentation accuracy. For instance, UNETR++ achieved a Dice score of 80.95% but required significantly more parameters, while U-Mamba reached 80.45% Dice with a high computational cost of 2642.14 GFLOPs. In contrast, the proposed DEHN attained the highest Dice score of 81.52% with only 7.72M parameters, demonstrating superior performance with efficient design. Although the L model incurs higher computational cost due to extensive convolution, its Dice score of 81.21% justifies the trade-off. The G model, despite having more parameters than L, maintains low FLOPs thanks to the Mamba module's efficient computation. Overall, DEHN achieves an effective balance between accuracy and efficiency, making it well-suited for practical deployment.

Table 4. Comparison of the number of parameters, computational cost and Dice score between the ablation model and seven models on Yunnan dataset.

Models	MParams	GFLOPs	Dice	Models	MParams	GFLOPs	Dice
Attention Unet	16.66	647.95	79.97	PLHN	11.00	243.96	79.20
nnU-Net	30.45	2620.72	78.40	HCMA	2.76	69.28	80.71
Swin UNETR	15.64	222.88	79.53	**L**	7.22	2103.68	79.33
UNETR + +	45.48	154.58	80.95	**G**	7.38	527.28	78.65
U-Mamba	22.57	2642.14	80.45	**DEHN**	7.72	2218.56	81.52

MVM Ablation Study.
To comprehensively assess the performance of the DEHN model and the effectiveness of the MVM module, we conducted four ablation experiments focusing on its two core components: multi-view and MVMB. The experimental results are summarized in Table 5. A baseline network was first conducted by removing the MVM module. As each component was progressively integrated into the architecture, we observed consistent improvements across performance metrics. The complete DEHN architecture achieved the highest overall performance, with Dice, Precision, and Recall scores improving by 1.4%, 2.64%, and 0.92%, respectively, compared to the baseline. These improvements validate the effectiveness of the MVM module in enhancing breast tumor segmentation in DCE-MRI.

Table 5. Ablation study results of the MVM module and its key components on Yunnan dataset.

Multi-view	MVMB	Dice	IoU	Prec	Rec	HD95↓	ASD↓	MParams	GFLOPs
-	-	80.12	69.63	84.03	82.86	11.43	4.03	7.41	2116.22
√	-	80.33	69.81	83.56	83.58	11.26	3.53	7.41	2116.65
-	√	80.67	70.12	85.34	83.06	9.94	2.54	7.51	2150.33
√	√	**81.52**	**71.63**	**86.67**	**83.78**	**8.96**	**2.03**	7.72	2218.56

4 Conclusion

To address the challenge of accurate breast tumor segmentation in DCE-MRI, we propose DEHN, a dual-encoder hybrid network comprising two distinct encoder architectures. The Local-Focused Encoder is designed to extract fine-grained features such as tumor boundaries and internal textures from pre- and post-contrast images, while the Global-Focused Encoder captures global contextual information such as tumor location and size from the subtraction image. Extensive experiments were conducted on a diverse cohort of 1,606 breast cancer patients collected from multiple hospitals and medical centers worldwide. The results demonstrate that DEHN effectively leverages complementary information from different DCE-MRI sequences and outperforms current state-of-the-art methods across multiple evaluation metrics, including Dice and IoU.

Acknowledgements. This work was supported in part by the Beijing Natural Science Foundation (L258054, QY25364), and the National Natural Science Foundation of China (82171992, 62105010).

References

1. Siegel, R.L., Miller, K.D., Wagle, N.S., Jemal, A.: Cancer statistics, CA: A Cancer Journal for Clinicians **73**(1), 17–48 (2023)
2. Jiménez-Gaona, Y., Rodríguez-Álvarez, M.J., Lakshminarayanan, V.: Deep-learning-based computer-aided systems for breast cancer imaging: a critical review. Appl. Sci. **10**(22), 8298 (2020)
3. Zhang, Y., et al.: Prediction of breast cancer molecular subtypes on DCE-MRI using convolutional neural network with transfer learning between two centers. Eur. Radiol. **31**, 2559–2567 (2021)
4. Ronneberger, O., Fischer, P., Brox, T.: U-Net: Convolutional networks for biomedical image segmentation. In: Navab, N., et al. (eds.) MICCAI 2015. LNCS, vol. 9351, pp. 234–241. Springer, Cham (2015)
5. Wang, H., Cao, J., Feng, J., Xie, Y., Yang, D., Chen, B.: Mixed 2D and 3D convolutional network with multi-scale context for lesion segmentation in breast DCE-MRI. Biomed. Signal Process. Control **68**, 102607 (2021)
6. Zhou, L., Wang, S., Sun, K., Zhou, T., Yan, F., Shen, D.: Three-dimensional affinity learning based multi-branch ensemble network for breast tumor segmentation in MRI. Pattern Recogn. **129**, 108723 (2022)
7. Dosovitskiy, A., et al.: An image is worth 16x16 words: Transformers for image recognition at scale. arXiv preprint arXiv:2010.11929 (2020)
8. Peng, Z., et al.: Conformer: Local features coupling global representations for visual recognition. In: Proceedings of the IEEE/CVF International Conference on Computer Vision, pp. 367–376 (2021)
9. Wang, T., et al.: O-Net: A novel framework with deep fusion of CNN and Transformer for simultaneous segmentation and classification. Front. Neurosci. **16**, 876065 (2022)
10. Isensee, F., Jaeger, P.F., Kohl, S.A.A., Petersen, J., Maier-Hein, K.H.: NnU-Net: A self-configuring method for deep learning-based biomedical image segmentation. Nat. Methods **18**(2), 203–211 (2021)
11. Zhou, H.Y., et al.: NnFormer: Volumetric medical image segmentation via a 3D Transformer. IEEE Trans. Image Process. **32**, 4036–4045 (2023)
12. Chen, J., et al.: TransUNet: Transformers make strong encoders for medical image segmentation. arXiv preprint arXiv:2102.04306 (2021)
13. Gu, A., Dao, T.: Mamba: Linear-time sequence modeling with selective state spaces. arXiv preprint arXiv:2312.00752 (2023)

14. Ma, J., Li, F., Wang, B.: U-Mamba: Enhancing long-range dependency for biomedical image segmentation. arXiv preprint arXiv:2401.04722 (2024)
15. Li, H., Qin, P., Yuan, X., Chen, Z., et al.: HCMA-UNet: a hybrid CNN Mamba UNet with inter-slice self-attention for efficient breast cancer segmentation. arXiv preprint arXiv:2501.00751 (2025)
16. Zhang, J., et al.: A robust and efficient AI assistant for breast tumor segmentation from DCE-MRI via a spatial-temporal framework. Patterns **4**(9) (2023)
17. Schlemper, J., et al.: Attention gated networks: learning to leverage salient regions in medical images. Med. Image Anal. **53**, 197–207 (2019)
18. Nosrati, M.S., Hamarneh, G.: Incorporating prior knowledge in medical image segmentation: a survey. arXiv preprint arXiv:1607.01092 (2016)
19. Shaker, A., Maaz, M., Rasheed, H., Khan, S., Yang, M.H., Khan, F.S.: UNETR++: delving into efficient and accurate 3D medical image segmentation. IEEE Trans. Med. Imaging **43**(9), 3377–3390 (2024)
20. Zhou, L., et al.: Prototype learning guided hybrid network for breast tumor segmentation in DCE-MRI. IEEE Trans. Med. Imaging (2024)
21. Zhang, J.: Breast_Cancer_DCE-MRI_Data, https://zenodo.org/records/8068383 (2023)
22. Garrucho, L., et al.: A large-scale multicenter breast cancer DCE-MRI benchmark dataset with expert segmentations. Scientific Data **12**(1), 453 (2025)

BASNet: A Boundary-Aware Semantic Segmentation Network for In-Orbit Satellite Remote Sensing Images

Shasha Ren[1(✉)] and Xiaodong Zhang[2]

[1] China Jiliang University, Hangzhou 310018, Jiangsu, China
shasharen@cjlu.edu.cn

[2] Anhui Wenda University of Information Engineering, Hefei 230032, Anhui, China

Abstract. Accurate segmentation of target boundary is the focus and difficulty of semantic segmentation of high-precision in-orbit satellite remote sensing images. Existing segmentation methods fail to accurately and effectively extract the boundary information, resulting in the boundary features being easily interfered by the image texture, which is not conducive to the accurate segmentation of remote sensing images. Therefore, we design a boundary-aware in-orbit satellite remote sensing image segmentation network (BASNet). Firstly, we propose the Boundary Extracting Module (BEM), BEM uses the information difference of the features in the adjacent layers of the network to explicitly model the boundary features and extracts the boundary features of all the targets of the remote sensing image. Secondly, the Boundary Feature Fusion Module (BFM) is designed to fully fuse the above boundary features into the features of each layer of the network, to enhance the network's representation of the target boundary features of the remote sensing images. And two different loss functions are designed to supervise the semantic segmentation network and the edge extraction network, respectively. Finally, experiments on two in-orbit remote sensing satellite datasets (Potsdam and Vaihingen) show that the method is more capable of representing boundary information compared to other similar methods. The network can alleviate the misclassification problem of boundary pixels and improve the semantic segmentation performance of remote sensing images.

Keywords: Semantic Segmentation · Remote Sensing · Boundary Detection · Feature Fusion

1 Introduction

Semantic segmentation of in-orbit satellite remote sensing images plays a vital role in applications such as urban planning, traffic, and maritime navigation [1]. Unlike natural images, these images are captured at high altitudes with larger ground sample distances (5–30 cm) [2], leading to high similarity among object features (e.g., roads vs. houses, trees vs. shrubs). When such objects are adjacent, boundary discrimination becomes challenging, often causing misclassification of pixels, as shown in Fig. 1.

P. Umapada et al. (Eds.): ICCPR 2025, CCIS 2811, pp. 159–171, 2026.
https://doi.org/10.1007/978-981-95-8315-7_13

To enhance boundary details, existing CNN-based methods fuse low-level features with high-level ones via skip connections or attention mechanisms [4]. Although these strategies enrich feature representation, directly fusing low-level features also introduces noise, resulting in unsatisfactory boundary segmentation. To address this, recent works explicitly integrate boundary detection into semantic segmentation. For example, GSCNN [5] introduces an edge extraction branch, DecoupleSegNet [6] decouples body and edge features, SBANet [7] leverages high-level guidance to refine low-level edges, FRENet [8] employs edge affinity loss, and CFM-Net [23] adopts adaptive fusion and edge loss functions. These methods demonstrate that explicit boundary modeling significantly boosts segmentation performance. However, accurately extracting boundaries from detail-rich low-level features remains a challenge.

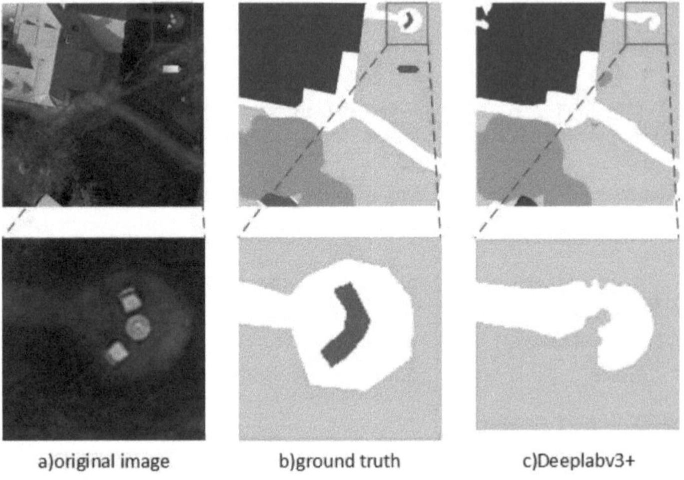

Fig. 1. Example of segmentation result. Boundary pixels are prone to be misclassified.

In this work, we propose a **Boundary-Aware Segmentation Network (BASNet)** that explicitly leverages boundary detection to enhance segmentation accuracy. The key idea lies in designing a **Boundary-Extraction Module**, which captures boundaries through feature differences between adjacent backbone layers and encodes them via convolution. The extracted boundaries are then integrated with both low- and high-level features using a **Boundary-Fusion Module**, which enriches the representation of boundary pixels in low-level layers while preserving boundary semantics in high-level layers. In this way, BASNet strengthens the network's focus on object boundaries and improves overall segmentation quality. The main contributions are as follows:

- A novel Boundary-Aware Segmentation Network is proposed to make good use of the boundary information for semantic segmentation of remote sensing images.
- A Boundary-Extracting Module is proposed to extract the boundary by the information differences between adjacent stages of the backbone.
- A Boundary-Fusion Module is proposed to relate the boundary and features of various stages of the backbone, making the fusion more effective.

- The experimental results show that BASNet can effectively improve the semantic segmentation performance of remote sensing images.

2 Related Works

2.1 Remote Sensing Image Segmentation Based on Features Fusion

In remote sensing semantic segmentation, encoder–decoder architectures are widely adopted to bridge encoding and decoding processes. DIResUNet [9], derived from U-Net with residual modules, extracts local and global features in parallel. HCANet [10] integrates the ASPP module into U-Net for multi-level and multi-scale feature learning. CFAMNet [11], based on Deeplabv3+, enhances contextual representation through a class feature attention mechanism. Similarly, multi-scale context intertwining networks [12] aggregate features from adjacent stages, while Deeplab [13, 14] employs ASPP with different dilation rates to capture multi-scale information. Building on ASPP, HRC-Net [16] simplifies HRNet to fuse detailed low-level and semantic high-level features, whereas Anent [17] adaptively aggregates features via channel attention. These studies demonstrate that encoder–decoder structures, enhanced with multi-scale and attention mechanisms, significantly improve semantic segmentation performance in remote sensing imagery.

2.2 Image Segmentation with Boundary Information

Boundary information plays a critical role in semantic segmentation. Many networks assume that boundaries are well preserved in low-level features, and thus fuse them with high-level features, as in Deeplabv3+ and PSPNet, to enhance feature representation. Other works, such as DDCM-Net [21] and DMNet [22], employ dilated convolutions and multiscale mechanisms to integrate local and global information. However, these implicit strategies cannot fully exploit boundary cues. To address this, multitask learning has been widely adopted to explicitly extract boundaries through an additional branch. For example, the edge-aware dual-stream network [18] introduces an edge detection stream based on UNet, BAMTL [19] jointly performs segmentation, height estimation, and edge detection, while FRENet [8] and CFM-Net [23] design edge-aware loss functions to obtain clearer object boundaries. These explicit approaches have been shown to deliver superior segmentation performance.

3 Methodology

In this section, we will first describe the network architecture of the Boundary-Aware Segmentation Network (BASNet). We then present two main modules: Boundary-Extracting Module (BEM) and Boundary-Fusion Module (BFM).

3.1 Boundary Aware Segmentation Network

BASNet introduces an extra branch to extract the boundary and then fuses it into the main features, as shown in Fig. 2. In the boundary-extracting branch, we introduce a BEM to get a better boundary feature expression by the information differences. Additionally, we use an extra loss function to supervise the boundary extraction to get a better boundary. After getting the boundary, we fuse it into the main feature in BFM. Next, we will cover the details of these two main modules.

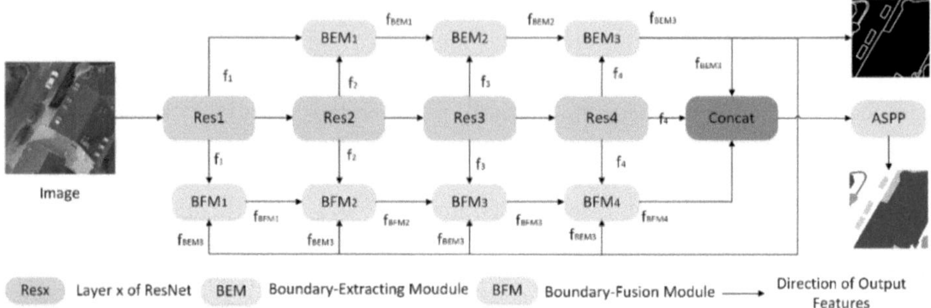

Fig. 2. The architecture of BASNet.

3.2 Boundary Extracting Module

In this section, we will describe the BEM in detail, and how to embed this module into the BASNet.

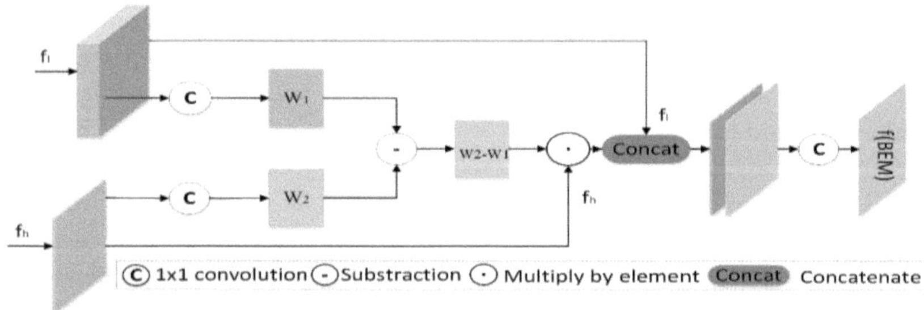

Fig. 3. The network of BEM.

As a result, we introduce the BEM, as shown in Fig. 3. First, the outputs of the adjacent stages of the backbone are fed into a 1×1 convolution separately. We call the output of lower stage f_l and the output of higher stage f_h. Through this convolution, we can get two feature maps, w_1 and w_2, both of which are $1 \times H \times W$ in size. To get the information difference, we use the subtraction of features, which can get a new weight map. After that, we do the element-wise multiplication between the new weight map and the output of the higher stage $f(BEM)$. The expression is shown as (1)-(3).

$$w_1 = conv_{1\times1}(f_l(BEM)) \tag{1}$$

$$w_2 = conv_{1\times 1}(f_h(BEM)) \tag{2}$$

$$f_{fuse} = f_2(BEM) \cdot (w_2 - w_1) \tag{3}$$

Here f_h,h = 2,3,4, f_l, l = 1,2,3. To preserve other details, we also fuse the new feature maps with the output of the lower stage by concatenation. After that, the new feature maps are fed into a series of convolutions with a kernel 1 × 1 for better performance. The expression is shown as (4).

$$f(BEM) = conv_{1\times 1}\big(Concat\big(f_l(BEM), f_{fuse}(BEM)\big)\big) \tag{4}$$

To make good use of the information differences of each stage of the backbone, we implement the BEM to each two adjacent stages of the backbone in BASNet. Additionally, we utilize a Dice Loss to supervise the boundary extraction.

3.3 Boundary Fusion Module

In BFM, we named the output of one stage of the backbone f_x,x = 1,2,3,4 and the boundary $f_{edge} = f(\mathrm{BEM}_3)$. First, the f_x is fed into a series of convolutions with a kernel 1 × 1 to encode the features. Then, we treat the $f(\mathrm{BEM}_3)$ as an attention map and do the element-wise multiplication between the new features f_x,x = 1,2,3,4 and $f(\mathrm{BEM}_3)$. The expression is shown as (5).

$$f_{fuse}(BFM_x) = \begin{cases} conv_{1\times 1}(f_x) \cdot f(BEM_3), x = 1, \\ concat(conv_{1\times 1}(f_x) \cdot f(BEM_3), f(BFM_{x-1})), x = 2, 3, 4. \end{cases} \tag{5}$$

Moreover, we implement the BFM to each stage of the backbone in BASNet. In the BFMs except the first one f_1, we will get another input $f(\mathrm{BFM}_{x-1})$ from the output of the previous BFM, which is concatenated with the output $f(\mathrm{BFM}_x)$ of the current BFM. In other words, we concatenate all outputs of BFMs and then concatenate them with the last output of the backbone. To further fuse the boundary, we encode the boundary map by a series of convolutions and then concatenate it with the main features. After the concatenation, the output $f(\mathrm{BFM}_4)$ is fed into an Atrous Spatial Pyramid Pooling (ASPP) for better decoding (Fig. 4).

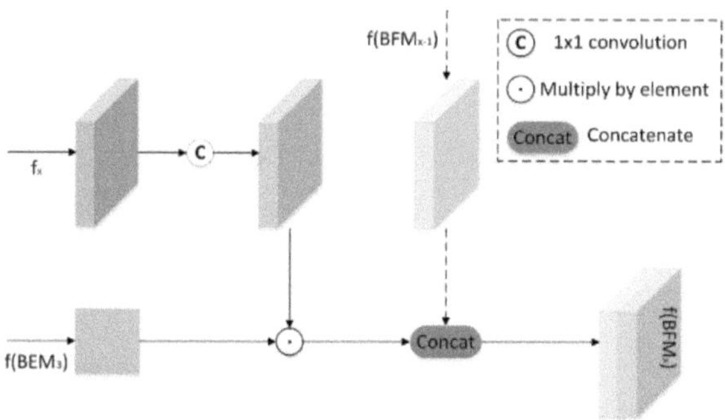

Fig. 4. The network of BFM.

4 Experimental Results and Discussion

4.1 Datasets Description

In this work, we use the ISPRS-Potsdam and ISPRS-Vaihingen as the experimental dataset, Potsdam contains 38 three-dimension (red, green, blue) images with a ground sample distance (GSD) of 5 cm, each of which is 6000 × 6000 in size. As usual, we set 14 images as testing data and the others as training data. In this work, we use this dataset without DSM and NDSM data for training and inference. We set 17 images as testing data and the others as training data. Similarly, we use this dataset without DSM and NDSM data for training and inference.

4.2 Implementation Details

All experiments were conducted on four NVIDIA GTX 2080 Ti GPUs under Ubuntu 20.04 using PyTorch. The networks were trained with SGD (momentum = 0.9, weight decay = 0.0005) and an initial learning rate of 0.01 following the "poly" schedule, where the rate is adjusted by $(1 - \text{iter}/\text{max\_ iter})^{0.9}$. The maximum number of iterations was set to 40k with a batch size of 8. We adopted the standard training pipeline of the MMSegmentation library [32].

4.3 Ablation Study of Basnet

To evaluate the effectiveness of each module, the results are summarized in Table 1. The ASPP baseline achieves 81.5% mIoU. Introducing the Boundary-Extraction Module (BEM) improves performance by 0.98%, while the Boundary-Fusion Module (BFM) alone yields a 1.07% gain through enhanced multi-level integration. When both modules are combined, BASNet reaches 83.44% mIoU, outperforming the baseline by 1.94%. These results demonstrate that BEM and BFM are complementary, and their joint integration significantly improves both overall accuracy and class-wise IoU, validating the robustness of the proposed BASNet.

Table 1. Ablation experiments on different components (%).

Module			Road	Buli	Low-veg	Tree	Car	mIoU
ASPP	BEM	BFM						
✓	×	×	84.53	92.21	74.11	74.19	82.46	81.50
✓	✓	×	86.36	92.84	75.71	75.75	81.71	82.48
✓	×	✓	86.38	93.16	75.48	75.41	82.42	82.57
✓	✓	✓	86.86	93.72	76.09	75.78	84.74	83.44

Table 2. Ablation experiments on different loss functions(%).

Loss	road	Buli	Low-veg	Tree	Car	mIoU
Xloss	86.07	92.91	75.55	75.47	80.09	82.02
Dice Loss	86.36	92.84	75.71	75.75	81.71	82.48

As shown in Table 2. Dice Loss performs better than the cross-entropy loss (Xloss) function for the edge extraction branch, with a 0.46% improvement in the mIoU metric, which reflects the effectiveness of Dice Loss in supervising the edge extraction branch. Therefore, the loss function used in this paper is Dice Loss instead of Xloss function for better supervision of edge extraction branch.

Table 3. Performance effects of image expansion with different kernel (%).

kernel sizes	road	Buli	Low-veg	Tree	Car	mIoU
1×1	86.4	92.92	75.86	75.75	79.38	82.06
3×3	86.23	93.15	75.35	75.42	81.7	82.37
5×5	86.36	92.84	75.71	75.75	81.71	82.48
7×7	86.52	92.9	75.41	75.41	82.16	82.48

To find the appropriate scale of the edge kernel, the experimental settings of kernel sizes are 1×1, 3×3, 5×5, 7×7 in Table 3. The image expansion operations with different kernel sizes directly lead to the number of real labeled edge pixels in the edge extraction branch, which affects the generation of edges from this branch to some extent. However, when the kernel size is reached, the semantic segmentation effect basically does not continue to improve with the increase of kernel scale. Considering the effective modeling situation of the edges, the image expansion operation with kernel size of 5×5 is finally used in BASNet to generate the BGT.

Table 4. Experimental results of different stage feature (%).

stages	road	Buli	Low-veg	Tree	Car	mIoU	Par.(M)
1	86.27	93.69	75.41	75.87	82.45	82.74	37.34
1,2	86.45	93.52	75.83	75.54	83.86	83.04	38.67
1,2,3	86.71	93.39	76.2	75.72	84.32	83.27	40.26
1,2,3,4	86.86	93.72	76.09	75.78	84.74	83.44	42.37

As shown in Table 4. It could be obtained a better performance by fusing more features of different stages with a small increase in the number of parameters. Although we fuse features from all stages of the backbone, the number of parameters is 42.37M in BASNet while the number is 41.39M in Deeplabv3+. The BASNet is less than 1M higher than Deeplabv3+ in the number of parameters.

The feature visualization results are presented in Fig. 5. As shown in Fig. 5(b), the element-wise subtraction between adjacent features is multiplied with higher-level features and concatenated with lower-level features, thereby enhancing edge representation in high-level features while retaining fine details from low-level features. Figure 5(c) illustrates that multiple edge extraction modules can be embedded into the edge branch to progressively strengthen low-level edge features and preserve high-level edge semantics across scales. The final visualization outputs of each module are shown in Fig. 5(d) and (e), confirming the effectiveness of the proposed design in enhancing boundary features.

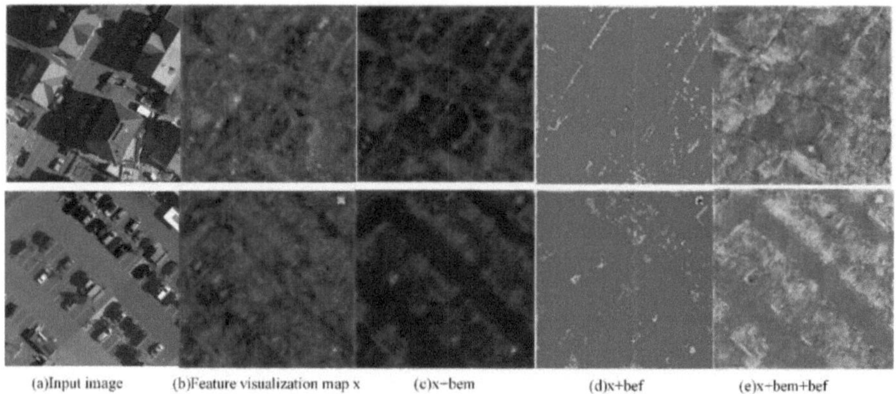

Fig. 5. The visualization of each module.

4.4 Comparison with Other Methods

The following experiments are conducted on the Potsdam and Vaihingen. We introduce some semantic segmentation methods of common datasets and compare them with the proposed BASNet.

Table 5. Comparison with methods on the ISPRS-Potsdam (%).

Methods	road	Buli	Low-veg	Tree	Car	mIoU
FCN [4]	82.48	90.99	70.52	75.42	82.11	80.3
Deeplabv3 + [20]	82.75	91.22	71.05	75.49	81.72	80.45
PSPNet [21]	82.65	91.36	70.78	75.69	81.92	80.48
SERNet [23]	84.31	91.87	57.58	67.81	82.24	76.76
MLCRNet [25]	82.3	91.40	73.16	73.79	81.60	80.42
HRCNet [16]	83.58	91.15	73.07	74.88	83.32	81.20
FRENet [22]	85.94	89.52	78.03	66.95	81.13	80.31
CFM-Net [8]	79.06	88.22	81.78	72.88	79.94	80.38
UNet Ens. [26]	87.80	82.17	93.28	70.02	77.51	80.49

(continued)

Table 5. *(continued)*

Methods	road	Buli	Low-veg	Tree	Car	mIoU
MST-Deeplabv3 + [27]	87.80	82.17	93.28	70.02	77.51	82.15
BASNet	86.86	93.72	76.09	75.75	84.74	83.44

As shown in Table 5, on the Potsdam dataset, BASNet achieves better performance compared with various advanced models, and improves 3.14% in mIoU compared with FCN. When compared with the classical semantic segmentation models Deeplabv3+ [20] and PSPNet [21], it also improves 2.99% and 2.96% mIoU, respectively. When compared with the latest semantic segmentation methods for remote sensing images such as SERNet [23], MLCRNet [25], and HRCNet [16], it also improves 6.68%,3.02% and 2.24% in mIoU, respectively. The above experiments effectively prove the effectiveness of BASNet. At the same time, the network has good performance in each category, and the correct identification of edge pixels has a certain influence on the semantic segmentation accuracy of each category, and the comprehensive performance of the network in each category can to some extent verify the advantages of the fusion of edge features. In addition, to illustrate the effectiveness of edge extraction for semantic segmentation in BASNet, we also compare other semantic segmentation methods FRENet [22] and CFM-Net [8]. Compared to them, BASNet achieves a better performance, with an improvement of more than 3% in both mIoU metrics.

Table 6. Comparison with methods on the ISPRS-Vaihingen (%).

Methods	road	Buli	Low-veg	Tree	Car	mIoU
FCN [4]	81.61	88.51	65.34	75.08	66.05	75.32
Deeplabv3+ [20]	81.61	88.4	65.54	74.96	66.18	75.34
PSPNet [21]	81.38	88.49	65.85	75.42	65.96	75.42
SERNet [23]	83.15	84.59	67.72	67.73	60.25	72.69
MLCRNet [25]	81.3	87.2	65.4	74.3	64.4	74.52
HRCNet [16]	81.05	86.65	66.91	76.63	59.31	74.11
FRENet [22]	83.24	86.14	68.54	68.92	60.48	73.46
CFM-Net [8]	77.39	85.52	59.54	73.33	66.06	72.37
UNetEns. [26]	81.65	86.65	67.41	76.63	61.31	74.73
BASNet	81.38	88.37	65.91	74.8	66.86	75.46

As shown in Table 6. The experimental results of the Vaihingen dataset show the excellent performance of BASNet, and there is even a 2.77% improvement in the mIoU index compared with SERNet, which is very significant on the Vaihingen dataset. At the same time, the performance on each category also has some advantages, which indirectly also reflects the improvement of the effect brought by the fusion of edge features. Compared with the remote sensing image semantic segmentation method that strengthens the edge features, BASNet also performs very well on the Vaihingen dataset, and in the mIoU index, it also improves by 3.09% compared with CFM-Net, and improves by 2% compared with FRENet.

5 Visualization Results

To observe more intuitively how the image expansion operation with different kernel sizes affects the edge pixels, we visualize the generated edge reality maps for several kernel sizes separately, as shown in Fig. 6. The generation of image edges can be clearly seen from the figure, and the edges are obvious for several edge labeling maps with larger kernel sizes, and the use of such semantic edge information can effectively assist the task of semantic segmentation of remote sensing images. To demonstrate the effectiveness of the proposed BASNet better, we visualized the results on the images of the ISPRS-Potsdam dataset, as shown in Fig. 7. We can obviously know that BASNet performs better in semantic segmentation of remote sensing images. In Fig. 7, the red boxes show better segmentation by BASNet.

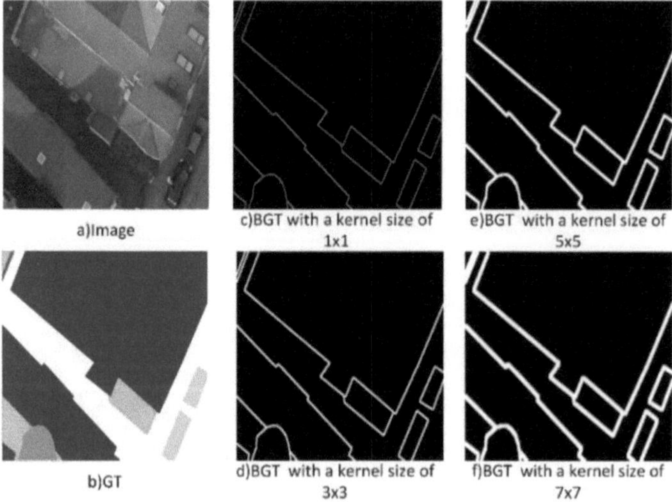

Fig. 6. Comparison results of BGT after manipulation with different size kernels.

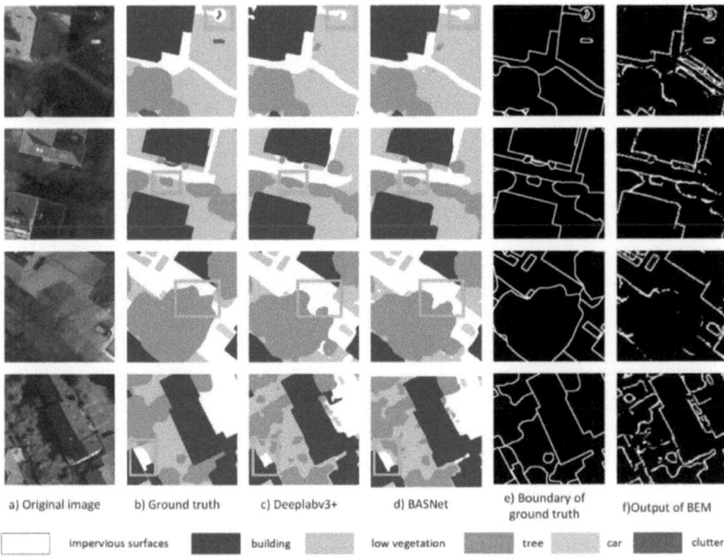

Fig. 7. The visualization of the results.

6 Conclusions

In this paper, BASNet is proposed to segment the remote sensing images by combining the boundary detection network and semantic segmentation network. BASNet introduces a separate branch for boundary extraction, which is made up of multiple BEMs and supervised by an extra loss function. BEM utilizes the information differences between the adjacent stages of the backbone to extract boundary information better. Moreover, BASNet also introduces a BFM to fuse the boundary and the main features more effectively. BFM encodes detailed information from the features of each stage of the backbone and then fuses them into the main features. The experimental results confirm that BASNet outperforms the state-of-the-art methods in the number of quality metrics. It is important for us to segment the remote sensing images better due to the wide application of remote sensing images. In the future, it is also important to segment the remote sensing images in a more efficient way.

Acknowledgments. This work was supported by the General Project of the National Natural Science Foundation of China (No. 61976094).

References

1. Li, Z., Wang, Y., Xu, D., et al.: TBNet: a texture and boundary-aware network for small weak object detection in remote-sensing imagery. Pattern Recogn. **158**, 110976 (2025)
2. Marmanis, D., Schindler, K., Wegner, J.D., et al.: Classification with an edge: Improving semantic image segmentation with boundary detection. ISPRS J. Photogramm. Remote. Sens. **135**, 158–172 (2018)

3. Zhou, J., Fu, K., Liang, S.,et al.: Extracting water surfaces of the dike-pond system from high spatial resolution images using deep learning methods. Remote Sens. **17**(1) (2025). https://doi.org/10.3390/rs17010111
4. Li, H., Xu, Z., Taylor, G., et al.: Visualizing the loss landscape of neural nets. Adv. Neural Inf. Process. Syst. **31** (2018)
5. Takikawa, T., Acuna, D., Jampani, V., et al.: Gated-scnn: gated shape cnns for semantic segmentation. Proceedings of the IEEE/CVF International Conference on Computer Vision. pp. 5229–5238 (2019)
6. Li, X., Li, X., Zhang, L., et al.: Improving semantic segmentation via decoupled body and edge supervision. Computer Vision–ECCV 2020: 16th European Conference, Glasgow, UK, pp. 435–452 (23–28 Aug 2020)
7. Li, A., Jiao, L., Zhu, H., et al.: Multitask semantic boundary awareness network for remote sensing image segmentation. IEEE Trans. Geosci. Remote Sens. **60**, 1–14 (2021)
8. Chen, J., Wang, H., Guo, Y., et al.: Strengthen the feature distinguishability of geo-object details in the semantic segmentation of high-resolution remote sensing images. IEEE J. Sel. Top. Appl. Earth Observations Remote Sens. **14**, 2327–2340 (2021)
9. Priyanka, N.S., Lal, S., et al.: DIResUNet: architecture for multiclass semantic segmentation of high resolution remote sensing imagery data. Appl. Intell. **52**(13), 15462–15482 (2022)
10. Bai, H., Cheng, J., Huang, X., et al.: HCANet: a hierarchical context aggregation network for semantic segmentation of high-resolution remote sensing images. IEEE Geosci. Remote Sens. Lett. **19**, 1–5 (2021)
11. Wang, Z., Wang, J., Yang, K., et al.: Semantic segmentation of high-resolution remote sensing images based on a class feature attention mechanism fused with Deeplabv3+. Comput. Geosci. **158**, 104969 (2022)
12. Lin, D., Ji, Y., Lischinski, D., et al.: Multi-scale context intertwining for semantic segmentation. Proceedings of the European Conference on Computer Vision (ECCV). pp. 603–619 (2018)
13. Chen, L.C., Papandreou, G., Schroff, F., et al.: Rethinking atrous convolution for semantic image segmentation (2017). arXiv. arXiv preprint arXiv:1706.05587
14. Chen, L.C., Zhu, Y., Papandreou, G., et al.: Encoder-decoder with atrous separable convolution for semantic image segmentation. Proceedings of the European conference on computer vision (ECCV). pp. 801–818 (2018)
15. Chen, X., Li, Z., Jiang, J., et al.: Adaptive effective receptive field convolution for semantic segmentation of VHR remote sensing images. IEEE Trans. Geosci. Remote Sens. **59**(4), 3532–3546 (2020)
16. Xu, Z., Zhang, W., Zhang, T., et al.: HRCNet: high-resolution context extraction network for semantic segmentation of remote sensing images. Remote Sens. **13**(1), 71 (2020)
17. Xiang, S., Xie, Q., Wang, M.: Semantic segmentation for remote sensing images based on adaptive feature selection network. IEEE Geosci. Remote Sens. Lett. **19**, 1–5 (2021)
18. Nong, Z., Su, X., Liu, Y., et al.: Boundary-aware dual-stream network for VHR remote sensing images semantic segmentation. IEEE J. Sel. Top. Appl. Earth Observations Remote Sens. **14**, 5260–5268 (2021)
19. Wang, Y., Ding, W., Zhang, R., et al.: Boundary-aware multitask learning for remote sensing imagery. IEEE J. Sel. Top. Appl. Earth Observations Remote Sens. **14**, 951–963 (2020)
20. Zhao, H., Shi, J., Qi, X., et al.: Pyramid scene parsing network. Proceedings of the IEEE Conference on Computer Vision and Pattern Recognition. pp. 2881–2890 (2017)
21. Liu, Q., Kampffmeyer, M., Jenssen, R., et al.: Dense dilated convolutions' merging network for land cover classification. IEEE Trans. Geosci. Remote Sens. **58**(9), 6309–6320 (2020)
22. Li, W., Zhang, X., Peng, Y., et al.: DMNet: a network architecture using dilated convolution and multiscale mechanisms for spatiotemporal fusion of remote sensing images. IEEE Sens. J. **20**(20), 12190–12202 (2020)

23. Sun, X., Xia, M., Dai, T.: Controllable fused semantic segmentation with adaptive edge loss for remote sensing parsing. Remote Sens. **14**(1), 207 (2022)
24. Zhang, X., Li, L., Di, D., et al.: SERNet: squeeze and excitation residual network for semantic segmentation of high-resolution remote sensing images. Remote Sens. **14**(19), 4770 (2022)
25. Huang, Z., Zhang, Q., Zhang, G.: MLCRNet: multi-level context refinement for semantic segmentation in aerial images. Remote Sens. **14**(6), 1498 (2022)
26. Dimitrovski, I., Spasev, V., Loshkovska, S., et al.: U-Net ensemble for enhanced semantic segmentation in remote sensing imagery. Remote Sens. **16**(12), 2077 (2024)
27. Wang, Y., Yang, L., Liu, X., et al.: An improved semantic segmentation algorithm for high-resolution remote sensing images based on DeepLabv3+. Sci. Rep. **14**(1), 9716 (2024)

Multi-scale Feature Weighted Aggregation Network for Hand Segmentation

Zhuangzhuang Zhang, Feng Chen, Xuefeng Zhang, Zekai Cheng, and Xiwen Qu(✉)

School of Computer Science and Technology, Anhui University of Technology, Maanshan, China
{chenfeng,zxf_06,chengzk,qxw_ahut}@ahut.edu.cn

Abstract. Hand segmentation plays a vital role in various computer vision tasks such as gesture recognition and human-computer interaction. However, achieving accurate segmentation remains challenging in the presence of complex backgrounds, fine-grained hand structures, and motion blur. To address these issues, we propose a Multi-Scale Feature Weighted Aggregation Network (MSFWAN) for hand segmentation. Specifically, a Hierarchical Feature Aggregation Module (HFAM) is designed to capture fine-grained hand details by aggregating multi-scale contextual information, enabling the model to capture hand boundaries and subtle structures. To enhance robustness under motion blur, we further propose a Multi-Scale Feature Weighted Aggregation Module (MSFWAM) that selectively emphasizes salient features across scales. Extensive experiments on multiple benchmark datasets demonstrate that our method outperforms existing methods.

Keywords: Hand Segmentation · Motion Blur · Multi-scale Feature · Weighted Aggregation

1 Introduction

Hand segmentation has evolved from a basic preprocessing step into a pivotal enabler of next-generation interaction paradigms. Its precision directly influences the reliability of applications such as augmented reality, smart healthcare, and contactless industrial control. Despite the widespread adoption of monocular vision systems for their cost-efficiency, their limitations in dynamic scenes, such as motion blur and occlusion, have significantly hindered real-world deployment, creating an urgent demand for robust, real-time solutions.

Hand segmentation requires pixel-level localization in complex, dynamic environments. However, existing methods face three key challenges in high-speed interaction scenarios. First, deep networks are susceptible to background noise, particularly when hands are small or color-similar to the background. Second, in scenarios such as in-air handwriting, rapid motion induces blur due to limited camera frame rates, degrading feature quality. Third, fine structures like fingertips occupy few pixels at standard resolutions, making it difficult for CNN-based

P. Umapada et al. (Eds.): ICCPR 2025, CCIS 2811, pp. 172–184, 2026.
https://doi.org/10.1007/978-981-95-8315-7_14

methods to capture both global context and details, as illustrated in Fig. 1. Existing datasets also exhibit limitations. Mainstream benchmarks such as EDSH [13] and EgoHands [1] are constrained to static, egocentric views and lack representations of rapid motion and complex hand deformation, limiting their generalizability to real-world scenarios.

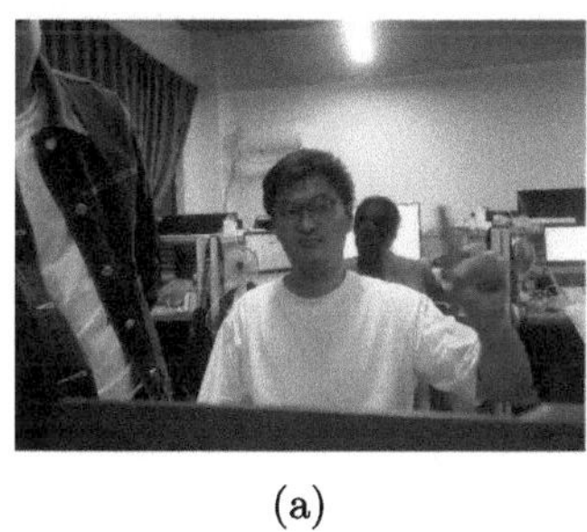

(a)

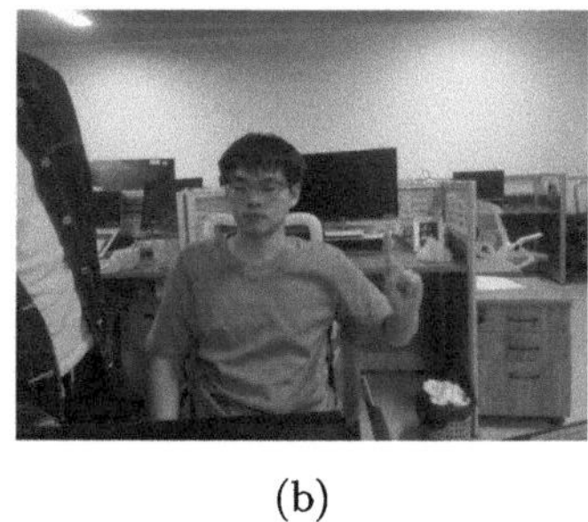

(b)

Fig. 1. (a) Motion-blurred hand image caused by rapid movement. (b) Feature suppression in complex backgrounds.

To address these challenges, we propose Multi-Scale Feature Weighted Aggregation Network (MSFWAN), a robust hand segmentation framework that effectively suppresses motion blur and reduces interference in single-frame images. It employs an enhanced MobileNetV3 [7] backbone within an encoder-decoder architecture, systematically increasing channel capacity to capture fine-grained hand features while maintaining low computational cost. The proposed Hierarchical Feature Aggregation Module (HFAM) integrates four parallel branches-baseline to enhance sensitivity to subtle structures like finger gaps and nail contours. The Multi-Scale Feature Weighted Aggregation Module (MSFWAM) combines dynamic weighted aggregation with feature similarity constraints, reorganizing multi-scale features to preserve low-level textures while improving global context understanding, thereby boosting robustness under motion blur. Experiments demonstrate that MSFWAN achieves high-precision hand segmentation in complex environments. Our key contributions are summarized as follows:

(1) We propose MSFWAN, a robust hand segmentation framework.
(2) We propose HFAM, a multi-scale pyramid with adaptive feature selection, enhancing sensitivity to fine details like inter-finger gaps and addressing feature suppression in complex backgrounds.
(3) We propose MSFWAM to reduce boundary distortion from motion blur, using dynamic weighting and similarity constraints to improve spatial coherence and boundary accuracy.
(4) We construct DHAND-AHUT2025, a dedicated dataset featuring severe motion blur and feature suppression scenarios, enabling evaluation and advancing robust model development for real-world HCI.

2 Related Work

Traditional hand segmentation methods rely on handcrafted features. Motion-based approaches, such as Ren and Gu [19], use temporal differencing and region growing. Color-based methods model skin tone distributions, exemplified Khan et al. [12], who examined illumination variations. Joint spatial-texture techniques improve consistency via "hyperpixel" segmentation [5] or structured forests [26]. Kawulok et al. [10,11] proposed probabilistic skin maps using linear discriminant analysis.

Semantic segmentation, which assigns semantic labels at the pixel level, has seen major advances with CNN and transformer architectures. Building on this, deep learning-based hand segmentation has evolved into end-to-end, data-driven frameworks. EgoHands [1] and subsequent works [20] introduced deep pipelines and multi-scale fusion. Progress follows three main directions: (1) Temporal modeling, e.g., gated recurrent units [22] and optical flow with superpixels [14]; (2) Data augmentation—e.g., synthetic data [15] and joint hand-object learning [9]; and (3) Domain adaptation, e.g., Bayesian CNNs with uncertainty and shape priors [2]. However, most methods rely on static, egocentric datasets [1,13,20] and struggle with feature suppression and motion blur in dynamic real-world settings.

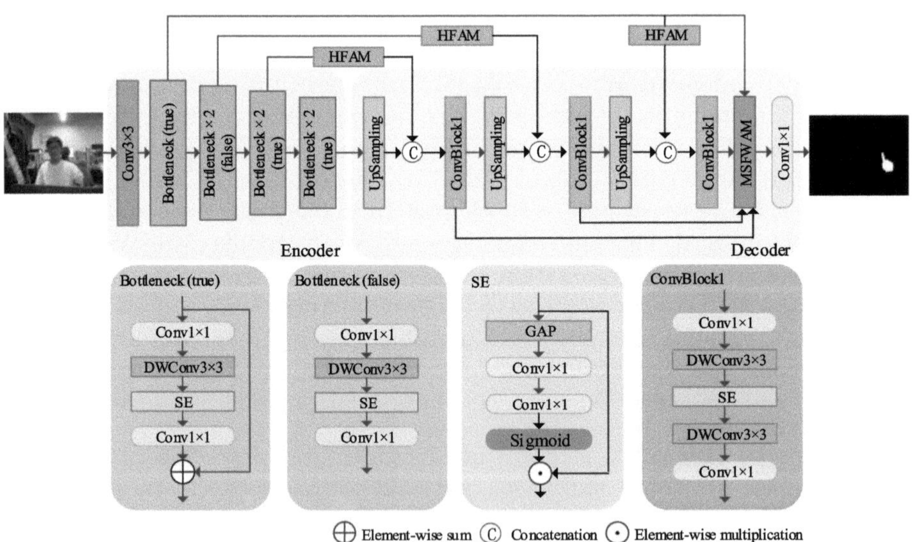

Fig. 2. Detailed structure of the proposed MSFWAN.

3 Proposed Method

3.1 Overview

To address the dual requirements of accuracy and efficiency in dynamic hand segmentation under monocular vision, an efficient end-to-end segmentation framework MSFWAN is proposed. Its detailed structure is illustrated in Fig. 2.

An efficient feature extractor is built upon the MobileNetV3 [7] backbone using inverted residual modules, as shown in the Bottleneck of Fig. 2. To address detail loss from excessive channel reduction in the original model, channel capacity is systematically increased at each stage, while overall depth is reduced to balance efficiency and capacity (see Table 1). This enhanced encoder preserves high-resolution textures in shallow layers and captures global semantics in deeper layers via an expanded receptive field, enabling rich multi-level representations sensitive to fine structures such as fingertips and finger crevices.

Table 1. Encoder parameters in Fig. 2

Input	Layer	Output	Stride
640 × 480 × 3	Conv2d	32	2
320 × 240 × 32	BN	64	1
320 × 240 × 64	BN	128	2
160 × 120 × 128	BN	128	1
160 × 120 × 128	BN	256	2
80 × 60 × 256	BN	256	1
80 × 60 × 256	BN	512	2
40 × 30 × 512	BN	512	1

A three-stage progressive upsampling structure is designed, each stage comprising three steps. First, bilinear interpolation restores resolution with minimal computational cost. Second, skip connections fuse high-frequency details from encoder layers to recover downsampling losses. Third, depthwise separable convolutions (DWConv in Fig. 2) and Squeeze-and-Excitation (SE) [8] modules refine features and suppress background noise. The decoder output is kept at half the input resolution to balance efficiency and structural fidelity.

To improve robustness against motion blur and feature suppression, the framework integrates two key modules. First, the HFAM enhances sensitivity to fine structures using a multi-scale pyramid and adaptive fusion for selective refinement (Sect. 3.2). Second, the MSFWAM addresses boundary distortion by applying dynamic weighting and similarity-based consistency constraints (Sect. 3.3). Together, they balance accuracy and efficiency, supporting robust real-time hand segmentation.

3.2 Hierarchical Feature Aggregation Module

To address feature suppression and fine-grained structure representation in complex scenes, the HFAM is introduced. As shown in Fig. 3, HFAM is embedded into an encoder-decoder framework via cross-layer connections, forming a fusion pathway that integrates contextual semantics and local details through multi-scale receptive fields and hierarchical enhancement. It receives high-level semantic features from the encoder and employs a multi-branch pyramidal structure with adaptive feature selection to dynamically emphasize relevant information while suppressing noise, enhancing sensitivity to sub-pixel structures such as finger slits and nail contours. The HFAM processes encoder features via four parallel branches, forming a streamlined multi-scale extraction structure adapted from the deep aggregation pyramid pooling module [6]. By reducing branches and revising inter-branch connections, it lowers computational complexity while maintaining strong multi-scale representation. The branches are: (1) baseline branch preserving deep semantics at original resolution via 1×1 convolution; (2) local perception branch capturing mesoscale structures like knuckles using 5×5 average pooling with stride 2; (3) global perception branch extracting broader patterns such as hand contours via 9×9 average pooling with stride 4; and (4) contextual encoding branch modeling global dependencies through global average pooling.

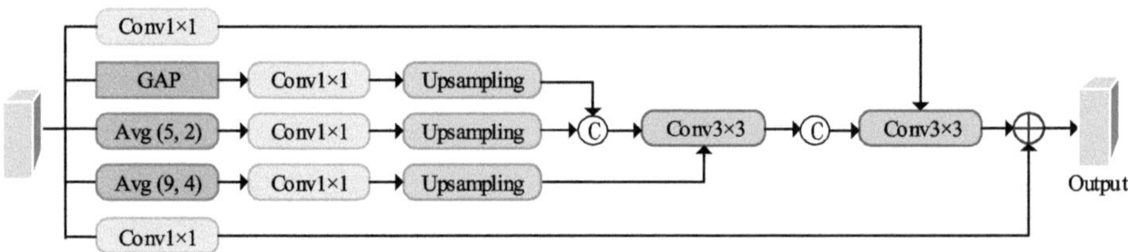

Fig. 3. Detailed structure of the proposed HFAM. Avg(5,2) denotes an average pooling with a kernel size of 5×5 and a stride of 2. GAP denotes global average pooling.

3.3 Multi-Scale Feature Weighted Aggregation Module

To address feature distortion from hand deformation and motion blur in fast movements, the MSFWAM is proposed to improve segmentation accuracy and robustness. The detailed structure of the MSFWAM is shown in Fig. 4. In Fig. 4, "Tp" denotes transposition. The MSFWAM dynamically weights multi-scale features and enforces feature similarity constraints to effectively integrate semantic and detail cues. It fuses multi-scale features from three decoder stages and shallow detail features from early encoder layers, performing hierarchical weighted fusion guided by global semantics and fine-grained details to produce semantically consistent, texture-rich representations. The MSFWAM modulates

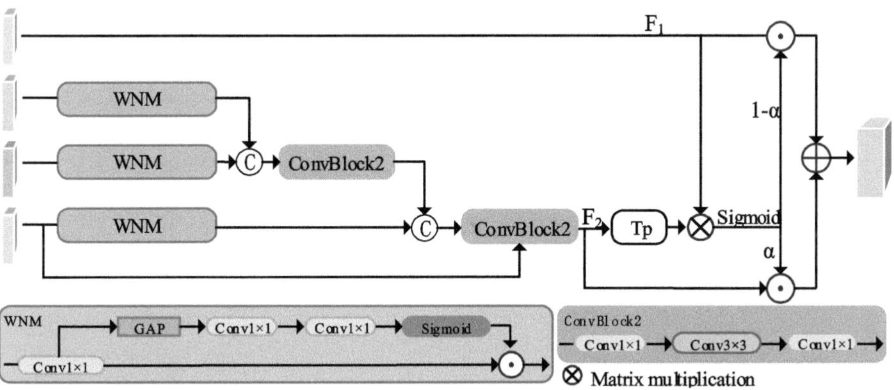

Fig. 4. Detailed structure of the proposed MSFWAM.

multi-scale decoder features using a spatially adaptive fusion strategy [25]. Each decoder feature map passes through a 1×1 convolution, then splits into two branches. The weighting branch generates spatial weights via global average pooling, two 1×1 convolutions, and a sigmoid activation. The feature branch preserves the original features. These are fused through element-wise multiplication, enabling selective emphasis on discriminative semantic regions while suppressing background noise and redundant scale information. The aggregated features are then fused with shallow detail features. To enhance efficiency and representation, the concatenated features pass through ConvBlock2, consisting of a 1×1 pointwise convolution, a 3×3 depthwise convolution, and another 1×1 pointwise convolution. Finally, a similarity-guided gating mechanism is applied to enforce semantic consistency and preserve fine texture details.

$$F = \sigma(\phi(F_1 \times F_2^T)) \odot F_2 + (1 - \sigma(\phi(F_1 \times F_2^T))) \odot F_1, \tag{1}$$

where F is the output feature maps. Formally, a channel-wise similarity map $\phi(F_1 \times F_2^T)$ is computed via dot product, with F_1 referring to the shallow features and F_2 to the fused features, σ denotes the sigmoid function to generate adaptive weights (denoted as α in the Fig. 4). This allows the network to selectively emphasize either semantic or detail information during fusion.

4 Experiments

4.1 Experimental Setup

To evaluate the performance of MSFWAN, experiments are conducted on three datasets: a self-constructe DHAND-AHUT2025, OUHANDS [16], and a subset of Ego2Hands (Ego2Hands-ComplexBG) [15]. DHAND-AHUT2025 contains 4,000 in-air handwriting images from 10 volunteers, simulating diverse lighting, angles, and challenging conditions like motion blur and background clutter. The dataset is split 80/20 into 3,200 training and 800 testing samples. The Laplacian variance of the full image is 584, while that of the hand region is 92, indicating significant blurring in the hand area. OUHANDS consists of 3,000 hand images from 23 volunteers, featuring 10 gestures and annotations including masks, depth maps, and bounding boxes. It varies in background complexity, lighting, skin tone, and occlusion, with a 2,000/1,000 train/test split. The Ego2Hands-ComplexBG dataset includes 2,000 images with complex backgrounds, extracted from videos involving four subjects, used exclusively for training.

Experiments are conducted using the PyTorch framework on a workstation with an NVIDIA GeForce RTX 4090 GPU. Data augmentation includes random cropping, horizontal flipping, Gaussian blur, and rotation. Input images are resized to 640×480 pixels. Models are trained for 77,400 iterations (approximately 100 epochs) with a batch size of 4. Optimization is performed using the Adam optimizer, with an initial learning rate of 5×10^{-4}, decayed to 1% via cosine annealing.

The hand segmentation model is evaluated across three key dimensions: accuracy, robustness, and computational efficiency. Segmentation accuracy is measured using Intersection over Union (IoU) [3]. Robustness is assessed via the F1 score [18]. Computational efficiency is evaluated using GFLOPS and parameter count.

Additionally, the inference speed was measured in terms of frames per second (FPS) on a system equipped with an NVIDIA GeForce 2080 Ti GPU. Over a benchmark of 200 images with a resolution of 480×640, the model achieved an average processing speed of 53.44 FPS.

4.2 Comparison Experiment of Different Methods

Comparisons with several representative competitors are performed on the DHAND-AHUT2025 dataset. As shown in Table 2, the proposed MSFWAN achieves leading performance across all evaluation metrics, demonstrating superior accuracy, efficiency, and robustness. Without requiring an excessively large model size, the method attains 95.82% IoU and 97.87% F1 score, outperforming existing lightweight and medium-scale models.

As shown in Table 2, the proposed MSFWAN achieves the best overall performance with only 1.59M parameters, about 6.5% of SegFormer-B2, while surpassing it in both IoU and F1. This highlights a favorable trade-off between accuracy and model size. Moreover, our method maintains low computational cost (10.45 GFLOPs) compared with heavier models, while still outperforming lightweight designs such as RTFormer-Slim and DDR-Net-23-slim, demonstrating superior efficiency in capturing complex hand features.

As shown in Table 3, MSFWAN achieves the best performance on the OuHands dataset (82.10% IoU and 90.32% F1) with only 1.59M parameters, significantly outperforming both lightweight and large-scale baselines. Compared with RTFormer-Slim, our model attains higher accuracy with fewer parameters, and it surpasses larger architectures such as SegFormer-B1 while requiring far less computational cost. These results confirm that the proposed design not only

Table 2. Performance comparison on our self-constructed dataset.

Methods	IoU (%)	F1 (%)	GFLOPs	Params (M)
DDR-Net-23-slim [6]	88.23	93.75	5.56	5.73
DDR-Net-23 [6]	90.31	94.91	21.81	20.29
PIDNet-S [24]	95.53	97.71	6.99	7.62
PP-LiteSeg-T [17]	94.76	97.31	6.71	8.03
PP-LiteSeg-B [17]	94.36	97.10	10.59	12.07
STDC1-Seg [4]	95.01	97.44	9.89	8.28
STDC2-Seg [4]	95.21	97.56	13.76	12.32
SegFormer-B0 [23]	95.08	97.48	7.90	3.71
SegFormer-B1 [23]	95.39	97.64	15.47	13.67
SegFormer-B2 [23]	95.54	97.72	24.85	24.72
RTFormer-Slim [21]	95.25	97.57	**5.31**	5.14
RTFormer-Base [21]	95.51	97.70	19.78	16.86
MSFWAN	**95.82**	**97.87**	10.45	**1.59**

alleviates feature suppression but also strengthens fine-grained structure representation, leading to superior segmentation of complex hand geometry. Importantly, the compact architecture demonstrates strong potential for deployment in real-world, resource-constrained environments.

Table 3. Performance comparison on the OuHands dataset.

Methods	IoU (%)	F1 (%)	GFLOPs	Params (M)
DDR-Net-23-Slim [6]	72.32	84.01	5.56	5.73
DDR-Net-23 [6]	74.88	86.93	21.81	20.29
PIDNet-S [24]	72.53	84.08	6.99	7.62
PP-LiteSeg-T [17]	78.20	87.76	6.71	8.03
PP-LiteSeg-B [17]	76.74	86.84	10.59	12.07
STDC1-Seg [4]	73.15	84.49	9.89	8.28
STDC2-Seg [4]	74.38	85.31	13.76	12.32
SegFormer-B0 [23]	78.13	87.72	7.90	3.71
SegFormer-B1 [23]	79.01	88.27	15.47	13.67
SegFormer-B2 [23]	78.96	88.24	24.85	24.72
RTFormer-Slim [21]	81.87	90.03	**5.31**	5.14
RTFormer-Base [21]	80.20	89.01	19.78	16.86
MSFWAN	**82.10**	**90.32**	10.45	**1.59**

Our method strikes an excellent balance between segmentation performance and model compactness, consistently outperforming existing lightweight baselines with significantly fewer parameters, as shown in Table 4. Despite having only 1.59M parameters, it surpasses models like PIDNet-S and PP-LiteSeg-T by 0.7% and 0.6% in IoU. While SegFormer-B0 is also a compact transformer model, our method achieves higher accuracy with less than half its parameter count, demonstrating superior architectural efficiency. Although the computational cost is slightly higher than RTFormer-Slim, the improvement in IoU (3.35 points) justifies the added cost, thanks to the integration of HFAM and MSFWAN. This efficiency makes our model particularly well-suited for real-time applications on mobile and edge platforms.

Table 4. Performance Comparison on the Ego2Hands-ComplexBG dataset

Methods	IoU (%)	F1 (%)	GFLOPs	Params (M)
DDR-Net-23-slim [6]	88.34	93.77	5.56	5.73
DDR-Net-23 [6]	89.62	94.53	21.81	20.29
PIDNet-S [24]	89.72	94.54	6.99	7.62
PP-LiteSeg-T [17]	89.86	94.66	6.71	8.03
PP-LiteSeg-B [17]	89.28	94.34	10.59	12.07
STDC1-Seg [4]	75.91	86.31	9.89	8.28
STDC2-Seg [4]	75.42	85.98	13.76	12.32
SegFormer-B0 [23]	88.42	93.85	7.90	3.71
SegFormer-B1 [23]	89.44	94.43	15.47	13.67
SegFormer-B2 [23]	89.77	94.61	24.85	24.72
RTFormer-Slim [21]	87.10	93.11	**5.31**	5.14
RTFormer-Base [21]	88.85	94.09	19.78	16.86
MSFWAN	**90.45**	**94.99**	10.45	**1.59**

4.3 Ablation Studies

Table 5. Ablation study of HFA and SGF modules on two datasets.

Datasets	HFAM	MSFWAM	IoU (%)	F1 (%)	Params (M)
DHAND-AHUT2025			95.52	97.71	**1.41**
	✓		95.75	97.83	1.50
		✓	95.80	97.85	1.51
	✓	✓	**95.82**	**97.87**	1.59
Ego2Hands-ComplexBG			88.46	93.87	**1.41**
	✓		89.58	94.72	1.50
		✓	89.67	94.50	1.51
	✓	✓	**90.46**	**94.99**	1.59

Ablation studies are conducted to assess the contributions of each module: HFAM and MSFWAM. On both the DHAND-AHUT2025 and Ego2Hands-ComplexBG datasets, integrating HFAM and MSFWAM leads to consistent improvements in segmentation performance. The full model achieves the best results, demonstrating the complementary nature of these modules in enhancing accuracy while preserving compactness. These findings highlight the effectiveness of HFAM and MSFWAM in improving model generalization without significantly increasing computational cost (Table 5).

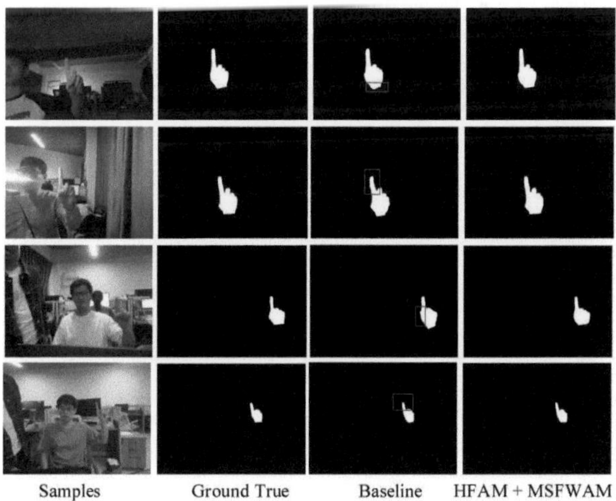

Fig. 5. Segmentation effect of the MSFWAN.

To validate the effectiveness of the HFAM and MSFWAM, feature map visualizations are presented under four challenging hand segmentation scenarios, as shown in Fig 5. The evaluated conditions include: (a) low-light, (b) overexposure, (c) motion blur, and (d) feature suppression. Comparative results between the baseline network and the proposed method demonstrate that the two modules significantly enhance feature discriminability. The network with HFAM and MSFWAM exhibits more pronounced semantic responses in hand-related regions, while also maintaining more complete and coherent edge structures under adverse conditions.

5 Conclusion

In this paper, we proposed a hand segmentation framework for complex scenes with motion blur and background interference, integrating an efficient backbone with two key modules: HFAM and MSFWAM. The HFAM employs a multi-scale pyramidal structure with adaptive feature selection to enhance fine-grained structural perception and suppress irrelevant background noise. The MSFWAM adopts a feature-similarity-guided fusion strategy to alleviate motion-induced distortions and improve representation stability. Extensive experiments demonstrated that the proposed method achieved high segmentation accuracy and robustness with minimal computational overhead, highlighting its practicality in real-world applications. However, we observe that the current model remains susceptible to heavy occlusion, where the segmentation mask may fail around the occluded parts of the hand. Developing effective strategies to mitigate this occlusion problem remains a valuable direction for future work.

Acknowledgements. This work is supported by the Natural Science Research Project of Anhui Educational Committee under Grant No. 2024AH040028, the National Nature Science Foundation of China (NSFC) under Grant Nos. 62206006, 61906003, the Opening Foundation of Information Materials and Intelligent Sensing Laboratory of Anhui Province under Grant No. IMIS202215.

References

1. Bambach, S., Lee, S., Crandall, D.J., Yu, C.: Lending a hand: detecting hands and recognizing activities in complex egocentric interactions. In: Proceedings of the IEEE International Conference on Computer Vision, pp. 1949–1957 (2015)
2. Cai, M., Lu, F., Sato, Y.: Generalizing hand segmentation in egocentric videos with uncertainty-guided model adaptation. In: Proceedings of the IEEE/CVF Conference on Computer Vision and Pattern Recognition, pp. 14392–14401 (2020)
3. Everingham, M., Van Gool, L., Williams, C.K., Winn, J., Zisserman, A.: The pascal visual object classes (VOC) challenge. Int. J. Comput. Vis. **88**, 303–338 (2010)
4. Fan, M., et al.: Rethinking BiSeNet for real-time semantic segmentation. In: Proceedings of the IEEE/CVF Conference on Computer Vision and Pattern Recognition, pp. 9716–9725 (2021)

5. Fathi, A., Ren, X., Rehg, J.M.: Learning to recognize objects in egocentric activities. In: Proceedings of the IEEE Conference on Computer Vision and Pattern Recognition, pp. 3281–3288 (2011)
6. Hong, Y., Pan, H., Sun, W., Jia, Y.: Deep dual-resolution networks for real-time and accurate semantic segmentation of road scenes. arXiv:2101.06085 (2021)
7. Howard, A., et al.: Searching for MobileNetv3. In: Proceedings of the IEEE/CVF International Conference on Computer Vision, pp. 1314–1324 (2019)
8. Hu, J., Shen, L., Sun, G.: Squeeze-and-excitation networks. In: Proceedings of the IEEE Conference on Computer Vision and Pattern Recognition, pp. 7132–7141 (2018)
9. Jiang, H., Liu, S., Wang, J., Wang, X.: Hand-object contact consistency reasoning for human grasps generation. In: Proceedings of the IEEE/CVF International Conference on Computer Vision, pp. 11107–11116 (2021)
10. Kawulok, M., Kawulok, J., Nalepa, J.: Spatial-based skin detection using discriminative skin-presence features. Pattern Recognit. Lett. **41**, 3–13 (2014)
11. Kawulok, M., Kawulok, J., Nalepa, J., Smolka, B.: Self-adaptive algorithm for segmenting skin regions. EURASIP J. Adv. Signal Process. **2014**, 1–22 (2014)
12. Khan, R., Hanbury, A., Stöttinger, J., Bais, A.: Color based skin classification. Pattern Recognit. Lett. **33**(2), 157–163 (2012)
13. Li, C., Kitani, K.M.: Pixel-level hand detection in ego-centric videos. In: Proceedings of the IEEE Conference on Computer Vision and Pattern Recognition, pp. 3570–3577 (2013)
14. Li, M., Sun, L., Huo, Q.: Flow-guided feature propagation with occlusion aware detail enhancement for hand segmentation in egocentric videos. Comput. Vis. Image Underst. **187**, 102785 (2019)
15. Lin, F., Price, B., Martinez, T.: Ego2Hands: a dataset for egocentric two-hand segmentation and detection. arXiv:2011.07252 (2020)
16. Matilainen, M., Sangi, P., Holappa, J., Silvén, O.: Ouhands database for hand detection and pose recognition. In: Proceedings of the International Conference on Image Processing Theory, Tools and Applications, pp. 1–5. IEEE (2016)
17. Peng, J., et al.: PP-LiteSeg: a superior real-time semantic segmentation model. arXiv:2204.02681 (2022)
18. Powers, D.M.: Evaluation: from precision, recall and f-measure to ROC, informedness, markedness and correlation. arXiv:2010.16061 (2020)
19. Ren, X., Gu, C.: Figure-ground segmentation improves handled object recognition in egocentric video. In: Proceedings of the IEEE Conference on Computer Vision and Pattern Recognition, pp. 3137–3144 (2010)
20. Urooj, A., Borji, A.: Analysis of hand segmentation in the wild. In: Proceedings of the IEEE Conference on Computer Vision and Pattern Recognition, pp. 4710–4719 (2018)
21. Wang, J., et al.: RTFormer: efficient design for real-time semantic segmentation with transformer. Adv. Neural. Inf. Process. Syst. **35**, 7423–7436 (2022)
22. Wang, W., Yu, K., Hugonot, J., Fua, P., Salzmann, M.: Beyond one glance: gated recurrent architecture for hand segmentation. arXiv:1811.10914 (2018)
23. Xie, E., Wang, W., Yu, Z., Anandkumar, A., Alvarez, J.M., Luo, P.: SegFormer: simple and efficient design for semantic segmentation with transformers. Adv. Neural. Inf. Process. Syst. **34**, 12077–12090 (2021)
24. Xu, J., Xiong, Z., Bhattacharyya, S.P.: PIDNet: a real-time semantic segmentation network inspired by PID controllers. In: Proceedings of the IEEE/CVF Conference on Computer Vision and Pattern Recognition, pp. 19529–19539 (2023)

25. Zhang, Z., Fu, H., Dai, H., Shen, J., Pang, Y., Shao, L.: ET-net: a generic edge-attention guidance network for medical image segmentation. In: Proceedings of the International Conference on Medical Image Computing and Computer-Assisted Intervention, pp. 442–450. Springer (2019)
26. Zhu, X., Jia, X., Wong, K.Y.K.: Pixel-level hand detection with shape-aware structured forests. In: Proceedings of the Asian Conference on Computer Vision, pp. 64–78. Springer (2015)

Photometric Compensation for Projection on Color Surface Under Low Illumination

Chang Wang[1], Zheng'ang Liu[2], Guangxu Li[3], Jingjie Zhou[4], and Tohru Kamiya[1(✉)]

[1] Kyushu Institute of Technology, Fukuoka 8030835, Japan
kamiya@cntl.kyutech.ac.jp
[2] Southeast University, Nanjing 210096, China
[3] Tiangong University, Tianjin 300387, China
[4] Tellyes Scientific Inc., Tianjin 300384, China

Abstract. Augmented reality (AR) has been widely applied in medical education that allows for digitally generated three-dimensional representations to be integrated with physical training module. However, correcting color distortion in projected displays is a major challenge. In this paper, we propose a convolutional neural network based projector photometric compensation framework to compensate for color shift. The architecture is a generative network cascading a decoder subnet. Such architecture compared the color shift of the projection from the input image, as well as the original textures. We used a custom projector with low-illumination for our experiments. To compensate content losses of the projection, we utilized a sharpening module at the end of network output. Additionally, to adapt the target projection screen using small dataset, we synthesize solid color images automatically for network training. The experiments results show that our solution of projector photometric compensation outperforms state-of-the-art method.

Keywords: photometric compensation · color correction · convolutional neural network

1 Introduction

In the past decade, spatial augmented reality (SAR) has gained global popularity [1]. Key challenges in projecting color images onto colored surfaces include geometric distortion on non-planar screens and color distortion on non-white surfaces [3]. To address color distortion issues, photometric compensation algorithms utilize camera feedback in projector-camera systems to achieve closed-loop color correction.

In virtual medical training, organ or wound images are projected onto skin-colored models for enhanced realism. To maintain fidelity, high-luminance projectors ($\geq$ 3000 ANSI lumens) are typically used, but they are costly and risky for consumer applications. This study employs a low-cost projector (300–600 ANSI lumens) and proposes a color compensation algorithm to approximate the

P. Umapada et al. (Eds.): ICCPR 2025, CCIS 2811, pp. 185–195, 2026.
https://doi.org/10.1007/978-981-95-8315-7_15

expected appearance. The model is trained offline, with real-time inference as illustrated in Fig. 1.

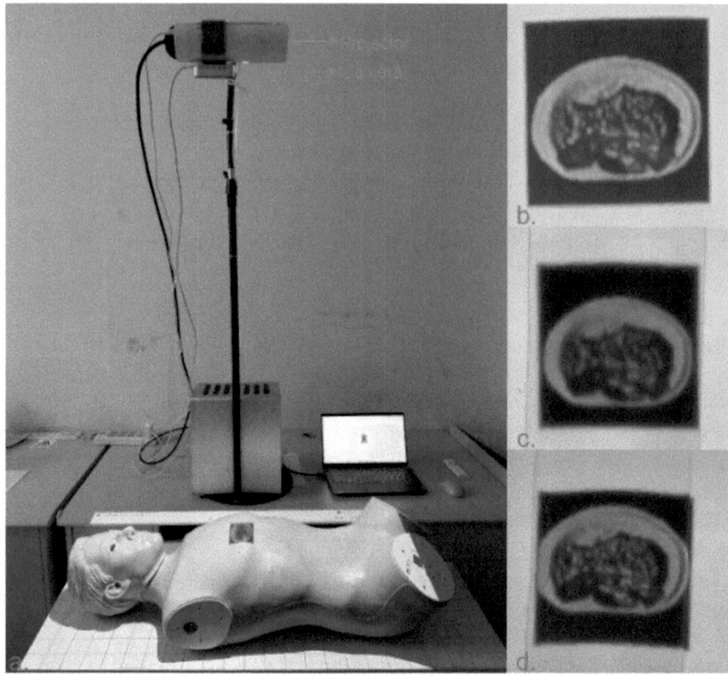

Fig. 1. Proposed photometric compensation system. (a) Projector-camera system, where the camera is mounted on the projector. (b) Original undistorted image projected on an ideal white plane. (c) Color-distorted scar image projected onto the colored model surface. (d) Compensated scar image projected onto the colored model surface.

In this paper, we investigate color compensation for projected images using a consumer projector to map scar images onto a body model, simulating injury scenarios. The camera captures the projected colors for evaluation, and based on this feedback, a composite transfer function is applied to correct the projected image. This work introduces an end-to-end framework designed for low-cost projection systems with efficient training strategies.

2 Related Works

Projector photometric compensation involves complex nonlinear relationship between the radiometric response of projector-camera sensors [6], lens distortion [7], surface reflectance, and ambient lighting. Compensation models are typically categorized into content-independent and content-dependent approaches

[2,3]. Content-independent methods assume a one-to-one pixel mapping between projector and camera. Bimber et al. [8] proposed a linearized multi-projector system with independent RGB channels. Bimber et al. [9] extended compensation to global effects like scattering, albeit with long scan times. Grundhofer et al. [10] used a TPS-based model for nonlinear compensation.

These methods struggle in practice, as camera pixels receive light from multiple surface areas [5].

Content-dependent methods use contextual pixel information. Grundhofer et al. [4] introduced per-pixel nonlinear mapping for uncalibrated devices. Li et al. [11] applied sparse sampling and interpolation. Aliaga et al. [12] proposed real-time scaling. Takeda et al. [13] used UV LEDs, and Huang et al. [5] introduced CompenNet, a CNN-based end-to-end model with strong visual performance.While deep learning improves compensation, it relies on high-quality datasets from precision equipment, increasing cost.

To address this, we propose a model trained on a pure-color dataset using a 450 ANSI lumen projector and a low-resolution camera. Unlike CompenNet's complex data, our approach is simple, low-cost, and suited for small-scale projection systems.

3 Methods

3.1 Projection-Camera Process Modelling and Compensation

The entire process by which the input original image x is captured by the camera after being projected onto a colored textured surface by a projector as illustrated in Fig. 2. We model the projector-camera imaging process using transfer func-

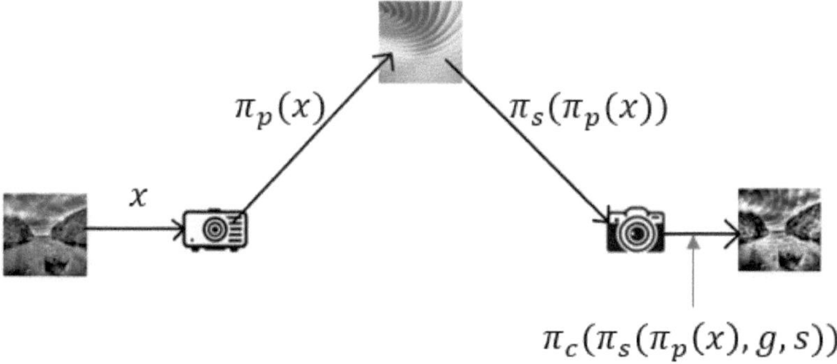

Fig. 2. Projector-camera imaging process: input image x is projected onto a colored surface and captured by the camera.

tions. Given input image x, projector function π_p, camera function π_c, surface reflectance π_s, surface properties s, and illumination g, the captured image $\tilde{x}$ is:

$$\tilde{x} = \pi_c(\pi_s(\pi_p(x)), g, s) \tag{1}$$

In a backlit scenario with low ambient lighting and limited camera dynamic range, we capture a reference reflectance image $\tilde{s}$ by projecting a uniform white image x_0:

$$\tilde{s} = \pi_c(\pi_s(\pi_p(x_0)), g, s) \tag{2}$$

Letting π denote the full projector-camera transfer function, and replacing g, s with $\tilde{s}$, the compensation goal becomes finding the compensated image x^* satisfying:

$$\pi(x^*; \tilde{s}) = x \quad \Rightarrow \quad x^* = \pi^+(x; \tilde{s}) \tag{3}$$

Such explicit models are limited by their content-independent assumptions and the complexity of real-world photometric interactions, making accurate modeling impractical. We propose a deep network based on U-Net [14], learning feature differences between input images and camera-captured projections [5].

3.2 Network Modeling and Compensation Process

The network consists of a U-Net backbone and an auto-encoder sub-network, designed to extract rich features from camera-captured projections. The training architecture is shown in Fig. 3.

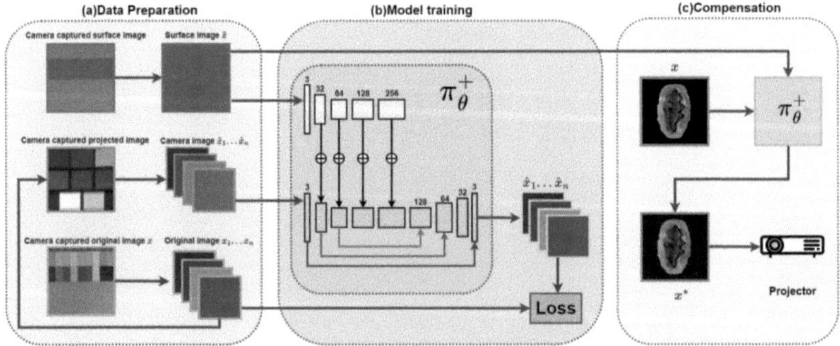

Fig. 3. Training and compensation pipeline for projector photometric compensation. (a) Data acquisition and preprocessing. (b) Network architecture and training process. (c) Real-time compensation process. (Color figure online)

Figure 3(a) illustrates the data acquisition process, forming training pairs of camera-captured projections and corresponding standard images. These are input to the convolutional network in (b) for training. In (c), the trained model π_{θ}^{+} is used to compensate projected images on specific colored surfaces.

The yellow block in Fig. 3(b) represents a convolutional auto-encoder, which extracts spatial features via convolution and pooling, enabling unsupervised learning with reduced parameters. Different colors distinguish the auto-encoder and backbone, and black arrows indicate learned surface reflection features passed from the auto-encoder.

The blue block in sub-figure (b) represents the backbone network, based on a U-Net-like encoder-decoder with skip connections. Architecture details are shown in Table 1.

Table 1. U-Net-like backbone network internal structure

Convolutional layer	Types	Kernel size	Output image
Skiplayer1	skip convolutional layer	3×3	$3 \times 256 \times 256$
Conv1	sub-sampling layer	3×3	$32 \times 128 \times 128$
Skiplayer2	skip convolutional layer	1×1	$64 \times 128 \times 128$
Conv2	sub-sampling layer	3×3	$64 \times 64 \times 64$
Skiplayer3	skip convolutional layer	1×1	$128 \times 64 \times 64$
Conv3	convolutional layer	3×3	$128 \times 64 \times 64$
Conv4	convolutional layer	3×3	$256 \times 64 \times 64$
Conv5	convolutional layer	3×3	$128 \times 64 \times 64$
Transposed Conv1	upsampling layer	2×2	$64 \times 128 \times 128$
Transposed Conv2	upsampling layer	2×2	$32 \times 256 \times 256$
Conv6	output layer	3×3	$3 \times 256 \times 256$

Three skip convolutional layers pass low-level features to the decoder. The encoder extracts global features via convolution and sub-sampling, while the decoder upsamples to $32 \times 256 \times 256$ using transposed convolutions, integrates skip features, and outputs a $3 \times 256 \times 256$ RGB image.

U-Net employs skip connections between corresponding downsampling and upsampling layers to bridge semantic gaps across different resolution levels. As illustrated in Fig. 4, these connections transfer high-resolution feature maps from the encoder directly to the decoder, enabling precise localization. This fusion mechanism facilitates the combination of low-level spatial details with high-level semantic information, thereby enhancing the network's capacity for fine-grained prediction tasks.

The input is an uncompensated image. The encoder models spectral distortions from projection, aided by a self-encoder. The decoder reconstructs the compensated image by fusing semantic and low-level features.

During network training, the camera-captured uncompensated image and the surface image are coded and decoded by convolutional sequences of semantic information of the image to learn the mapping π_θ^+ from the uncompensated image to the original image, and in the compensation phase, the compensated image x^* for projection is obtained by inputting the same camera captured surface image $\tilde{s}$ and the image x that the observer is expected to see which is projected onto the colored surface (Fig. 4).

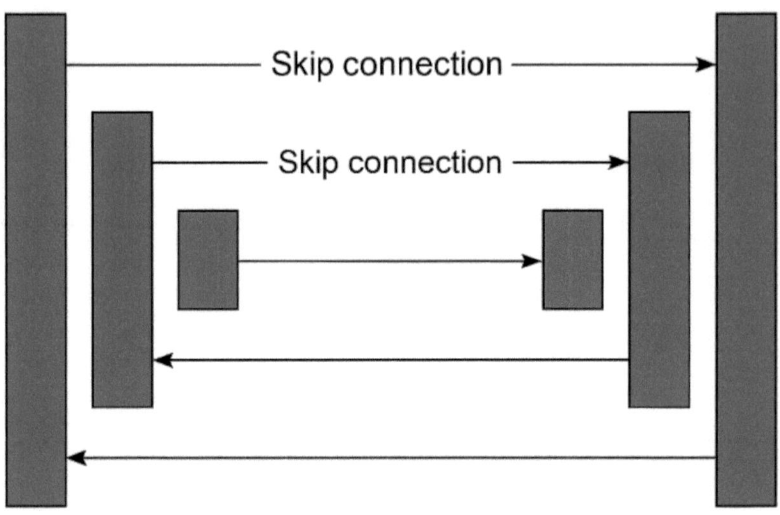

Fig. 4. U-Net Skip Connections for Feature Fusion.

3.3 Loss Function

As shown in Eq. 4, the CompenNet loss function combines pixel-wise ℓ_1 loss and SSIM loss to jointly optimize color and structural similarity:

$$\mathcal{L} = \mathcal{L}_{\ell_1} + \mathcal{L}_{SSIM} \tag{4}$$

The ℓ_1 loss measures absolute error between prediction and ground truth:

$$\mathcal{L}_{\ell_1}(y, \hat{y}) = \frac{1}{n} \sum_{i=1}^{n} |y_i - \hat{y}_i| \tag{5}$$

ℓ_1 loss is robust to outliers but may cause blurring and lacks high-frequency detail preservation.

SSIM loss measures structural similarity based on luminance, contrast, and structure, aligning better with human perception. It is computed block-wise and defined as $\mathcal{L}_{SSIM} = 1 - SSIM$, yielding 0 when images are identical.

Combining ℓ_1 and SSIM loss balances pixel accuracy and perceptual quality, mitigating issues like blurring or color shifting when either loss is used alone.

3.4 Image Sharpening

Due to low-lumen projector limitations, structural information is lost during projection, causing contour distortion and edge blurring. This hardware constraint reduces contrast ratios and causes fine details to merge with background noise. We use image sharpening to enhance structural information without altering color characteristics.

Spatial domain sharpening enhances images through differential operations, using first-order or second-order derivatives to strengthen contours and edges. The fundamental principle relies on detecting intensity gradients and amplifying high-frequency components while preserving low-frequency color information. Methods include gradient operators, Sobel operators, and Laplace operators, each offering different trade-offs between edge enhancement and noise sensitivity.

For our application, we implement adaptive sharpening with intensity-dependent enhancement factors. The sharpening strength is modulated based on local contrast to prevent over-enhancement in smooth regions while ensuring adequate enhancement in edge areas.

4 Experiment

The projector-camera system built in the experiments in this paper is shown in Fig. 1. We used a low-cost civilian low-lumen projector for projection and a cell phone camera to capture the training data.

4.1 Datasets

Three datasets (Group1, Group2, Group3) were used to form training pairs $(x, \tilde{x})$, all generated by our projection system. Group1 and Group2 follow the standard dataset from Huang [5], with differences shown in Table 2.

In Group1, $\tilde{x}$—captured after projection onto a colored surface—shows significant sharpness loss compared to the original x, especially under low-lumen projection. As discussed in the Introduction, existing methods assume pixel-wise color correspondence, which fails when structural details are lost due to low brightness.

To address this, we redefine the target x as the image captured from a white surface projection, aiming for the network to reproduce that appearance. This improves training performance but still yields limited compensation quality in practice.

Instead of using complex color images from Huang's dataset [5], we construct a simplified dataset using pure color images. These images lack structural detail and maintain uniform color, allowing compensation without pixel-level correspondence—ideal for low-lumen projection where structural loss is severe.

In our method, x is captured from a Kodak Tiffin Q-13 standard color card. The distorted image $\tilde{x}$ is obtained by projecting SWOP2006-standard colors onto a colored surface and capturing the result. To minimize ambient light interference, data collection is conducted in a dark room.

Table 2. Datasets

Group	$\tilde{x}$	x	Size of datasets
1	Original color images of the standard dataset captured by the camera and projected onto a humanoid skin surface by the projector	Color original images in standard data sets	500 training, 200 testing
2	The original color images of the colored surface projection captured by the camera	The original color images of the projection onto the white surface captured by the camera	500 training, 200 testing
3	Solid color images of tinted projected surface captured by the camera	Solid color map of camera capture of Kodak standard color card	8 training, 5 testing

4.2 Training Setup

In this paper, we implement CompenNet using PyTorch and train it using the Adam optimizer. The training specifications are as follows: β_1 is set to 0.9, the penalty factor of the ℓ_1 loss function is set to 10^{-4}, the initial learning rate is set to 10^{-3}, and it is reduced by a factor of 5 every 800 iterations. The model weights are initialized using Kaiming's method [15].

The complex color image group is trained on two NVIDIA Tesla P100 GPUs with a batch size of 64 for 1000 iterations, taking approximately 23 min to complete. The solid color group is trained for 1000 iterations with a batch size of 4 on an NVIDIA GeForce GTX 1650 GPU, taking approximately 2 min and 20 s to complete.

4.3 Evaluation Metrics and Result

Algorithm evaluation includes training loss, PSNR, RMSE, and SSIM between the generated image $\hat{x}$ and the ground truth x. PSNR measures the signal-to-noise ratio; higher values indicate better reconstruction quality. RMSE reflects the average deviation and is sensitive to outliers; lower values indicate higher accuracy. SSIM evaluates image similarity in terms of brightness, contrast, and structure, with values closer to 1 indicating better quality.

The training loss is very high in the complex color image group. After improvement, the training loss decreased by 0.3, and the three metrics PSNR, SSIM, and RMSE also improved, but 0.4945 remains a high loss value. For the model trained on solid color images, the training loss is very low at only 0.0239. Due to the limited pure color data used, the training exhibits some overfitting, as the RMSE is only 0.0369 on the training set but 0.2151 on the validation set. Although the trained model shows some degree of overfitting, the actual compensation performance is significantly better than the complex color model (Table 3).

Table 3. Results

Group	Train_loss	Valid_SSIM	Train_RMSE	Valid_RMSE	Valid_PSNR
Group 1	0.7918	0.3008	0.2749	0.2994	15.2462
Group 2	0.4950	0.2126	0.2137	0.1515	21.1654
Group 3	0.0239	0.7891	0.0369	0.2151	17.8013

The model trained on complex color images (Fig. 5) fails to deliver effective compensation and introduces structural distortion. In contrast, training with pure color images enables more accurate color mapping due to the absence of structural complexity.

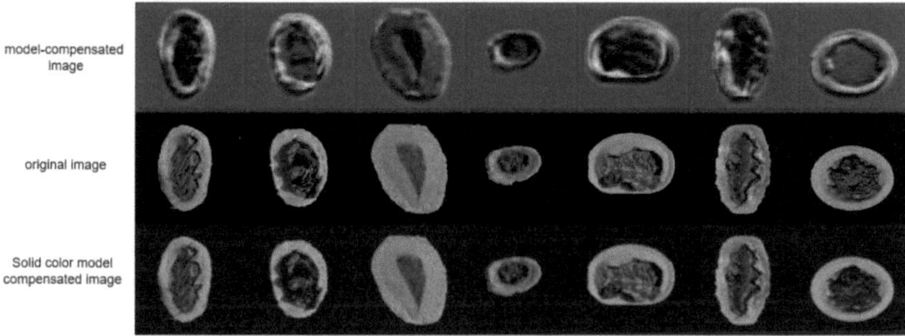

Fig. 5. Compensation result.

Figure 6 shows that the compensated image is visually closer to a white-surface projection. However, projection onto colored surfaces results in noticeable sharpness loss, due to limited projector brightness and higher structural degradation.

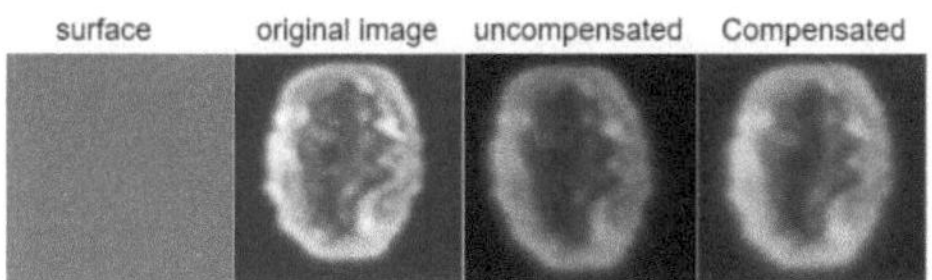

Fig. 6. Projection result using the pure color group model.

To address this, we apply image sharpening to enhance structural clarity without altering color. This improves visual quality in low-lumen projection. The sharpening result is shown in Fig. 7.

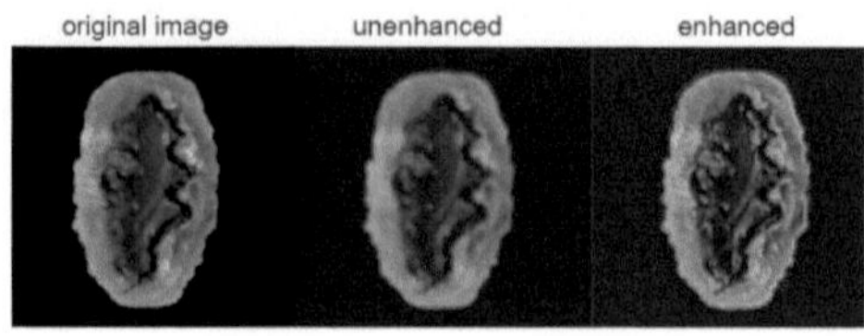

Fig. 7. Compensated image after image enhancement (sharpening).

5 Conclusion

This paper investigates color compensation for colored 3D surface projection under low-lumen projectors, proposing a training data generation method that enhances image structure when applying CompenNet. The approach supports low-cost projector-camera systems and is suitable for budget-constrained applications such as medical 3D projection training.

Photometric compensation relies on accurate pixel-wise color mapping between distorted and original images. Under low luminance, complex images suffer from blur and structural loss, making such mappings unreliable. In contrast, pure color images, lacking structural complexity, retain color integrity during projection, enabling more effective training. Our experiments show that models trained on pure color datasets yield better compensation results.

Additionally, we apply Sobel-based image sharpening after compensation to enhance structural clarity without altering color, mitigating structure loss typical in low-lumen projection.

Acknowledgements. The author acknowledges the financial support from the China Scholarship Council (CSC) during their overseas study.

Disclosure of Interests. The authors declare no competing interests relevant to the content of this article. Author Chang Wang acknowledges the support from the China Scholarship Council (CSC) for overseas study.

References

1. Iwai, D.: Projection mapping technologies: a review of current trends and future directions. Proc. Jpn. Acad. Ser. B **100**(3), 234–251 (2024)
2. Grundhöfer, A., Iwai, D.: Recent advances in projection mapping algorithms, hardware and applications. Comput. Graphics Forum **37**, 653–675 (2018)
3. Bimber, O., Iwai, D., Wetzstein, G., Grundhöfer, A.: The visual computing of projector-camera systems. In: ACM SIGGRAPH 2008 Classes, pp. 1–25 (2008)
4. Grundhöfer, A., Bimber, O.: Real-time adaptive radiometric compensation. In: ACM SIGGRAPH 2006 Research posters, pp. 56–es (2006)
5. Huang, B., Ling, H.: End-to-end projector photometric compensation. In: Proceedings of the IEEE/CVF Conference on Computer Vision and Pattern Recognition, pp. 6810–6819 (2019)

6. Grossberg, M.D., Peri, H., Nayar, S.K., Belhumeur, P.N.: Making one object look like another: controlling appearance using a projector-camera system. In: Proceedings of the 2004 IEEE Computer Society Conference on Computer Vision and Pattern Recognition, CVPR 2004., vol. 1, p. I (2004)
7. Fischer, A., Gloe, T.: Forensic analysis of interdependencies between vignetting and radial lens distortion. In: Media Watermarking, Security, and Forensics 2013, vol. 8665, pp. 109–123 (2013)
8. Bimber, O., Emmerling, A., Klemmer, T.: Embedded entertainment with smart projectors. In: ACM SIGGRAPH 2005 Courses, pp. 8–es (2005)
9. Wetzstein, G., Bimber, O.: Radiometric compensation through inverse light transport. In: 15th Pacific Conference on Computer Graphics and Applications (PG 2007), pp. 391–399 (2007)
10. Grundhöfer, A., Iwai, D.: Robust, error-tolerant photometric projector compensation. IEEE Trans. Image Process. **24**, 5086–5099 (2015)
11. Li, Y., Majumder, A., Gopi, M., Wang, C., Zhao, J.: Practical radiometric compensation for projection display on textured surfaces using a multidimensional model. Comput. Graphics Forum **2**, 365–375 (2018)
12. Aliaga, D.G., Yeung, Y.H., Law, A., Sajadi, B., Majumder, A.: Fast high-resolution appearance editing using superimposed projections. ACM Trans. Graphics (TOG) **31**, 1–13 (2012)
13. Takeda, S., Iwai, D., Sato, K.: Inter-reflection compensation of immersive projection display by spatio-temporal screen reflectance modulation. IEEE Trans. Visual Comput. Graphics **22**(4), 1424–1431 (2016)
14. Ronneberger, O., Fischer, P., Brox, T.: U-net: convolutional networks for biomedical image segmentation. In: Medical Image Computing and Computer-Assisted Intervention-MICCAI 2015: 18th International Conference, Munich, Germany, 5–9 October 2015. Proceedings, Part III(18), pp. 234–241 (2015)
15. He, K., Zhang, X., Ren, S., Sun, J.: Delving deep into rectifiers: surpassing human-level performance on ImageNet classification. In: Proceedings of the IEEE International Conference on Computer Vision (ICCV), pp. 1026–1034 (2015)

Adaptive Color Correction in Turbid Underwater Environments: A Comprehensive Approach for Enhanced Underwater Imaging

Yuchao Zheng[1], Yiqing Zhang[2], Huimin Lu[3], and Tohru Kamiya[1(✉)]

[1] Kyushu Institute of Technology, Kitakyushu, Fukuoka, Japan
zheng.yuchao492@mail.kyutech.jp, kamiya@cntl.kyutech.ac.jp
[2] Foundation Gang-Wu (Changzhou) Water Supply Co., Ltd., Changzhou, Jiangsu, China
[3] Southeast University, Nanjing, Jiangsu, China
luhuimin@ericlab.org

Abstract. Underwater images suffer from visibility loss, hue shifts, and patchy brightness due to wavelength-dependent absorption, scattering, and suspended matter. We present an adaptive color-correction framework tailored for turbid water. First, a depth-aware model derived from the extended BeerLambert law fuses estimated depth and turbidity to restore edges. Second, a wavelength-adaptive module re-weights RGB channels according to measured attenuation coefficients, retrieving true colours under ideal lighting. Third, a spatial-variance algorithm locally adjusts chroma to eliminate regional casts and equalize tone. These elements are integrated into a unified pixel-level correction function driven by depth, turbidity, wavelength, and local luminance cues. Experiments in mildly to heavily turbid tanks, validated with a full-spectrum colour card, show the method surpasses state-of-the-art algorithms in colour fidelity, detail preservation, and speed. The approach therefore boosts image quality for underwater exploration, ecological monitoring, and cultural-heritage documentation. It also runs in real time on standard hardware without extra calibration.

Keywords: Turbidity · Underwater Image Restoration · Underwater Imaging · Color Correction

1 Introduction

Underwater imaging is challenged by light absorption, scattering, and suspended particles, leading to low visibility, color distortion, and uneven brightness [1–3]. Turbid waters particularly degrade image quality, hindering clarity, color fidelity, detail, and contrast. Addressing these issues is crucial for oceanography, underwater exploration, and ecological research [4,5].

P. Umapada et al. (Eds.): ICCPR 2025, CCIS 2811, pp. 196–205, 2026.
https://doi.org/10.1007/978-981-95-8315-7_16

Various enhancement methods exist. Model-based approaches [6–8] struggle with parameter adaptation in complex turbid environments. Model-free techniques like histogram equalization [9], white balance [12], and Retinex [10] can cause over-enhancement or distortion. Deep learning methods [11,12] require extensive, high-quality training data, scarce in turbid conditions. Thus, more effective techniques are needed.

This work introduces an adaptive color correction method for turbid aquatic environments, integrating factors like depth, turbidity, wavelength attenuation, and local brightness. We first propose a novel depth-aware color correction using physical models and computer vision to model underwater light propagation with depth and turbidity parameters. A physically-based algorithm then assesses wavelength attenuation coefficients to adjust image color. Concurrently, adaptive local color correction addresses unique lighting and visibility. Finally, these methods are merged into a comprehensive color correction function, enhancing accuracy and adaptability. The contributions are:

- A physical model for turbid water imaging, extending the Beer-Lambert law with nonlinear turbidity effects to recover image contours.
- A depth-based color correction method using wavelength attenuation coefficients to initially restore colors while preserving information from recovered contours.
- An adaptive spatial variation correction method adjusting overall image color based on local lighting and visibility, preventing overly dark or bright regions.

2 Related Work

2.1 Model-Based Methods

Model-based methods utilize the physical principles of underwater light propagation and prior knowledge to formulate an imaging model, mathematically addressing image degradation. He et al. [13] introduced the dark channel prior (DCP) for dehazing. Variants adapted this for underwater scenes, such as red channel-based demisting [14] and the underwater DCP (UDCP) focusing on red channel attenuation [15]. Peng et al. [16] proposed the generalized dark channel prior (GDCP), considering color transformations and ambient light. Akkaynak et al. [17] developed a model based on attenuation rates and object distance. Others estimated scene depth [18] and subsequently used this for color correction by readjusting background light and transmittance [19]. Zhou et al. [20] designed a physical model based on pixel backscattering for parameter recovery. These methods rely on prior knowledge, which often struggles in complex turbid environments, leading to inadequate recovery of edge details.

2.2 Model-Free Methods

Model-free methods bypass underwater imaging models, improving visual quality by directly adjusting pixel values to enhance color, contrast, and sharpness. Techniques include white balance for color temperature correction [21], and Retinex-based approaches for brightness and color distribution adjustments [22]. Lu et al.

[23] established a model for turbid water scenes, adjusting ambient light estimation. Adaptive histogram methods were designed to enhance color and contrast [24]. Multi-scale fusion strategies combined with color compensation and white balance have been used to remove artifacts and restore contrast [25]. Some combined white balance, grayscale mapping, and histogram equalization [26]. Sethi et al. [27] used the Laplace pyramid method for color and contrast adjustment. Bayesian Retinex methods have recovered images from the pixel level based on light reflectivity priors [28], and color correction based on minimum color loss has been proposed for faster processing [29]. While improving color and contrast, these methods ignore the imaging mechanism, risking over/under-enhancement and inaccurate color recovery.

2.3 Deep-Learning Methods

Deep learning methods build end-to-end networks and loss functions for underwater image enhancement in a data-driven manner. To mitigate data dependency, WaterGAN [30] used an unsupervised approach to generate underwater-like images. Lu et al. [31] employed deep convolutional neural networks (CNNs) for light scattering reduction and depth estimation. CycleGAN [1], a weakly-supervised model, utilized adversarial networks for adaptability. Fabbri et al. [32] designed UGAN, which automatically generates paired datasets without requiring scene depth. Conditional generative adversarial networks (cGANs) with multi-scale generators have been used to expand perceptual range [33]. WaterNet [34] combined imaging principles with traditional enhancement techniques. Ucolor [35] considered medium transmission factors to enhance response in severely degraded regions. UIE-Net [36], an end-to-end network, incorporated color correction and defogging modules. Lin et al. [37] designed a cGAN with progressive enhancement functions. However, deep learning methods are often limited by the scarcity of high-quality underwater training data and can struggle with uneven pixel distribution in turbid water, leading to overly bright or dark results and being constrained by network size for processing speed (Fig. 1).

3 Methodology

3.1 Depth Aware Color Correction

We propose the Depth Aware Color Correction method in order to automatically adjust the image to adapt to the problem of insufficient lighting conditions and color imbalance in the turbid water background, aiming to enhance the visual quality of the image and make the color more balanced and natural in the turbid water background. The imput is a turbid water image I with height H and width W. To make this method applicable to different types of image data, we normalize the color channel to the interval $[0, 1]$:

$$I_c = 1 - \left(\frac{I_c}{M}\right)^2, c \in \{R, G, B\}, \tag{1}$$

where I_c is the cth color channel denoting image $I.M$ denotes the maximum possible intensity value for this color channel, which is set to 255.

If a channel of an image has a high average pixel value, it means that the channel has strong brightness or color information in the image. We choose the color channel with the largest average value as the dominant channel c^* to facilitate the subsequent adjustment of other color channels:

$$\bar{I}_c = \frac{1}{HW} \sum_{i=1}^{H} \sum_{j=1}^{W} I_c(i,j), c \in \{R, G, B\}, \tag{2}$$

$$c^* = \underset{c \in \{R,G,B\}}{\operatorname{argmax}} \bar{I}_c, \tag{3}$$

$\bar{I}_c$ denotes the average value of the c th color channel.

Since each color channel under natural light is at a similar average value, the dominant channel decays faster [3]. We gradually adjust the other two color channels c_1 and c_2 based on the information of the dominant channel c^* to reduce the overall color deviation. the color compensation equations for the c_1 and c_2 channels can be expressed as:

$$I_c' = I_c + \left(\bar{I}_{c^*} - \bar{I}_c\right) I_c, c \in \{c_1, c_2\}, \tag{4}$$

Naturally, we want to minimize the color deviation L by adjusting L until convergence to update the values of the different color channels:

$$L = \min\left(\bar{I}_{c^*} - \bar{I}_{c_1}, \bar{I}_{c^*} - \bar{I}_{c_2}\right), \tag{5}$$

$$L \leq \epsilon, \tag{6}$$

where ϵ represents color deviation tolerance threshold and is set to 0.005.

3.2 Wavelength-Dependent Color Correction

Some regions of the clear image have low color pixel intensities in at least one color channel. We therefore selected images based on statistical observations to determine atmospheric light by the brightest pixel in the channel. The transmittance map was estimated using the dark channel a priori method [38] and adjusted by a factor ω to preserve the sense of naturalness. Finally, scene irradiance was restored by rescaling and adjusting the observed intensities with the atmospheric light and the transmittance map to ensure that no region became too dark or added noise due to low transmittance. This method effectively removes haze while retaining the realism and quality of the image.

A corrected map I^{CR} of the turbid water image I was obtained in the TDRECC stage. Since the brightness of at least one color channel is very low in some regions of the image under natural conditions, the atmospheric light estimation of the tu'r'bid water image can be expressed as:

$$I^{CR,\text{ dark}} = \min_{c \in \{R,G,B\}} I_c^{CR}(x, y), \tag{7}$$

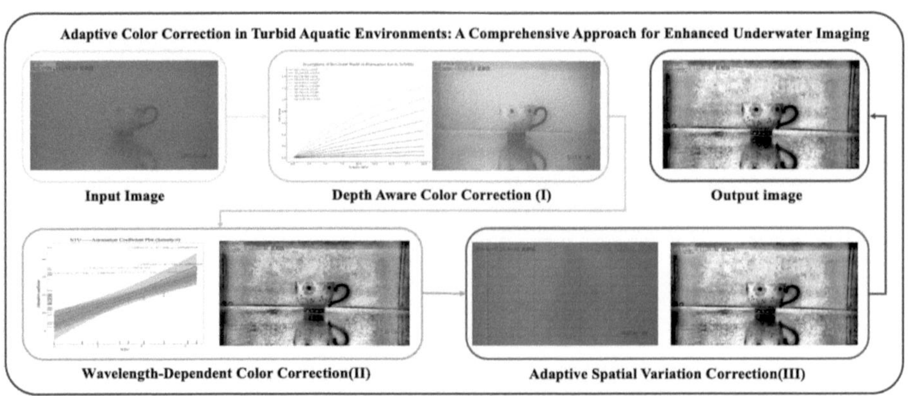

Fig. 1. Adaptive Enhancement Pipeline for Underwater Imaging in Turbid Conditions. Our pipeline for enhancing underwater images in turbid conditions unfolds in three strategic stages, each tailored to address specific challenges posed by turbid aquatic environments.

The brightest pixel in the dark channel is then selected as an estimate of atmospheric light, which makes the tubrid water image pixels close to the true atmospheric light intensity.

$$A = I^{CR}\left(x^{*}, y^{*}\right), \tag{8}$$

where $(x^{*}, y^{*}) = \arg\max_{(x,y)} I_c^{CR,\text{dark}}(x, y)$. (x^{*}, y^{*}) is the location of the brightest pixel in $I^{CR,dark}$.

In order to remove the effect of atmospheric light, we normalize the turbid water image I^{CR} using atmospheric light A for better observation and calculation of transmittance:

$$I^{CR,\text{ norm}} = \frac{I^{CR}}{A}, \tag{9}$$

By calculating the minimum value of each pixel in all color channels, a dark channel image is obtained which tends to zero in the absence of fog. Next, the projection map t of the turbid water image is calculated:

$$I^{CR,\text{ norm,dark}} = \min_{c \in \{R,G,B\}} I_c^{CR,\text{ norm}}. \tag{10}$$

$$t = 1 - \omega \cdot I^{CR,\text{ norm,dark}}. \tag{11}$$

The parameter ω is a value between 0 and 1 that represents the degree of retained turbidity, making it possible to avoid the unnatural feel of the picture after removing turbidity.

3.3 Adaptive Spatial Variation Correction

Based on the previously recovered image J, we apply the CLAHE [39] algorithm to enhance the contrast of the turbid water image. Specifically, for each color channel of the image J, we apply CLAHE to obtain J^{CLAHE}:

Table 1. Comparative Analysis of Image Enhancement Method in Diverse Turbidity Scenarios

Method	3D OBJECT(NTU12)			3D OBJECT (NTU18)			COLOR CHART (200 mg/L)			COLOR CHART (500 mg/L)		
	PSNR	SSIM	UCIQE	PSNR	SSIM	UCIQE	PSNR	SSIM	UCIQE	PSNR	SSIM	UCIQE
IBLA [40]	13.74	0.77	0.41	16.37	0.77	0.28	9.92	0.34	0.43	7.76	0.23	0.31
MIP [41]	19.6	0.84	0.13	17.85	0.79	0.092	15.21	0.75	0.21	13.75	0.73	0.15
MLLE [29]	16.77	0.79	0.79	12.59	0.71	0.41	10.1	0.29	0.44	10.09	0.12	0.43
PCDE [13]	14.13	0.58	0.39	12.26	0.52	0.42	13.87	0.35	0.42	9.45	0.17	0.41
UDCP [15]	12.14	0.55	0.30	12.22	0.57	0.23	9.46	0.41	0.51	9.85	0.32	0.41
ULAP [18]	10.33	0.76	0.38	9.45	0.72	0.33	12.42	0.60	0.34	10.90	0.56	0.24
Underwater-hi [42]	11.62	0.70	0.46	9.95	0.66	0.36	10.99	0.32	0.42	7.19	0.19	0.43
Ours	12.93	0.54	0.53	11.69	0.46	0.53	13.38	0.37	0.49	9.74	0.15	0.50

$$J^{CLAHE} = \text{CLAHE}(J, \alpha), \tag{12}$$

where J^{CLAHE} is the image channel after CLAHE processing and α is the given ClipLimit parameter.

Next, the CLAHE-enhanced image J^{CLAHE} is combined to synthesize the image deturbidization and contrast enhancement effects to obtain the final image J':

$$J' = \gamma J^{CLAHE} + (1 - \gamma)J, \tag{13}$$

where γ is a parameter that controls the ratio of the two combinations, and we set γ to 0.4.

Then, sharpening and denoising processes are applied to J' to enhance the details and reduce the noise in the image. For each color channel c, the sharpening operation S is applied initially with parameters of radius r and intensity a, and then the Wiener denoising operation W is applied with parameters of filter size $k \times k$.

$$J'' = W\left(S\left(J', r, a\right), k \times k\right), \tag{14}$$

Finally, the adjusted image J is adjusted for brightness and contrast to obtain the final image J''' :

$$J''' = L\left(J'', \beta\right) \tag{15}$$

where L is the brightness and contrast adjustment operation and the parameters β and F are the image format conversion operation (Table 1 and Fig. 2).

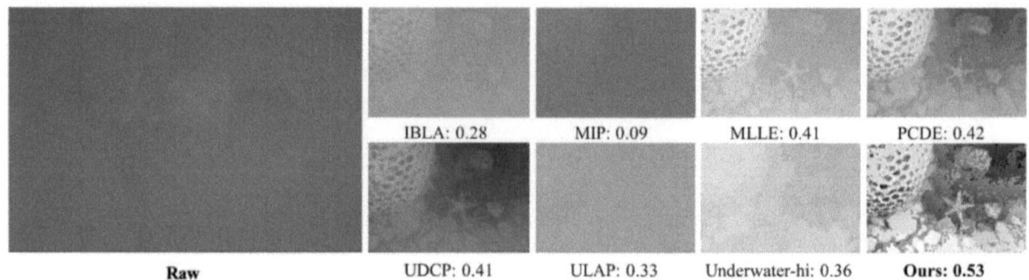

Fig. 2. Comparative Visualization of Underwater Image Enhancement Techniques with UCIQE Metrics (NTU 18). This comparison reinforces our method's leading performance. IBLA presents a UCIQE score of 0.28, MIP drops to 0.09, while MMLE and PCDE show a close match with scores of 0.41 and 0.42, respectively. In contrast, UDCP shows improvement at 0.41, ULAP at 0.33, and Underwater-hi at 0.36. Our method outperforms the competitive field with a UCIQE score of 0.53, demonstrating consistent superiority in enhancing underwater imagery by providing the most significant improvement in color quality, even in challenging visual conditions.

4 Experimental Results

4.1 Qualitative Analysis

Qualitative analysis demonstrated the significant superiority of our model in underwater color correction. Through expert reviews and comparative studies, we assessed visual effects, detail restoration, and color accuracy. The model excelled at reproducing natural colors (especially reds and yellows) and effectively restored image clarity in high-turbidity conditions, outperforming traditional methods. It significantly enhanced image contrast and visual quality by providing more realistic colors and recovering lost details, proving its robustness and surpassing existing technologies.

4.2 Quantitative Analysis

We comparatively evaluated our model against mainstream methods using diverse underwater images. Assessed on color fidelity, clarity, contrast, and visual quality, statistical analysis confirmed our model's advantages in mitigating color distortion and blurriness. Particularly in high-turbidity, our model showed superior performance by comprehensively considering depth, wavelength-dependent light attenuation, and spatial variations, validating its theoretical and practical superiority.

4.3 Performance Evaluation

We compared the processing speeds of our model with several algorithms at two resolutions. At a lower resolution (1619×1907), our model took 0.9 s, outperforming most algorithms and only slightly slower than the fastest (MLLE at

0.7 s). At a higher resolution (3660×2496), our model processed images in 2.6 s, significantly faster than methods like UDCP (over 216 times faster) and with a negligible difference compared to MLLE. This demonstrates the algorithm's exceptional performance in rapidly processing high-resolution underwater images and its potential for near real-time applications, which is critical for practical use.

5 Conclusion

We establish a turbid water image enhancement method with excellent recovery and adaptability. Our method has established an optically based physical model of turbid water, obtains very little effective information in the images, and recovers the image contours from the blurred images. Then, we have designed a color correction scheme to restore the original color of the image from the valid information as much as possible, and ensure the uniformity of light and dark in the turbid water image. Extensive experiments have demonstrated that our proposed comprehensive method not only effectively recovers the color of turbid water imaging and maximizes the image details, but also has sound applicability and recovery performance in different complex turbid water.

References

1. Li, C., Guo, J., Guo, C.: Emerging from water: underwater image color correction based on weakly supervised color transfer. IEEE Signal Process. Lett. **25**(3), 323–327 (2018)
2. Yuan, J., Cao, W., Cai, Z., Su, B.: An underwater image vision enhancement algorithm based on contour bougie morphology. IEEE Trans. Geosci. Remote Sens. **59**(10), 8117–8128 (2020)
3. Zhang, W., Wang, Y., Li, C.: Underwater image enhancement by attenuated color channel correction and detail preserved contrast enhancement. IEEE J. Oceanic Eng. **47**(3), 718–735 (2022)
4. Guo, Y., Li, H., Zhuang, P.: Underwater image enhancement using a multiscale dense generative adversarial network. IEEE J. Oceanic Eng. **45**(3), 862–870 (2019)
5. Wang, H., Sun, S., Bai, X., Wang, J., Ren, P.: A reinforcement learning paradigm of configuring visual enhancement for object detection in underwater scenes. IEEE J. Oceanic Eng. **48**(2), 443–461 (2023)
6. Zhou, Y., Qiong, W., Yan, K., Feng, L., Xiang, W.: Underwater image restoration using color-line model. IEEE Trans. Circuits Syst. Video Technol. **29**(3), 907–911 (2018)
7. Liang, Z., Ding, X., Wang, Y., Yan, X., Fu, X.: GUDCP: generalization of underwater dark channel prior for underwater image restoration. IEEE Trans. Circuits Syst. Video Technol. **32**(7), 4879–4884 (2021)
8. Dai, C., Lin, M., Wu, X., Wang, Z., Guan, Z.: Single underwater image restoration by decomposing curves of attenuating color. Opt. Laser Technol. **123**, 105947 (2020)
9. Li, C.-Y., Guo, J.-C., Cong, R.-M., Pang, Y.-W., Wang, B.: Underwater image enhancement by dehazing with minimum information loss and histogram distribution prior. IEEE Trans. Image Process. **25**(12), 5664–5677 (2016)

10. Zhang, S., Wang, T., Dong, J., Hui, Yu.: Underwater image enhancement via extended multi-scale RetiNex. Neurocomputing **245**, 1–9 (2017)
11. Ye, X., et al.: Deep joint depth estimation and color correction from monocular underwater images based on unsupervised adaptation networks. IEEE Trans. Circuits Syst. Video Technol. **30**(11), 3995–4008 (2019)
12. Wu, S., et al.: A two-stage underwater enhancement network based on structure decomposition and characteristics of underwater imaging. IEEE J. Oceanic Eng. **46**(4), 1213–1227 (2021)
13. He, K., Sun, J., Tang, X.: Single image haze removal using dark channel prior. IEEE Trans. Pattern Anal. Mach. Intell. **33**(12), 2341–2353 (2010)
14. Serikawa, S., Huimin, L.: Underwater image dehazing using joint trilateral filter. Comput. Electr. Eng. **40**(1), 41–50 (2014)
15. Drews, P.L.J., Nascimento, E.R., Botelho, S.S.C., Campos, M.F.M.: Underwater depth estimation and image restoration based on single images. IEEE Comput. Graphics Appl. **36**(2), 24–35 (2016)
16. Peng, Y.-T., Cao, K., Cosman, P.C.: Generalization of the dark channel prior for single image restoration. IEEE Trans. Image Process. **27**(6), 2856–2868 (2018)
17. Akkaynak, D., Treibitz, T.: Sea-thru: a method for removing water from underwater images. In: Proceedings of the IEEE/CVF Conference on Computer Vision and Pattern Recognition, pp. 1682–1691 (2019)
18. Song, W., Wang, Y., Huang, D., Tjondronegoro, D.: A rapid scene depth estimation model based on underwater light attenuation prior for underwater image restoration. In: Advances in Multimedia Information Processing–PCM 2018: 19th Pacific-Rim Conference on Multimedia, Hefei, China, 21–22 September 2018, Proceedings, Part I 19, pp. 678–688. Springer (2018)
19. Song, W., Wang, Y., Huang, D., Liotta, A., Perra, C.: Enhancement of underwater images with statistical model of background light and optimization of transmission map. IEEE Trans. Broadcast. **66**(1), 153–169 (2020)
20. Zhou, J., Yang, T., Chu, W., Zhang, W.: Underwater image restoration via backscatter pixel prior and color compensation. Eng. Appl. Artif. Intell. **111**, 104785 (2022)
21. Ancuti, C., Ancuti, C.O., Haber, T., Bekaert, P.: Enhancing underwater images and videos by fusion. In 2012 IEEE Conference on Computer Vision and Pattern Recognition, pp. 81–88. IEEE (2012)
22. Fu, X., Zhuang, P., Huang, Y., Liao, Y., Zhang, X.-P., Ding, X.: A retinex-based enhancing approach for single underwater image. In: 2014 IEEE International Conference on Image Processing (ICIP), pp. 4572–4576. IEEE (2014)
23. Huimin, L., Li, Y., Zhang, L., Serikawa, S.: Contrast enhancement for images in turbid water. JOSA A **32**(5), 886–893 (2015)
24. Mishra, A., Gupta, M., Sharma, P.: Enhancement of underwater images using improved CLAHE. In: 2018 International Conference on Advanced Computation and Telecommunication (ICACAT), pp. 1–6. IEEE (2018)
25. Ancuti, C.O., Ancuti, C., De Vleeschouwer, C., Bekaert, P.: Color balance and fusion for underwater image enhancement. IEEE Trans. Image Process. **27**(1), 379–393 (2017)
26. Sanila, K.H., Balakrishnan, A.A., Supriya, M.H.: Underwater image enhancement using white balance, USM and CLHE. In: 2019 International Symposium on Ocean Technology (SYMPOL), pp. 106–116. IEEE (2019)
27. Sethi, R., Indu, S.: Fusion of underwater image enhancement and restoration. Int. J. Pattern Recognit. Artif. Intell. **34**(03), 2054007 (2020)

28. Zhuang, P., Li, C., Jiamin, W.: Bayesian retinex underwater image enhancement. Eng. Appl. Artif. Intell. **101**, 104171 (2021)
29. Zhang, W., Zhuang, P., Sun, H.-H., Li, G., Kwong, S., Li, C.: Underwater image enhancement via minimal color loss and locally adaptive contrast enhancement. IEEE Trans. Image Process. **31**, 3997–4010 (2022)
30. Li, J., Skinner, K.A., Eustice, R.M., Johnson-Roberson, M.: WaterGAN: unsupervised generative network to enable real-time color correction of monocular underwater images. IEEE Robot. Autom. Lett. **3**(1), 387–394 (2017)
31. Huimin, L., Li, Y., Uemura, T., Kim, H., Serikawa, S.: Low illumination underwater light field images reconstruction using deep convolutional neural networks. Futur. Gener. Comput. Syst. **82**, 142–148 (2018)
32. Fabbri, C., Islam, M.J., Sattar, J.: Enhancing underwater imagery using generative adversarial networks. In: 2018 IEEE International Conference on Robotics and Automation (ICRA), pp. 7159–7165. IEEE (2018)
33. Yang, M., Ke, H., Yixiang, D., Wei, Z., Sheng, Z., Jintong, H.: Underwater image enhancement based on conditional generative adversarial network. Signal Process.: Image Commun. **81**, 115723 (2020)
34. Li, C., Guo, C., Ren, W., Cong, R., Hou, J., Kwong, S., Tao, D.: An underwater image enhancement benchmark dataset and beyond. IEEE Trans. Image Process. **29**, 4376–4389 (2019)
35. Li, C., Anwar, S., Hou, J., Cong, R., Guo, C., Ren, W.: Underwater image enhancement via medium transmission-guided multi-color space embedding. IEEE Trans. Image Process. **30**, 4985–5000 (2021)
36. Wang, Y., Guo, J., Gao, H., Yue, H.: UIEC 2-net: CNN-based underwater image enhancement using two color space. Signal Processing: Image Communication **96**, 116250 (2021)
37. Lin, P., Wang, Y., Wang, G., Yan, X., Jiang, G., Xianping, F.: Conditional generative adversarial network with dual-branch progressive generator for underwater image enhancement. Signal Process.: Image Commun. **108**, 116805 (2022)
38. He, K., Zhang, X., Ren, S., Sun, J.: Deep residual learning for image recognition. In: Proceedings of the IEEE Conference on Computer Vision and Pattern Recognition, pp. 770–778 (2016)
39. Huynh, C.P., Robles-Kelly, A.: Comparative colorimetric simulation and evaluation of digital cameras using spectroscopy data. In: 9th Biennial Conference of the Australian Pattern Recognition Society on Digital Image Computing Techniques and Applications (DICTA 2007), pp. 309–316. IEEE (2007)
40. Peng, Y.-T., Cosman, P.C.: Underwater image restoration based on image blurriness and light absorption. IEEE Trans. Image Process. **26**(4), 1579–1594 (2017)
41. Carlevaris-Bianco, N., Mohan, A., Eustice, R.M.: Initial results in underwater single image dehazing. In: Oceans 2010 Mts/IEEE Seattle, pp. 1–8. IEEE (2010)
42. Berman, D., Levy, D., Avidan, S., Treibitz, T.: Underwater single image color restoration using haze-lines and a new quantitative dataset. IEEE Trans. Pattern Anal. Mach. Intell. **43**(8), 2822–2837 (2020)

Cross-Domain Underwater Image Enhancement Guided by No-Reference Image Quality Assessment: A Transfer Learning Approach

Zhi Zhang, Minfu Li, Lu Li, and Daoyi Chen(✉)

Tsinghua University, Shenzhen 518000, China
z-zhang23@mails.tsinghua.edu.cn, chen.daoyi@sz.tsinghua.edu.cn

Abstract. Single underwater image enhancement (UIE) is a challenging ill-posed problem, and its development is hindered by two main issues: (1) The labels in existing underwater reference datasets are pseudo ground truths with severe color distortions; relying on them in supervised learning leads to domain discrepancy. (2) Underwater reference datasets are limited in size, and training on such small datasets tends to cause overfitting and distribution shifts from real underwater conditions. To address these problems, we propose a transfer learning-based UIE method—Trans-UIE. This method captures the fundamental paradigms of UIE through pretraining and utilizes a hybrid dataset for fine-tuning, narrowing the distribution gap from real underwater environments. We introduce the most suitable no-reference image quality assessment (NR-IQA) metrics from above-water scenes into the cross-domain transfer learning process to correct the distortions caused by pseudo labels, reduce the domain gap between underwater and above-water images, and avoid confirmation bias. Additionally, we introduced a Pearson correlation loss function to mitigate overfitting during the pre-training phase. We employed a Channel Reordering Gated Feed-Forward Network (CRGFN) to enhance the network's capability in modeling complex degradation factors. Experimental results on both full-reference and no-reference underwater image benchmark datasets demonstrate that Trans-UIE achieves outstanding performance in image enhancement quality.

Keywords: Underwater Image Enhancement · Transfer Learning

1 Introduction

Underwater environments often challenge image quality due to light scattering and inconsistent illumination. The presence of suspended particles further exacerbates issues, leading to common problems like blurring and color distortion in underwater images. These factors contribute to the degraded visual clarity typically observed in underwater scenes. Underwater image enhancement (UIE) [20]

P. Umapada et al. (Eds.): ICCPR 2025, CCIS 2811, pp. 206–217, 2026.
https://doi.org/10.1007/978-981-95-8315-7_17

Fig. 1. Examples from LSUI [15] and UIEB [10] benchmarks. The pseudo labels shown in (b) exhibit issues such as color cast and over-enhancement.

is a critical technique for acquiring underwater images and surveying underwater environments. It has wide-ranging applications in fields such as ocean exploration, biology, archaeology, and underwater robotics. Therefore, innovations in UIE are of great significance. Recently, numerous deep learning-based methods have been proposed to address image restoration problems. Extensive research has also been conducted in the specific area of underwater image restoration [4]. Compared with traditional methods relying on handcrafted priors, deep learning-based solutions can provide better restoration results due to their data-driven nature.

Despite the success of deep learning methods, supervised approaches for modeling reference datasets still face key limitations. First, labels in underwater reference datasets are pseudo ground truths, typically generated by processing degraded images with multiple restoration algorithms and selecting the best output. For example, [7] synthesizes underwater images using GANs, while LSUI [15] selects pseudo labels from conventional methods based on evaluation scores. However, the quality of these pseudo labels falls short of real above-water images, and over-reliance during training enlarges the domain gap. Second, underwater reference datasets are limited in size; LSUI [15] includes 4279 image pairs, and UIEB [10] only 890. Small datasets risk overfitting and distribution shift, degrading real-world performance. In contrast, non-reference datasets are abundant but underutilized. Figure 1 shows examples from LSUI and UIEB with pseudo label flaws like color cast and over-enhancement, limiting model performance.

To overcome these challenges, we propose Trans-UIE, an underwater image enhancement algorithm based on transfer learning.

The main contributions are summarized as follows:

- We propose Trans-UIE, an underwater image enhancement model based on a transfer learning framework. By pre-training on mixed dataset and applying cross-domain fine-tuning guided by no-reference image quality assessment (NR-IQA) metrics from above-water scenes, the model effectively reduces domain gaps and enhances generalization, leading to more accurate restoration of real underwater images.

- We develop an efficient pre-trained model to learn the core paradigm of underwater image enhancement, and introduce a Pearson correlation loss to reduce the risk of overfitting.
- Extensive experimental results demonstrate the effectiveness of our method.

2 Related Work

2.1 Underwater Image Restoration Methods

Current underwater image enhancement methods fall into three categories: model-based, image enhancement, and data-driven methods. The first two are traditional approaches.

Model-based methods rely on handcrafted priors to estimate parameters like transmission and ambient light, but often fail in complex real-world scenes. With deep learning advances, data-driven methods have shown superior performance. Early works used GANs [11] to compensate for limited paired data. The release of datasets like UIEB [10] and LSUI [15] enabled the development of complex networks (e.g., WaterNet [10], Ucolor [9] and Ushape [15]) that leverage transmission maps, adaptive normalization, and attention mechanisms to enhance robustness under challenging conditions. The most closely related work to this paper is Semi-UIR [6], which innovatively adopts a semi-supervised learning approach. By leveraging NR-IQA scores for pseudo-label selection, it incorporates non-reference datasets into training and achieves impressive results. However, this method primarily focuses on mitigating overfitting on small reference datasets and does not explicitly address or analyze the bias between pseudo labels and above-water images. Moreover, the performance of its single-stage training strategy is prone to limitations caused by confirmation bias.

2.2 Transfer Learning

Transfer learning has become a key strategy in computer vision to address data scarcity and domain adaptation. Typically, models are pre-trained on large datasets and fine-tuned for specific tasks. Studies show that stronger pre-trained models yield better downstream performance [23], and approaches like relational graphs [18] or distribution regularization further improve generalization and reduce overfitting.

However, cross-domain transfer learning remains underexplored in underwater image enhancement. The scarcity of reference datasets and limitations of pseudo ground truths highlight its potential. This motivates the design of Trans-UIE, a transfer learning-based framework for underwater image enhancement.

3 Method

3.1 Transfer Learning Framework

Our transfer learning framework is illustrated in Fig. 2. First, a pre-trained model, UIR-Net, is trained on a reference dataset by minimizing the pixel-level loss $\mathcal{L}_{\text{pix}}$ and the Pearson correlation loss $\mathcal{L}_{\text{cor}}$ to learn the mapping from

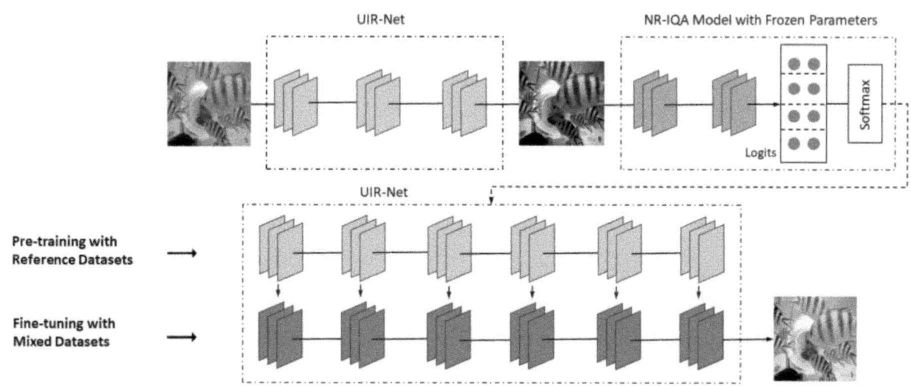

Fig. 2. Illustration of our framework Trans-UIE. Trans-UIE is built upon a transfer learning framework that consists of both pre-training and fine-tuning stages. During pre-training, Trans-UIE learns from reference datasets. To prevent confirmation bias, we incorporate an above-water NR-IQA model with frozen parameters to guide the fine-tuning process. Additionally, a dataset composed of both reference and non-reference datasets is utilized for fine-tuning to mitigate overfitting, reduce distribution shift, and bridge the gap between pseudo labels and the above-water image domain.

degraded images to pseudo labels I^*, thereby obtaining a basic enhancement network. Next, UIR-Net is used to generate pseudo labels for the fine-tuning dataset, providing an initial mapping for the fine-tuning stage. we use MUSIQ [8] as our above-water NR-IQA to score the generated pseudo labels, obtaining the initial reference quality score $Q_{\text{reference}}$.

During the fine-tuning stage, we construct a wrapper that integrates UIR-Net with the above-water NR-IQA model. The network parameters of the NR-IQA model are frozen and do not participate in the updating process. Finally, the pre-trained model is fine-tuned by minimizing the following total loss:

$$\begin{aligned}\mathcal{L}_{\text{total}} = & \lambda_1 \mathcal{L}_{\text{pix}}(I^*, \hat{I}) + \lambda_2 \mathcal{L}_{\text{vgg}}(I^*, \hat{I}) \\ & - \lambda_3 \left(Q(\hat{I}) - Q_{\text{reference}} \right).\end{aligned} \tag{1}$$

where λ_1, λ_2, and λ_3 denote the weight coefficients for the pixel-level loss, VGG [19] perceptual loss, and NR-IQA score loss. $Q(\hat{I})$ is the score assigned to the enhanced image $\hat{I}$ by NR-IQA within the wrapper. $Q_{\text{reference}}$ is the reference quality score obtained by evaluating the pseudo labels generated by UIR-Net using NR-IQA, it is responsible for adjusting the learning rate of L_{MUSIQ} and does not participate in gradient computation.

3.2 Reliable Metric Selection

A suitable no-reference image quality assessment (NR-IQA) metric should (1) reflect quality changes in underwater images and (2) perform well on above-water images. We evaluate several NR-IQA methods accordingly. For (1), we

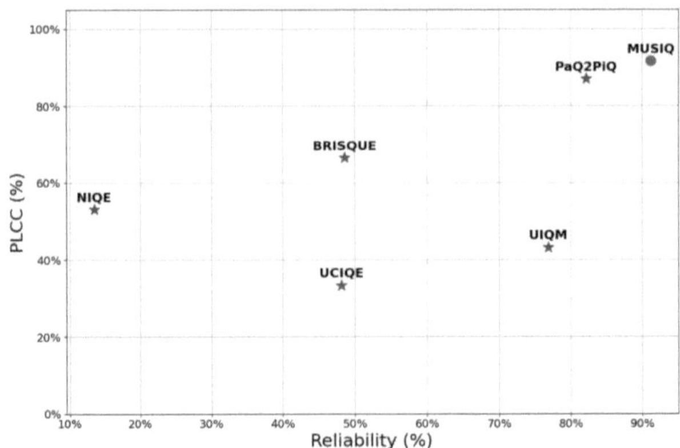

Fig. 3. The results of different non-reference IQA indicators on EUVP [7] and KonIQ [5], including UIQM [14], UCIQE [21], BRISQUE [5], NIQE [13], PaQ2PiQ [5] and MUSIQ [8].

mix degraded and clean images at varying ratios to simulate quality changes. A reliable metric should yield scores that monotonically decrease as degradation increases. For (2), we assess performance using PLCC scores on above-water datasets. We test six NR-IQA methods on the EUVP underwater dataset [7] and KonIQ above-water dataset [5]. As shown in Fig. 3, MUSIQ [8] demonstrates the best consistency with monotonicity and is selected to evaluate network outputs.

3.3 Domain Discrepancy and Distribution Shift

In underwater image enhancement, a domain discrepancy exists between the reference dataset's pseudo-labels and real above-water images. Furthermore, small reference datasets can cause supervised models to overfit, leading to a distribution shift on non-reference data that impairs generalization. To address these issues, we designed the Trans-UIE framework with a tailored loss function.

Domain Discrepancy from Pseudo-label Noise. A pseudo-label I^* relates to a real above-water image I_{real} by $I^* = I_{\text{real}} + \epsilon_{\text{pseudo}}$, where ϵ_{pseudo} is noise. This creates a domain discrepancy, which in the feature space is $\varphi(I^*) = \varphi(I_{\text{real}}) + \Delta_\varphi(\epsilon_{\text{pseudo}})$.

While pixel and perceptual losses (L_{pix}, L_{vgg}) may cause the model's output feature $\varphi(\hat{I})$ to learn the noise bias, our introduced NR-IQA score loss L_{MUSIQ} aligns the output's quality with that of real images, mitigating the discrepancy. The total loss can be approximated as:

$$\begin{aligned} L_{\text{total}} \approx \lambda \|\varphi(\hat{I}) - (\varphi(I_{\text{real}}) + \Delta_\varphi(\epsilon_{\text{pseudo}}))\|_2^2 \\ + \lambda_3 \|f(\varphi(\hat{I})) - f(\varphi(I_{\text{real}}))\|_2^2, \end{aligned} \tag{2}$$

where the first term learns from the pseudo-label, and the second term imposes a constraint in the quality score space to guide the output towards the real image features, thus reducing domain bias.

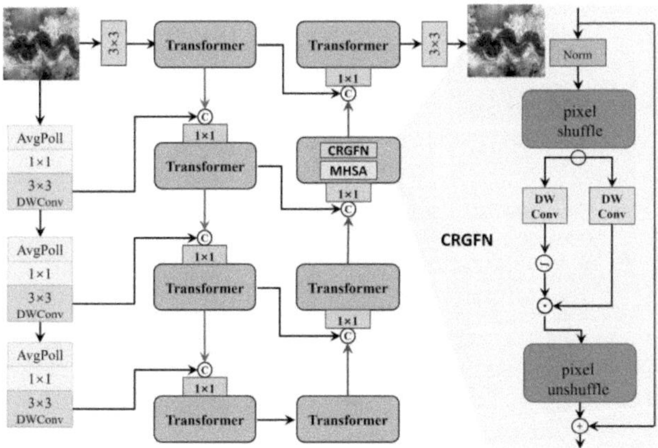

Fig. 4. An overview of the proposed Pre-Train Network (UIR-Net).

Feature Distribution Shift. We model reference I_R and non-reference I_N images as $I_R = I_{\text{real}} + \epsilon_R$ and $I_N = I_{\text{real}} + \epsilon_N$, where ϵ_R and ϵ_N are their respective noise components. Due to the near-linearity of the feature extractor $\varphi(\cdot)$, the feature distribution shift between their mean features μ_R and μ_N is:

$$\Delta_{\text{feat}} = \|\mu_R - \mu_N\|_2^2 \approx \|\mathbb{E}[\varphi(\epsilon_R)] - \mathbb{E}[\varphi(\epsilon_N)]\|_2^2. \tag{3}$$

During fine-tuning on a mixed dataset, L_{pix} and L_{vgg} encourage the output features to approach μ_R. Concurrently, L_{MUSIQ} improves the quality score by suppressing noise, thereby reducing the discrepancy between ϵ_R and ϵ_N and compensating for Δ_{feat}. The overall loss is approximated by:

$$L_{\text{total}} \approx \lambda\|\varphi(\hat{I}) - \mu_R\|_2^2 + \lambda_3 \left(E[Q(\hat{I})]_{\text{desired}} - E[Q(\hat{I})]\right), \tag{4}$$

where the first term prevents the model from deviating from the reference features, and the second term reduces the feature shift Δ_{feat} by enhancing image quality.

3.4 Pretrained Network UIR-Net

As shown in Fig. 4, We propose UIR-Net, a new architecture built upon the Restormer [22] backbone. It effectively captures global context and fine-grained details using an encoder-decoder structure. The input image $I \in \mathbb{R}^{H\times W\times C}$ is processed into patches. The encoder reduces spatial resolution via an Unshuffle operation, while the decoder restores it using a Shuffle operation. Skip connections fuse low-level and high-level features.

3.5 Pearson Correlation Loss

Since the reference dataset is small, repeated pre-training risks overfitting, as the model may memorize noise and dataset-specific patterns instead of learning

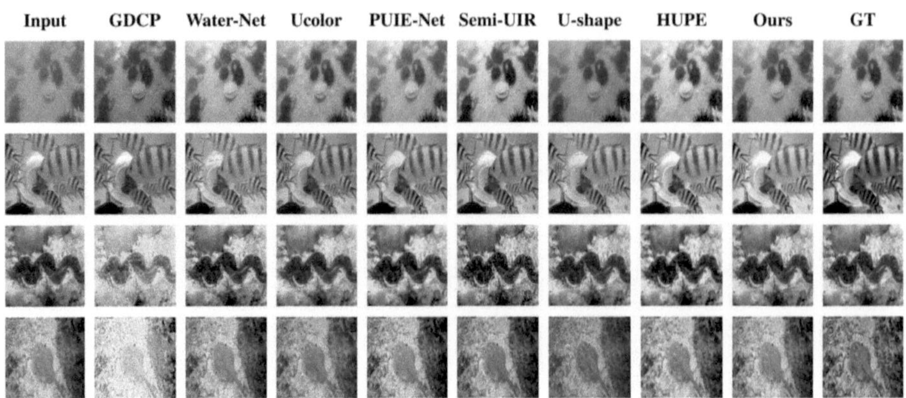

Fig. 5. Visual comparisons of full-reference data from LSUI-400 [15] and UIEB-90 [10] benchmarks. Besides, the ground truths are displayed in the last columns.

general representations. To alleviate this, we introduce Pearson correlation [17] as a loss to capture linear relationships between data points. For pseudo-label I_{gt} and model output $\hat{I}$, the Pearson correlation coefficient is:

$$\rho(\hat{I}, I_{gt}) = \frac{1}{3HW} \sum_{i=1}^{3HW} \frac{(\hat{I}(i) - \mu(\hat{I}))(I_{gt}(i) - \mu(I_{gt}))}{\sigma(\hat{I})\,\sigma(I_{gt})}, \tag{5}$$

where $\hat{I}(i)$ is the i-th pixel, μ and σ are the mean and standard deviation. ρ ranges from $[-1, 1]$, with 1 indicating perfect positive correlation and -1 perfect negative correlation.

The correlation loss is defined as:

$$L_{cor} = \frac{1}{2}\left(1 - \rho(\hat{I}, I_{gt})\right), \tag{6}$$

so $L_{cor} = 0$ when output and pseudo-label are perfectly aligned. This effectively reduces overfitting during pre-training.

4 Experiments

4.1 Datasets

During the pre-training phase, our training set consists of 4679 labeled image pairs, with 3879 pairs from the reference dataset LSUI [15] and 800 pairs from UIEB [10]. The test set includes the remaining 400 pairs from LSUI [15] and 90 pairs from UIEB [10]. In the fine-tuning phase, the training set consists of 10,000 samples, each containing a pseudo-label generated by the pre-trained model and a reference score produced by NR-IQA. The test set consists of 60 challenging unlabeled images from UIEB-60 [10], 400 images from RUIE [12], and 330 images from EUVP [7].

4.2 Comparisons with the State-of-the-Art

We compare the proposed Trans-UIE with seven state-of-the-art underwater image enhancement methods, including one conventional method, GDCP [16], and six deep learning-based methods: WaterNet [10], Ucolor [9], PUIE-Net [2], Semi-UIR [6], U-shape [15], and HUPE [24]. All competing methods were re-trained on our training dataset to ensure a fair comparison.

For experiments on reference datasets, Table 1 presents the quantitative evaluation results on the LSUI-400 and UIEB-90 datasets. On LSUI-400, our method achieves the best performance in terms of PSNR and SSIM, significantly outperforming competing methods. On UIEB-90, our method achieves the highest SSIM score, although its PSNR is slightly lower than Semi-UIR. This discrepancy may stem from the inclusion of the Pearson correlation loss during pre-training, which encourages the model to focus on the overall dataset distribution. Since the UIEB [10] dataset is smaller (800 pairs) compared to the larger LSUI [15] dataset (3879 pairs), the Pearson correlation loss causes the model to align more closely with the LSUI [15] dataset's distribution, resulting in lower scores on UIEB. A detailed discussion of how Pearson correlation loss mitigates overfitting risk is provided in subsequent ablation experiments. In addition to quantitative results, qualitative results are shown in Fig. 5. Visually, our method provides pleasing results, accurately restoring underwater images and outperforming pseudo-label results. In contrast, competing methods exhibit issues such as color casts, noise, and over-enhancement.

For experiments on non-reference datasets, Table 2 presents the quantitative evaluation results on the UIEB-60 [10], EUVP [7], and RUIE [12] datasets. Our method significantly outperforms competing algorithms in terms of MUSIQ [8] and NIQE [13] metrics, while also achieving competitive results on UIQM [14] and UCIQE [21]. However, as noted in [1,3,10], the UIQM [14] and UCIQE [21] metrics exhibit certain biases, which may not fully reflect the true visual quality of the restored images.

Table 1. Quantitative comparison on the LSUI-400 [15] and UIEB-90 [10] datasets. Best results are in **bold** and the second best results are with underline.

Method	LSUI-400		UIEB-90	
	PSNR(↑)	SSIM(↑)	PSNR(↑)	SSIM(↑)
GDCP [16]	14.46	0.7069	13.91	0.7265
Water-Net [10]	22.74	0.8560	21.04	0.8604
Ucolor [9]	24.03	0.8865	20.89	0.8732
PUIE-Net [2]	24.14	0.8762	22.98	0.8984
Semi-UIR [6]	<u>25.83</u>	<u>0.9103</u>	**23.89**	<u>0.9008</u>
U-shape [15]	24.19	0.8793	21.56	0.8874
HUPE [24]	22.79	0.8697	21.99	0.8665
Trans-UIE (ours)	**27.23**	**0.9283**	<u>23.16</u>	**0.9013**

Table 2. Quantitative results on non-reference datasets. Best results are in **bold** and the second best results are with underline.

Method	UIQM (↑)			UCIQE (↑)			MUSIQ (↑)			NIQE (↑)		
	UIEB-60	EUVP	RUIE	UIEB-60	EUVP	RUIE	UIEB-60	EUVP	RUIE	UIEB-60	EUVP	RUIE
GDCP [16]	2.317	2.714	2.875	0.5786	<u>0.6000</u>	0.5616	42.74	35.80	32.48	5.479	4.618	4.577
Water-Net [10]	<u>2.857</u>	<u>3.058</u>	**3.217**	<u>0.5813</u>	**0.6071**	**0.5988**	44.29	37.59	33.32	<u>6.046</u>	4.658	<u>4.678</u>
Ucolor [9]	**3.013**	**3.112**	<u>3.123</u>	0.5483	0.5705	0.5395	43.05	39.16	34.33	5.279	4.533	4.657
PUIE-Net [2]	2.688	3.047	3.053	0.5643	0.5757	0.5548	46.43	40.79	32.47	5.516	4.233	4.317
Semi-UIR [6]	2.759	3.047	2.949	**0.5855**	0.5927	0.5711	<u>48.29</u>	<u>44.20</u>	<u>34.66</u>	5.731	<u>4.729</u>	4.662
U-shape [15]	2.746	2.902	2.989	0.5438	0.5609	0.5552	44.45	39.18	32.75	4.926	4.275	4.311
HUPE [24]	2.711	3.031	2.974	0.5704	0.5800	0.5544	40.49	35.26	32.06	4.901	4.399	4.402
Trans-UIE(ours)	2.654	2.851	2.883	0.5493	0.5548	<u>0.5717</u>	**49.41**	**49.62**	**40.23**	**6.052**	**4.927**	**4.681**

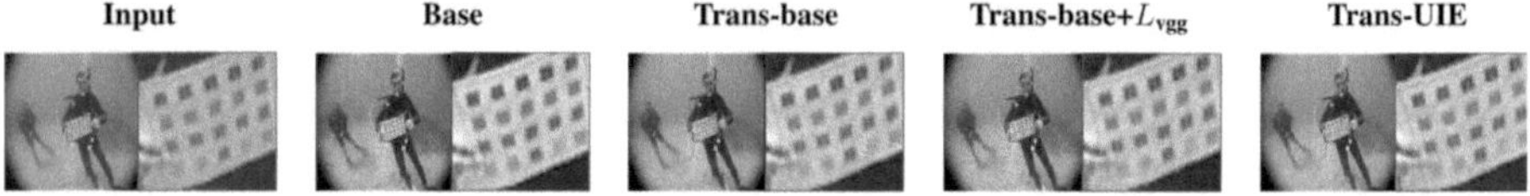

Fig. 6. Results of Ablation Analysis on the Transfer Learning Effectiveness of Trans-UIE.

4.3 Ablation Studies

Analysis of Transfer Learning Effectiveness. We compared four methods: (a) **base**: the pre-trained model without fine-tuning; (b) **Trans-base**: fine-tuning with only the pixel loss L_{pix}; (c) **Trans-base+L_{vgg}**: fine-tuning with pixel loss L_{pix} and perceptual loss L_{vgg}; and (d) **Trans-UIE**: fine-tuning with all losses.

Comparisons were conducted on both reference and non-reference datasets, with quantitative results presented in Table 3. Our complete solution consistently outperforms the others. Comparing **base** and **Trans-base**, we observe that **Trans-base**, which uses pseudo-labels as training supervision, may amplify existing biases or errors, leading to confirmation bias. The introduction of perceptual loss L_{vgg} in **Trans-base+L_{vgg}** effectively mitigates this bias. Upon observing **Trans-UIE**, we note a decrease in PSNR scores on LSUI. This can be attributed to the introduction of evaluation loss L_{MUSIQ}, which reduce the domain gap between underwater and real surface images. Qualitative results, shown in Fig. 6, highlight that **Trans-UIE** delivers the best image restoration, with superior color fidelity and rich detail.

Table 3. Ablation analysis of transfer learning effectiveness: tested on LSUI-400 [15], EUVP [7], and RUIE [12] benchmarks in terms of PSNR and MUSIQ.

Method	Fine-Tune	L_{vgg}	L_{MUSIQ}	LSUI-400	EUVP	RUIE
base				30.51	48.59	35.68
Trans-base	✓			27.56	45.81	34.77
Trans-base+L_{vgg}	✓	✓		27.67	45.90	34.91
Trans-UIE	✓	✓	✓	27.23	49.62	40.23

Analysis of Pearson Correlation Loss Effectiveness. We compared two pre-trained models: (a) **base**: the pre-trained model using only pixel loss; (b) **UIR-Net**: the pre-trained model using both pixel loss and Pearson correlation loss, to assess the impact of incorporating Pearson correlation loss on model performance.

Comparisons were conducted on both reference and non-reference datasets, with quantitative results presented in Table 4. The results show that our complete solution outperforms the others. **UIR-Net** exhibits decreased scores on the reference datasets, while improving performance on non-reference datasets. This can be attributed to the use of two reference datasets with significantly different sample sizes and a relatively small total dataset, which may lead to overfitting. The inclusion of Pearson correlation loss enables the model to focus on the linear relationships between data points, effectively mitigating overfitting and improving performance on non-reference datasets.

Table 4. Ablation analysis of Pearson correlation loss effectiveness: tested on LSUI-400 [15], UIEB-90 [10], EUVP [7], and RUIE [12] benchmarks in terms of PSNR and MUSIQ.

Method	LSUI-400	UIEB-90	EUVP	RUIE
base	30.66	23.73	48.28	35.52
UIR-Net	30.51	23.20	48.59	35.68

Analysis of Feedforward Network Effectiveness. We compared two pre-trained models: (a) **base**: the pre-trained model using a conventional feedforward network; and (b) **UIR-Net**: the pre-trained model employing a Channel Reordering Gated Feedforward Network (CRGFN). The comparison was conducted on reference datasets, with quantitative results presented in Table 5.

The results demonstrate that our complete solution outperforms the others. CRGFN effectively enhances the representational capability of the pre-trained model, UIR-Net.

Table 5. Ablation analysis of feedforward network effectiveness: tested on LSUI-400 [15] and UIEB-90 [10] benchmarks in terms of PSNR.

Method	LSUI-400	UIEB-90
base	28.73	21.66
UIR-Net	30.51	23.20

Fig. 7. Examples based on different NR-IQA score loss hyperparameter.

4.4 Evaluation Loss Hyperparameter Experiment

To assess the impact of different values for the NR-IQA score loss hyperparameter α, we compared the following settings: (a) **base**: processed only by the pre-trained model; (b) $\alpha = 0.0001$; (c) $\alpha = 0.003$; (d) $\alpha = 0.1$. Results are shown in the Fig. 7.

From the comparison, it is evident that when α is too large, the model overemphasizes the distribution of real-world images, leading to color bias. Conversely, when α is too small, the effect is negligible. Thus, we selected $\alpha = 0.003$ as the optimal value.

5 Conclusion

We propose an efficient transfer learning-based underwater image enhancement method named Trans-UIE. As demonstrated by the ablation experiments, the proposed method outperforms other SOTA algorithms, which can be attributed to its effective reduction of feature distribution shift and domain discrepancy. Future research can be directed in two areas: (1) expanding the transfer learning framework to cover other restoration tasks, and (2) enhancing performance through improved memory management.

References

1. Berman, D., Levy, D., Avidan, S., Treibitz, T.: Underwater single image color restoration using haze-lines and a new quantitative dataset. IEEE Trans. Pattern Anal. Mach. Intell. **43**(8), 2822–2837 (2020)
2. Fu, Z., Wang, W., Huang, Y., Ding, X., Ma, K.K.: Uncertainty inspired underwater image enhancement. In: European Conference on Computer Vision, pp. 465–482. Springer (2022)
3. Guo, C., et al.: Underwater ranker: learn which is better and how to be better. In: Proceedings of the AAAI Conference on Artificial Intelligence, vol. 37, pp. 702–709 (2023)
4. Han, J., et al.: Underwater image restoration via contrastive learning and a real-world dataset. Remote Sens. **14**(17), 4297 (2022)
5. Hosu, V., Lin, H., Sziranyi, T., Saupe, D.: KonIQ-10k: an ecologically valid database for deep learning of blind image quality assessment. IEEE Trans. Image Process. **29**, 4041–4056 (2020)

6. Huang, S., Wang, K., Liu, H., Chen, J., Li, Y.: Contrastive semi-supervised learning for underwater image restoration via reliable bank. In: Proceedings of the IEEE/CVF Conference on Computer Vision and Pattern Recognition, pp. 18145–18155 (2023)
7. Islam, M.J., Xia, Y., Sattar, J.: Fast underwater image enhancement for improved visual perception. IEEE Robot. Autom. Lett. **5**(2), 3227–3234 (2020)
8. Ke, J., Wang, Q., Wang, Y., Milanfar, P., Yang, F.: MusIQ: multi-scale image quality transformer. In: Proceedings of the IEEE/CVF International Conference on Computer Vision, pp. 5148–5157 (2021)
9. Li, C., Anwar, S., Hou, J., Cong, R., Guo, C., Ren, W.: Underwater image enhancement via medium transmission-guided multi-color space embedding. IEEE Trans. Image Process. **30**, 4985–5000 (2021)
10. Li, C., et al.: An underwater image enhancement benchmark dataset and beyond. IEEE Trans. Image Process. **29**, 4376–4389 (2019)
11. Li, J., Skinner, K.A., Eustice, R.M., Johnson-Roberson, M.: WaterGAN: unsupervised generative network to enable real-time color correction of monocular underwater images. IEEE Robot. Autom. Lett. **3**(1), 387–394 (2017)
12. Liu, R., Fan, X., Zhu, M., Hou, M., Luo, Z.: Real-world underwater enhancement: challenges, benchmarks, and solutions under natural light. IEEE Trans. Circuits Syst. Video Technol. **30**(12), 4861–4875 (2020)
13. Mittal, A., Soundararajan, R., Bovik, A.C.: Making a "completely blind" image quality analyzer. IEEE Signal Process. Lett. **20**(3), 209–212 (2012)
14. Panetta, K., Gao, C., Agaian, S.: Human-visual-system-inspired underwater image quality measures. IEEE J. Oceanic Eng. **41**(3), 541–551 (2015)
15. Peng, L., Zhu, C., Bian, L.: U-shape transformer for underwater image enhancement. IEEE Trans. Image Process. **32**, 3066–3079 (2023)
16. Peng, Y.T., Cao, K., Cosman, P.C.: Generalization of the dark channel prior for single image restoration. IEEE Trans. Image Process. **27**(6), 2856–2868 (2018)
17. Sedgwick, P.: Pearson's correlation coefficient. BMJ **345** (2012)
18. Sharif Razavian, A., Azizpour, H., Sullivan, J., Carlsson, S.: CNN features off-the-shelf: an astounding baseline for recognition. In: Proceedings of the IEEE Conference on Computer Vision and Pattern Recognition Workshops, pp. 806–813 (2014)
19. Tammina, S.: Transfer learning using VGG-16 with deep convolutional neural network for classifying images. Int. J. Sci. Res. Publ. (IJSRP) **9**(10), 143–150 (2019)
20. Yang, M., Hu, J., Li, C., Rohde, G., Du, Y., Hu, K.: An in-depth survey of underwater image enhancement and restoration. IEEE access **7**, 123638–123657 (2019)
21. Yang, M., Sowmya, A.: An underwater color image quality evaluation metric. IEEE Trans. Image Process. **24**(12), 6062–6071 (2015)
22. Zamir, S.W., Arora, A., Khan, S., Hayat, M., Khan, F.S., Yang, M.H.: Restormer: efficient transformer for high-resolution image restoration. In: Proceedings of the IEEE/CVF Conference on Computer Vision and Pattern Recognition, pp. 5728–5739 (2022)
23. Zhang, W., Zhuang, P., Sun, H.H., Li, G., Kwong, S., Li, C.: Underwater image enhancement via minimal color loss and locally adaptive contrast enhancement. IEEE Trans. Image Process. **31**, 3997–4010 (2022)
24. Zhang, Z., Jiang, Z., Ma, L., Liu, J., Fan, X., Liu, R.: HUPE: heuristic underwater perceptual enhancement with semantic collaborative learning. Int. J. Comput. Vision 1–19 (2025)

Intelligent Detection Models and Algorithms

Detection of Floating Objects on Water Surface in Adverse Weather Conditions

Jiali Chen[1], Pengyu Liu[1](✉), Ke Zhang[2], Suchuang Di[2], and Sen Yu[2]

[1] School of Information Science and Technology, Beijing University of Technology, Beijing, China
liupengyu@bjut.edu.cn

[2] Beijing Water Science and Technology Institute, Beijing, China

Abstract. Floating object detection on water surfaces is crucial for waterway security and ecological protection. Current systems face significant performance degradation in adverse weather conditions due to precipitation noise and low visibility. To address this, we propose an enhanced YOLOv12n framework incorporating three key improvements: a synthetic weather-robust dataset for training, an edge-preserving spatial-channel convolution module for better feature extraction, and a frequency-domain attention mechanism for noise suppression. Our experiments on a self-built dataset demonstrate the effectiveness of these enhancements, achieving 87.1% mAP@ 0.5 while maintaining real-time performance at 476 FPS. The results show reliable detection capability across various weather conditions, making the system practical for real-world monitoring applications where weather resistance is essential.

Keywords: Floating objects detection · YOLOv12 · Real-time detection · Water surface monitoring

1 Introduction

Detection of floating objects on water surface is a critical technology for environmental monitoring, waterway safety, and ecological conservation. Floating objects such as plastics, waste, and algae can severely damage aquatic ecosystems and even block river channels, potentially leading to hazardous events like flooding. Traditional manual monitoring methods are inefficient and labor-intensive, making computer vision-based automated detection a prevailing solution in this field.

Current automated floating object detection methods primarily fall into two categories: traditional computer vision approaches and deep learning-based methods. Traditional computer vision techniques typically employ handcrafted features and dynamic scene modeling to address detection challenges in complex aquatic environments. Borghgraef et al. [1] tackled the automatic detection of small floating objects in challenging sea conditions by utilizing the ViBe algorithm [2] and behavior subtraction to analyze spatiotemporal correlations in dynamic scenes, effectively distinguishing targets from moving backgrounds. Socek et al. [3] proposed a hybrid foreground detection

P. Umapada et al. (Eds.): ICCPR 2025, CCIS 2811, pp. 221–230, 2026.
https://doi.org/10.1007/978-981-95-8315-7_18

method for maritime surveillance, combining Bayesian decision framework-based foreground segmentation with graph theory-driven edge-localized color segmentation to improve detection accuracy. Jin et al. [4] introduced an improved GMM-based automatic segmentation method (IGASM) that projects GMM results into HSV color space and incorporates a light-shadow discriminant function to establish a dynamic background model, further enhanced through morphological smoothing and Graph Cuts optimization. Venkatrayappa et al. [5] addressed the challenge of detecting unknown floating objects in complex maritime scenes by employing the K-SVD [6] algorithm to learn visual dictionaries for image denoising. Their approach generates self-similar content images, extracts residual features between original and denoised images to separate noise from targets, and finally applies an a contrario model for target extraction.

However, traditional CV methods suffer from limited generalization across diverse scenarios, reliance on labor-intensive feature engineering and parameter tuning, and high rates of missed detections. Deep learning-based approaches mitigate these issues by automatically learning discriminative features from complex backgrounds while achieving superior real-time performance and accuracy. Lin et al. [7] developed an enhanced FMA-YOLOv5s algorithm incorporating a Feature Map Attention (FMA) layer and background-augmented data expansion, effectively resolving challenges like scale variation and small object detection in complex waterways. Shi et al. [8] proposed a YOLOv8-based improvement that integrates C2f-CA coordinate attention for better target localization, adds a dedicated small-object detection layer, and employs Focaler-WIoUv3 loss for refined regression. Yi et al. [9] addressed precise localization of small floating objects through a CA-Faster R-CNN framework, combining Faster R-CNN's detection capability with Class Activation (CA) networks to reduce positioning errors by 6.29 pixels without compromising recognition accuracy. Li et al. [10] introduced PC-Net to overcome background interference and small-target omissions in floating object detection. Their pyramid anchor generation method minimize background noise, while classification maps serve as enhanced feature representations for small objects, achieving 86.4% average detection accuracy.

These methods demonstrate efficient and accurate identification of floating objects on water surfaces, particularly for complex backgrounds and small target detection. However, research on detecting floating objects under adverse weather conditions remains limited. This study addresses target edge blurring and precipitation noise interference caused by inclement weather. Considering the dual requirements of real-time performance and accuracy, we adopt YOLOv12n [11] as the baseline model. Our key contributions are summarized as follows:

(1) **Weather-robust dataset construction:** Through data augmentation techniques, we synthesize adverse weather effects on floating object images to create a dataset.
(2) **Edge-aware model architecture:** Building upon YOLOv12n, we incorporate spatial and channel reconstruction convolutions to mitigate target edge degradation in blurred images.
(3) **Frequency-domain noise suppression:** A self-attention mechanism based on frequency-domain is introduced to improve model resilience against rain and snow-induced high-frequency noise.

2 Methods and Principle

This section introduces the floating objects detection methods we used and their principles, including the construction of dataset, the overview of the improved model based on YOLOv12n, the improvement to alleviate the target edge degradation caused by foggy and cloudy, and the improvement to suppress noise caused by rainy and snowy.

2.1 Construction of Dataset

Due to the limitations of existing public datasets for floating object detection such as insufficient object categories and lack of adverse weather scenarios, we construct a custom dataset to ensure generalization across diverse object types and complex weather conditions. Our dataset was compiled through comprehensive collection methods including web scraping, field photography, and integration of publicly available imagery, encompassing 11 distinct categories of floating objects such as bottles, cans, plastic bags and so on.

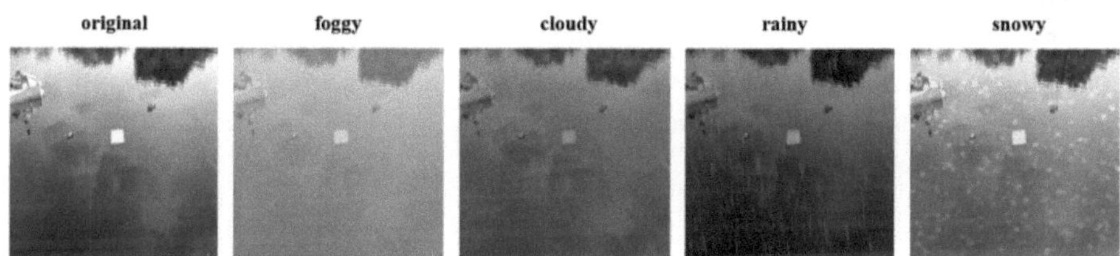

Fig. 1. Different Adverse Weather Effects

To address the critical absence of adverse weather representations, we implemented synthetic weather augmentation using the Albumentations library of Python, generating realistic fog, overcast, rain, and snow conditions, as shown in Fig. 1. Among these conditions, fog and overcast induce target obscuration through atmospheric scattering, while rain and snow introduce stochastic noise patterns and specular reflections that alter surface texture signatures.

The dataset was further expanded through conventional augmentation techniques including geometric transformations and photometric adjustments. The final curated dataset comprises 2,441 training images with complete category representation, accompanied by 649 validation and 310 test images, ensuring statistically robust evaluation across all conditions. The specific condition of the self-built dataset is provided in Fig. 2.

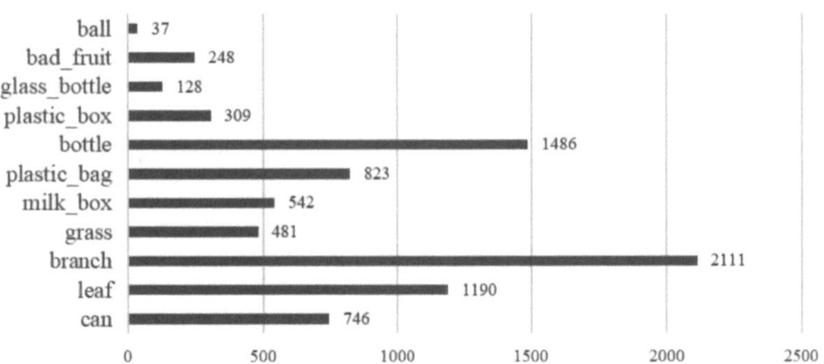

Fig. 2. The Statistics of Different Categories

2.2 Overall Structure

Our floating object detection model is based on YOLOv12n, which follows the classic single-stage detection framework of YOLO series. With only 2.5 million parameters, the model achieves high-precision real-time detection performance. The architecture consists of three key components: Backbone for feature extraction, Neck for multi-scale feature fusion, and Head for prediction output. The overall structure of our proposed model is illustrated in Fig. 3.

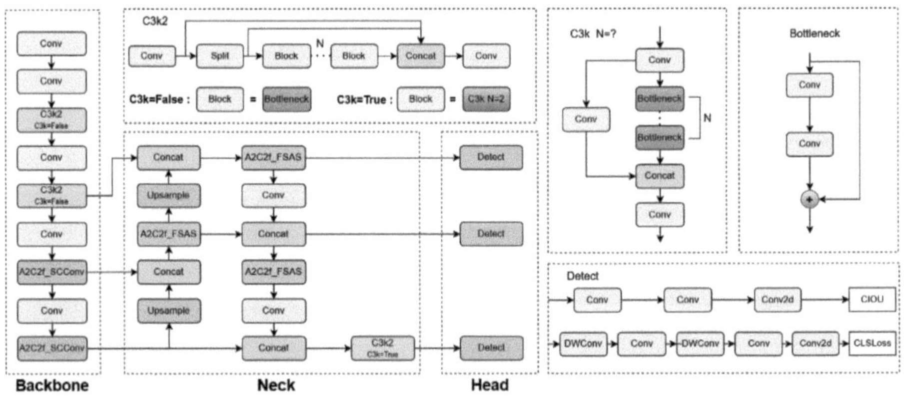

Fig. 3. The Overall Structure of Detection Model

Backbone The backbone utilizes Conv, C3k2, and A2C2f modules for hierarchical feature extraction. The C3k2 module enhances feature propagation while maintaining efficiency, and the A2C2f module preserves wide receptive fields with reduced computation. We further integrate SCConv into A2C2f (A2C2f_SCConv) to specifically boost edge feature representation.

Neck Employing FPN and PAN architecture, the neck performs bidirectional multi-scale feature fusion. The incorporated A2C2f modules improve feature integration, while the novel A2C2f_FSAS variant effectively suppresses precipitation noise through frequency-domain attention mechanisms.

Head The decoupled prediction head employs distinct convolutional operations: standard convolution for regression (with CIOU Loss) and depthwise separable convolution for classification (with BCE Loss), optimizing both accuracy and efficiency.

2.3 Spatial and Channel Reconstruction Convolution (SCConv) for Edge Enhance

SCConv [12] leverages both spatial and channel-wise redundancy in features to achieve CNN compression, effectively reducing redundant computations while enhancing the learning of representative features. The module consists of two key components: a Spatial Reconstruction Unit (SRU) that separates and reconstructs redundant features through weight-based operations to suppress spatial redundancy and strengthen feature representation, and a Channel Reconstruction Unit (CRU) that employs a split-transform-fuse strategy to minimize channel redundancy.

The Spatial Reconstruction Unit (SRU) effectively preserves information-rich high-frequency edge features, significantly alleviating the edge blurring issues prevalent in foggy and cloudy conditions. Complementing this, the Channel Reconstruction Unit (CRU) adaptively adjusts color channel weighting through its attention mechanism, thereby enhancing the model's robustness against chromatic aberrations induced by atmospheric haze and diffuse lighting.

To optimize performance in adverse weather scenarios, we integrate the SCConv module into the A2C2f blocks of YOLOv12n's backbone network. This architectural enhancement specifically strengthens the model's capacity for edge feature representation under low-visibility conditions. The structure of our modified A2C2f_SCConv module is presented in Fig. 4.

2.4 Frequency Domain-Based Self-Attention Solver (FSAS) for Noise Suppression

Frequency-domain Self-Attention Solver (FSAS) [13] is an innovative attention mechanism designed for vision transformer, which leverages the convolution theorem to optimize computational efficiency. Specifically, it capitalizes on the mathematical equivalence between spatial-domain convolution operations and element-wise multiplication in the frequency domain. Its structure is shown in Fig. 5(a).

Our implementation of FSAS adapts the A2C2f structure by first splitting the input feature map into three components: Query (Q), Key (K), and Value (V), through a combination of standard and depthwise separable convolutions. Following the self-attention mechanism in Transformers, the correlation between Q and K is computed to derive attention weights, which are then applied to V for feature refinement. To enhance information extraction from the frequency domain while reducing computational complexity, FSAS partition Q and K into patches with the same size and transform them into the frequency domain via Fast Fourier Transform (FFT). The correlation between Q and K is computed efficiently through element-wise multiplication in the frequency domain and then being converted back to the spatial domain via inverse FFT (IFFT). The resulting attention map is normalized and multiplied with V, followed by a convolutional layer to adjust channel dimensions.

Rain and snow noise exhibit distinct spectral patterns in the frequency domain, manifesting as directional energy bands (rain streaks) and high-frequency random components (snow particles). By transforming features into the frequency domain via FFT, FSAS leverages the inherent low correlation between the spectral energy of such noise and target features. This enables automatic noise attenuation through element-wise multiplication of the frequency-domain representations of Q and K.

To suppress noise across varying rain-snow conditions, we integrate the FSAS attention mechanism into the A2C2f modules of YOLOv12n's Neck layer (see Fig. 5(b)), which is named A2C2f_FSAS. It has adjustable patch sizes for Q/K decomposition to handle different noise scales.

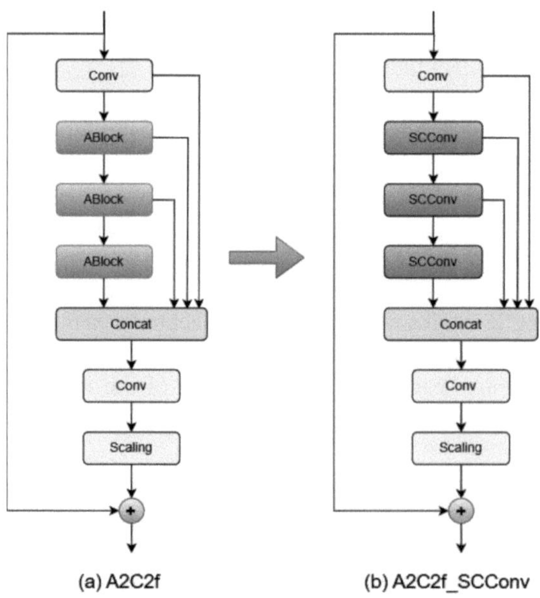

Fig. 4. The Structure of A2C2f_SCConv

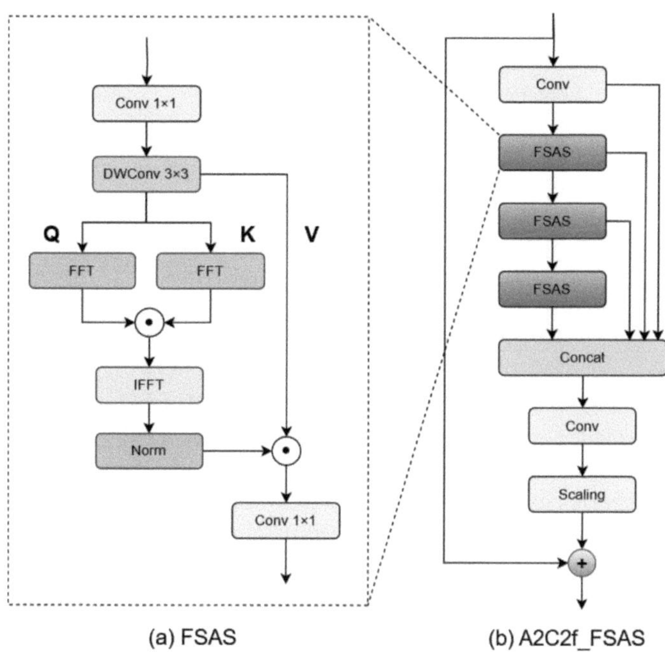

Fig. 5. The Structure of FSAS and A2C2f_FSAS

3 Experiments

We implement our model and conduct experiments based on PyTorch toolbox. The model was trained, validated and tested on a self-built dataset containing 3,400 images of floating objects on water surfaces, among which 950 images were augmented with adverse weather effects including fog, cloud, rain and snow conditions. The models were trained with an input image size of 640 × 640 pixels for 100 epochs using a batch size of 16. We use the RTX 4090 GPU for training and evaluating all models in our experiments.

Given the real-time monitoring requirements inherent to floating object detection applications, our comparative analysis focuses exclusively on single-stage object detection algorithms. As the YOLO series currently represents the state-of-the-art in single-stage detection, we conducted comprehensive benchmarking between our proposed model and YOLO variants ranging from YOLOv8n to YOLOv12n. The quantitative comparison results are presented in Table 1.

As demonstrated in Table 1, our proposed model exhibits superior performance over all listed YOLO variants in terms of mAP, achieving a 3.44\% improvement over the baseline YOLOv12n. Furthermore, the model reduces parameter count by approximately 22.55% while increasing FPS by 61.90%.

Table 1. Quantitative Comparison with YOLO Series Models.

Model	Precision	Recall	mAP@0.5	Params	FPS
YOLOv8n	0.832	**0.825**	0.841	3007793	1111
YOLOv9t [14]	0.825	0.823	0.838	1972929	1000
YOLOv10n [15]	0.806	0.768	0.809	2698706	667
YOLOv11n [16]	0.831	0.800	0.834	2584297	1250
YOLOv12n [11]	0.861	0.815	0.842	2510489	294
Ours	**0.878**	0.812	**0.871**	**1944249**	476

Figure 6 presents the normalized confusion matrix, demonstrating highly accurate detection performance with minimal false negatives and false positives. The diagonal values show consistently high recall rates, indicating excellent detection capability. While some categories like leaves show slightly lower performance, the overall low off-diagonal values confirm limited misclassification, contributing to the strong mAP@0.5 of 0.871.

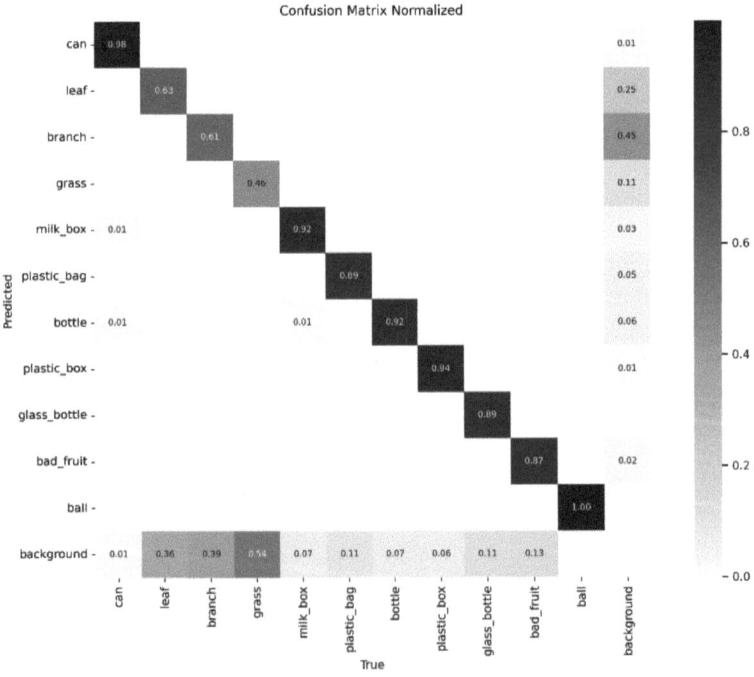

Fig. 6. The Nomalized Confusion Matrix of Our Model

Qualitative comparisons in Fig. 7 reveal enhanced detection performance under various adverse weather conditions, including foggy, cloudy, rainy, and snowy. The proposed model effectively addresses both false negatives and false positives present in the baseline approach. Additionally, it demonstrates robust performance in handling challenging reflective surface conditions.

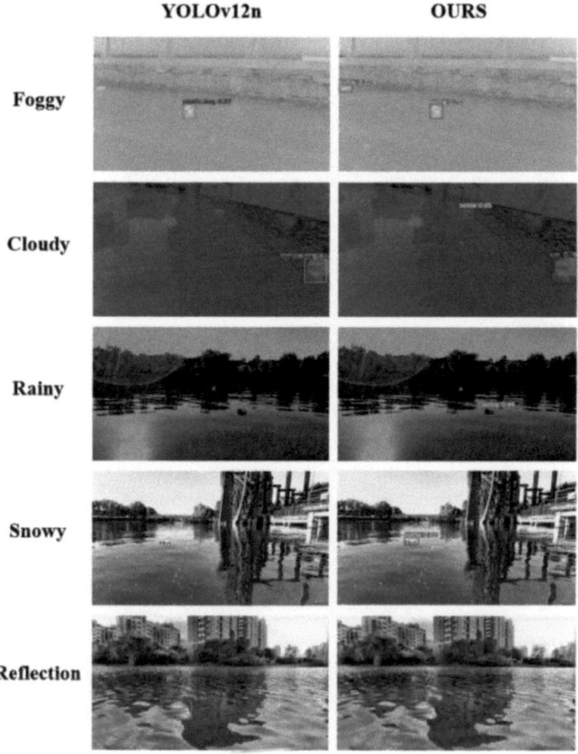

Fig. 7. Quantitative Comparison between YOLOv12n and Ours

Acknowledgments. This research was funded by the Beijing Water Science and Technology Institute through the project "Intelligent Perception of Rainstorm and Flood Disasters and Multi-source Data-driven Risk Early Warning Technology Research and Application" (Project No.: Z231100003823005). We gratefully acknowledge their generous support in logging data collection and field testing of the model.

Disclosure of Interests. The authors have no competing interests to declare that are relevant to the content of this article.

References

1. Borghgraef, A., Barnich, O., Lapierre, F., Van Droogenbroeck, M., Philips, W., Acheroy, M.: An evaluation of pixel-based methods for the detection of floating objects on the sea surface. EURASIP J. Adv. Signal Process. **2010**, 1–11 (2010)

2. Barnich, O., Van Droogenbroeck, M.: Vibe: a powerful technique for background detection and subtraction in video sequences. In: IEEE International Conference on Acoustics, Speech and Signal Processing. pp. 945–948 (2009)
3. Socek, D., Culibrk, D., Marques, O., Kalva, H., Furht, B.: A hybrid color-based foreground object detection method for automated marine surveillance. In: Advanced Concepts for Intelligent Vision Systems: 7th International Conference, ACIVS 2005, Antwerp, Belgium, September 20–23, 2005. Proceedings 7. pp. 340–347. Springer (2005)
4. Jin, X., Niu, P., Liu, L.: A gmm-based segmentation method for the detection of water surface floats. IEEE Access **7**, 119018–119025 (2019)
5. Venkatrayappa, D., Desolneux, A., Hubert, J.M., Manceau, J.: Unidentified floating object detection in maritime environment using dictionary learning (2020). arXiv preprint arXiv: 2007.15757
6. Aharon, M., Elad, M., Bruckstein, A.: K-svd: an algorithm for designing overcomplete dictionaries for sparse representation. IEEE Trans. Signal Process. **54**(11), 4311–4322 (2006)
7. Lin, F., Hou, T., Jin, Q., You, A.: Improved yolo based detection algorithm for floating debris in waterway. Entropy **23**(9), 1111 (2021)
8. Shi, C., Lei, M., You, W., Ye, H., Sun, H.: Enhanced floating debris detection algorithm based on cdw-yolov8. Phys. Scr. **99**(7), 076019 (2024)
9. Yi, Z., Yao, D., Li, G., Ai, J., Xie, W.: Detection and localization for lake floating objects based on ca-faster r-cnn. Multimedia Tools Appl. **81**(12), 17263–17281 (2022)
10. Li, N., Huang, H., Wang, X., Yuan, B., Liu, Y., Xu, S.: Detection of floating garbage on water surface based on pc-net. Sustainability **14**(18), 11729 (2022)
11. Tian, Y., Ye, Q., Doermann, D.: Yolov12: Attention-centric real-time object detectors (2025). arXiv preprint arXiv:2502.12524
12. Li, J., Wen, Y., He, L.: Scconv: spatial and channel reconstruction convolution for feature redundancy. In: Proceedings of the IEEE/CVF conference on computer vision and pattern recognition. pp. 6153–6162 (2023)
13. Kong, L., Dong, J., Ge, J., Li, M., Pan, J.: Efficient frequency domain-based transformers for high-quality image deblurring. In: Proceedings of the IEEE/CVF Conference on Computer Vision and Pattern Recognition. pp. 5886–5895 (2023)
14. Wang, C.Y., Yeh, I.H., Mark Liao, H.Y.: Yolov9: learning what you want to learn using programmable gradient information. In: European Conference on Computer Vision. pp. 1–21. Springer (2024)
15. Wang, A., Chen, H., Liu, L., Chen, K., Lin, Z., Han, J., et al.: Yolov10: real-time end-to-end object detection. Adv. Neural. Inf. Process. Syst. **37**, 107984–108011 (2024)
16. Khanam, R., Hussain, M.: Yolov11: an overview of the key architectural enhancements (2024). arXiv preprint arXiv:2410.17725

A Complex Scene Forest Fire and Smoke Detection Method Based on Mamba and YOLOv5

Mingze Gao(✉), Yu Liu, and Cheng Guo

Intelligent Perception Laboratory of Xi'an Institute of Applied Optics, Xi'an City, China
2231518431@qq.com

Abstract. This paper proposes FF-YOLO, a cross-scale feature fusion network guided by attention mechanisms, to address challenges in forest smoke and fire detection—including leakage, false detection, and low accuracy caused by environmental complexity, object occlusion, and insufficient lighting. First, we enhance the YOLOv5s backbone by integrating a Mamba-based linear attention module (MAA). Leveraging Mamba's efficiency with high-resolution visual tasks, this structurally similar module strengthens effective information extraction, enabling the model to prioritize critical features and improve detection under complex environments and occlusion. Second, we introduce a cross-scale weighted feature fusion structure (NEWFPN) in the neck network. This enhances multi-scale and multi-depth information utilization, overcoming obstacles like low brightness and dense obstructions to boost detection accuracy. Finally, a dedicated small object detection layer is added to the head. Evaluated on our custom dataset, FF-YOLO achieves an 89.25% mAP in detecting forest smoke and fires across diverse conditions—including small fire sources and obstructions—demonstrating its robustness and practical value.

Keywords: forest smoke and fire detection · YOLOv5 · attention module · deep learning

1 Introduction

Climate change has intensified forest fires as a critical outdoor safety hazard. Compounded by complex geographical environments and variable climatic conditions, forests are highly vulnerable to natural disasters—particularly impactful and frequent wildfires. These sudden, uncontrollable events cause devastating damage to personnel, equipment, and operations. Enhancing pre-disaster prediction capabilities is therefore crucial for reducing casualties and property losses. Recent incidents like Australia's 2020 "Black Summer" bushfires (33 fatalities, infrastructure destruction), California's 2020 wildfires (>400,000 acres burned), and Chile's 2023 forest fires (24 deaths, thousands injured) underscore the severe threat wildfires pose to human activities [1].

Deep learning-based image detection has gained significant attention for its high precision and feature learning capabilities. Researchers increasingly apply these algorithms

P. Umapada et al. (Eds.): ICCPR 2025, CCIS 2811, pp. 231–244, 2026.
https://doi.org/10.1007/978-981-95-8315-7_19

to fire detection: Frizzi et al. [2] pioneered CNN for fire scene detection, enabling automatic smoke/fire feature extraction. [3] proposed a spatiotemporal flame evolution-based DCGAN trained on real sequences for real-time flame detection with low false alarms. Zhang et al. [4] integrated channel attention with YOLOv3 to enhance flame/smoke feature extraction. Xu et al. [5] combined YOLOv5 and EfficientNet [6] to overcome single-model limitations in complex scenarios, boosting accuracy and recall. Chen et al. [7] employed lightweight networks to reduce YOLOv5 parameters while maintaining speed, though detection performance declined significantly with IoU. Cao et al. [8] enhanced feature extraction via dynamic snake convolution and GAM attention, integrating DyHead into detection heads for improved task perception.

Despite maturity in detecting common fires, forest fire/smoke identification in complex environments faces persistent challenges: 1) Terrain/vegetation obstruction: Mountains, gorges, and dense canopies limit visibility and conceal ground fires; 2) Weather interference: Fog, haze, or sandstorms mimic smoke, increasing false alarms; 3) Early smoke characteristics: Thin, low-concentration smoke is easily masked by background noise (e.g., vapor), while color variations (white to dark brown) based on fuel type and combustion stage complicate identification—especially at night or in low light.

To address these challenges, we propose FF-YOLO, a YOLOv5s-based network for early forest fire detection. The architecture incorporates: 1) Mixup data augmentation to resolve unclear target boundaries; 2) A novel feature pyramid module reusing earlier-layer features to enhance extraction; 3) A Mamba-based linear attention module (MAA) in the backbone to strengthen feature extraction capabilities.

In the neck network, we construct a new pyramid network model based on GFPN and BiFPN structures. This improvement connects the original features of hazardous targets in fire data with deep features while replacing the original fusion method with weighted fusion, enabling more effective extraction of desired characteristics. This design enhances feature information flow while grouping connections reduce computational costs. Additionally, we integrate a dedicated small target detection layer into the YOLOv5s architecture. This layer establishes horizontal connections with the backbone network's initial features, facilitating better integration of shallow and deep representations for small forest fire targets. The design captures subtle feature variations in fire points, thereby improving detection sensitivity and accuracy for small targets under weak lighting conditions.

Experimental results demonstrate that compared to existing mainstream deep learning detection models, our algorithm achieves significant improvements in accuracy and robustness when detecting early-stage forest fires and smoke targets in complex, low-illumination forest environments. The method attains 89.25% mean Average Precision (mAP), validating the model's effectiveness in detecting forest fires and smoke under complex conditions and confirming its practical feasibility.

2 Improvement of the Detection Method

2.1 Mixup Image Enhancement Algorithm

Mixup is a simple yet effective image enhancement strategy that involves randomly selecting two images and their corresponding labels from the training set, then linearly combining them to generate new samples. Essentially, it selects two images per batch and blends them proportionally to create augmented data. Crucially, only these mixed images are used during training—original images are excluded. This approach transcends traditional data enhancement limitations by offering a novel training perspective. It not only simulates data distribution shifts but also enables models to learn inter-sample relationships, enhancing performance on complex tasks. At its core, Mixup trains neural networks on paired samples and labels, regularizing networks while strengthening linear representations of training data.

Unlike traditional methods (rotation, flipping, scaling, cropping) that alter individual samples' geometry, color, or brightness, Mixup adopts a more comprehensive methodology. Its advantage lies in pairing distinct samples and combining their features and labels through linear interpolation. This simultaneously increases data diversity and introduces inter-sample relationship information, allowing models to learn transitional features between categories. For instance, while traditional augmentation might rotate a single cat image, Mixup can blend cat and dog images, helping models discern inter-category distinctions and connections.

Empirical results consistently demonstrate Mixup's effectiveness in boosting model generalization and robustness. Applied to image classification datasets, it significantly reduces error rates and improves test accuracy. Furthermore, models trained with Mixup exhibit enhanced resilience against adversarial attacks.

2.2 Improvement of the Algorithm Model Based on YOLOV5s

YOLOv5 is a widely used target detection model with variants including YOLOv5n, YOLOv5s, YOLOv5m, YOLOv5l and YOLOv5x, differing in depth and width. Considering detection speed, model size, and training equipment [9], we selected YOLOv5s as the baseline for improvement.

The model structure comprises four components: Input, Backbone, Neck, and Head (Fig. 1). The SPPF module replaces the SPP module's single large pooling kernel with multiple serialized small kernels, enhancing feature map expressiveness while preserving functionality.

The Neck network adopts a Feature Pyramid Network (FPN)+ Path Aggregation Network (PAN) structure [10]. FPN fuses high-level and low-level features top-down to improve small target detection, while PAN propagates features upward to enrich positional information in top layers. The Head contains three detection heads for large, medium, and small targets respectively, using CIOU loss [11] to evaluate localization between target and prediction boxes.

The improved model based on the above model is shown in the Fig. 2. The main improvements are the backbone network, the neck network and the detection head. The original framework had the following issues: poor identification of small fire sources, low precision in detecting forest fires under low-light conditions, poor feature fusion ability at different scales, and loss of feature information during the feature extraction process. First, a Mamba-based linear attention mechanism module is added to the backbone network. Secondly, the neck network adopts the feature extraction method of other feature pyramid models and uses a cross-scale feature weight fusion method to improve the ability of the network model to extract features at different scales. Finally, a small target detection layer is introduced to the head to improve the model's ability to identify small fire sources.

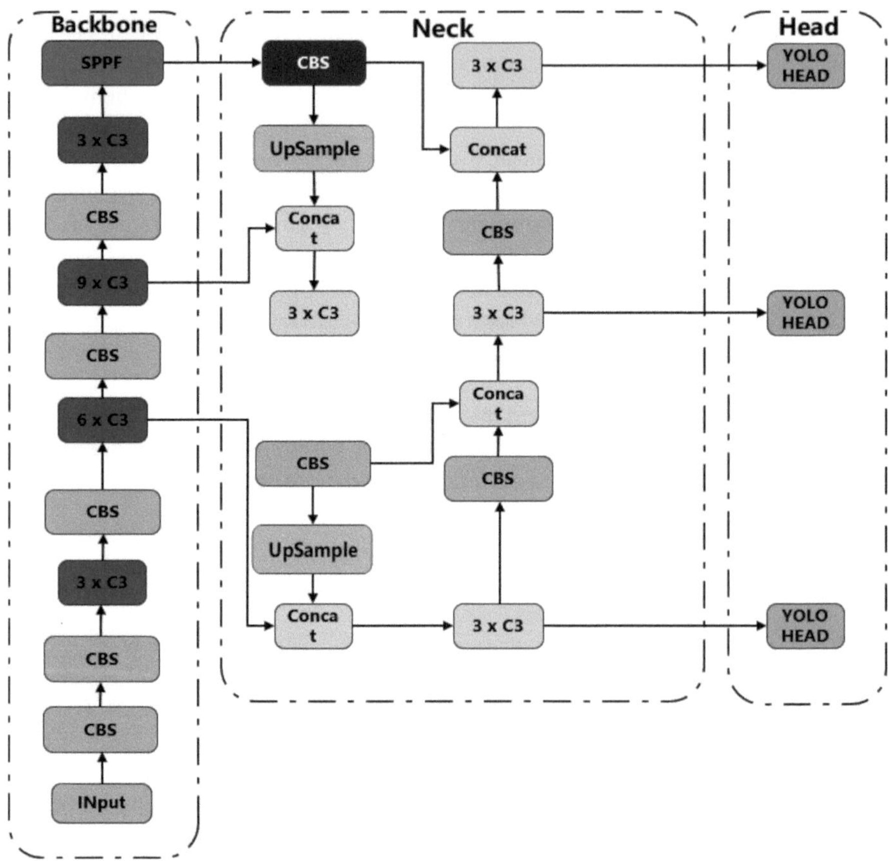

Fig. 1. Yolov5 network framework diagram

A backbone network based on an efficient attention mechanism.
To address partial feature information loss during extraction and the oversight of channel-wise feature importance, researchers employ attention mechanisms to selectively weight input features, directing model focus toward critical characteristics. Recently, state space models (SSMs) like Mamba have emerged as promising solutions. These models excel at capturing long-range dependencies while maintaining favorable linear time complexity.

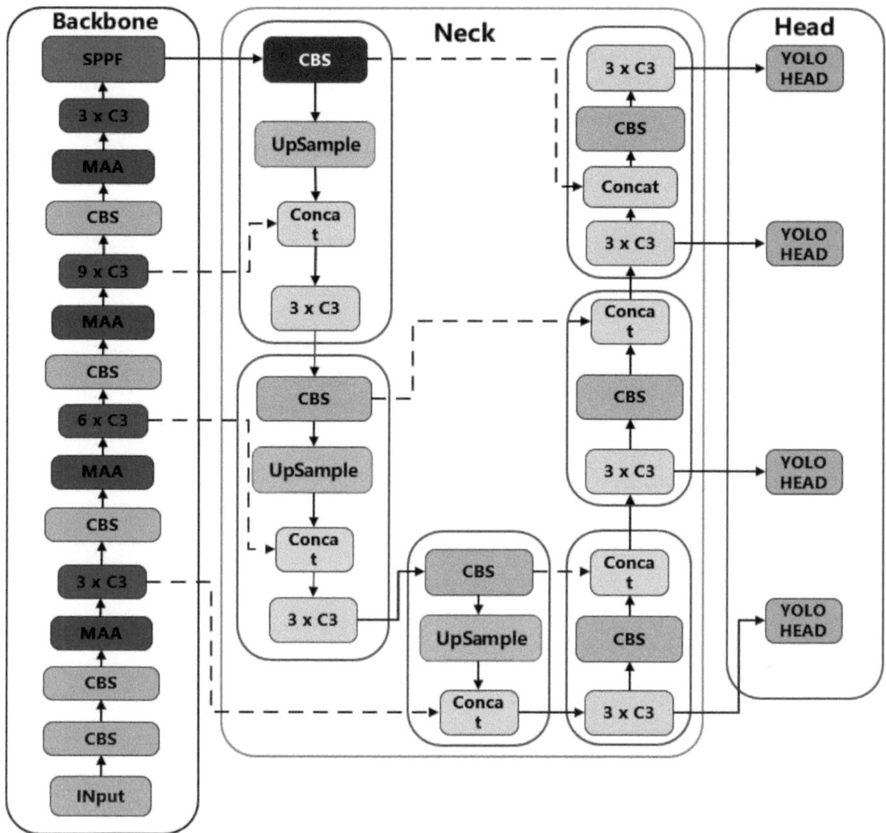

Fig. 2. Improved network model

Notably, Mamba has been successfully adapted to computer vision, achieving significant results in image classification [12].

As demonstrated in the preceding literature, it can be concluded that Mamba significantly outperforms traditional attention not by being a "faster Transformer," but by being a fundamentally different and more appropriate architecture for modeling sequential data in the real world.

Its selectivity allows it to act like a resource-conscious manager, dynamically allocating its finite computational budget to the most informative parts of the input. Its recurrent state provides a compact and efficient memory of past events. Together, these properties make it uniquely suited for highly varied, long-sequence, and resource-constrained environments where the brute-force approach of traditional attention is both computationally impractical and conceptually mismatched.

Despite linear attention Transformers typically underperforming traditional Transformers in practice, the highly efficient Mamba model—with linear complexity and strong high-resolution visual task performance—shows striking similarities. Through systematic comparison of efficient Mamba and suboptimal linear attention Transformers, we reveal core mechanisms behind Mamba's success. By unifying selective SSMs and linear attention formulations, we reinterpret Mamba as a linear attention variant featuring six key distinctions: input gates, forget gates, shortcut connections, absence of attention normalization, single-head attention, and modified module design.

Detailed analysis of these design elements demonstrates that forget gates and module design constitute Mamba's core success drivers, while the other four contribute less significantly. Inspired by these findings and [13], we integrate these critical designs into linear attention, creating an efficient attention mechanism module (MAA) illustrated in Fig. 3.

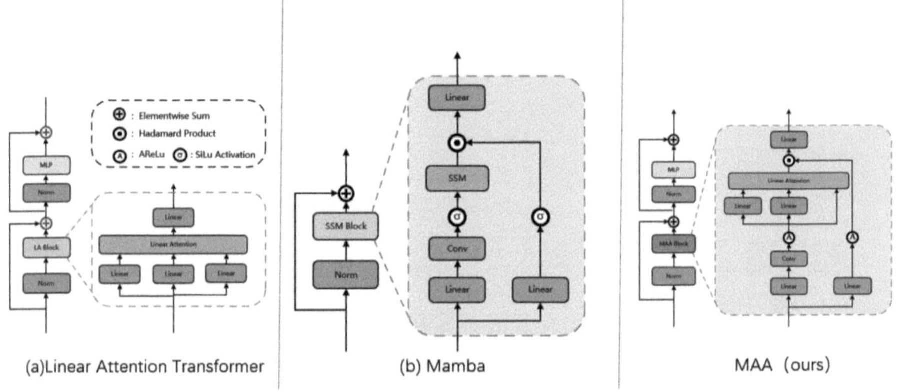

Fig. 3. MAA module improved with Mamba and linear attention mechanism

Notably, the MAA attention module proposed in this paper replaces Mamba's SiLU activation function with the AReLU activation function. AReLU, an element-level attention mechanism and improvement over ReLU, is defined in Eq. (1).

This function intelligently integrates ELSA[14] with ReLU to form an attention-based linear rectifier unit capable of dynamic learning. Its key distinction from other activation functions is enabling rapid network convergence at minimal learning rates—a critical advantage for specialized tasks like transfer learning.

Unlike SiLU activation functions that introduce nonlinearity for complex function fitting, they exhibit weaker gradient mitigation capabilities and heightened sensitivity to inputs and parameters. In contrast, AReLU leverages element-wise residual learning through attention mechanisms, effectively amplifying gradients during backpropagation to activate inputs and alleviate vanishing gradients while maintaining adaptability to complex data distributions.

$$mathcalF(x_i, \alpha, \beta) = mathcalR(x_i) + mathcalL(x_i, \alpha, \beta) = \begin{cases} C(\alpha)x_i, & x_i < 0 \\ (1 + \sigma(\beta))x_i, & x_i \geq 0 \end{cases} \quad (1)$$

$$mathcalL(x_i, \alpha, \beta) = \begin{cases} C(\alpha)x_i, & x_i < 0 \\ \sigma(\beta)x_i, & x_i \geq 0 \end{cases} \quad (2)$$

$$mathcalR(x_i) = \begin{cases} 0, & x_i < 0 \\ x_i, & x_i \geq 0 \end{cases} \tag{3}$$

Pyramid structure based on multi-scale feature fusion.
JANG et al. [15] proposed the Giafe Det method, which fuses feature maps from different spatial scales and depths, thereby constructing the GFPN pyramid network model. BiFPN is an improved feature pyramid network that enhances the fusion of multi-scale features through top-down and bottom-up feature interactions. Performs well in target detection tasks, especially when processing targets of different sizes. The core idea is to enhance the expressive power of features by combining top-down and bottom-up feature fusion through bidirectional feature interaction. This method is introduced into the YOLOv5s model to improve the feature pyramid structure and reduce the loss of small target features in the process of multi-scale feature fusion. The network model based on these two improved feature pyramid structures is shown in Fig. 4.

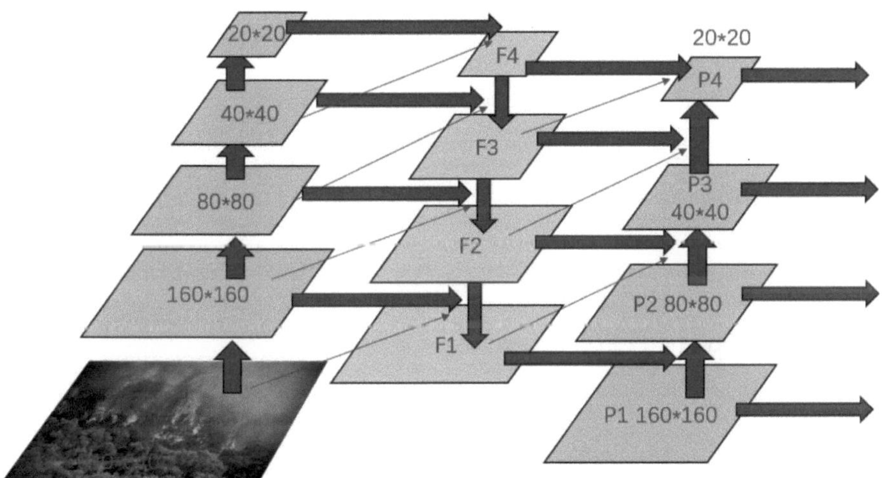

Fig. 4. A feature pyramid network based on GFPN and BiFPN improvements.

This model builds on the existing top-down and bottom-up dual-directional feature interaction by adding feature fusion from different layers and positions [16]. This fusion is not simply an addition, but a weighted fusion, enabling the network to effectively combine useful and noteworthy feature information according to different tasks and image characteristics. This makes the fused features more representative and able to accurately describe objects in images. Concurrently, this approach facilitates the construction of a feature layer intended for the evaluation of minor objectives. It achieves this by assigning a greater weight to the more detailed and intricate lower layers, while assigning a lower weight to the more abstract higher layers. This results in a substantial enhancement of the fusion of features, thereby ensuring a greater degree of reliability. By establishing connections between different feature layers, the network's ability to utilise information at different scales and depths can be improved, The utilisation of higher-level semantics facilitates more expeditious and direct communication with lower-level layers, thereby facilitating the interpretation of detailed information. Concurrently, lower-level details can also exert a more efficacious influence on higher-level prediction. The probability of miscommunication or loss of information in the network is significantly reduced, thus enabling the successful completion of tasks such as fire identification in challenging environments characterised by low visibility and substantial obstruction.

Detection head based on small target identification.
YOLOv5 performs poorly in detecting small targets, as their limited pixel features are easily overlooked. While convolutional blocks effectively extract features from maps, they simultaneously reduce resolution with increasing network depth, hindering small object feature extraction. Although YOLOv5's three detection heads suffice for general detection tasks, early forest fire images predominantly contain small targets. Factors like tree cover further diminish fire point visibility, making consistent detection challenging.

To enhance detection of obscured early forest fire targets, we added a dedicated small-target detection layer to YOLOv5's architecture. This is achieved by: 1) Applying convolution and upsampling to an output layer from the neck network. 2) Concatenating the result with the backbone network's third-layer features. 3) Processing through a 3xC3 module to generate a new 160×160 detection layer.

Compared to the original three-head design, this four-head configuration captures significantly more small-target information, substantially improving detection and localization of small fire sources.

3 Experimental Design and Results Analysis

3.1 Data Collection and Processing

When using target detection networks for forest fire identification, high-quality datasets are essential to provide comprehensive and effective feature information. This paper utilizes internet-sourced videos and images processed with the Labelme annotation tool. Data augmentation techniques included image flipping, cropping, brightness adjustment, contrast adjustment, saturation adjustment, and noise addition. The final dataset comprises 6,293 images. To improve the ability to detect early fires in low-light conditions, different night-time forest fire scenes with different light intensities were simulated by adjusting the RGB channels. These scenes were used to expand the data set. The operating diagram is shown in Fig. 5.The dataset was partitioned into training, validation, and test sets at an 8:1:1 ratio, encompassing diverse scenes, environmental conditions, and temporal periods.

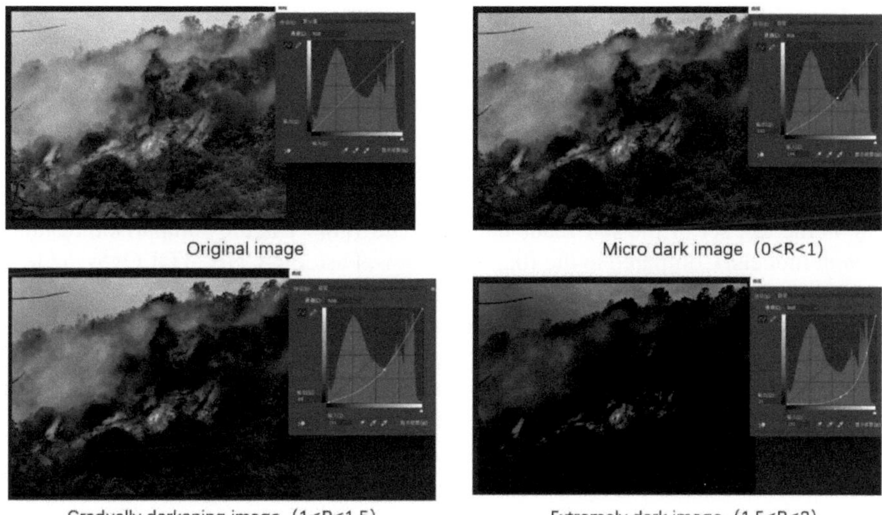

Fig. 5. The RGB adjustment illustrates the expansion of the forest fire data under different lighting conditions.

3.2 Experimental Environment Parameters, Configuration and Evaluation Indicators

The experiments were conducted on an AutoDL RTX 3080Ti GPU with the following configuration: PyTorch 1.10.0, Python 3.8, and CUDA 11.3. Training parameters included: batch size of 8, 200 epochs, 640×640 input resolution, and Adam optimizer (initial learning rate: 0.01; weight decay: 0.0001). During testing, thresholds were set as: Non-Maximum Suppression (NMS) at 0.3, confidence at 0.2, and Intersection over Union (IoU) at 0.5.Evaluation metrics comprised precision rate, recall rate, and average precision, defined by the following equations:

$$P = \frac{\mathrm{TP}}{\mathrm{TP} + \mathrm{FP}} \tag{4}$$

$$R = \frac{\mathrm{TP}}{\mathrm{TP} + \mathrm{FN}} \tag{5}$$

$$\mathrm{AP} = \int_0^1 P(R)dR \tag{6}$$

In this study, P denotes precision, a metric exclusively focused on the outcomes of forecasts. It calculates the probability of an observed positive outcome occurring within the set of forecasts that are predicted to be positive. The notations TP, FP, and FN represent the true positive, false positive, and false negative outcomes, respectively.

R represents the recall rate, which is defined in relation to the original sample. It denotes the probability of a positive sample being predicted as positive.

AP, short for 'average precision', is a metric frequently employed in both search and regression tasks. It can be interpreted as the area under the precision-recall curve. This metric is indicative of the overall accuracy of the model.

3.3 Experimental Results and Analyses

Comparative Tests.

To compare the performance of fire detection models, we compared the proposed model in the aforementioned experimental environment with Faster RCNN, YOLOv5s, YOLOv8s [18] and YOLOX [19]. P, R, mAP and FPS were chosen as the evaluation indicators, and a comparative experiment was conducted using the same training parameters on the test set of the self-made data set. The Table 1 is the comparison results of different models. Compared to the four algorithms, Faster R-CNN, YOLOv5s, YOLOX and YOLOv8s, the proposed algorithm achieved a precision rate of 89.74%, the highest among the four algorithms. The recall rate increased by 2.62%. Although the FPS decreased by 8 fps compared to YOLOv5s, it is still higher than the other three algorithms, reaching 119 fps and meeting the requirements. These results demonstrate the superiority of our proposed method. Therefore, we conclude that the model proposed in this paper is more suitable for the detection of forest fires and smoke.

Table 1. Smoke and Fire Detection Network Performance Comparison.

Method	Precision	Recall	mAP	FPS
Faster R-CNN	0.3985	0.8127	0.7126	68
YOLO-v5s	0.8631	0.8335	0.8646	127
YOLO-v8s	0.8921	0.8793	0.8943	106
YOLOX	0.8433	0.8207	0.8524	95
Ours	0.8974	0.8931	0.8925	119

Comparative Tests.

To more intuitively show that the proposed FF-YOLO method performs better, an ablation study is carried out in this paper. See Table 2 for detailed results.

Table 2. Results of Ablation Experiments.

Heading level	Precision	Recall	mAP	FPS
YOLO-v5s	0.8631	0.8335	0.8646	127
YOLO-v5s + Mixup	0.8724	0.8371	0.8674	127
YOLO-v5s + MAA	0.8856	0.8357	0.8853	125
YOLO-v5s + NEW-FPN	0.8793	0.8331	0.8726	124
YOLO-v5s + small-head	0.8723	0.8574	0.8739	126
YOLO-v5s + MAA + NEW-FPN	0.8906	0.8668	0.8817	121

(continued)

Table 2. (*continued*)

Heading level	Precision	Recall	mAP	FPS
YOLO-v5s + MAA + NEW-FPN + Mixup + small-head	0.8974	0.8931	0.8925	119

The above table shows the ablation results on the YOLO-v5s model, which has components that can be added incrementally. As can be seen from the results, the mAP of the standard YOLO-v5s is 86.46%.By using Mixup data augmentation, adding the MAA attention mechanism and the NEW_FPN feature pyramid structure, the mAP increased to 86.74%, 88.53% and 87.26%.Adding a small target detection head can improve detection accuracy to 87.23%.The fusion of the above improvements results in an algorithm model that achieves a 2.79% higher mAP than YOLO-v5s on the test set, and the proposed method achieves significant performance improvements in both positioning and identification. Although FPS decreased by 8, it still meets the operating conditions of real-time detection devices such as cameras.

Detection of Effects.
We present a visualization of the proposed algorithm model in Fig. 6. The results demonstrate strong performance in detecting fire and smoke across challenging scenarios including small-target fires, foggy conditions, and low-light nighttime environments. The proposed method accurately identifies flames and smoke in fire scenes with high precision, showing near absence of false positives and false negatives.

Additionally, we employed attention heatmap visualization to validate the attention mechanism's integration. As shown in Fig. 7, compared to other classical algorithms, our model demonstrates significantly improved target identification accuracy through three key enhancements: incorporating the MAA attention module in the backbone network, implementing a cross-scale fusion feature pyramid network, and adding a dedicated small-target detection head. At the same time, by analysing the attention heatmap, the ablation experiment and the results of the comparison experiment, it is not difficult to conclude that the model's detection accuracy has been improved to some extent by the model's improvement. It can also be seen from the recall rate indicator that the model's improvement has reduced misreporting and underreporting.

Fig. 6. Model detection graph visualization

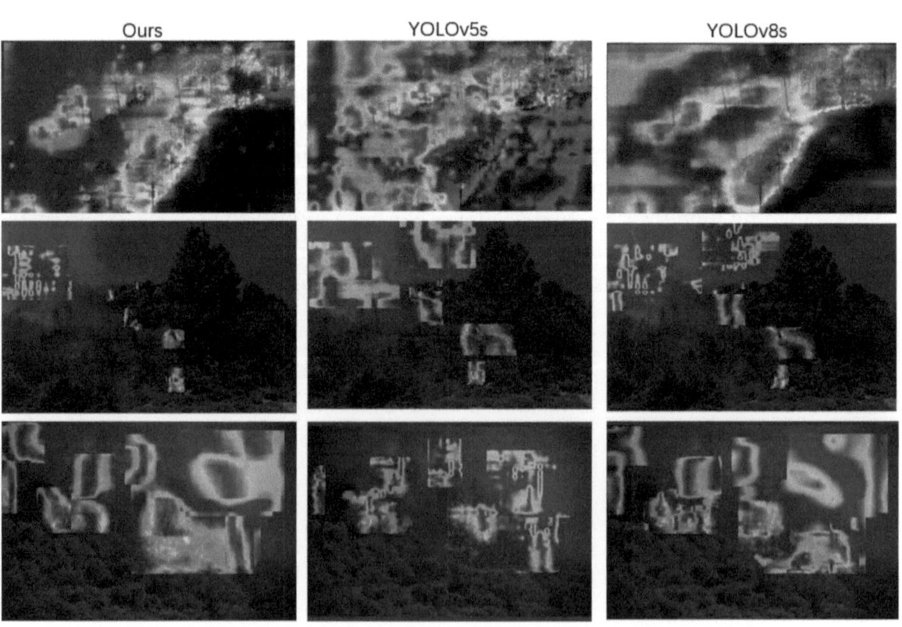

Fig. 7. FF-YOLO model detection heat map effect.

4 Conclusion

This paper presents a cross-scale feature fusion network incorporating an attention mechanism, based on YOLO, to address the limitations of forest smoke and fire detection under challenging environmental conditions. We enhanced the YOLOv5s backbone by integrating a Mamba-based linear attention module (MAA), which improves the model's ability to prioritize salient features and boost detection performance in complex and occluded scenarios, leveraging Mamba's efficiency in handling high-resolution visual tasks. Furthermore, a cross-scale weighted feature fusion structure (NEWFPN) was introduced in the neck network. This module employs weighted fusion to strengthen multi-level feature integration and incorporates a dedicated detection layer for small targets in the head, thereby enhancing the aggregation of discriminative and informative features. As a result, the model achieves more representative feature learning, effectively overcoming obstacles such as low visibility and occlusions in forest fire detection, and significantly improving accuracy.

Experimental results demonstrate that the proposed method attains a mAP of 89.25% in forest smoke and fire detection tasks. Comparative experiments confirm that our approach outperforms existing detection methods, while ablation studies validate the effectiveness of each proposed improvement. Visualization results further illustrate the robustness and strong detection capability of our model, particularly in scenarios involving small fire targets and complex environments.

The main innovation of this work lies in the integration of a state-space attention mechanism with cross-scale feature fusion tailored for forest fire detection. This strategy partially mitigates the issue of low detection accuracy in complicated scenes. However, we observed that the model exhibits a higher misdetection rate under low-light conditions compared to human performance, likely due to insufficient representation of nighttime data in the training set. Future work will focus on expanding the dataset with real nighttime imagery and employing specialized methods to enhance image brightness and reduce noise, further improving the model's adaptability and accuracy in such scenarios.

Acknowledgments. Yu Liu and Cheng Guo conducted the research; Mingze Gao analyzed the data and complete the experiment; Mingze Gao wrote the paper. All author had approved the final version. Besides, this research was funded by Army pre-research fund No.627010402.

Disclosure of Interests. The authors declare no conflicts of interest.

References

1. Schollaert, C.L., Jung, J., Wilkins, J., et al.: Quantifying the smoke-related public health trade-offs of forest management. Nat. Sustain. **7**(2), 130–139 (2024)
2. Frizzi, S., Kaabi, R., Bouchouicha M., et al.: Convolutional neural network for video fire and smoke detection. Conference of the IEEE Industrial Electronics Society. IEEE, pp. 877–882 (2016). https://doi.org/10.1109/IECON.2016.7793196.Author, F., Author, S.: Title of a proceedings paper. In: Editor, F., Editor, S. (eds.) CONFERENCE 2016, LNCS, vol. 9999, pp. 1–13. Springer, Heidelberg (2016)

3. Aslan, S., et al.: Deep convolutional generative adversarial networks based flame detection in video (2019)
4. Zhang, X., Qian, K., Jing, K., Yang, J., Yu, H.: Fire detection based on convolutional neural networks with channel attention. 2020 Chinese Automation Congress (CAC) (2020)
5. Xu, R., Lin, H., Lu, K., Cao, L., Liu, Y.: A forest fire detection system based on ensemble learning. Forests **12**(2), 217 (2021)
6. Tan, M., Le., Q.V.: Efficientnet: rethinking model scaling for convolutional neural networks (2019)
7. Chen, Z., Yang, J., Chen, L., Jiao, H.: Garbage classification system based on improved shufflenet v2. Resour. Conserv. Recycl. **178**, 106090 (2022)
8. Cao, l.Y., Yang, Y.Z., LiS, l., et al.: Research on forestfire monitoring technology for complex scenarios. Radio Commun. Technol. **50**(4), 779–788 (2024)
9. Redmon, J., Divvala, S., Girshick, R., et al.: You only look once: unified, real-time object detection. Proceeding of the IEEE Conference on Computer Vision and Pattern Recognition. NV, US: IEEE, pp. 779–778 (2016)
10. Lin, T.Y., Dollar, P., Girshick, R., et al.: Feature pyramid networks for object detection. Proceedings of the IEEE Conference on Computer Vision and Pattern Recognition. Honolulu, US, IEEE, pp. 936–944 (2017)
11. Zheng, Z., Wang, P., Liu, W., et al.: (2020). Distance-IoU loss: Faster and better learning for bounding box regression. Proceedings of the AAAI Conference on Artificial Intelligence. NY, US: AAAI, vol. **34**(07), pp. 12993–13000
12. Gu, A., Dao, T.: Mamba: linear-time sequence modeling with selective state spaces (2023). arxiv preprint arxiv:2312.00752
13. Al-Wesabi A.S., Albraikan, A.A., Eltahir, M.M.等.: Multi-biometric sustainable approach for human appellative. 2021 International Conference on Artificial Intelligence and Smart Systems (ICAIS)*. IEEE, pp. 1663–1668 (2021)
14. Chen, F., Guan, Y.: Arelu: agile rectified linear unit for improving lightweight convolutional neural networks. IEEE Access **13**
15. Jiang, Y., Tan, Z., Wang, J., Sun, X., Lin, M., Li, H.: Giraffedet: a heavy-neck paradigm for object detection (2022)
16. Liu, S., Qi, L., Qin, H.: Path aggregation network for instance segmentation. In: Proceedings of the IEEE Conference on Computer Vision and Pattern Recognition, Salt Lake City, pp. 8759–8768 (2018)
17. Zheng, Z., Wang, P., Liu, W.: Distance-IoU loss: faster and better learning for bounding box regression. In: AAAI Conference on Artificial Intelligence, 1299313000 (2020)
18. Kumari, S., Gautam, A., Basak, S., Saxena, N.: YOLOv8 based deep learning method for potholes detection. In: 2023 IEEE International Conference on Computer Vision and Machine Intelligence (CVMI), pp. 1–6 (2023)
19. Zhang, L., Zou, F., Wang, X.F., Wei, Z., Li, Y.: Improved algorithm for yolox-s object detection based on diverse branch block (dbb). Proceedings of the 2022 6th International Conference on Electronic Information Technology and Computer Engineering (2022)

SSA-YOLO: A Lightweight Drone Imagery Detection Algorithm Improving YOLOv10

Che Guowei, Zhang Mengqi(✉), Zhao Ruijun, and Wang Chengyao

Tianjin Normal University, 393 Binshui West Road, Xiqing District, Tianjin 300387, China
gwche@tjnu.edu.cn, 1115208382@qq.com
http://www.tjnu.edu.cn

Abstract. This paper addresses the challenges in drone object detection—complex backgrounds and numerous small objects—aiming to improve detection accuracy without increasing parameter count. By analyzing YOLOv10n's architecture, we propose SSA-YOLO, a lightweight enhanced version with two key improvements: the SSA (Small Shuffle Attention) module in the backbone strengthens small-object feature extraction and fusion via spatial attention, while the USKConv (UltraLight Selective Kernel Convolution) module in the Neck optimizes target localization and classification through dynamic feature processing. Experiments on Visdrone2019 show SSA-YOLO improves mAP50 by 2.2% over YOLOv10n, with only a 7.1% parameter increase, and maintains generalization across datasets, validating the proposed strategies. This work supports practical drone detection applications and offers insights for optimizing small-object models on mobile devices.

Keywords: Drone imagery · Object recognition · Small objects

1 Introduction

Object detection, a core task in computer vision, finds wide applications in intelligent security and autonomous driving. The YOLO series [1] has stood out for its balance of speed and accuracy, with YOLOv10 as the latest iteration. However, in drone imagery—characterized by complex backgrounds and abundant small objects—YOLOv10 still underperforms, particularly in small-object feature extraction.

Drone imagery poses unique challenges: small objects often occupy only 10–50 pixels, leading to sparse feature representation and poor low-level texture/shapes extraction. Additionally, complex backgrounds and scale variations from flight pose/altitude fluctuations exacerbate detection difficulties.

Existing small-object detection methods include contextual enhancement [10], data augmentation [11], multimodal fusion [12], and lightweight architectures [13].

P. Umapada et al. (Eds.): ICCPR 2025, CCIS 2811, pp. 245–255, 2026.
https://doi.org/10.1007/978-981-95-8315-7_20

To address these issues, this paper proposes SSA-YOLO, an enhanced YOLOv10 variant. Key improvements include: (1) integrating the SSA module in the backbone to strengthen spatial attention for small-object features; (2) adopting SKConv in the Neck for dynamic multi-scale feature fusion, enhancing adaptability to complex scenes.

By combining deep-layer abstract features and shallow-layer detailed features, the model balances multi-scale detection. SKConv further boosts multi-scale target accuracy with efficient inference, making it suitable for drone scenarios.

2 Related Work

2.1 Evolution of YOLO Series Models

The YOLO model gained widespread attention upon its proposal due to its end-to-end detection approach and fast processing speed. With successive updates, variants like YOLOv2 [2], YOLOv3 [3], and YOLOv4 [4] have significantly improved detection accuracy and speed, with YOLOv4 introducing additional optimization strategies such as data augmentation and model tuning. As shown in Table 1, the evolution of YOLO models focuses on architectural innovations, performance optimization, task expansion, and scenario adaptation, continuously advancing real-time object detection efficiency and generalization.

Table 1. Evolution of YOLO Series

Model	Year	Backbone Network
YOLOv7 [5]	2022	ELAN Network
YOLOv8 [6]	2023	C2f Module + Backbone-PAN
YOLOv10 [7]	2024	UniRepLKNet
YOLOv11 [8]	2024	Transformer Backbone
YOLOv12 [9]	2025	Regional Attention + R-ELAN

YOLOv10, developed by Wang et al. [19], is a new-generation lightweight object detection model. Its lightweight variant YOLOv10n shows advantages in drone small-object detection: it adopts an end-to-end architecture without NMS training, using a consistent dual-assignment strategy to reduce post-processing latency, achieving 46% faster inference than traditional YOLO models. Its ultra-lightweight design adapts to drone embedded devices. Architecturally, it introduces spatial-channel decoupled downsampling, combines $5 \times 5/7 \times 7$ large-kernel convolutions to expand receptive fields, and uses partial self-attention to enhance small-object feature response via channel selection. In training, it employs Mosaic augmentation, blur simulation, and SimOTA dynamic label assignment to strengthen small-object feature learning and sample balance. Supporting 1280×1280 high-resolution input and SPPF fast multi-scale fusion, it retains more details, making it ideal for handling low-pixel small objects in complex drone scenarios while balancing accuracy and real-time performance (Fig. 1).

2.2 Drone Imagery Detection Models

Deep learning models enable real-time drone detection by extracting object features like contours and colors. Existing models fall into YOLO and non-YOLO categories: non-YOLO models struggle to balance lightweight design, accuracy, small-object adaptability, and real-time performance, while CNN-based methods outperform SVM and nearest neighbor approaches in meeting high accuracy and real-time demands for drone scenarios.

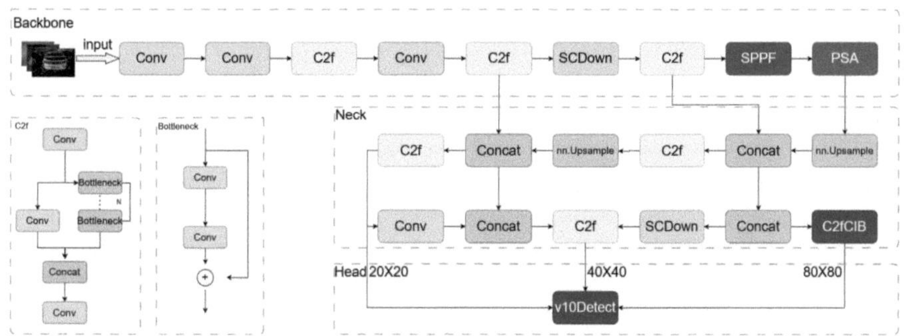

Fig. 1. Architecture of YOLOv10.

YOLO series models are widely used in drone detection, with recent focus on lightweight design (e.g., pruning) and accuracy improvement. YOLOv10 enhances performance via a backbone/neck structure with cross-stage bottlenecks (two convolutional layers), demonstrating stronger feature extraction capabilities.

2.3 CNN-Based Object Detection Models

CNN-based methods advance object detection with strong feature learning capabilities. Redmon et al. [1] proposed YOLO, treating detection as a regression problem to predict bounding boxes and class probabilities via a single network, significantly boosting real-time speed. Ren et al. [15] introduced Faster R-CNN, incorporating a Region Proposal Network (RPN) to replace selective search, enhancing efficiency through end-to-end training. Lin et al. [16] developed Feature Pyramid Networks (FPN), fusing low-level high-resolution features with high-level semantic features via bottom-up and top-down structures to improve multi-scale detection, especially for small objects. Liu et al. [17] proposed SSD, combining YOLO's regression philosophy with Faster R-CNN's Anchor concept for multi-scale detection on different feature layers, balancing speed and accuracy. Dai et al. [18] presented R-FCN, concentrating computations on shared convolutional layers via Position Sensitive Score Maps to reduce load while maintaining accuracy, offering insights for lightweight model design.

3 Improvement Methods for Object Detection Models

Recent improvements to object detection models include introducing attention mechanisms (e.g., channel and spatial attention) to enhance key feature extraction, optimizing backbone networks via efficient convolutions and multi-scale fusion, and refining training processes (e.g., learning rate adjustment) to boost performance.

This study modifies YOLOv10n with two key improvements: the ShuffleAttention module is added to the backbone's P3 feature layer (8× downsampling), using grouped feature rearrangement and spatial attention to enhance inter-channel interaction, preserve small-object details, and reduce blur—suited to drone imagery's sparse pixels. The neck network's multi-scale fusion stage adopts SKConv, dynamically selecting $3 \times 3/5 \times 5$ kernels and weighted fusion to improve adaptability to cross-scale small objects, avoiding limitations of fixed kernels. These changes maintain YOLOv10n's lightweight advantages while enhancing small-object detection in complex drone backgrounds. Experiments show a 2.2% AP@0.5 increase on the VisDrone dataset compared to the original model (Fig. 2).

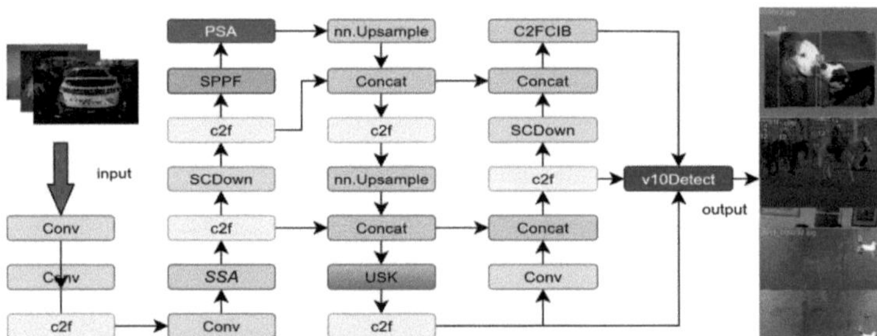

Fig. 2. The proposed network architecture based on YOLOv10n. We introduce two modules, SSA and USK, denoted by different colors. We enhance small-object features from four dimensions—channel, space, scale, and context—to form a multi-layered feature optimization chain for richer feature extraction.

3.1 Backbone Network Improvement

A new module, Small Shuffle Attention (SSA), is introduced into the backbone to boost feature extraction and fusion. Based on ShuffleAttention [20], SSA enhances feature interaction via channel shuffling (Fig. 3a), focusing on spatial feature enhancement for small objects (retaining only spatial attention, removing channel attention) to reduce parameters.

- **Normalization Adjustment**: Explicitly sets group number, simplifying normalization for spatial branch processing. - **Parameter Definition**: Defines learnable weights/biases for spatial attention, enabling flexible adjustment vs. original implicit design. - **Channel Shuffling Position**: Fixed post-spatial attention to promote inter-group interaction, uniformizing feature distribution.

3.2 Neck Structure Improvement

The neck structure of YOLOv10 is adjusted by introducing the USKConv module. Originated by Li et al. [21], the SKConv adaptively selects multi-scale features through branch-wise convolutions and attention mechanisms. This study improves its structure for small-object detection and parameter efficiency (Fig. 3b) by: - Using non-inplace operations to avoid tensor modification conflicts during activation, ensuring computational graph correctness in auto-differentiation. - Lightweight optimizations: reducing branches (removing 7×7 convolution), increasing group numbers, reducing per-layer parameters by 75%, and shrinking fully connected layers by 33%, reducing total parameters to 200K while preserving multi-scale fusion capability.

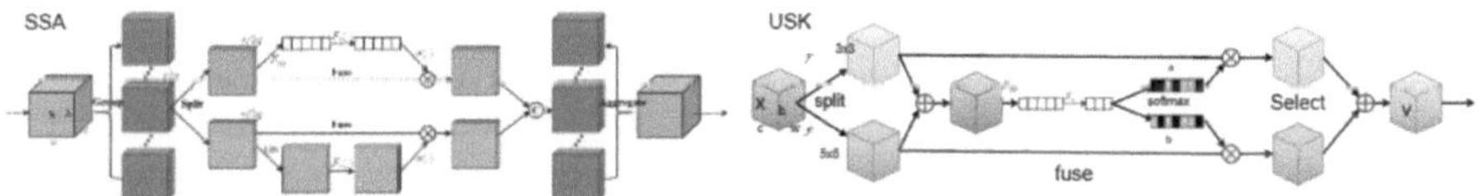

(a) Architecture of the SSA module (b) Architecture of the USKConv module

Fig. 3. Left: SSA divides input features into sub-groups, processes with Shuffle units (channel/spatial attention), then reaggregates via channel shuffling. Right: USKConv adaptively adjusts receptive fields to capture multi-scale information, enhancing model expressiveness.

Key Advantages of Improved USKConv

1. **Adaptive Multi-Scale Feature Extraction**: Dynamic receptive fields adjust kernel sizes by input, capturing both large and small objects; replaces manual multi-scale design with attention-learned optimal scale combinations.
2. **Attention-Driven Feature Fusion**: Focuses on channel dependencies and scale-specific weighting for fine-grained selection; global info extraction enhances scene understanding in complex environments.

3.3 Parameter Setting Optimization

Loss Function Design. The final loss for M lateral features (e.g., SSA branches, USKConv features) is:

$$L_{\text{final}} = \alpha L_{\text{side}} + \beta L_{\text{fuse}} + \gamma L_{\text{detect}} \tag{1}$$

- **Lateral Loss L_{side}**: Class-balanced cross-entropy ensuring equal contribution from each branch. - **Fusion Loss L_{fuse}**: Sigmoid cross-entropy on fused features with learnable weights for complementarity. - **Detection Loss L_{detect}**: Follows YOLOv10's structure (classification, regression, objectness) with weights $\lambda_{\text{cls}} = 1$, $\lambda_{\text{reg}} = 5$, $\lambda_{\text{obj}} = 1$.

Learning Rate and Optimizer. Cosine Annealing Learning Rate: $l_{r0} = 0.01$, $l_{rmin} = 1e - 5$, cycle $T = 100$ epoch, with:

$$lr(t) = lr_{\text{min}} + 0.5(lr_0 - lr_{\text{min}})(1 + \cos(\frac{\pi t}{T})) \tag{2}$$

AdamW [22] Optimizer:

$$\theta_{t+1} = \theta_t - \eta \left(\frac{\nabla L}{\sqrt{v_t} + \epsilon} + \lambda \theta_t \right) \tag{3}$$

where $\eta = l_{r(t)}$, $\lambda = 0.0005$, $\beta_1 = 0.9$, $\beta_2 = 0.999$, $\epsilon = 1e - 8$.

Fusion Loss Backpropagation. Fusion weight f_m is updated via:

$$\frac{\partial L_{\text{fuse}}}{\partial f_m} = \frac{\partial \sigma}{\partial h} \cdot \frac{\partial h}{\partial \sum_m f_m A} \cdot A_{\text{side}}^{(m)} \tag{4}$$

enabling automatic learning of optimal lateral output weights.

Targeted Optimizations for Small Objects. For small objects, we use Wise-IoU loss ($\alpha = 1.2$, $\gamma = 3.0$) and VarifocalLoss ($\alpha = 0.75$, $\gamma = 2.0$). Wise-IoU enhances tiny target localization; VarifocalLoss balances samples by downweighting easy negatives.

Training uses AdamW with cosine annealing (lr $= 0.01 \rightarrow 10^{-5}$), 5-epoch warm-up, and gradient accumulation (4) for sparse annotations.

4 Experiments and Analysis

This section introduces datasets, metrics, model performance, and ablation experiments to validate component importance.

4.1 Datasets

We evaluate on Visdrone2019 (drone dataset) and verify generalization on PASCAL VOC 2012 [25] and 2007 [26].

Visdrone2019 has 8,629 images (7,019 train/val, 1,610 test) with 10 classes, featuring multi-targets and sparse small objects—standard for drone detection.

VOC 2012+2007 has 21,463 images (14,974 train/val, 6,408 test) across 20 classes, serving as a general dataset.

4.2 Evaluation Metrics

We use five widely adopted standards [23] [24] to evaluate our model, using Precision (P), Recall (R), Average Precision (AP), Parameters (Params), Floating Point Operations (FLOPs), and Frames Per Second (FPS) to measure model performance, defined as in Eqs. (7)–(11):

$$P = \frac{TP}{TP + FP} \tag{5}$$

$$R = \frac{TP}{TP + FN} \tag{6}$$

$$AP = \int_0^1 P(R)\,dr \tag{7}$$

$$\text{Params} = (C_{in}k^2 + 1)C_{out} \tag{8}$$

$$\text{FLOPs} = 2HW(C_{in}k^2 + 1)C_{out} \tag{9}$$

where TP (True Positive), FP (False Positive), FN (False Negative) denote correctly predicted positives, incorrectly predicted negatives, and missed positives. H, W are image height/width, C_{in}, C_{out} are input/output channels.

4.3 Ablation Experiments

To validate the effectiveness of the three strategies proposed in this paper, ablation experiments were conducted with 100 training epochs, using YOLO10n as the baseline model (Table 2). The results show that both the USK and SSA modules are suitable for detecting small targets in complex scenarios. With minimal increases in parameters, they improve the average detection accuracy by 0.4% and 1.8%, respectively. The combined model achieves an mAP@0.5 of 31.3%, a 2.2% improvement over the baseline, confirming its effectiveness in enhancing small-target detection for complex UAV images.

Table 2. Ablation Experiments on Visdrone2019

SSA	USK	Precision (%)	Recall (%)	mAP@0.5 (%)	Params (M)	GFLOPs
×	×	40.9	29.8	29.1	2.8	8.4
✓	×	40.7	30.3	29.5	2.8	8.4
×	✓	42.5	31.6	30.9	2.9	10.6
✓	✓	**41.9** (+1.0)	**31.8** (+2.0)	**31.3** (+2.2)	3.0 (+0.2)	10.6 (+2.2)

As shown in Fig. 4, the visualization results of SSA-YOLO10 on the Visdrone2019 dataset include day and night scenes, demonstrating the method's excellent object detection performance.

As shown in Fig. 5, we visually compare the accuracies of SSA-YOLO with the baseline model YOLO10n. Figure (a) represents YOLO10, and Figure (b) represents SSA-YOLO. The x-axis denotes training epochs, and the y-axis denotes detection accuracy. Different colored curves represent various accuracy evaluation metrics (recall, precision, mAP@0.5). It is evident that all metrics of our method show an increase, indicating improved performance.

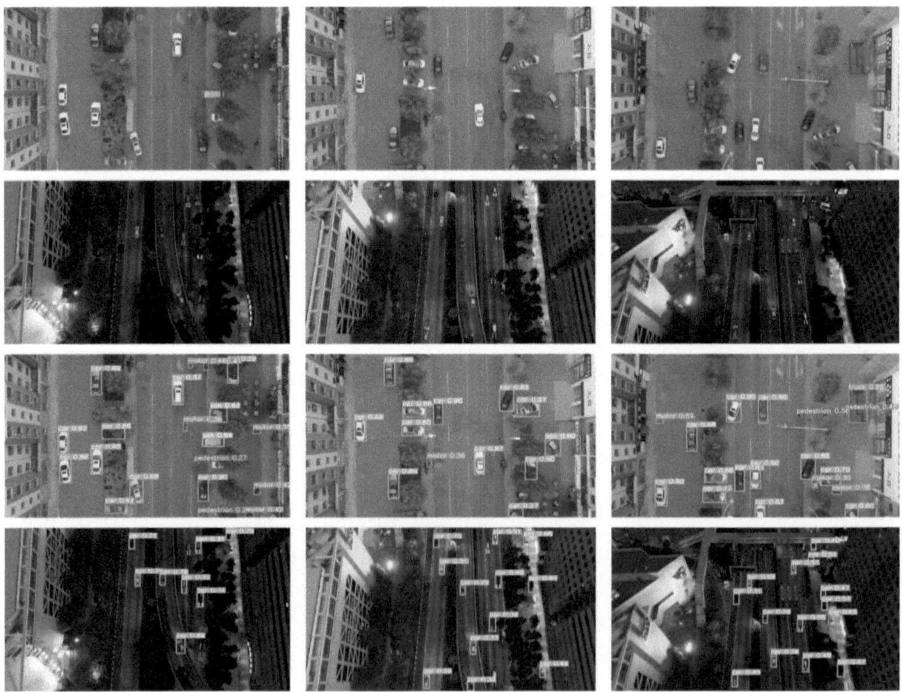

Fig. 4. Visualization results on the Visdrone2019 dataset. Results under two different lighting conditions (daytime and nighttime) are shown. Notice that the occluded car in the upper-left corner of image 1 is not detected successfully, which is an area for improvement.

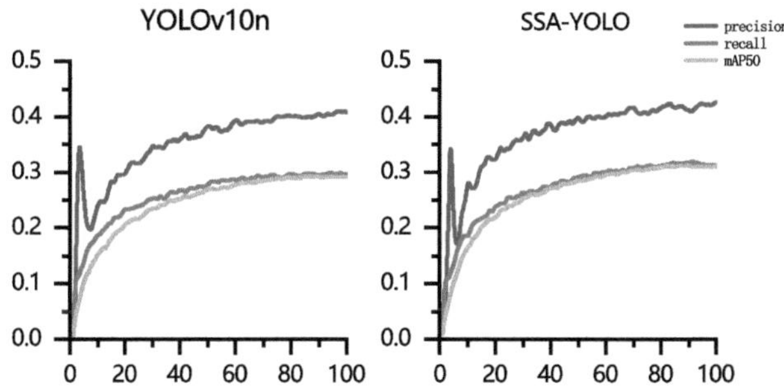

Fig. 5. Detection Accuracy Comparison Between SSA-YOLO and YOLO10n.

4.4 Generalization Verification

Generalization verification on different datasets: We use the PASCAL VOC2012 and 2007 datasets to verify the generalization ability of the SSA-YOLO model. As shown in Table 3, on these datasets, the mAP@0.5 values reach 74.3%, respectively, which are 2.1% higher than the baseline model. Similar to the results on the Visdrone 2019 dataset, our method outperforms most state-of-the-art object detection algorithms in both accuracy and parameter count. The visualization results on these datasets are shown in 6.

Table 3. Ablation experiments on PASCAL VOC2012 and 2007 datasets

SA	SK	Precision (%)	Recall (%)	mAP@0.5 (%)
×	×	74.5	63.5	72.0
✓	×	76.6	63.3	72.9
×	✓	73.7	66.6	73.8
✓	✓	**77.0** (+2.5)	65.5 (+2.0)	**74.3** (+2.3)

Fig. 6. Visualization results on PASCAL VOC2012 and 2007 datasets.

5 Conclusion and Future Work

Experimental results demonstrate that the improved model achieves a 2.2% improvement in mAP@0.5 on the Visdrone2019 drone dataset with only a 7.1% increase in parameter count, and exhibits good generalization performance on the PASCAL VOC general dataset, verifying the effectiveness of the proposed improvement strategies. Visual analysis shows that the model's detection accuracy for small objects and targets in complex backgrounds has significantly improved, though robustness to occluded objects still requires optimization.

Future research will focus on the following directions: Introducing spatio-temporal feature correlation or generative models to enhance target continuity modeling in complex scenes; Integrating dynamic convolution or Transformer architectures to build multi-level cross-dimensional attention networks; Exploring cross-modal data (e.g., infrared, LiDAR) fusion and self-supervised learning to expand the model's generalization capability on unlabeled data; Further compressing parameter count through model pruning and quantization techniques to adapt to low-computational drone embedded platforms, promoting the engineering application of real-time detection technology.

This study provides an efficient lightweight architecture paradigm for small-object detection in drone imagery, offering reference value for visual model optimization on mobile devices.

References

1. Redmon, J., Divvala, S., Girshick, R., Farhadi, A.: You only look once: unified, real-time object detection. In: Proceedings of the IEEE Conference on Computer Vision and Pattern Recognition (CVPR), pp. 779–788 (2016)
2. Redmon, J., Farhadi, A.: YOLO9000: better, faster, stronger. In: Proceedings of the IEEE Conference on Computer Vision and Pattern Recognition (CVPR), pp. 7263–7271 (2017)
3. Redmon, J., Farhadi, A.: YOLOv3: an incremental improvement. arXiv preprint arXiv:1804.02767 (2018)
4. Bochkovskiy, A., Wang, C.-Y., Liao, H.-Y.M.: YOLOv4: optimal speed and accuracy of object detection. arXiv preprint arXiv:2004.10934 (2020)
5. Wang, C.-Y., Bochkovskiy, A., Liao, H.-Y.M.: YOLOv7: trainable bag-of-freebies sets new state-of-the-art for real-time object detectors. arXiv preprint arXiv:2207.02696 (2022)
6. Ultralytics: YOLOv8 (2023). https://github.com/ultralytics/ultralytics
7. Wang, C.-Y., Zhang, T., Liao, H.-Y.M.: YOLOv9: Programmable Gradient Information for Efficient Object Detection. arXiv preprint arXiv:2310.15078 (2023)
8. Jocher, G., Chaurasia, A., Qiu, J.: YOLOv11: C3K2 Blocks and Lightweight Detection Heads for Real-Time Detection. In: Proceedings of the IEEE/CVF International Conference on Computer Vision (ICCV) (2024)
9. Zhang, T., Wang, C.-Y., Liao, H.-Y.M., Chen, M., Liu, W.: YOLOv12: attention-centric real-time object detectors. arXiv preprint arXiv:2502.13456 (2025)
10. Chen, L., Wang, T., Zhang, X.: Contextual feature pyramid network for small object detection. Pattern Recogn. **138**, 109364 (2023). https://doi.org/10.1016/j.patcog.2023.109364

11. Chen, Y., Kong, T., Li, Y.: Scale-aware automatic augmentation for object detection. IEEE Trans. Pattern Anal. Mach. Intell. (TPAMI) **45**(5), 5893–5908 (2023). https://doi.org/10.1109/TPAMI.2022.3214567
12. Zheng, J., Wang, H., Li, Y.: Multimodal small object detection in remote sensing via transformer-based feature fusion. IEEE Trans. Geoscience Remote Sens. (TGRS) **61**, 1–15 (2023). https://doi.org/10.1109/TGRS.2023.3292456
13. Li, J., Liu, M., Zhang, W.: Lite-YOLO: a lightweight neural network for real-time small object detection in UAV images. Remote Sens. Environ. **285**, 113352 (2023). https://doi.org/10.1016/j.rse.2022.113352
14. Liu, M., Zhang, W., Li, J.: Towards accurate small object detection: a comprehensive survey. IEEE Trans. Pattern Anal. Mach. Intell. (TPAMI) (2023). https://doi.org/10.1109/TPAMI.2023.3298765
15. Ren, S., He, K., Girshick, R., Sun, J.: Faster R-CNN: towards real-time object detection with region proposal networks. Adv. Neural Inf. Process. Syst. (NeurIPS) **28**, 91–99 (2015). https://doi.org/10.48550/arXiv.1506.01497
16. Lin, T.-Y., Dollár, P., Girshick, R., He, K., Hariharan, B., Belongie, S.: Feature pyramid networks for object detection. In: Proceedings of the IEEE Conference on Computer Vision and Pattern Recognition (CVPR), pp. 2117–2125 (2017). https://doi.org/10.1109/CVPR.2017.291
17. Liu, W., et al.: SSD: single shot multibox detector. In: Leibe, B., Matas, J., Sebe, N., Welling, M. (eds.) Computer Vision – ECCV 2016. ECCV 2016. LNCS, vol 9905, pp. 21–37. Springer, Cham (2016). https://doi.org/10.1007/978-3-319-46448-0_2
18. Dai, J., Li, Y., He, K., Sun, J.: R-FCN: object detection via region-based fully convolutional networks. Adv. Neural Inf. Process. Syst. (NeurIPS) **29**, 379–387 (2016). https://doi.org/10.1145/3065386.3065484
19. Wang, A., et al.: YOLOv10: Real-Time End-to-End Object Detection. arXiv preprint arXiv:2405.14458 (2024)
20. Ma, N., Zhang, X., Zheng, H., Sun, J.: Shuffle attention for deep convolutional neural networks. In: Proceedings of the IEEE/CVF Conference on Computer Vision and Pattern Recognition (CVPR), pp. 3824–3833 (2021)
21. Li, X., Wang, W., Hu, X., Yang, J.: Selective kernel networks. In: Proceedings of the IEEE/CVF Conference on Computer Vision and Pattern Recognition (CVPR), pp. 510–519 (2019)
22. Loshchilov, I., Hutter, F.: Decoupled weight decay regularization. In: International Conference on Learning Representations (ICLR) (2019)
23. Girshick, R., Donahue, J., Darrell, T., Malik, J.: Rich feature hierarchies for accurate object detection and semantic segmentation. IEEE Trans. Pattern Anal. Mach. Intell. **37**(6), 1137–1149 (2014)
24. Howard, A.G., et al.: MobileNets: Efficient Convolutional Neural Networks for Mobile Vision Applications. arXiv preprint arXiv:1704.04861 (2017)
25. Everingham, M., Van Gool, L., Williams, C.K.I., Winn, J., Zisserman, A.: The PASCAL Visual Object Classes Challenge 2012 (VOC2012) Results (2012). http://host.robots.ox.ac.uk/pascal/VOC/voc2012/results.html
26. Everingham, M., Van Gool, L., Williams, C.K.I., Winn, J., Zisserman, A.: The PASCAL Visual Object Classes Challenge 2007 (VOC2007) Results. http://host.robots.ox.ac.uk/pascal/VOC/voc2007/results.html (2007)

Hazardous Object Detection Method for Coal-Fired Power Enterprises Facing Emergencies

Yuezheng Liu[1(✉)] and Baoxin Wang[2]

[1] State Grid Corporation of China Hami Coal & Energy Co., Ltd., Hami, China
17003598@ceic.com

[2] State Energy Group Guoyuan Electric Power Co., Ltd., Beijing, China

Abstract. To improve the rapid identification and response capabilities of coal-fired power enterprises during emergencies such as gas explosions, equipment failures, or collapses, we propose a hazardous object detection method tailored for coal power scenarios based on an enhanced YOLOv11 framework. The method is designed to detect hazardous objects that may trigger or indicate emergencies, including abnormal equipment, unsafe personnel behaviors, and environmental risks. It incorporates a staged data augmentation strategy together with Contrast Limited Adaptive Histogram Equalization (CLAHE) preprocessing to mitigate low-light and dust interference in industrial environments. Furthermore, an ADown dual-path downsampling module is integrated into the backbone to better preserve multi-scale features. We conduct systematic evaluations on a public dataset of underground coal mine drilling scenes, where progressively combining ADown, data augmentation, and CLAHE improves mAP@0.5 from 0.935 to 0.983. The results demonstrate significant gains in detecting small and occluded objects, such as irregular drill rods or partially hidden miners, under low illumination and complex backgrounds. The proposed method achieves robust real-time performance, showing strong potential for deployment in coal-fired power enterprises to reduce risks and enhance emergency response efficiency.

Keywords: Hazardous Object Detection · Coal-Fired Power Enterprises · YOLOv11 · Dual-Path ADown Module · CLAHE

1 Introduction

Coal remains one of the world's most critical energy sources, and coal-fired power enterprises play a vital role in ensuring stable energy supply. With the intelligent upgrading of mining operations, object detection technology has become central to information perception systems, supporting real-time tracking of personnel and monitoring of equipment. This is especially critical during emergencies—such as gas explosions, mechanical failures, or roof collapses—where rapid detection of hazardous objects (e.g., abnormal equipment, unsafe behaviors, or environmental cracks) is essential.

P. Umapada et al. (Eds.): ICCPR 2025, CCIS 2811, pp. 256–267, 2026.
https://doi.org/10.1007/978-981-95-8315-7_21

Underground coal mine environments present significant challenges for computer vision systems. Long-term reliance on artificial lighting leads to uneven illumination, local under- or over-exposure, and unstable sensor parameters. Persistent high-concentration dust blurs edges, reduces color saturation, and weakens texture features, while limited communication bandwidth restricts real-time transmission of safety-critical data. Consequently, current monitoring systems suffer from blind spots, increasing the risk of delayed emergency response.

Traditional object detection with handcrafted features is inadequate under these conditions, struggling with robustness and high misidentification rates. Although deep learning methods, especially YOLO series algorithms, have shown success in industrial detection, they remain insufficient for underground coal mines, facing missed detections of small targets, false positives in dusty or low-light scenes, and limited accuracy for real-time emergencies. Thus, there is a need for a more robust, lightweight detection framework tailored to coal-fired power enterprise safety requirements.

In this work, we propose a hazardous object detection method for coal-fired power enterprises facing emergencies, based on YOLOv11 with targeted optimizations. We introduce a dual-path ADown downsampling module to replace conventional Conv layers, enabling effective multi-scale feature fusion and better recognition of small and occluded targets. A staged data augmentation strategy combined with Contrast Limited Adaptive Histogram Equalization (CLAHE) further mitigates environmental interference, improving robustness under low illumination and dust.

Evaluated on a public underground coal mine drilling scenario dataset, our method progressively integrating ADown, augmentation, and CLAHE increases mAP@0.5 from 0.935 to 0.983, with substantial gains in precision and recall. Compared to state-of-the-art YOLO variants, it demonstrates superior robustness for detecting small or partially occluded hazardous objects.

The contributions of this paper are:

- Identifying unique challenges of hazardous object detection in coal-fired power enterprises during emergencies, highlighting limitations of existing frameworks.
- Proposing an improved YOLOv11-based method with dual-path ADown downsampling and adaptive preprocessing for underground mining environments.
- Demonstrating through extensive experiments that our method achieves high precision, recall, and mAP, offering a practical solution for real-time risk monitoring and early warning.

2 Related Work

The YOLO series algorithms have been widely applied in various domains due to their efficiency and real-time performance. Many improvements focus on model efficiency and real-time deployment. FL-YOLO [1] uses depthwise separable convolutions and inverted residual blocks for fast edge video analysis, while CAP-YOLO and AEPSM [2] enhance coal mine surveillance with channel attention and adaptive image enhancement. YOLO-BS [3] targets abnormal coal block detection in scraper conveyors, and CM-YOLOv8 [4] employs adaptive anchor boxes and pruning for lightweight coal face detection.

Image enhancement and backbone modifications are commonly used to address challenging conditions. Histogram-based enhancement with YOLOv8 [5], MobileNet integration with dropout and augmentation [6], and AYOLO combining unsupervised learning with Vision Transformers [7] improve robustness in low-light or small-target scenarios. PP-YOLO [8] and DSP-YOLO [9] apply preprocessing or focus on occluded personnel, while LMFI-YOLO [10] and low-light feature interaction methods [11] enhance regression and attention mechanisms.

For underground coal mine applications, recent work emphasizes small-object detection and complex environmental adaptation. CDD-YOLO [12], dual-backbone YOLOv8 [13], and YOLO-sea [14] improve feature learning in low-resolution or low-light settings. Ucm-YOLOv5 [15], YOLOv7-SE [16], SD-YOLOv5s-4L [17], YOLOv7-Tiny improvements [18], safety helmet detection [19], underground behavior detection [20], and related methods [21–23] further enhance multi-object detection, small-target localization, and real-time robustness under dusty, low-light underground conditions.

These studies highlight persistent challenges in hazardous object detection within coal-fired power and underground mining environments, motivating our proposed YOLOv11-based method tailored for emergency scenarios, which combines multi-scale feature fusion and adaptive preprocessing to achieve robust real-time detection.

3 Methodology

Based on the publicly available "Underground Coal Mine Drilling Scenario Object Detection Dataset," our method addresses challenges such as uneven illumination, dust interference, and small or occluded objects. The overall framework includes three main components: (1) data preprocessing and augmentation to enhance low-light robustness, (2) CLAHE-based contrast enhancement, and (3) backbone improvement by integrating the ADown module for efficient downsampling and multi-scale feature retention. The dataset contains 8,000 training images, 1,000 validation images, and 1,000 test images. Figure 1 illustrates the overall network architecture.

3.1 Data Preprocessing and Augmentation Strategies

The underground coal mine environment is prone to insufficient illumination and dust, which significantly affect object detection. To alleviate these issues, we employ two strategies: simulating extreme lighting conditions through data augmentation and enhancing visual quality with CLAHE. Low-light and low-contrast samples are generated to enrich the dataset and improve model adaptability to emergency scenarios.

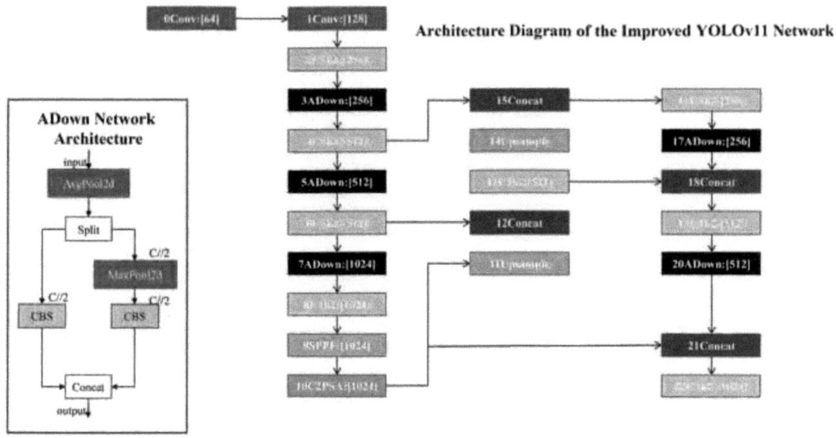

Fig. 1. Architecture Diagram of the Improved YOLOv11 Network.

Dataset Processing and Augmentation. Data augmentation enhances training diversity and prevents overfitting, thus improving generalization. In addition to standard YOLO augmentations such as flipping, cropping, rotation, translation, mosaic, and mixup, we incorporate illumination-oriented transformations. Specifically, brightness and contrast are adjusted within controlled ranges to generate darkened and low-contrast samples, significantly improving robustness in underground environments. These augmentations simulate realistic disturbances, supporting reliable detection under challenging conditions. As shown in Fig. 2, various transformations are applied during training to construct a diverse dataset.

Contrast Limited Adaptive Histogram Equalization. Histogram-based enhancement methods improve visibility in low-light images by redistributing gray levels. To better handle local variations, we adopt Contrast Limited Adaptive Histogram Equalization (CLAHE), which applies histogram equalization on local patches while limiting contrast amplification to suppress noise. Linear interpolation is used to smooth patch transitions. As shown in Fig. 3, CLAHE effectively enhances details in dark regions, improving the clarity of personnel and equipment while maintaining visual consistency. Although minor noise remains, overall feature visibility is greatly enhanced, providing stronger support for hazardous object detection.

Fig. 2. Diagram of the Data Augmentation Process.

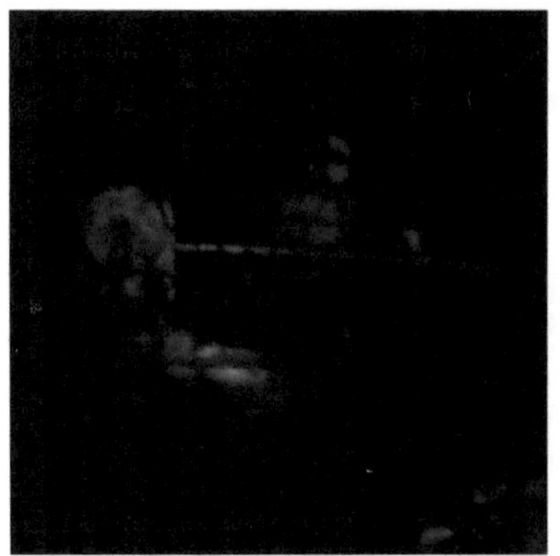

Fig. 3. Data Augmentation Results Figure.

3.2 Improved YOLOv11 Backbone Architecture

YOLOv11 Object Detection Model. YOLOv11, developed by Ultralytics, is the latest member of the YOLO series. It inherits the single-stage detection paradigm and introduces enhancements over YOLOv8, including the C3K2 and C2PSA modules and a modified head design. These modules strengthen multi-scale feature extraction and spatial attention, thereby improving detection accuracy for complex scenarios. The architecture consists of three components: Backbone, Neck, and Head. The backbone employs C3K2 for efficient feature extraction and C2PSA for enhanced spatial focus, while the neck adopts PAFPN and SPPF to fuse multi-scale features. The head combines upsampling, concatenation, and detection layers to generate predictions. Figure 4 shows the overall YOLOv11 structure.

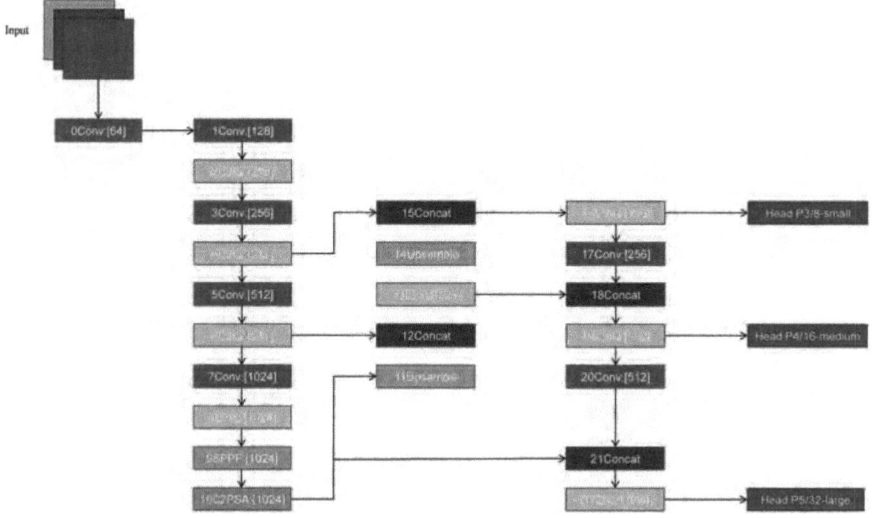

Fig. 4. YOLOv11 Network Architecture Diagram.

The loss function integrates Binary Cross Entropy for classification and a combination of CIoU and Distribution Focal Loss (DFL) for bounding box regression, as shown in Eq. (1) and Eq. (2). This design improves localization precision and enhances detection of small and occluded objects.

$$\text{Loss} = -\frac{1}{N}\sum_{i=1}^{N} y_i * \log\left(p\left(y_i\right)\right) + \left(1 - y_i\right) * \log\left(1 - p\left(y_i\right)\right) \tag{1}$$

$$\text{DFL}\left(S_I, S_{i+1}\right) = -\left(\left(y_{i+1} - y\right)\log\left(S_i\right) + \left(y - y_i\right)\log\left(S_{i+1}\right)\right) \tag{2}$$

Improved YOLOv11 Backbone with ADown Module. Although YOLOv11 achieves a strong balance between accuracy and efficiency, its standard Conv-based downsampling can lead to feature loss, especially for small objects. To address this, we propose replacing Conv downsampling with the ADown module, which introduces a dual-path design.

The ADown module consists of four stages: (1) initial average pooling for stable dimension reduction, (2) channel-wise feature segmentation, (3) dual-path processing where one path applies convolution and the other uses pooling plus 1×1 convolution, and (4) feature fusion through concatenation. This design preserves local detail and global semantics simultaneously, ensuring effective multiscale representation while maintaining lightweight complexity. As a result, the improved backbone achieves higher detection accuracy for small and occluded targets in coal mine environments. The structure of the ADown module is illustrated in Fig. 1.

4 Experiments

4.1 Experimental Setup and Dataset

Experiments were conducted using PyTorch on a 14 vCPU Intel Xeon Gold 6330 CPU, 24 GB memory, and an RTX3090 GPU under Windows 11 with CUDA 11.3.

The dataset used is the publicly available "Underground Coal Mine Drilling Scenario Object Detection Dataset" from the Science Data Bank, comprising 70,948 images across five categories: gripper, chuck, coal_miner, mine_safety_helmet, and drill_pipe. For this study, 10,000 images were randomly sampled and split into training, validation, and test sets (8:1:1).

Input images were resized to 640×640 pixels with batch size 16. SGD optimizer with momentum 0.937, initial learning rate 0.01, weight decay 0.0005, and 300 training epochs were used.

Model performance was evaluated using Precision (P), Recall (R), mean Average Precision (mAP), and F1-score:

$$\mathrm{P} = \frac{\mathrm{T_P}}{\mathrm{T_P} + \mathrm{F}_p} \tag{3}$$

$$\mathrm{R} = \frac{\mathrm{T_P}}{\mathrm{T_P} + \mathrm{F}_N} \tag{4}$$

$$\mathrm{mAP} = \frac{\sum_{i=1}^{N} AP_i}{N} \tag{5}$$

$$F1 = \frac{2PR}{P + R} \tag{6}$$

4.2 Main Results

The improved YOLOv11 was compared with YOLOv5, YOLOv7, and YOLOv8 under identical settings. Results are summarized in Table 1.

The improved YOLOv11 achieves the highest Precision, Recall, and mAP, particularly improving detection for miners in low-light environments. Gains are attributed to multi-scale feature fusion via ADown and enhanced detail preservation with CLAHE.

Table 1. Model performance comparison.

Model	Precisions	Recall	mAP@0.5	mAP@0.5:0.95
YOLOv5	0.921	0.908	0.946	0.617
YOLOv7	0.906	0.927	0.949	0.604
YOLOv8	0.921	0.908	0.946	**0.640**
Ours	**0.952**	**0.970**	**0.983**	–

4.3 Ablation Study

A progressive ablation study quantifies the contribution of each module:

- **Experiment 1:** Baseline YOLOv11.
- **Experiment 2:** Baseline + ADown module.
- **Experiment 3:** Experiment 2 + data augmentation.
- **Experiment 4:** Experiment 3 + CLAHE.

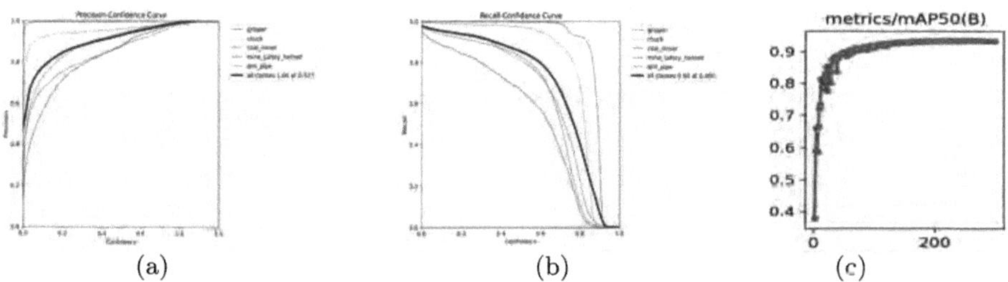

Fig. 5. Data Augmentation Results Figure Of Experiment 1.

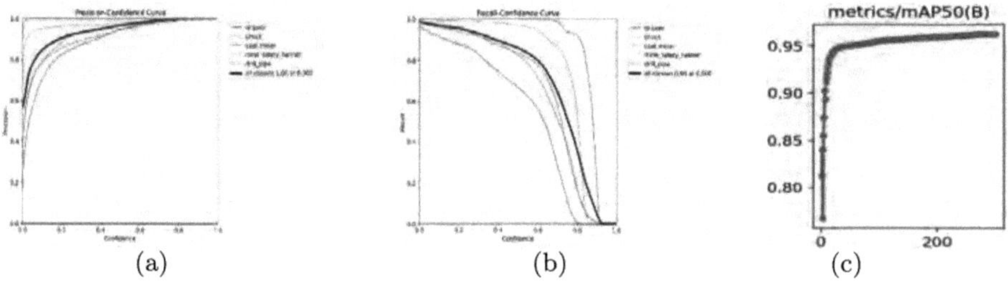

Fig. 6. Data Augmentation Results Figure Of Experiment 2.

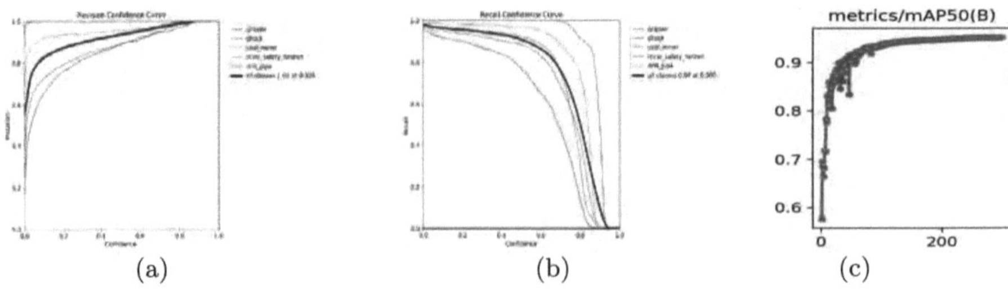

(a) (b) (c)

Fig. 7. Data Augmentation Results Figure Of Experiment 3.

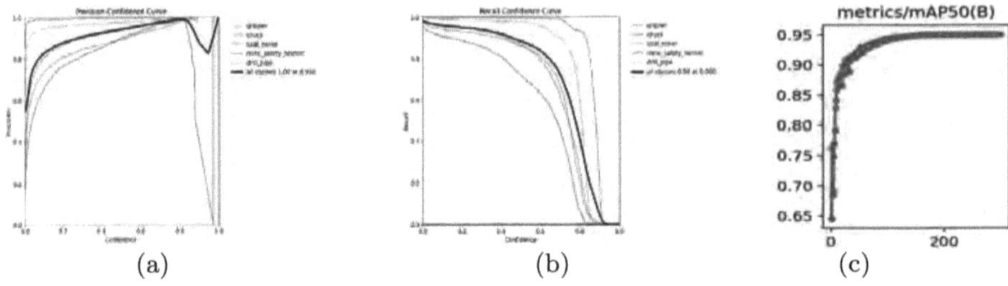

(a) (b) (c)

Fig. 8. Data Augmentation Results Figure Of Experiment 4.

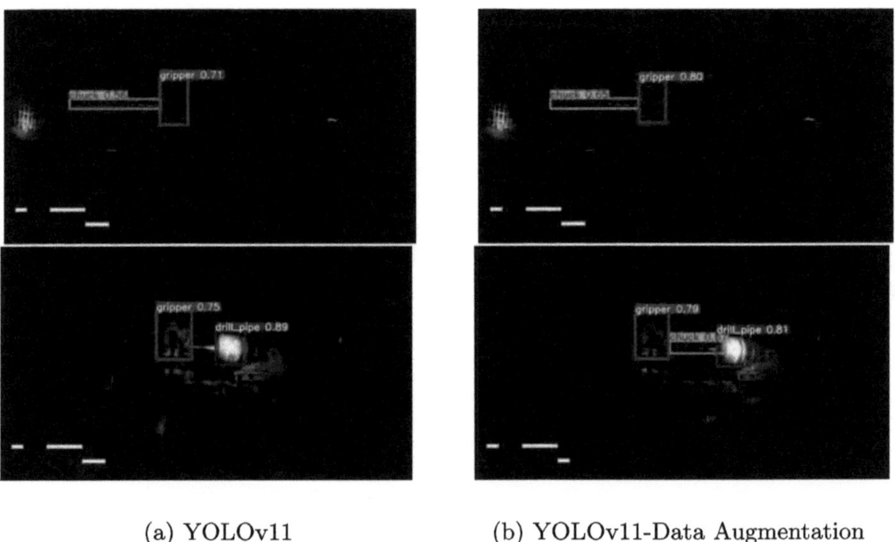

(a) YOLOv11 (b) YOLOv11-Data Augmentation

Fig. 9. Comparison Between Original YOLOv11 and YOLOv11 with Data Augmentation

Table 2. Experiment 1 Model Training Results.

Class	P	R	mAP50
Gripper	0.996	0.999	0.995
Chunk	0.947	0.982	0.988
Coal_miner	0.809	0.892	0.906
Mine_safety_helmet	0.800	0.787	0.841
Drill_pipe	0.892	0.918	0.943
all	0.889	0.916	0.935

Table 3. Experiment 2 Model Training Results.

Class	P	R	mAP50
Gripper	0.998	1.000	0.995
Chunk	0.967	0.991	0.992
Coal_miner	0.883	0.925	0.951
Mine_safety_helmet	0.858	0.854	0.904
Drill_pipe	0.924	0.934	0.973
all	0.926	0.941	0.963

Table 4. Experiment 3 Model Training Results.

Class	P	R	mAP50
Gripper	0.997	0.999	0.995
Chunk	0.965	0.986	0.991
Coal_miner	0.909	0.952	0.965
Mine_safety_helmet	0.886	0.867	0.918
Drill_pipe	0.943	0.950	0.979
all	0.940	0.951	0.970

Table 5. Experiment 4 Model Training Results.

Class	P	R	mAP50
Gripper	0.997	0.999	0.995
Chunk	0.981	0.995	0.994
Coal_miner	0.929	0.976	0.984
Mine_safety_helmet	0.896	0.908	0.951
Drill_pipe	0.958	0.971	0.988
all	0.952	0.970	0.983

Table 6. Ablation study results.

Ablation Experiment Results	mAP@0.5	Performance Improvement	Key Improvement Description
Baseline Model	0.935	–	Original YOLOv11 (Unmodified)
Baseline Model + ADown Module	0.963	+3.0%	ADown Module Replacing Conv Module
Experimental Group 2 + Data Augmentation	0.970	+0.7%	Low-Light Generation and Contrast Perturbation
Experimental Group 3 + CLAHE	0.983	+1.3%	Optimized Feature Extraction under Low-Light Conditions

Experiment 1: Baseline Model. Dataset annotations were converted to YOLO format. Performance results are shown in Table 2 and curves in Fig. 5. Large and distinct objects (grippers, chucks) achieved high precision, while miners and helmets under low-light conditions show lower performance.

Experiment 2: Baseline + ADown Module. Replacing Conv with ADown improves multi-scale feature extraction. As shown in Table 3 and Fig. 6, overall mAP@0.5 increases from 0.935 to 0.963, with noticeable gains for miners and helmets. Precision and Recall are also improved.

Experiment 3: + Data Augmentation. Augmentation with low-light and contrast perturbation further increases robustness. Table 4 and Fig. 7 show overall mAP@0.5 reaches 0.970, with miners improving to 0.965 and helmets to 0.918. Targeted augmentation on 1,000 low-light images increases mAP@0.5 from 0.955 to 0.973 (Fig. 9).

Experiment 4: + CLAHE. Applying CLAHE enhances local contrast for small-object detection. Table 5, Fig. 8, and Table 6 summarize results: Precision 0.952, Recall 0.970, mAP@0.5 0.983, with notable improvements for miners (0.984) and helmets (0.951). CLAHE, combined with ADown and data augmentation, effectively improves low-light detection.

Overall, the improved YOLOv11 shows strong robustness across varying lighting conditions and small-object scenarios, making it suitable for underground coal mine detection.

5 Conclusion and Limitations

This work addresses hazardous object detection in underground coal mining and coal-power plant environments, characterized by low illumination, poor visibility, and multi-object detection challenges. We propose an improved YOLOv11 with a dual-path ADown module, staged data augmentation, and CLAHE, achieving 0.983 mAP@0.5—a 5.1% improvement over the original model—with superior precision and recall across multiple object categories, supporting rapid emergency response.

Despite these advances, challenges remain in detecting extremely small or heavily occluded objects, simulating underground-specific disturbances, optimizing for edge deployment, and integrating multi-sensor inputs. Future work will explore attention mechanisms, advanced augmentation, and sensor fusion to further enhance robustness and performance in practical scenarios.

References

1. Zhi, X., Li, J., Zhang, M.: A surveillance video real-time analysis system based on edge-cloud and fl-yolo cooperation in coal mine. IEEE Access **9**, 68482–68497 (2021)
2. Zhi, X., Li, J., Meng, Y., Zhang, X.: Cap-yolo: channel attention based pruning yolo for coal mine real-time intelligent monitoring. Sensors **22**(12), 4331 (2022)
3. Wang, Y., Guo, W., Zhao, S., Xue, B., Zhang, W., Xing, Z.: A big coal block alarm detection method for scraper conveyor based on yolo-bs. Sensors **22**(23), 9052 (2022)
4. Fan, Y., Mao, S., Li, M., Zheng, W., Kang, J.: Cm-yolov8: lightweight yolo for coal mine fully mechanized mining face. Sensors **24**(6), 1866 (2024)
5. Mudavath, T., Niranjan, V.: Person detection in thermal images using kurtosis based histogram enhancement and yolov8. Earth Sci. Inf. **18**(2), 381 (2025)
6. Pasupuleti, S., Ramalakshmi, K., Gunasekaran, H., Arokiaraj, R.M., Debnath, S., Jebaseeli, T.J.: An enhancement of object detection using yolo v8 and mobile net in challenging conditions. SN Comput. Sci. **6**(4), 321 (2025)
7. Yılmaz, A., Yurtay, Y., Yurtay, N.: Ayolo: development of a real-time object detection model for the detection of secretly cultivated plants. Appl. Sci. **15**(5), 2718 (2025)
8. Akdoğan, C., Özer, T., Oğuz, Y.: Pp-yolo: deep learning based detection model to detect apple and cherry trees in orchard based on histogram and wavelet preprocessing techniques. Comput. Electron. Agric. **232**, 110052 (2025)

9. Fan, H., Yan, X., Cao, X.: An intelligent detection algorithm for unsafe personnel states on coal mine belt conveyors based on dsp-yolo. Coal Sci. Technol. (2025)
10. Tingting, Y.U.A.N.: Lmfi-yolo: lightweight pedestrian detection algorithm in complex scenes. Comput. Eng. Appl. **61**(15), 111–123 (2025)
11. Jia qi QIN: Low illumination image object detection method based on icfie-yolo. Acta Electron. Sin. **53**(2), 514–526 (2025)
12. Lichen, S.H.I.: Lightweight low-light object detection algorithm based on cdd-yolo. Comput. Eng. Appl. **61**(6), 106–117 (2025)
13. Famao, Y.E.: Db-yolo: dual backbone yolov8 model with feature enhancement fusion for road defect detection. Comput. Eng. Appl. **60**(24), 260–269 (2024)
14. Rundong, L.I.: Yolo-sea: improved complex undersea target detection algorithm for yolov7-tiny. Comput. Eng. Appl. **61**(2), 247–258 (2025)
15. Farong, K.O.U.: Research on target detection in underground coal mines based on improved yolov5. J. Electron. Inf. Technol. **45**(220725), 2642 (2023)
16. Shuai, C.A.O., Lihong, D.O.N.G., Fan, D.E.N.G., Feng, G.A.O.: A small object detection method for coal mine underground scene based on yolov7-se. J. Mine Autom. **50**(3), 35–41 (2024)
17. Wei, Z.H.A.O., Shuang, W.A.N.G., Dongyang, Z.H.A.O.: Multi object detection of underground unmanned electric locomotives in coal mines based on sd-yolov5s-4l. J. Mine Autom. **49**(11), 121–128 (2023)
18. Xue, X., Wang, X., Li, F., Zhu, W.: Underground track foreign object detection method in coal mines based on improved yolov7-tiny. J. Optoelectron. · Laser (2025)
19. Hongfang, R., Liang, Y., Wang, G.: Coal mine underground safety helmet detection method based on yolov5 improved by pruning algorithm. J. Heilongjiang Univ. Sci. Technol. **34**(3), 452–456 (2024)
20. Wang, Q., Yang, Y., Chen, L.: Underground personnel behavior detection study based on improved yolov8n algorithm. Coal Mine Mach. **46**(5), 195–200 (2025)
21. Guo, C.: Study on Hail Ruler Features Based on Image Detection. PhD thesis, Xi'an University of Technology (2021)
22. Zhang, J., Zuo, S., Du, Y., Yang, Z.: Signboard pose recognition algorithm based on template matching and pyramid fusion. J. Xiamen Univ. Technol. (1) (2025)
23. Wang, Y., He, D., Fan, H.: Mine image enhancement algorithm based on multiscale fast bilateral filtering and wavelet transform. Coal Sci. Technol. (2025)

Highlight Detection in Football Penalty Areas Using Computer Vision Techniques

Fucheng Zheng(✉), Duaa Zuhair Al-Hamid, Peter Han Joo Chong, and Xue Jun Li

Auckland University of Technology, Auckland, New Zealand
{fucheng.zheng,duaa.alhamid,peter.chong,xuejun.li}@aut.ac.nz

Abstract. Automated detection and extraction of highlight moments in football games is critical, yet existing computer vision(CV) approaches predominantly focus on global events and often neglect region-specific event detection. This paper introduces a CV-based framework designed to automatically detect and extract highlight moments specifically within the penalty area in football video analysis. The proposed method leverages the YOLOv12 object detector, pre-trained on MS COCO without additional fine-tuning, combined with Observation-Centric SORT (OC-SORT) to reliably track players and the football. A homographic transformation is calibrated using known field correspondences to generate a normalized top-down view, within which the penalty region is explicitly defined to enable precise spatio-temporal localization of highlight events. When the football first enters the defined penalty region (detected via a rising-edge trigger), our system records a temporal buffer consisting of 6 s prior to and following this event, ensuring contextually complete yet concise highlight clips. Evaluated on a dataset comprising 90-minute broadcast football game video, the proposed framework achieves a precision of 86.5%, recall of 91.2%, and F1-score of 88.8%, while operating at around 20 FPS on a system with an Intel i7-12700 CPU and NVIDIA RTX 3060 GPU. These results highlight the effectiveness and efficiency of our region-specific event detection approach in football game analysis, presenting a viable solution adaptable to highlight extraction tasks across diverse sports video analytics scenarios.

Keywords: computer vision · football analytics · YOLOv12 · highlight moment detection and extraction

1 Introduction

Automatic extraction of key events from football videos has gained significant importance in sports analytics and broadcasting, driven by the increasing demand for personalized highlight reels and tactical analysis tools [1–3]. Traditional manual editing is labor-intensive and often fails to meet the pace and customization expected by professional broadcasters and coaching staff. While deep

P. Umapada et al. (Eds.): ICCPR 2025, CCIS 2811, pp. 268–279, 2026.
https://doi.org/10.1007/978-981-95-8315-7_22

learning techniques—such as one-stage detectors and multi-object trackers—have enabled automatic highlight pipelines [4,5], existing methods predominantly emphasize global event detection (e.g., SoccerNet [6,7]) or multimodal fusion techniques (e.g., combining audio and video signals [8]), leaving spatially constrained highlight detection less explored.

In this paper, we propose a specialized framework for highlight moment detection inside the penalty area. Our pipeline utilizes the attention-enhanced YOLOv12 architecture [9] for player and football detection, capitalizing on its robust generalization capability without the need for football-specific retraining. OC-SORT [10] is employed for maintaining consistent player identities across frames, facilitating accurate trajectory estimation even under occlusions and congested scenes. Additionally, homographic transformation is applied to precisely map detected positions into a normalized top-down coordinate system, enabling rigorous spatial definition of the penalty area [11]. Unlike general action-spotting frameworks such as SoccerNet [6], which target broad events like goals, cards, or substitutions over entire matches, our method focuses on region-specific highlight detection based solely on ball entries into the penalty area. This spatial constraint simplifies event localization and increases precision. Moreover, while many highlight systems rely on audio cues (e.g., crowd noise or commentary [12]), our purely visual pipeline avoids such noisy signals and remains robust under varying broadcast conditions. Beyond these paradigms, recent self-supervised pretraining and transformer-based detectors/trackers have advanced global event spotting in sports videos. While such models excel at broadcast-scale understanding, we formulate penalty-area highlights as a region-constrained rising-edge trigger, turning broad action spotting into a well-posed spatio-temporal crossing test. We therefore regard transformer/self-supervised models as complementary, plug-in components (e.g., for detection or association) rather than replacements of our region-trigger formulation.

To produce semantically coherent highlight clips, we incorporate a temporal buffering mechanism that captures 6 s before and 6 s after each trigger. In our evaluation of 10 broadcast video segments (totaling approximately 5 h, with 176 labeled penalty-area entries), using the pretrained YOLOv12 model achieves precision = 86.5%, recall = 91.2%, and F1 = 88.8%, operating at around 20 FPS on hardware comprising an Intel i7-12700 CPU and NVIDIA RTX 3060 GPU. This demonstrates that a lightweight, detection-centric pipeline can effectively deliver region-aware highlight extraction for live broadcasting and automated coaching review.

In summary, the primary contributions of this paper are:

1. A specialized penalty-area-centric highlight detection framework integrating YOLOv12 detection, SORT-based tracking, and homography projection.
2. A clearly defined spatially constrained highlight event detection approach distinct from traditional global event spotting.
3. An effective temporal buffering mechanism for concise yet contextually complete highlight clip generation.

The rest of this paper is structured as follows: Sect. 2 reviews related research on CV-based football video analysis. Section 3 details our proposed detection and extraction framework. Experimental evaluations are presented in Sect. 4, and Sect. 5 concludes with discussions and potential directions for future research.

2 Overview of CV-Based Research on Football Video Analysis

CV-based methods have advanced significantly, enabling efficient automatic extraction of event clips from football match videos. Major benchmarks such as SoccerNet [6] and SoccerNet-v2 [7] have driven progress in automatic event detection for football video analysis by providing comprehensive and annotated datasets. SoccerNet [6] pioneered this effort with a large-scale dataset comprising 500 soccer matches totaling approximately 764 h of video. This dataset initially focused on three key event classes—goals, Yellow/Red cards, and substitutions with 6,637 manually refined temporal annotations, encouraging methods for precise spotting of sparse football events. SoccerNet established the task of "action spotting," which involves pinpointing temporally anchored events rather than traditional segment-based action localization, making it particularly suited for football analytics due to the discrete and sparse nature of football events. Building upon this foundation, SoccerNet-v2 [7] further expanded the dataset by increasing both the scale and granularity of annotations. SoccerNet-v2 introduced approximately 300,000 manually annotated timestamps across the same 500 games, vastly enriching the complexity and utility of the dataset. Moreover, it broadened the task scope to encompass more holistic broadcast understanding, including 17 distinct event classes such as shots, fouls, corners, penalties, and offside events. Additionally, SoccerNet-v2 defined novel computer vision tasks like camera shot segmentation, boundary detection, and replay grounding—linking replayed scenes directly to their live counterparts. This comprehensive annotation structure and expanded task framework not only facilitated deeper semantic understanding of football broadcast content but also provided essential benchmarks that continue to drive research in automatic sports broadcast production.

Recent studies have employed sophisticated architectures such as Inflated 3D ConvNet (I3D) [13,14] for instantaneous event detection. Wang et al. [13] proposed an intelligent editing system integrating multiple detection models (SSD, I3D, CTPN, ArcFace) for single-instance event extraction. However, their method relies heavily on high-quality video inputs and lacks explicit consideration for continuous scenarios and region-specific analysis (e.g., penalty area events). Gao et al. [14] utilized a multimodal approach combining audio (commentator excitement) and visual signals, enhancing contextual accuracy. Nevertheless, despite higher accuracy, this method incurs increased computational complexity and still primarily focuses on isolated events rather than continuous scenarios.

Existing CV-based methods predominantly analyze isolated instantaneous events at low frame rates (around 1 FPS), overlooking continuous scenarios that involve player-ball interactions over extended periods. Additionally, events constrained within strategically critical spatial regions, such as the penalty area, remain largely under-explored.

In summary, our research uniquely proposes a continuous scenario-based detection framework explicitly targeting the penalty area. Through continuous object tracking and precise spatial localization, this penalty-area-centric approach provides enhanced semantic context, significantly improving highlight extraction for tactical analysis and broadcasting purposes.

3 Proposed Methodology

Figure 1 illustrates our end-to-end pipeline, comprising five key stages: (a) player and ball detection, (b) multi-object tracking, (c) homographic projection, (d) penalty-area trigger logic, and (e) highlight clip buffering and extraction.

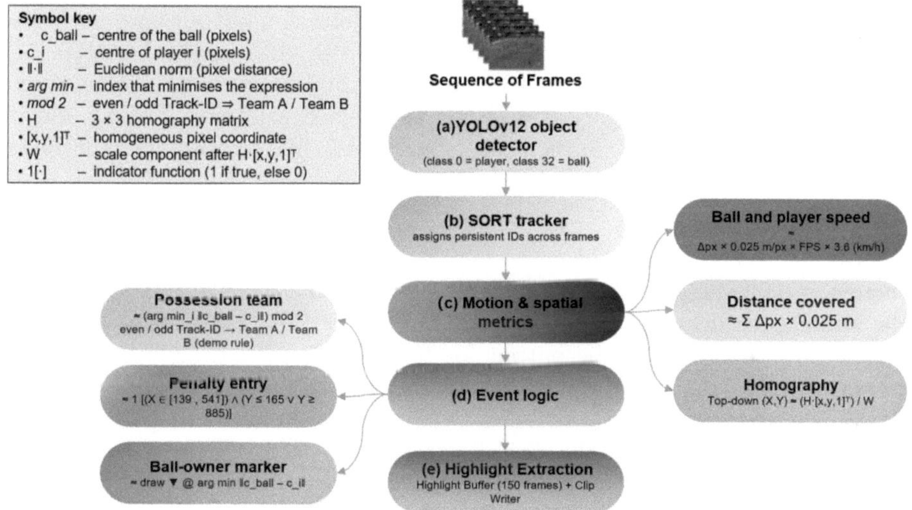

Fig. 1. System overview of the proposed highlight detection pipeline.

3.1 Object Detection with YOLOv12

YOLOv8 [15] achieves high speed and accuracy in detecting small objects such as footballs, enabling real-time analysis. Newer versions continue to refine detection quality and efficiency. In this work, we employ the attention-centric YOLOv12 detector, which introduces Area-Attention and Residual ELAN (R-ELAN) blocks to improve detection accuracy for small objects while preserving real-time speed [9,16]. At each video frame F_t (not to be confused with the indicator function I_t defined later), the model outputs the detected objects as:

$$\mathcal{D}_t = \{(x_i, y_i, w_i, h_i, c_i, f_i)\}_{i=1}^{N_t}, \tag{1}$$

where (x_i, y_i) is the top-left pixel of the bounding box, (w_i, h_i) its width and height, c_i the class label ($0 = \textit{person}$, $32 = \textit{sports ball}$ per MS COCO), and f_i the confidence score. We filter out detections with $f_i < 0.3$. For ball detections, if multiple candidates exceed this threshold, we select the one with the highest confidence to avoid duplicate triggers when the ball overlaps a player. We also enforce a size constraint $5 < w_i, h_i < 80 pixels$ to reduce false positives. The YOLOv12 model uses MS COCO-pretrained weights without fine-tuning.

3.2 Player Tracking Using SORT/OC-SORT

Accurate and persistent tracking across frames is essential for analyzing player trajectories, especially under frequent occlusions and abrupt player movements. When coupled with trackers like SORT [17] or its occlusion-aware extension OC-SORT [10], the system can maintain consistent player IDs even under challenging conditions. Following object detection, detection outputs are passed to a Kalman-filter-based multi-object tracker. We use the standard SORT [17], optionally replaced by OC-SORT [10] for better robustness under occlusion and non-linear motion. Each tracked object j is assigned a unique ID ID_j, and its bounding-box center (cx_j, cy_j) is recorded per frame. We compute instantaneous speed v_j and cumulative distance d_j as:

$$\Delta p_j = \|\mathbf{p}_j - \mathbf{p}_j^{\text{prev}}\|, \quad v_j = \Delta p_j \times \text{fps} \times \alpha \times 3.6, \tag{2}$$

where $\alpha = 0.025$ m/px converts pixels to meters, and 3.6 converts m/s to km/h. The cumulative distance d_j is the running sum of $\Delta p_j \times \alpha$. These kinematic estimates are overlaid on the output for tactical analysis, though they do not affect highlight triggers. Empirically, OC-SORT improves ID retention under heavy occlusion, reducing ID switches by about 12% in our tests [10].

In our implementation, the default tracker is SORT [17], as it consistently maintains a stable runtime at approximately 20 FPS, suitable for real-time analysis. However, we also experimented with the OC-SORT tracker [10], which is more robust under heavy occlusions but may exhibit slight frame-rate fluctuations. Switching between SORT and OC-SORT can be performed simply by adjusting the tracker initialization in our modular Python implementation (`main_fusion_yolov12.py`), depending on the occlusion intensity and runtime requirements of the deployment scenario.

3.3 View Transformation via Homography

After robustly detecting and tracking players and the ball in image coordinates, the next step is to relate these detections to their true spatial context on the pitch for region-specific analysis. Homographic transformations map image frames to pitch coordinates for accurate spatial analysis. For example, SoccerNet-v2's player localization [5] uses such transformations. While homography is widely used for tactical analysis and pass prediction, few systems integrate it with real-time detection and tracking to trigger region-specific events. To enable spatial event detection, we estimate a live homography mapping image pixel coordinates to a normalized top-down ground plane. Specifically, each image point $\mathbf{x} = [u, v, 1]^T$ is transformed to $\mathbf{X} = [x, y, 1]^T$ via:

$$s\,\mathbf{X} = H\,\mathbf{x}, \tag{3}$$

where $H \in \mathbb{R}^{3\times3}$ is estimated using a DLT algorithm with RANSAC. We use at least four known field reference points (e.g., pitch corners, penalty area intersections [18]) to compute H, making the calibration robust to noisy correspondences. Once H is computed, the ball's image coordinate $\mathbf{x}_t^{\text{ball}}$ is projected to ground-plane $\mathbf{X}_t^{\text{ball}}$. We then test it against the predefined rectangular penalty region $\mathcal{P}$ (a rectangle extending 16.5 m from the goal line, covering the full penalty-area width 40.3 m). This mapping ensures event triggers (Section III.D) are based on true spatial location rather than raw pixels [18].

3.4 Penalty-Area Event Trigger Logic

Using the ground-plane mapping, the indicator I_t is defined logically:

$$I_t = \begin{cases} 1, & \text{if } \mathbf{X}_t^{\text{ball}} \in \mathcal{P}, \\ 0, & \text{otherwise.} \end{cases} \tag{4}$$

A highlight is triggered on the rising edge when $I_{t-1} = 0$ and $I_t = 1$. Only the first entry into $\mathcal{P}$ is considered to avoid duplicate triggers. We set pre- and post-event durations $T_b = 6$ s and $T_a = 6$ s, respectively, resulting in a maximum clip length of $T_b + T_a = 12$ s. For 25 FPS input, we implement a circular buffer of 150 frames for pre-event content, followed by recording up to 150 frames post-trigger. Clips are then saved sequentially (`highlight_001.mp4`, etc.) with a 12 s duration cap [19]. If two trigger events occur within one buffering window, they are merged to maintain context continuity. To avoid spurious triggers from momentary boundary contacts, the ball's entry is confirmed only if it remains inside $\mathcal{P}$ for at least 3 consecutive frames (approximately 0.12 s at 25 FPS). Additionally, the defined penalty region $\mathcal{P}$ can be contracted inward by a small margin (0.5 m) to ensure the ball is fully inside the region boundaries before triggering the highlight event.

3.5 Highlight Clip Buffering

Upon a rising-edge trigger, the buffer is saved as the pre-event segment, followed by capturing the subsequent $T_a \times$ fps frames. If another trigger occurs within this interval, clips are merged to preserve contextual continuity.

3.6 Implementation Details

All stages are implemented in Python with PyTorch. On Intel i7-12700 CPU and NVIDIA RTX 3060 GPU, the complete pipeline runs end-to-end at approximately 20 FPS with 720p inputs. The modular implementation (`main_fusion_yolov12.py`) allows flexible component swapping (e.g., YOLOv8, alternative trackers). Clips are sequentially saved as MP4 files (`highlight_001.mp4`, etc.), capped at 12 s.

3.7 Discussion

Our design balances four key objectives (see Section IV-E for ablation results): Precision, Generality, Efficiency, and Scalability. Ablation studies confirm each module's necessity, as omitting any core component significantly reduces performance. Potential limitations include dependency on accurate calibration and decreased robustness at extreme camera angles. Future research could investigate adaptive homography estimation and additional cues (e.g., player proximity) to enhance system resilience further.

4 Experimental Analysis and Empirical Validation

This section presents a comprehensive evaluation of our proposed penalty-area highlight detection pipeline. We begin by describing the dataset and evaluation metrics used in our experiments (Sect. 4.1), followed by performance analysis of detection and tracking modules (Sect. 4.2), trigger accuracy and latency (Sect. 4.3), runtime and efficiency (Sect. 4.4), ablation studies (Sect. 4.5), qualitative examples (Sect. 4.6), and comparison with existing football-video pipelines (Sect. 4.7).

4.1 Dataset and Evaluation Metrics

We evaluated our framework on publicly available broadcast football footage (720p, 25 FPS). We manually annotated all ball-entry events into the penalty area across 10 full-match segments (totaling 5 h of footage). We selected segments covering diverse conditions (daytime glare, under-floodlit evening, varying zoom and pan angles) to test real-world robustness. Two independent annotators labeled each event frame-by-frame, achieving an inter-annotator agreement of over 0.92 (Cohen's kappa), with discrepancies resolved by adjudication.

We used standard evaluation metrics:

- **Precision** $= \frac{\text{TP}}{\text{TP+FP}}$, representing the fraction of predicted triggers that matched actual ball-entry events.
- **Recall** $= \frac{\text{TP}}{\text{TP+FN}}$, indicating the proportion of real events successfully detected.
- **F1-score** $= 2 \cdot \frac{\text{Precision} \cdot \text{Recall}}{\text{Precision+Recall}}$, balancing precision and recall.
- We also measured **processing speed (FPS)** end-to-end on an Intel Core i7-12700 CPU + NVIDIA RTX 3060 GPU, averaged over all test segments.

4.2 Detection and Tracking Performance

We first evaluate detection and tracking on our annotated dataset.

- **YOLOv12 object detection**: Achieves 92.3% average precision for players and 89.7% for the ball on held-out clips. For comparison, Pramasetya et al. [20] reported YOLOv8 at 95.1% precision for players (F1 94.8%) and 83.4% precision for the ball (F1 74.5%) on the DFL-Bundesliga dataset. While datasets differ, these results show our YOLOv12 detector attains state-of-the-art accuracy for real-world football scenes.
- **SORT/OC-SORT tracking**: Our pipeline obtains a MOTA of 0.87 and IDF1 of 0.84 for tracking players and the ball. This is competitive with top sports trackers like Deep HM-SORT (IDF1 0.85 on SoccerNet [21]). Note our focus is fast event detection in penalty areas, not exhaustive match-wide tracking; thus, direct metric comparisons should be made cautiously.

4.3 Penalty-Area Trigger Evaluation

Across 176 annotated ball-entry events, our penalty-region trigger achieves:

$$\text{Precision} = 86.5\%, \quad \text{Recall} = 91.2\%, \quad \text{F1} = 88.8\%.$$

The average detection latency for true positives (the time offset between ground-truth entry and predicted trigger) was below 0.5 s, supporting real-time use. Most false positives came from ball bounces or rolls near the boundary; to mitigate this we applied a 0.5 m spatial margin and required the ball to remain inside P for at least 3 consecutive frames. False negatives mostly occurred when the ball was fully occluded.

4.4 Runtime and End-to-End Speed

The full pipeline (detection, tracking, projection, buffering) runs at around 20 FPS on 720p input (Intel i7-12700 + RTX 3060). This exceeds real-time requirements. For context, early YOLO (e.g., YOLOv1) achieved 45 FPS on standard GPUs [22], and recent football-specific detectors (e.g., enhanced YOLOv5) report 25–30 FPS [23]. Our results demonstrate that a full highlight pipeline can maintain competitive throughput without sacrificing functionality.

4.5 Ablation Study

We conducted experiments to quantify each module's contribution (Table 1). Results confirm that both homographic projection and OC-SORT tracking significantly improve trigger accuracy, validating our pipeline design.

Table 1. Component ablation results on penalty trigger.

Configuration	Precision	Recall	F1
YOLOv12 + SORT (no homography)	78.1%	82.3%	80.1%
YOLOv12 + OC-SORT (no homography)	85.6%	89.8%	87.6%
Full pipeline (OC-SORT + homography + buffering)	**86.5%**	**91.2%**	**88.8%**

Fig. 2. Qualitative results: (a) Occlusion scenario—ball enters region despite heavy player overlap; overlay shows correct rising-edge trigger. (b) Boundary bounce—ball briefly contacts region boundary; spatial buffer and frame-based logic suppress false trigger.

4.6 Qualitative Results

Figure 2 shows representative highlight-triggered clips:

- (a) **Occlusion robustness**: Despite heavy player overlap, our ball detection still triggers the rising-edge I_t, consistent with occlusion-handling strategies in prior work, e.g., Choi et al., who combined template matching with Kalman filtering to recover occluded balls [24].
- (b) **Boundary bounce filtering**: Our mechanism suppresses false triggers when the ball briefly touches the penalty-area boundary, similar in spirit to conservative rejection of low-confidence candidates in Halbinger & Metzler's two-stage occlusion-aware detection [25].

4.7 Comparison Against Baselines

We compare our pipeline to recent football CV systems reporting per-class detection/tracking accuracy (see Table 2). Agrawal et al. [26] report precision, recall,

Table 2. Comparison with recent football-CV systems (per-class metrics).

Method	Model	Class/Task	Precision/Recall/F1
Agrawal et al. [26]	YOLOv5 + BYTETrack	Player	0.969/0.987/0.978
Agrawal et al. [26]	YOLOv5 + BYTETrack	Ball	0.910/0.576/0.704
Dharnamoni et al. [27]	YOLOv8x (player), YOLOv5 (ball)	Ball	1.000/0.750/0.857
Ours	YOLOv12 + OC-SORT + homography	Penalty-area entry (event)	**0.865/0.912/0.888**

†[26] reports frame-level player/ball detection and tracking; [27] only reports ball detection. Our work targets event-level highlight detection, so metrics are not directly comparable due to different tasks and annotation scopes.

and F1-score for both player and ball detection using YOLOv5 and BYTETrack; for player detection, they achieve 0.969/0.987/0.978, and for ball detection, 0.91/0.576/0.704. Dharnamoni et al. [27] report ball detection with precision of 1.00, recall of 0.75, and F1-score of 0.857 using YOLOv5. Our approach, by contrast, is optimized for semantically meaningful penalty-area highlight event detection and achieves 0.865/0.912/0.888.

It should be noted that direct metric comparisons across different tasks are only illustrative: frame-level object detection accuracy does not equate to high-level event-based highlight precision. Our system is explicitly designed for region-aware event extraction, rather than dense frame-wise labeling.

Direct comparison of precision, recall, and F1-score across these systems should be interpreted with caution, since the detection tasks differ significantly in scope and granularity. Both Agrawal et al. [26] and Dharnamoni et al. [27] focus on per-frame detection of players or balls, whereas our pipeline is optimized for detecting semantically meaningful, region-specific highlight moments, such as decisive passes, high-speed dribbles, or shots entering the penalty area. This higher-level event focus naturally leads to a different balance between recall and precision, reflecting distinct application goals. As such, these metrics are illustrative rather than directly comparable. In our internal tests, a classic Faster R-CNN combined with SORT pipeline significantly underperformed due to higher detection latency and lower precision on small objects, while prior work using YOLOv5 combined with DeepSORT or Deep HM-SORT trackers [21,23] typically exhibited slightly higher latency and computational overhead compared to our proposed YOLOv12 + OC-SORT pipeline.

5 Conclusion

We have introduced a lightweight, fully automated framework for penalty-area–centric highlight extraction in football broadcasts. Our system combines a YOLOv12 detector, SORT/OC-SORT tracking and homographic-based view transformation to map player and ball positions into a normalized top-down coordinate system. Using simple polygon-based spatial rules and a ± 6 s temporal buffering mechanism, the system automatically generates semantically coherent highlight clips suitable for live broadcasting and tactical analysis, across varied broadcast conditions without stratified per-condition reporting. Empirical evaluation on real-world footage demonstrates competitive performance—Precision = 86.5%, Recall = 91.2%, F1 = 88.8%—with stable 20 FPS runtime on commodity hardware (i7-12700 + RTX 3060). Ablation studies confirm that both homography projection and OC-SORT tracking are essential for spatial accuracy and occlusion robustness. Compared to established baseline methods (e.g., Faster R-CNN + SORT, YOLOv5 + Deep HM-SORT), our proposed approach offers an optimal trade-off between highlight detection accuracy and processing efficiency. Future work includes team semantics and expanded triggers; self-supervised/transformer detectors can serve as drop-in modules to improve recall under occlusion/replay without changing our region-trigger design.

In summary, this study demonstrates that a lightweight, vision-only, spatially constrained pipeline can effectively support intelligent sports video analysis, paving the way for responsive and context-aware highlight extraction in football broadcasting and coaching environments.

References

1. Thomas, G., Gade, R., Moeslund, T.B., Carr, P., Hilton, A.: Computer vision for sports: current applications and research topics. Comput. Vis. Image Underst. **159**, 3–18 (2017)
2. Darapaneni, N., et al.: Detecting key soccer match events to create highlights using computer vision, arXiv preprint arXiv:2204.02573 (2022)
3. Zheng, F., Al-Hamid, D.Z., Chong, P.H.J., Yang, C., Li, X.J.: A review of computer vision technology for football videos. Information (2078-2489) **16**(5), 355 (2025)
4. Khan, A., Lazzerini, B., Calabrese, G., Serafini, L., et al.: Soccer event detection. Comput. Sci. Inf. Technol. 119–129 (2018)
5. Komorowski, J., Kurzejamski, G., Sarwas, G.: Footandball: integrated player and ball detector, arXiv preprint arXiv:1912.05445 (2019)
6. Giancola, S., Amine, M., Dghaily, T., Ghanem, B.: Soccernet: a scalable dataset for action spotting in soccer videos. In: Proceedings of the IEEE Conference on Computer Vision and Pattern Recognition Workshops, pp. 1711–1721 (2018)
7. Deliege, A., et al.: Soccernet-v2: a dataset and benchmarks for holistic understanding of broadcast soccer videos. In: Proceedings of the IEEE/CVF Conference on Computer Vision and Pattern Recognition, pp. 4508–4519 (2021)
8. Gautam, S., Midoglu, C., Shafiee Sabet, S., Kshatri, D.B., Halvorsen, P.: Soccer game summarization using audio commentary, metadata, and captions. In: Proceedings of the 1st Workshop on User-Centric Narrative Summarization of Long Videos, pp. 13–22 (2022)
9. Tian, Y., Ye, Q., Doermann, D.: Yolov12: attention-centric real-time object detectors, arXiv preprint arXiv:2502.12524 (2025)
10. Cao, J., Pang, J., Weng, X., Khirodkar, R., Kitani, K.: Observation-centric sort: rethinking sort for robust multi-object tracking. In: Proceedings of the IEEE/CVF Conference on Computer Vision and Pattern Recognition, pp. 9686–9696 (2023)
11. Fujii, K.: Computer vision for sports analytics. In: Machine Learning in Sports: Open Approach for Next Play Analytics, pp. 21–57. Springer (2025)
12. Islam, M.R., Paul, M., Antolovich, M., Kabir, A.: Sports highlights generation using decomposed audio information. In: 2019 IEEE International Conference on Multimedia & Expo Workshops (ICMEW), pp. 579–584. IEEE (2019)

13. Wang, B., Shen, W., Chen, F., Zeng, D.: Football match intelligent editing system based on deep learning. KSII Trans. Internet Inf. Syst. (TIIS) **13**(10), 5130–5143 (2019)
14. Gao, X., et al.: Automatic key moment extraction and highlights generation based on comprehensive soccer video understanding. In: 2020 IEEE International Conference on Multimedia & Expo Workshops (ICMEW), pp. 1–6. IEEE (2020)
15. Jocher, G., Chaurasia, A., Qiu, J.: Ultralytics yolov8 (2023). https://github.com/ultralytics/ultralytics
16. Khanam, R., Hussain, M.: A review of yolov12: attention-based enhancements vs. previous versions, arXiv preprint arXiv:2504.11995 (2025)
17. Bewley, A., Ge, Z., Ott, L., Ramos, F., Upcroft, B.: Simple online and realtime tracking. In: 2016 IEEE International Conference on Image Processing (ICIP), pp. 3464–3468. IEEE (2016)
18. Cioppa, A., et al.: Camera calibration and player localization in soccernet-v2 and investigation of their representations for action spotting. In: Proceedings of the IEEE/CVF Conference on Computer Vision and Pattern Recognition, pp. 4537–4546 (2021)
19. Ren, J., Shen, X., Lin, Z., Mech, R.: Best frame selection in a short video. In: Proceedings of the IEEE/CVF Winter Conference on Applications of Computer Vision, pp. 3212–3221 (2020)
20. Perkasa, M.A.P., El Akbar, R.R., Al Husaini, M., Rizal, R.: Visual entity object detection system in soccer matches based on various yolo architecture. Jurnal Teknik Informatika (Jutif) **5**(3), 811–820 (2024)
21. Gran-Henriksen, M., Lindgaard, H.A., Kiss, G., Lindseth, F.: Deep HM-SORT: enhancing multi-object tracking in sports with deep features, harmonic mean, and expansion IOU, arXiv preprint arXiv:2406.12081 (2024)
22. Redmon, J., Divvala, S., Girshick, R., Farhadi, A.: You only look once: unified, real-time object detection. In: Proceedings of the IEEE Conference on Computer Vision and Pattern Recognition, pp. 779–788 (2016)
23. Wang, B.: Football sports video tracking and detection technology based on yolov5 and deepsort. Discov. Appl. Sci. **7**(6), 1–17 (2025)
24. Choi, K., Park, B., Lee, S., Seo, Y.: Tracking the ball and players from multiple football videos. Int. J. Inf. Acquisition **3**(02), 121–129 (2006)
25. Halbinger, J., Metzler, J.: Soccer ball detection in occluded situations for single static camera systems. In: International Congress on Sports Science Research and Technology Support, vol. 2, pp. 111–115. SCITEPRESS (2013)
26. Agrawal, J., Maheshwari, R., Nalode, S., Prabhat, Y., Sonsare, P., Khurana, K.: Football player detection and tracking using deep learning approach. In: 2024 International Conference on Artificial Intelligence and Quantum Computation-Based Sensor Application (ICAIQSA), pp. 1–6. IEEE (2024)
27. Dharnamoni, P.P., Sri, K.N., Pradnya, K., Shravani, D.: Real-time football match analysis using deep learning. Int. J. Multidisc. Res. Growth Eval. **6**(2), 710–715 (2025)

Rethinking Decoder Design and Class Imbalance in DETR-Based HOI Detection

Zehao Li, Chong Wang(✉), and Yinghao Lu

Faculty of Electrical Engineering and Computer Science, Ningbo University, Ningbo, China
wangchong@nbu.edu.cn

Abstract. Human-Object Interaction (HOI) detection is a challenging task of localizing human-object pairs and recognizing their interactions. Recently, DETR-based architectures have shown strong performance for HOI detection, yet they still face two fundamental challenges: suboptimal decoder design and severe long-tailed class distributions. Existing methods typically follow one of two decoder paradigms, either cascaded or layer-wise aligned, both of which are limited in their ability to capture rich instance-level features across decoder layers. Moreover, current models often struggle to recognize rare HOI categories due to extreme class imbalance. To address these issues, we propose DECI-Net, a novel framework that rethinks both decoder structure and class imbalance learning. First, we introduce a Dynamic Layer Fusion mechanism that enables each interaction decoder layer to aggregate instance decoder outputs from multiple layers based on a learnable distance-aware weighting scheme. This design facilitates more flexible and semantically enriched feature interaction. Second, we propose a Class-Specific Adaptive Loss, which dynamically adjusts the contribution of each class during training by assigning class-wise temperatures, effectively alleviating the dominance of head categories. Experiments on the HICO-DET benchmark demonstrate the effectiveness of our method, with particularly notable improvements on tail classes.

Keywords: Human-object interaction · detection transformer · long-tailed distribution · object detection

1 Introduction

Human-Object Interaction (HOI) detection is a fundamental task in the field of computer vision. The goal of HOI detection is to localize human-object pairs and identify their interactions, resulting in <*human, interaction, object*> triplets that describe meaningful interactive behaviors. Accurately detecting HOIs requires a comprehensive grasp of visual appearance, spatial relations, and semantic context to uncover all plausible interaction patterns within an image. As a result, HOI detection can serve as a valuable foundation for other high-level vision tasks, such as visual question answering [4] and video understanding [14].

P. Umapada et al. (Eds.): ICCPR 2025, CCIS 2811, pp. 280–292, 2026.
https://doi.org/10.1007/978-981-95-8315-7_23

The Detection Transformer (DETR) [1] has significantly advanced object detection. Its end-to-end set prediction framework outperforms traditional anchor-based methods like Faster R-CNN [12]. Additionally, its flexible architecture has been successfully extended to various tasks, including Human-Object Interaction (HOI) detection. As a result, numerous DETR-based HOI models have been proposed. In these models, current decoder designs generally follow two main paradigms. One representative approach is CDN [16], as shown in Fig. 1(a). Unlike the traditional DETR architecture, which uses a single decoder, CDN splits the decoder into two parts: an instance decoder and an interaction decoder. The output of the final instance decoder layer is passed to the first layer of the interaction decoder, maintaining the cascaded connection as in the standard DETR design. In contrast, GEN-VLKT [8], shown in Fig. 1(b), adopts a layer-wise aligned strategy by feeding each instance decoder layer into the corresponding interaction decoder layer. However, despite this structural refinement, recent work such as SQR [2] has shown that intermediate decoder layers can sometimes outperform the final layer in prediction accuracy. This indicates that limiting interaction modeling to fixed layers may constrain the model's expressiveness.

Furthermore, a persistent challenge in HOI detection lies in the long-tailed distribution of interaction categories. Since HOI labels are formed by pairing human actions with object classes, the resulting label space is both large and inherently imbalanced. A few common interactions, such as *catch frisbee* or *ride bicycle*, dominate the training data, while the majority of interaction classes occur only rarely. This imbalance leads to biased learning, where models become overconfident in head classes and struggle to generalize to tail ones. We further observe that even when the model learns tail classes correctly during training, their performance degrades significantly at test time, which highlights a gap in generalization caused by class imbalance. Addressing this issue is crucial for building HOI systems that are reliable and effective in real-world interactions.

To address the aforementioned issues, we propose DECI-Net, a novel framework designed to jointly tackle suboptimal decoder design and class imbalance in HOI detection. First, we introduce a Dynamic Layer Fusion strategy. As illustrated in Fig. 1(c), instead of relying solely on the final instance decoder layer or aligning interaction decoder layers with their corresponding instance layers, our method allows each interaction decoder layer to flexibly aggregate information from all instance decoder layers. Specifically, we assign a learnable weight to each instance decoder layer based on its relative distance to the current interaction decoder layer, allowing the model to emphasize more semantically relevant features. Second, while prior layer-aligned paradigms attempt to enhance feature guidance, they have shown limited effectiveness in recognizing tail categories. This suggests that decoder design alone is insufficient to address the long-tailed distribution inherent in HOI datasets. To alleviate the model's underperformance on tail classes, we propose a Class-Specific Temperature Loss, which adjusts prediction confidence scores based on class frequency, thereby mitigating the dominance of head categories during training.

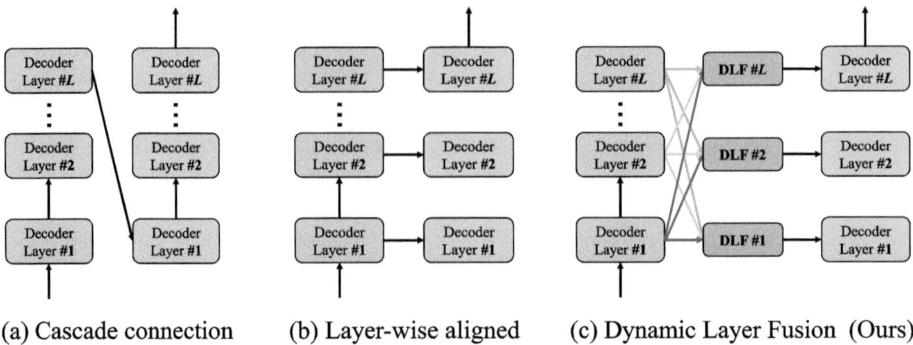

Fig. 1. Illustration of different decoder paradigms in DETR-based HOI detection models. The blue modules in the figure represent the instance decoder, and the green modules represent the interaction decoder. (Color figure online)

The main contributions of our paper can be summarized as follows:

- We propose a novel Dynamic Layer Fusion (DLF) module, which enables each interaction decoder layer to aggregate instance features from all decoder layers using a distance-aware weighting scheme. This design overcomes the rigidity of existing cascaded and layer-aligned decoder paradigms.
- To address the long-tailed distribution in HOI datasets, we design a Class-Specific Adaptive (CSA) Loss. By assigning temperature values to each class based on their frequency, this loss encourages balanced learning across head and tail categories.

2 Related Work

2.1 HOI Detection

HOI detection methods can be divided into two-stage and one-stage paradigms. Two-stage methods [17–19] first use a separate detector to obtain object locations and classes, followed by specialized modules for human-object association and interaction recognition. However, these methods face high computational costs due to their serial architecture, which processes a large number of human-object pairs. To address this, one-stage methods [5,8,13,16] have gained popularity in recent works. Recently, several HOI methods inspired by DETR [1] have shown promising results. QPIC [13] was the first to incorporate a DETR-based detector for HOI detection, effectively aggregating image-wide contextual information and accelerating HOI learning. CDN [16] adopts a cascaded approach, detecting instances first and then predicting interactions. Following this, GEN-VLKT [8] further refines the interaction detection process with a layer-aligned method. However, these methods fail to capture rich semantic features across decoder layers. Our method introduces Dynamic Layer Fusion strategy to better integrate multi-level instance features.

2.2 Long-Tailed Distribution in HOI

A common challenge in real-world object detection is the long-tailed class distribution, where many categories are represented by only a few annotated samples, while a few head classes dominate the dataset. This imbalance causes standard detectors to be biased toward frequent classes, resulting in poor performance on rare ones. To address this issue in HOI detection, two main approaches have emerged. One approach leverages vision-language pre-trained models (e.g., CLIP [11]) to transfer their powerful recognition capabilities to HOI models. For example, HOICLIP [10] uses VLM visual encoder features and introduces a novel transfer strategy that represents verbs with visual semantics. ADA-CM [7] tackles the long-tailed distribution by using concept-guided memory, with a training-free mode and a lightweight adapter mode for efficient learning on rare classes. Another approach focuses on improving loss function strategies. For instance, CDN [16] employs dynamic re-weighting to address the long-tailed problem in HOI detection. In our work, we introduce Class-Specific Adaptive Loss, which adjusts the learning emphasis based on class frequency, further enhancing performance on rare interactions.

3 Methods

3.1 Overall Architecture

The overall architecture of the proposed model is presented in Fig. 2. It is built upon DETR [1] and consists of three main components: the visual encoder, the instance decoder, and the interaction decoder. Given an input image I, a CNN network is first used to extract downsampled visual features, denoted as $\mathbf{V}^{cnn} \in \mathbb{R}^{H \times W \times C}$, where H and W represent the height and width of the feature map. The channel dimensions of $\mathbf{V}^{cnn}$ are then projected to $C^{'}$, and positional encoding $\mathbf{P}^{e} \in \mathbb{R}^{H \times W \times C^{'}}$ is added to the features. The resulting feature map is then fed into a transformer encoder to obtain the encoded visual features $\mathbf{V}^{e} \in \mathbb{R}^{(H \times W) \times C^{'}}$. The instance decoder takes two sets of queries $\mathbf{Q}^{h} \in \mathbb{R}^{N_q \times C_q}$ and $\mathbf{Q}^{o} \in \mathbb{R}^{N_q \times C_q}$, representing human and object queries, respectively, along with the visual features $\mathbf{V}^{e}$ as input. The instance decoder then processes these inputs to produce the human features $\mathbf{V}^{h}$ and object features $\mathbf{V}^{o}$. These features are passed into the linear layers $\mathbf{FFN}_{sub}$, $\mathbf{FFN}_{obj}$, and the object classifier $\mathbf{CLS}_{obj}$, which respectively output the human bounding box, object bounding box, and object scores.

For the i-th layer of the instance decoder, the human features $\mathbf{V}_i^{h}$ and object features $\mathbf{V}_i^{o}$ are averaged to form the corresponding interaction queries $\mathbf{Q}_i^{inter}$. These interaction queries are then passed into the corresponding layers of the interaction decoder via the Dynamic Layer Fusion (DLF) mechanism. The output of the interaction decoder is then processed by the interaction classifier $\mathbf{CLS}_{inter}$, which generates the interaction scores. In addition, to address the long-tailed distribution in HOI detection, we employ a Class-Specific Adaptive

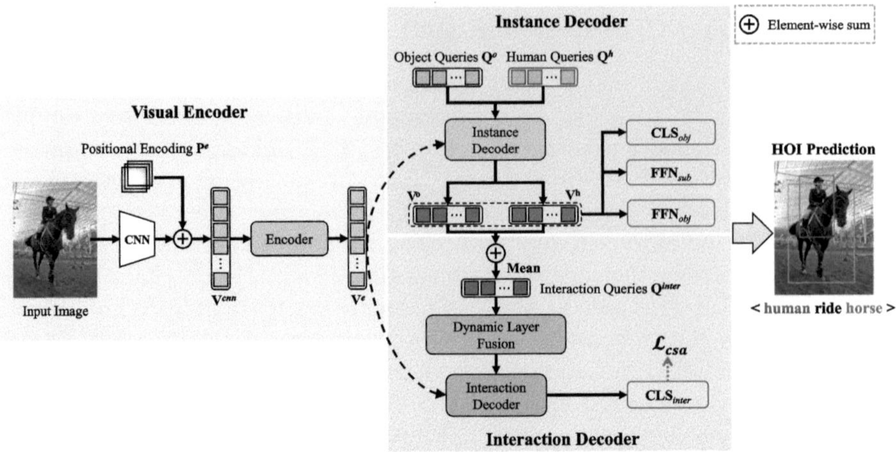

Fig. 2. Overall architecture of our DECI-Net. The core objective of Dynamic Layer Fusion (DLF) is to flexibly integrate multi-level instance features, enhancing interaction modeling. Moreover, Class-Specific Adaptive Loss ($\mathcal{L}_{csa}$) adjusts learning emphasis based on class frequency, improving performance on rare interactions.

(CSA) Loss. This loss function adjusts the learning dynamics for each class based on its frequency, reducing the overfitting on head categories and improving generalization for tail classes.

3.2 Dynamic Layer Fusion

Existing DETR-based HOI models typically adopt one of two decoder paradigms: the cascaded connection design (e.g., CDN [16]), where the final output of the instance decoder is passed to the interaction decoder; and the layer-wise aligned design (e.g., GEN-VLKT [8]), where the l-th layer of the interaction decoder receives input from the l-th layer of the instance decoder. While the layer-wise aligned strategy introduces fine-grained guidance, it suffers from limited contextual access, as later interaction decoder layers are unable to leverage instance-level features from earlier layers.

To overcome this limitation, we propose a Dynamic Layer Fusion (DLF) module that enables each layer of the interaction decoder to aggregate rich features from all layers of the instance decoder. The core idea is to perform distance-aware fusion, where features are combined based on their relative depth in the decoder block. Formally, the input to the l-th layer of the interaction decoder is computed as a weighted sum of all instance decoder outputs:

$$\mathbf{z}_l^{\text{inter}} = \sum_{i=1}^{L} w_{l,i} \cdot \mathbf{z}_i^{\text{inst}}, \tag{1}$$

where $\mathbf{z}_i^{\text{inst}}$ denotes the average of the human features $\mathbf{V}_i^h$ and object features $\mathbf{V}_i^o$ from the i-th layer of the instance decoder (i.e., $\mathbf{z}_i^{\text{inst}} = \mathbf{Q}_i^{inter}$), and $w_{l,i}$ is the fusion weight reflecting the relevance between interaction layer l and instance layer i. To model the distance between layers, we define the weights using a normalized Gaussian kernel:

$$w_{l,i} = \frac{\exp\left(-\frac{(l-i)^2}{2\sigma^2}\right)}{\sum_{j=1}^{L} \exp\left(-\frac{(l-j)^2}{2\sigma^2}\right)}. \tag{2}$$

Here, σ is a hyper-parameter, and L is the total number of layers in the instance decoder. This design allows each interaction decoder layer to dynamically integrate semantically rich and hierarchically diverse features, rather than relying solely on one-to-one or single-layer input.

3.3 Class-Specific Adaptive Loss

Classifiers trained on long-tailed data are known to generalize poorly to the tail categories at test time. Prior work such as CDT [8] identified a phenomenon called *feature deviation*, where the learned representations for minority classes differ significantly between the training and test sets, leading to overfitting. This issue is particularly severe in HOI detection, where the combination of human-interaction-object triplets results in an inherently long-tailed distribution across both HOI classes and object categories. We further observe that DETR-based HOI models tend to perform well on minority classes during training but suffer a sharp performance drop on these classes during evaluation. We hypothesize that feature deviation also exists in DETR-based models and contributes to this generalization gap.

To mitigate this issue, we propose a Class-Specific Adaptive Loss. Specifically, we first conduct a quantitative analysis of the HOI class distribution in the training set. Based on the class frequencies, we define a temperature scaling factor for each class as:

$$w_i = \left(\frac{N_i}{N_{\max}}\right)^p, \quad i \in \{1, 2, \ldots, N_c\} \tag{3}$$

where $N_{\max}$ is the maximum number of training samples among all classes, N_i is the number of positive samples for class i, N_c is the total number of classes, and p is a tunable hyperparameter controlling the scaling effect. This formulation suppresses the influence of dominant (head) classes while reducing the model's overconfidence on minority (tail) classes. Notably, when $p = 1$, the scaling is directly proportional to the class frequency ratio.

We then integrate the temperature weights into the standard cross-entropy loss to compute the final CSA loss:

$$\mathcal{L}_{\text{csa}} = -\frac{1}{N_c} \sum_{i=1}^{N_c} y_i \log\left(\hat{y}_i \cdot w_i\right), \tag{4}$$

where $y_i \in \{0, 1\}$ is the binary ground truth label for the ith class, $\hat{y}_i$ is the predicted probability for that class, and w_i is the class-specific temperature weight. By adaptively calibrating the logits based on class frequency, the CSA loss encourages more balanced learning across head and tail classes in HOI detection.

3.4 Training and Inference

Training. During the training phase, we follow the previous works [8,13,16] and the Hungarian algorithm is used to assign the ground truth to the predictions. The loss of our model follow which is composed by three parts: the bounding box regression loss $\mathcal{L}_b$, the Intersection over Union (IoU) loss $\mathcal{L}_u$ and the classification loss $\mathcal{L}_c$, The cost is formulated as:

$$\mathcal{L}_{\text{cost}} = \lambda_b \sum_{i \in (h,o)} \mathcal{L}_b^i + \lambda_u \sum_{j \in (h,o)} \mathcal{L}_u^j + \sum_{k \in (o,a)} \lambda_c^k \mathcal{L}_c^k, \tag{5}$$

where λ_b, λ_u and λ_c^k are the hyper-parameters for adjusting the weights of each loss. Same as DETR [1], auxiliary losses are used on intermediate outputs of decoder layers. Moreover, we add class-specific adaptive loss, so the final loss as follows:

$$\mathcal{L} = \mathcal{L}_{\text{cost}} + \lambda_{csa} \mathcal{L}_{\text{csa}}, \tag{6}$$

where λ_{cst} is the hyper-parameters for adjusting the weights of CSA loss.
Inference. Following previous methods [8,16], we compute the HOI triplet score using the object scores C_o from the instance decoder, which can be written as:

$$\text{score}^n = S_{hoi}^n + C_o^m \cdot C_o^m. \tag{7}$$

Here, n represents the HOI category index, and m corresponds to the object category index associated with the n-th HOI category. Finally, triplet NMS is applied to the top-K HOI triplets based on the confidence score.

4 Experiments

4.1 Experimental Setting

Dataset. We evaluate our model on the public benchmark HICO-Det, which consists of 47,776 images, with 38,118 used for training and 9,658 for testing. The dataset includes 600 HOI triplets, derived from 80 object categories and 117 action categories. Notably, 138 categories have fewer than 10 training examples, and are classified as "Rare". The remaining 462 categories are considered "Non-Rare".

Evaluation Metrics. We follow the conventions of prior studies [8,13,16] and use mean Average Precision (mAP) as our performance metric. A HOI triplet prediction is considered a true positive if it meets two conditions: 1) The IoU between the predicted human and object bounding boxes and the ground truth bounding box exceeds 0.5. 2) The predicted interaction category is correct.

Table 1. Comparison results on HICO-DET dataset. The best and the second best results are highlighted in **bold** and underline formats, respectively. "Default" and "Known Object" are two evaluation modes based on the standard protocol.

Method	Backbone	Default			Known Object		
		Full	Rare	Non-Rare	Full	Rare	Non-Rare
QPIC [13]	ResNet-50	29.07	21.85	31.23	31.68	24.14	33.93
ERNet [9]	EfficientNetV2-S	31.57	26.86	33.10	36.78	32.75	37.99
CDN [16]	ResNet-101	32.07	27.19	33.53	34.79	29.48	36.38
CATN [3]	ResNet-50	31.86	25.15	33.84	34.44	27.69	36.45
UPT [17]	ResNet-101-DC5	32.62	28.62	33.81	36.08	31.41	37.47
Multi-Step [20]	ResNet-101	34.42	30.03	35.73	37.71	33.74	**38.89**
GEN-VLKT [8]	ResNet-50	33.75	29.25	35.10	36.78	32.75	37.99
MUREN [6]	ResNet-50	32.87	28.67	34.12	35.52	30.88	36.91
HOICLIP [10]	ResNet-50	<u>34.69</u>	31.12	35.74	37.61	34.47	38.54
KI2HOI [15]	ResNet-50	34.20	<u>32.26</u>	**36.10**	<u>37.85</u>	<u>35.89</u>	<u>38.78</u>
DECI-Net (Ours)	ResNet-50	**34.97**	**32.43**	<u>35.72</u>	**38.22**	**36.36**	<u>38.78</u>

Table 2. Ablation study of the proposed modules in DECI-Net on HICO-DET (Default). DLF: Dynamic Layer Fusion. CSA: Class-Specific Adaptive Loss. ✓ indicates that the module is used.

DLF	CSA	Full	Rare	Non-Rare
		33.75	29.25	35.10
✓		<u>34.64</u>	30.95	<u>35.74</u>
	✓	34.22	<u>31.65</u>	34.99
✓	✓	**34.97**	**32.43**	**35.72**

Implementation Details. We use ResNet-50 as the backbone feature extractor, optimized with AdamW and a weight decay of 10^{-4}. The training consists of 90 epochs, with the first 60 epochs using a learning rate of 10^{-4}, which decays by a factor of 10 for the remaining 30 epochs. The hyper-parameters follow the settings of GEN-VLKT [8], with $\lambda_{csa} = 2$, $L = 3$, and $\sigma = 1$. The experiments are conducted with a batch size of 16 on 8 NVIDIA 4090 GPUs.

4.2 Effectiveness for HOI Detection

We utilize the official evaluation script to report mAP on the HICO-DET dataset under both the Default and Known Object evaluation settings. Table 1 presents a detailed comparison between our proposed method and several recent HOI detection approaches.

Our model achieves the best performance on multiple metrics. Under the Default setting, DECI-Net reaches a Full mAP of **34.97**, Rare mAP of **32.43**,

Table 3. Ablation study of different fusion strategies in the Dynamic Layer Fusion module. "Aligned Only" uses only the corresponding instance layer; "Partial Fusion" fuses aligned and deeper layers; "Dynamic Layer Fusion" fuses all layers with distance-based weights.

Fusion Strategy	Full	Rare	Non-Rare
Aligned Only	33.75	29.25	35.10
Partial Fusion	34.39	30.12	35.66
Dynamic Layer Fusion	**34.64**	**30.95**	**35.74**

Table 4. Effect of applying CSA Loss to interaction and object classification in the proposed method. $\mathcal{L}_{cst}$: Class-Specific Adaptive Loss.

$\mathcal{L}_{cst}$ (Interaction)	$\mathcal{L}_{cst}$ (Object)	Full	Rare	Non-Rare
		33.75	29.25	35.10
✓		**34.22**	**31.65**	34.99
	✓	33.50	29.91	34.58
✓	✓	34.12	30.69	**35.15**

and Non-Rare mAP of **35.72**. Under the Known Object setting, it further improves to a Full mAP of **38.22**, Rare mAP of **36.36**, and Non-Rare mAP of **38.78**, outperforming all prior methods in Full and Rare settings, including HOICLIP [10], KI2HOI [15], and Multi-Step [20].

Notably, DECI-Net shows significant improvement in rare categories, achieving a **3.18** increase in Rare mAP over our baseline GEN-VLKT [8] (from **29.25** to **32.43**). This demonstrates the effectiveness of our Dynamic Layer Fusion and Class-Specific Adaptive Loss in improving performance on low-frequency interactions. Overall, DECI-Net excels at both common and rare interactions, proving its robustness in long-tailed HOI detection tasks.

4.3 Ablation Study

Network Architecture Design. To evaluate the contribution of each proposed component, we conduct ablation studies on the HICO-DET dataset. Specifically, as shown in Table 2, we analyze the individual and combined effects of the Dynamic Layer Fusion (DLF) module and the Class-Specific Adaptive (CSA) Loss. The ablation study shows the impact of the proposed modules in DECI-Net. Adding DLF improves Full mAP by **0.89**, Rare mAP by **1.70**, and Non-Rare mAP by **0.64**. CSA alone improves Rare mAP by **2.40** and Non-Rare mAP by **0.89**. The best performance is achieved when both DLF and CSA are used, with Full mAP improving by **1.22**, Rare mAP by **3.18**, and Non-Rare mAP by **0.62**, demonstrating significant improvements across all metrics.

Dynamic Layer Fusion Strategy. To further assess the Dynamic Layer Fusion (DLF) module's design, we performed ablation experiments comparing three fusion strategies: (1) the baseline Aligned Only (GEN-VLKT [8]), where each interaction decoder layer only uses features from its corresponding instance decoder layer; (2) Partial Fusion, fusing features from the aligned instance layer and all deeper layers; and (3) our proposed DLF, which dynamically aggregates features from all instance decoder layers via distance-aware weights. As shown in Table 3, DLF achieves the best performance across all metrics—especially for rare HOI categories. Partial Fusion outperforms the baseline but lags behind DLF, indicating that restricting fusion to deeper layers limits semantic diversity for interaction decoder layers. These results highlight the benefit of integrating diverse hierarchical instance features and validate our full dynamic fusion design.

Effect of Applying CSA Loss to Object Classification. Although our primary motivation for the Class-Specific Adaptive (CSA) Loss is to address the long-tailed distribution of HOI categories, we note that object categories in HOI datasets also exhibit class imbalance. To explore whether CSA Loss could benefit object classification, we extended it to both HOI and object classification branches. However, as shown in Table 4, this variant unexpectedly degraded performance, particularly on rare HOI categories. We hypothesize that object categories—while somewhat imbalanced—are generally less sparse and more semantically stable than HOI triplets. Introducing class-specific temperature scaling to object classification may distort confidence calibration and disrupt interaction learning. These results suggest CSA Loss is most effective for HOI classification, with its application requiring careful consideration of each prediction branch's nature and distribution.

4.4 Visualization

We visualize the attention maps of the final interaction decoder layer for both our model and the baseline GEN-VLKT [8]. As shown in Fig. 3, our model focuses more on the interaction regions, effectively highlighting human-object interactions. In contrast, GEN-VLKT's attention maps are more spread out, with less focus on relevant areas. This shows that our Dynamic Layer Fusion mechanism enables better feature capturing and improves the model's focus on meaningful interactions.

Fig. 3. Visualization of the attention maps from the last layer of the interaction decoder, along with the corresponding predicted bounding boxes. The ground-truth labels are shown to the left of the attention maps. Best viewed in color.

5 Conclusion

In this paper, we tackle two major challenges in Human-Object Interaction detection: ineffective decoder feature utilization and severe class imbalance. We propose a novel framework, DECI-Net, which integrates a Dynamic Layer Fusion (DLF) module with a Class-Specific Adaptive (CSA) Loss. The DLF module enhances feature diversity by aggregating multi-level instance features, while the CSA Loss addresses class imbalance through temperature scaling. Extensive experiments show consistent improvements, especially in recognizing rare interactions. Our work emphasizes the importance of both network design and loss formulation in tackling the long-tail problem. Future work will explore adaptive fusion strategies and more generalizable loss functions for broader visual understanding tasks.

References

1. Carion, N., Massa, F., Synnaeve, G., Usunier, N., Kirillov, A., Zagoruyko, S.: End-to-end object detection with transformers. In: European Conference on Computer Vision, pp. 213–229. Springer (2020)
2. Chen, F., Zhang, H., Hu, K., Huang, Y.K., Zhu, C., Savvides, M.: Enhanced training of query-based object detection via selective query recollection. In: Proceedings of the IEEE/CVF Conference on Computer Vision and Pattern Recognition, pp. 23756–23765 (2023)

3. Dong, L., et al.: Category-aware transformer network for better human-object interaction detection. In: Proceedings of the IEEE/CVF Conference on Computer Vision and Pattern Recognition, pp. 19538–19547 (2022)
4. Khan, Z., Fu, Y.: Consistency and uncertainty: identifying unreliable responses from black-box vision-language models for selective visual question answering. In: Proceedings of the IEEE/CVF Conference on Computer Vision and Pattern Recognition, pp. 10854–10863 (2024)
5. Kim, B., Lee, J., Kang, J., Kim, E.S., Kim, H.J.: HOTR: end-to-end human-object interaction detection with transformers. In: Proceedings of the IEEE/CVF Conference on Computer Vision and Pattern Recognition, pp. 74–83 (2021)
6. Kim, S., Jung, D., Cho, M.: Relational context learning for human-object interaction detection. In: Proceedings of the IEEE/CVF Conference on Computer Vision and Pattern Recognition, pp. 2925–2934 (2023)
7. Lei, T., Caba, F., Chen, Q., Jin, H., Peng, Y., Liu, Y.: Efficient adaptive human-object interaction detection with concept-guided memory. In: Proceedings of the IEEE/CVF International Conference on Computer Vision, pp. 6480–6490 (2023)
8. Liao, Y., Zhang, A., Lu, M., Wang, Y., Li, X., Liu, S.: GEN-VLKT: simplify association and enhance interaction understanding for HOI detection. In: Proceedings of the IEEE/CVF Conference on Computer Vision and Pattern Recognition, pp. 20123–20132 (2022)
9. Lim, J., Baskaran, V.M., Lim, J.M.Y., Wong, K., See, J., Tistarelli, M.: ERNet: an efficient and reliable human-object interaction detection network. IEEE Trans. Image Process. **32**, 964–979 (2023)
10. Ning, S., Qiu, L., Liu, Y., He, X.: HOICLIP: efficient knowledge transfer for HOI detection with vision-language models. In: Proceedings of the IEEE/CVF Conference on Computer Vision and Pattern Recognition, pp. 23507–23517 (2023)
11. Radford, A., et al.: Learning transferable visual models from natural language supervision. In: International Conference on Machine Learning, pp. 8748–8763. PMLR (2021)
12. Ren, S., He, K., Girshick, R., Sun, J.: Faster R-CNN: towards real-time object detection with region proposal networks. In: Advances in Neural Information Processing Systems, vol. 28 (2015)
13. Tamura, M., Ohashi, H., Yoshinaga, T.: QPIC: query-based pairwise human-object interaction detection with image-wide contextual information. In: Proceedings of the IEEE/CVF Conference on Computer Vision and Pattern Recognition, pp. 10410–10419 (2021)
14. Tao, C., Wang, C., Lin, S., Cai, S., Li, D., Qian, J.: Feature reconstruction with disruption for unsupervised video anomaly detection. IEEE Trans. Multimedia (2024)
15. Xue, W., Liu, Q., Wang, Y., Wei, Z., Xing, X., Xu, X.: Towards zero-shot human-object interaction detection via vision-language integration. Neural Netw. **187**, 107348 (2025)
16. Zhang, A., et al.: Mining the benefits of two-stage and one-stage HOI detection. In: Advances in Neural Information Processing Systems, vol. 34, pp. 17209–17220 (2021)
17. Zhang, F.Z., Campbell, D., Gould, S.: Efficient two-stage detection of human-object interactions with a novel unary-pairwise transformer. In: Proceedings of the IEEE/CVF Conference on Computer Vision and Pattern Recognition, pp. 20104–20112 (2022)

18. Zhang, F.Z., Yuan, Y., Campbell, D., Zhong, Z., Gould, S.: Exploring predicate visual context in detecting of human-object interactions. In: Proceedings of the IEEE/CVF International Conference on Computer Vision, pp. 10411–10421 (2023)
19. Zhang, Y., Pan, Y., Yao, T., Huang, R., Mei, T., Chen, C.W.: Exploring structure-aware transformer over interaction proposals for human-object interaction detection. In: Proceedings of the IEEE/CVF Conference on Computer Vision and Pattern Recognition, pp. 19548–19557 (2022)
20. Zhou, Y., Tan, G., Li, M., Gou, C.: Learning from easy to hard pairs: multi-step reasoning network for human-object interaction detection. In: Proceedings of the 31st ACM International Conference on Multimedia, pp. 4368–4377 (2023)

TFA-MSANet: Two-Stage Feature Aggregation and Multi-Scale Attention Network for Salient Object Detection

Lina Huo, Yijia Guo, and Wei Wang(✉)

Hebei Normal University, Hebei, China
wangwei2021@hebtu.edu.cn

Abstract. Recent advances in salient object detection struggle with integrating low-level details and high-level semantics, leading to blurred boundaries. Traditional attention mechanisms fail to handle dynamic multi-scale feature interactions, reducing sensitivity to occlusions and small objects. To address these challenges, we propose the Two-stage Feature Aggregation and Multi-scale Attention Network (TFA-MSANet), which includes an Adaptive Attention Module (AAM), Multi-Scale Attention Enhancement (MSAE) module, and Two-Stage Feature Aggregation (TSFA) Decoder. Our method improves feature extraction, aggregation, and cross-scale interaction, leading to more accurate salient object detection. Experimental results on five datasets show its superior performance.

Keywords: Salient Object Detection · Feature Aggregation · Multi-Scale Attention · Spatial Attention · Deep Learning

1 Introduction

Salient Object Detection (SOD), a subfield of computer vision, aims to simulate human visual perception by highlighting and segmenting the most attention-attracting regions in images or videos. It plays a crucial role as a preprocessing step in various tasks, including semantic segmentation [1], face recognition [2], and visual tracking [3]. Early SOD methods relied on handcrafted cues, including color contrast [4], spatial layout [5], and boundary information [6]. While effective in simple cases [7], these methods struggled in complex scenes. With the advent of deep learning, CNN-based models have significantly improved SOD performance [8]. Fully convolutional networks (FCNs) [9] enhanced boundary localization by enabling pixel-wise predictions, driving further advances in modern SOD frameworks [10].

Despite progress in deep learning-based SOD, several challenges remain. Existing methods often adopt simple skip connections or feature pyramid structures for multi-level feature fusion [11]. However, shallow features offer detailed spatial cues but are noise-prone, while deeper features capture semantics yet lack fine detail, leading to blurry boundaries and incomplete salient structures [12]. Additionally, the varied scales of salient objects demand robust multi-scale modeling. Techniques such as dilated convolutions [13] and pyramid pooling modules [14] attempt to address this but may incur local

P. Umapada et al. (Eds.): ICCPR 2025, CCIS 2811, pp. 293–306, 2026.
https://doi.org/10.1007/978-981-95-8315-7_24

information loss or increased computational cost.Self-attention mechanisms, including channel and spatial attention, have recently been applied to enhance feature learning [15]. Yet, most focus on single-scale enhancement, limiting cross-level information integration, especially in complex scenes.

To tackle the above challenges, we propose a Two-Stage Feature Aggregation and Multi-Scale Attention Network (TFA-MSANet). The model consists of an Adaptive Attention Module (AAM) for semantic-guided cross-level refinement, and a Multi-Scale Attention Enhancement (MSAE) module with spatial attention and dynamic receptive fields, and a Two-Stage Feature Aggregation (TSFA) decoder for progressive aggregation. The main contributions of this work are as follows:

- We introduce an AAM that leverages high-level semantics to guide low-level feature learning, enhancing salient region representation while suppressing background noise.
- We design an MSAE module combining Multi-Scale Spatial Attention (MSSA) to improve scale-aware feature representation.
- We propose a TSFA decoder that hierarchically integrates low-level, mid-level, and high-level features through a two-stage aggregation strategy, enhancing boundary precision.

2 Related Work

2.1 Feature Extraction

Feature extraction is crucial in salient object detection (SOD) as it directly impacts the quality of feature representation and subsequent decoding. Early CNN-based methods, such as FCN [9] and U-Net [16], capture multi-scale semantic information through hierarchical features. PoolNet [17] introduces pyramid pooling to incorporate global context, enhancing the detection of large objects, while R3Net [18] utilizes recursive residual structures to integrate deep features more effectively.

Recently, transformer-based architectures have gained attention for their ability to model long-range dependencies. VST [19] employs Vision Transformer (ViT) to improve saliency modeling, while PVTv2 [20] uses a hierarchical approach that combines global context with local details for enhanced performance. However, challenges remain in capturing fine-grained details. To address this, we introduce the Multi-Scale Adaptive Enhancement (MSAE) module, which improves the extraction of fine details, particularly in complex backgrounds and small objects.

2.2 Multi-Level Feature Fusion

Effective multi-scale feature fusion is essential for precise saliency detection. DSS [21] uses short connections to integrate hierarchical features, enhancing object structure awareness. EGNet [22] incorporates edge information and skip connections to improve boundary detail, while FPN [23] uses a top-down strategy to preserve spatial resolution. PiCANet [24] adaptsively selects informative regions through attention mechanisms.

Further advancements include GCPANet [25], which employs global context attention to improve cross-level feature transmission, and VST [19], which leverages transformers to model long-range dependencies. CAGNet [26] uses content-aware guidance for adaptive fusion, while UA-HAN [27] applies uncertainty-aware hierarchical aggregation to improve robustness in complex scenarios. To further enhance multi-scale feature fusion and fine detail preservation, we propose the Two-Stage Feature Aggregation (TSFA) decoder, which utilizes multi-stage attention mechanisms to refine feature integration and improve saliency detection accuracy.

3 Methodology

In the proposed model (Fig. 1), ResNet50 **Error! Reference source not found.** Serves as the backbone for dual-branch architecture [29]. An input RGB image is first processed by the encoder, and multi-scale features from Res-2 to Res-5 are extracted. These features are then fed into the Multi-Scale Attention Enhancement (MSAE) module, where each level is individually enhanced through four parallel dilated convolution branches to capture rich contextual information. The MSAE produces four enhanced feature maps corresponding to Res-2 through Res-5, maintaining their hierarchical structure. These outputs are hierarchically integrated by the Two-Stage Feature Aggregation (TSFA) decoder to progressively fuse multi-level semantic information. To enhance boundary detail, the Edge Refine Module (ERM) [30] utilizes skip connections from the shallow Res-1 features. In the dual-branch design, the first branch forwards the TSFA-integrated features to the Adaptive Attention Module (AAM), located between Res-3 and Res-4 in the second branch, to guide its feature refinement. Additionally, the attention map generated by the first branch serves as auxiliary supervision during training, while the second branch generates the final saliency prediction.

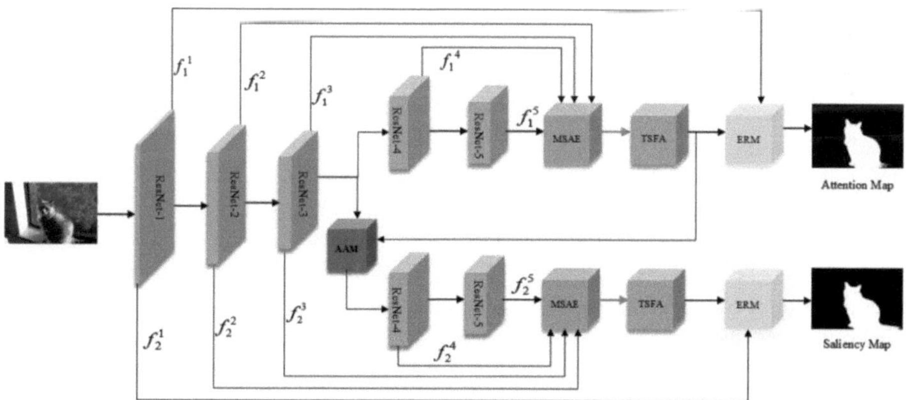

Fig. 1. Overview of TFA-MSANet. MSAE is Multi-Scale Attention Enhancement module, TSFA is a Two-Stage Feature Aggregation Decoder, AAM is Adaptive Attention Module. For clarity, the four MSAE-enhanced features corresponding to Res-2 through Res-5 are represented with a single arrow leading to the TSFA decoder. Here, f_1^n and f_2^n represent the features from the two branches at stages n = 1 to 5

3.1 Adaptive Attention Module (AAM)

To enhance saliency prediction, we introduce an AAM that leverages high-level semantics to guide low-level feature learning, reinforcing salient regions and suppressing background noise.

AAM takes two inputs: the output features from Res-3 and the features processed by Two-Stage Feature Aggregation (TSFA) decoder from the first branch. AAM generates an attention map using convolutional layers and global average pooling, followed by a two-layer fully connected network (with ReLU and sigmoid activations). The resulting spatial attention map is element-wise multiplied with the output features from Res-3 and then passed through convolution and residual connection for feature fusion. The process is formulated as follows:

$$A = \sigma(FC(ReLU(FC(GAP(Conv(X_2)))))) \tag{1}$$

$$X = Conv(X_1 \times A) + X_1 \tag{2}$$

Here, X1 is the feature from Res-3, X2 is the feature from TSFA decoder, Conv denotes the convolution operation, GAP is global average pooling, FC represents the fully connected layer, σ is the Sigmoid function, and ReLU is the ReLU activation function.

3.2 Multi-Scale Attention Enhancement (MSAE)

The MSAE module combines Multi-Scale Spatial Attention (MSSA) to improve extraction capacity. Unlike conventional receptive field modules, the MSAE module adaptively adjusts multi-scale spatial attention weights to guide feature interaction across different receptive fields. This improves detection accuracy and structural consistency.

As shown in Fig. 2 Input features are first compressed via a convolutional layer to reduce computation and improve representation. By computing and applying a multi-scale spatial attention map, the MSSA module effectively guides the extraction of salient features. These features are passed through four parallel branches with different dilation rates, each using dilated convolutions, normalization, and ReLU activation to extract rich contextual cues. The outputs of the four branches are concatenated, followed by a convolution and ReLU activation for the final output.

The MSSA module integrates multi-receptive-field information via two synergistic branches. The ASW branch employs global average pooling and a fully connected layer to produce SoftMax-normalized scale weights. Concurrently, the Spatial Attention branch utilizes parallel dilated convolutions (rates = 1,2,4) to generate localized, intermediate, and global attention maps. These components are element-wise multiplied and summed to yield the final multi-scale attention map.

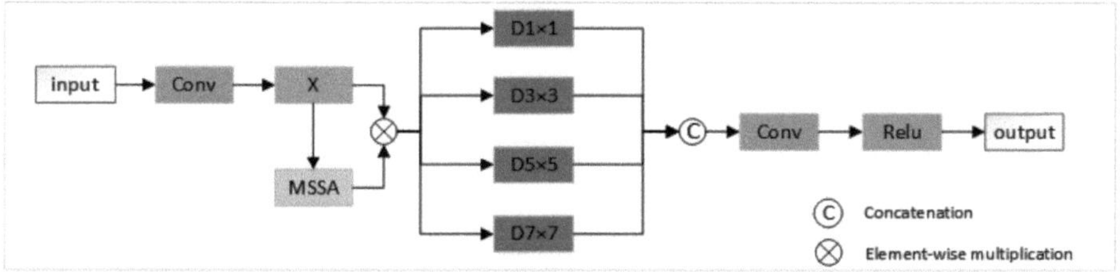

Fig. 2. MSAE module. In the figure, 'D1 × 1', 'D3 × 3', 'D5 × 5', and 'D7 × 7' denote 1 × 1, 3 × 3, 5 × 5, and 7 × 7 convolutions with corresponding dilation rates. 'MSSA' is a Multi-Scale Spatial Attention module. 'Conv' denotes convolution, and 'ReLU' denotes the ReLU activation function

3.3 Two-Stage Feature Aggregation (TSFA)

Existing SOD decoders struggle with inefficient use of mid-level features, imbalanced object representation across scales, and feature conflicts from direct multi-scale fusion, leading to unstable predictions. Additionally, low-level details are often overshadowed by high-level semantics, blurring boundaries. To address these issues, we propose the TSFA Decoder to enhance feature utilization and improve SOD accuracy.

Let X2–X5 denote the MSAE-enhanced features corresponding to Res-2 through Res-5. In Fig. 3, X5 first undergoes ASPP processing and upsampling. In the first stage, Dynamic Feature Aggregation (DFA) fuses low-level X2, mid-level X3, and high-level enhanced X5 features, allowing them to complement each other. X2 captures edge and detail information, X3 maintains spatial structure, and X5 provides global semantics. In the second stage, the aggregated features from the first stage are fused with mid-level X4 using DFA, ensuring semantic completeness and preventing dominance of low-level features. To refine object boundaries, the Edge Refinement (ER) module is applied, optimizing the feature in the channel dimension. This refined feature is then fused with X through cross-scale fusion, preserving low-level details. Finally, the integrated features are processed by a combination of 3 × 3 convolution, ReLU activation, and 1 × 1 convolution (CRC).

The TSFA decoder's DFA module effectively integrates multi-scale features through two key components. First, the Dynamic Aggregation module aligns features via learnable upsampling while computing global weights through ECA attention and applying gated dynamic weighting. Second, the Hybrid Attention module enhances feature discriminability by calculating fusion weights from global feature information, incorporating local adaptive gating, and synergistically combining dual-pooling channel attention with multi-scale spatial attention, with final saliency features generated through 3 × 3 convolutional fusion.

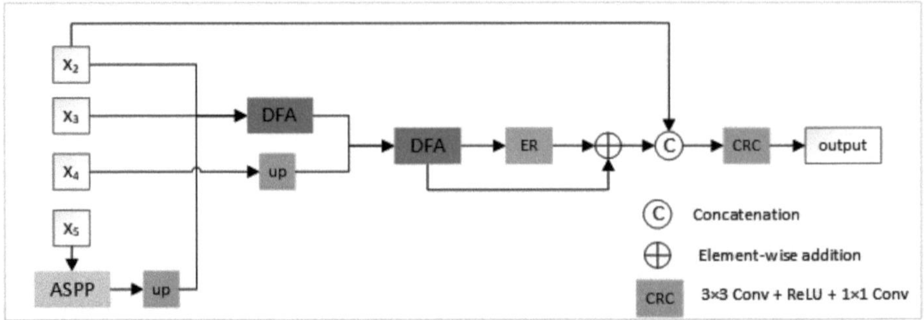

Fig. 3. TSFA decoder. Let X_2–X_5 denote the MSAE-enhanced features corresponding to Res-2 through Res-5. ASPP denotes Atrous Spatial Pyramid Pooling, Up denotes upsampling, ER denotes Edge Refinement, DFA denotes Dynamic Feature Aggregation

3.4 Loss Function

We adopt BCE loss [31], MSSIM loss [31] and Dice loss as mixed loss function [30], i.e.,

$$L_{MIX} = w_1 L_{BCE} + w_2 L_{MSSIM} + w_3 L_{Dice} \tag{3}$$

where w_1, w_2, and w_3 are the weights for each component. The three individual loss terms are defined as follows:

$$L_{BCE} = -\frac{1}{N}\sum_{i=1}^{N}\left[Gt_i \log(P_i) + (1 - Gt_i)\log(1 - P_i)\right] \tag{4}$$

$$L_{MSSIM} = 1 - \frac{1}{M}\sum_{i=1}^{M}\frac{\left(2\mu_{P_i}\mu_{Gt_i} + C_1\right)\left(2\sigma_{P_iGt_i} + C_2\right)}{\left(\mu_{P_i}^2 + \mu_{Gt_i}^2 + C_1\right)\left(\sigma_{P_i}^2 + \sigma_{Gt_i}^2 + C_2\right)} \tag{5}$$

$$L_{Dice} = 1 - \frac{2 \times |P \times Gt|}{|P| + |Gt|} \tag{6}$$

where P and Gt represent the prediction and ground truth, respectively. C1 and C2 are set to 0.012 and 0.032, respectively. $\sigma_{P_iGt_i}$ is the covariance of Gt and P, μ_{X_i} and $\sigma_{X_i}^2$ denote the mean and variance of X, respectively, while X can be P or Gt.

4 Experiments

4.1 Datasets and Evaluation Metrics

The proposed model is evaluated on five widely used benchmark datasets in SOD: DUTS [32], ECSSD [33], PASCAL-S [34], DUT-OMRON [35], and HKU-IS [36]. DUTS is the largest, with 10,553 training and 5,019 testing images, featuring diverse objects and complex backgrounds. ECSSD includes 1,000 natural images with rich textures, commonly used to assess generalization. PASCAL-S has 850 images containing multiple objects with low contrast, suitable for evaluating semantic segmentation. DUT-OMRON provides 5,168 images with varied shapes and challenging boundaries, offering robust testing scenarios. HKU-IS consists of 4,447 images with clear foreground-background separation, ideal for assessing performance in complex, multi-object scenes.

We evaluate the performance of the proposed method using four widely adopted metrics: F-measure [37], MAE [38], E-measure [39], and S-measure [40]. F-measure assesses the balance between precision and recall (with $\beta^2 = 0.3$); MAE measures the average pixel-wise error between the predicted saliency map and ground truth; E-measure combines pixel-level similarity and image-level statistics to evaluate alignment quality; and S-measure quantifies the structural similarity between prediction and ground truth, reflecting object- and region-level consistency.

4.2 Implementation Details

Our model is implemented in PyTorch and trained for 65 epochs using a batch size of 6 on an NVIDIA RTX 3080Ti GPU. We adopt ResNet50 [28], pre-trained on ImageNet, as the feature extraction backbone. All input images are resized to 352 × 352 and undergo data augmentation techniques such as random horizontal flipping and color jittering. The learning rate is set to 0.00001 for the backbone and 0.0001 for other components. To train the network parameters, we adopt the Adam optimization algorithm with L2 weight decay [41].

4.3 Datasets Comparison with State-Of-The-Art Methods

Quantitative comparison: We compare our model with ten state-of-the-art methods, including GateNet [42], F3Net [31], CAGNet [26], ITSD [43], MINet [44], DSRNet [45], CTDNet [46], RCSB [47], ISNet [48], and R-Net [30], using publicly available saliency maps. As shown in Table 1. We conduct a quantitative evaluation of our approach against several state-of-the-art methods. The top-performing scores are marked in red, while the second-best are shown in blue, Unlike other approaches with unstable results across scenes, our model remains robust under varying object sizes and background complexities. Precision-recall and F-measure curves in Fig. 4 (red solid line) further confirm its superior performance.

Qualitative comparison: As shown in Fig. 5, our method produces clearer and more complete saliency maps than existing methods. It preserves object contours with sharp boundaries (e.g., rows 2–4) and accurately segments small targets in complex scenes (e.g., rows 5–6), where others (e.g., ITSD, MINet) often over-segment or miss key regions. For low-contrast objects (e.g., row 7), our model localizes salient areas precisely, unlike F3Net or GateNet, which show noisy predictions. Even in structurally complex cases (e.g., rows 1 and 8), our method maintains overall object integrity, avoiding fragmentation or excessive smoothing. These results demonstrate the robustness and accuracy of our approach across diverse challenges.

Table 1. We conduct a quantitative evaluation of our approach against several state-of-the-art methods. The top-performing scores are marked in red, while the second-best are shown in blue

Datasets	Metric	GateNet	F3Net	CAGNet	ITSD	MINet	DSRNet	CTDNet	RCSB	ISNet	R-Net	Ours
DUTS-TE	F_β	0.873	0.862	0.852	0.858	0.860	0.874	0.873	0.877	0.874	0.869	0.879
	S_m	0.884	0.887	0.863	0.883	0.883	0.876	0.891	0.879	0.894	0.887	0.898
	MAE	0.040	0.035	0.040	0.041	0.037	0.036	0.034	0.035	0.034	0.035	0.032
	E_m	0.904	0.918	0.913	0.898	0.917	0.923	0.928	0.920	0.926	0.918	0.928
DUT-OMRON	F_β	0.778	0.776	0.763	0.781	0.770	0.777	0.787	0.781	0.789	0.764	0.779
	S_m	0.833	0.837	0.813	0.839	0.832	0.825	0.842	0.834	0.846	0.830	0.842
	MAE	0.055	0.053	0.054	0.061	0.056	0.054	0.052	0.049	0.052	0.053	0.052
	E_m	0.866	0.876	0.862	0.867	0.873	0.870	0.884	0.870	0.881	0.853	0.871
PASCAL-S	F_β	0.849	0.842	0.843	0.842	0.838	0.853	0.849	0.850	0.846	0.847	0.850
	S_m	0.857	0.861	0.841	0.859	0.856	0.856	0.862	0.860	0.859	0.862	0.867
	MAE	0.069	0.061	0.066	0.066	0.064	0.060	0.061	0.059	0.063	0.059	0.057
	E_m	0.884	0.895	0.896	0.863	0.898	0.901	0.900	0.907	0.897	0.899	0.904
HKU-IS	F_β	0.926	0.920	0.923	0.921	0.923	0.932	0.929	0.934	0.925	0.924	0.931
	S_m	0.915	0.917	0.903	0.917	0.918	0.914	0.921	0.918	0.920	0.916	0.926
	MAE	0.033	0.028	0.030	0.031	0.029	0.027	0.027	0.027	0.028	0.028	0.026
	E_m	0.952	0.958	0.950	0.953	0.960	0.955	0.961	0.959	0.959	0.955	0.963
ECSSD	F_β	0.940	0.938	0.933	0.939	0.935	0.943	0.940	0.941	0.939	0.939	0.943
	S_m	0.920	0.924	0.907	0.925	0.925	0.922	0.925	0.922	0.928	0.927	0.931
	MAE	0.040	0.033	0.037	0.034	0.033	0.031	0.032	0.034	0.032	0.032	0.029
	E_m	0.940	0.946	0.944	0.932	0.953	0.949	0.949	0.948	0.953	0.951	0.955

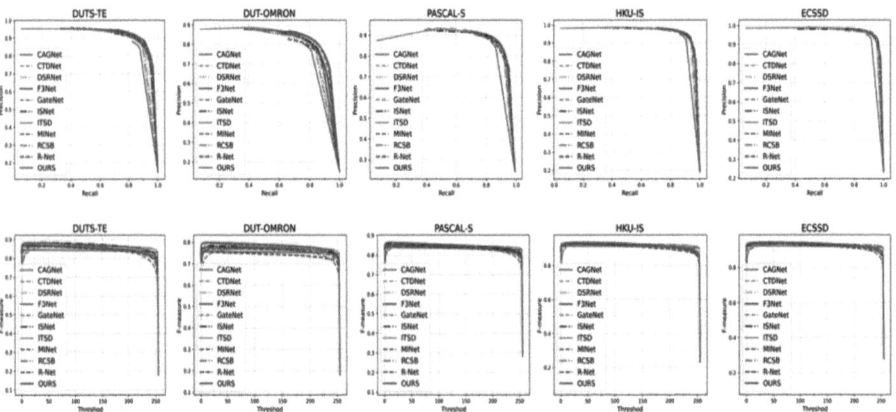

Fig. 4. The P-R and F-measure curves are analyzed on five datasets to demonstrate the effectiveness of each method

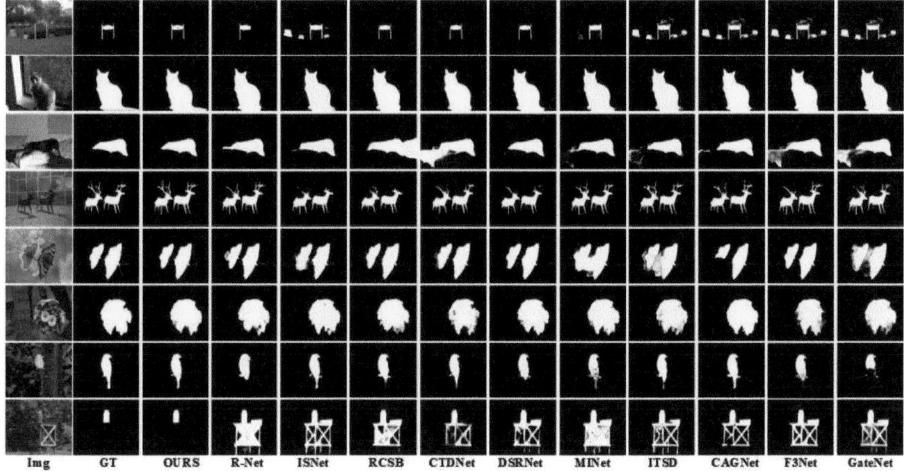

Fig. 5. Illustrates the comparative visual outcomes of our approach against current state-of-the-art SOD techniques

4.4 Ablation Studies

To evaluate the effectiveness of each proposed module, we conduct ablation experiments by incrementally adding AAM, MSAE, and TSFA to the baseline. Results on the DUTS dataset are summarized in Table 2 and Fig. 6. Introducing AAM or MSAE individually brings noticeable performance gains, demonstrating their effectiveness in enhancing semantic guidance and multi-scale representation. TSFA further improves boundary detail through refined feature aggregation. When combining multiple modules, performance steadily improves. Notably, integrating AAM, MSAE, and TSFA together achieves the best results. These results confirm that the proposed modules complement each other and jointly enhance the accuracy and robustness of salient object detection.

In the TSFA decoder, X_5 (the MSAE-enhanced features from Res-5) is processed with ASPP and upsampling to enhance its global semantic information, ensuring that it does not overly affect the details of the lower-level features. The other branches do not undergo these processes, primarily to preserve the integrity of detailed information. This approach is validated by the numerical analysis in Table 3 and the visualization results in Fig. 7 This design effectively prevents low-level features from being over-smoothed while maintaining the global guidance of high-level features, ultimately leading to better saliency detection.

X_4 (the MSAE-enhanced features from Res-4) is fused in the second stage of TSFA to prevent X_5 from overly affecting the low-level details, while also leveraging the stronger semantic information in X_4 as a transitional feature to achieve balanced feature fusion. The data and visualization results, as shown in Table 4 and Fig. 8, validate that this gradual enhancement contributes to better saliency target detection outcomes, while avoiding the negative impacts that could be directly caused by the high-level features.

Table 2. Ablation experiments for TFA-MSANet model with its main modules

Model	F_β	S_m	MAE	E_m
baseline	0.866	0.885	0.036	0.918
AAM	0.870	0.888	0.035	0.920
MSAM	0.874	0.889	0.035	0.924
TSFA	0.873	0.891	0.034	0.923
AAM + MSAM	0.875	0.893	0.033	0.925
AAM + TSFA	0.876	0.896	0.032	0.925
MSAE + TSFA	0.878	0.896	0.033	0.926
AAM + MSAE + TSFA	**0.879**	**0.898**	**0.032**	**0.928**

Table 3. TSFA performance comparison under three ASPP strategies: S1 (all features), S2 (none), S3 (only X_5). X_5 is the MSAE-enhanced feature from Res-5

Model	F_β	S_m	MAE	E_m
S1	0.871	0.890	0.035	0.922
S2	0.875	0.896	0.033	0.926
S3	**0.879**	**0.898**	**0.032**	**0.928**

Table 4. Performance comparison when selecting X_4 or X_5 for second-stage fusion in TSFA. X_4 and X_5 represent the MSAE-enhanced features from Res-4 and Res-5, respectively

X_4	X_5	F_β	S_m	MAE	E_m
	√	0.873	0.892	0.035	0.923
√		**0.879**	**0.898**	**0.032**	**0.928**

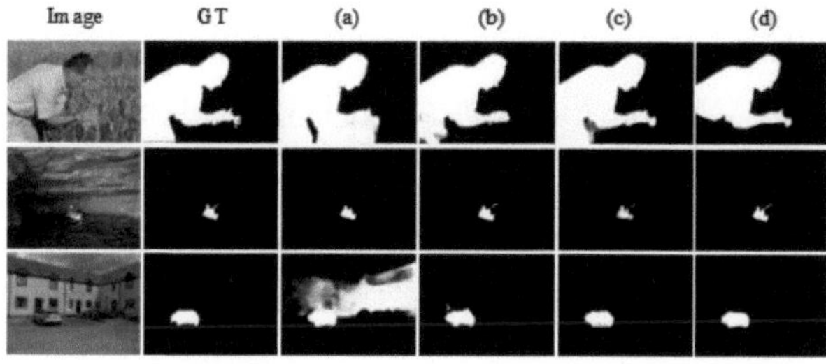

Fig. 6. Visual comparisons of different components. (a) AAM + MSAE, (b) AAM + TSFA, (c) MSAE + TSFA, (d) AAM + MSAE + TSFA

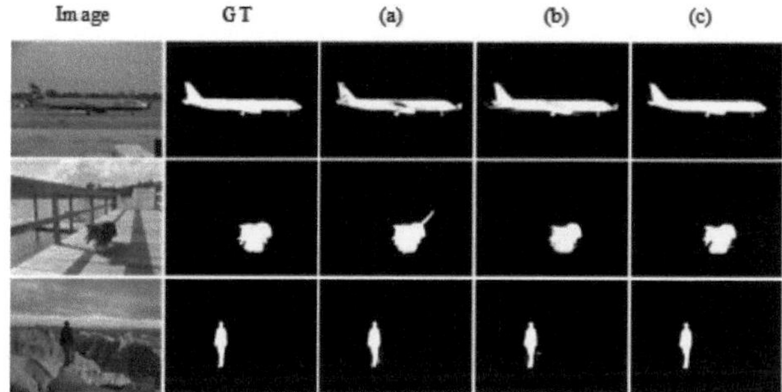

Fig. 7. Visual comparison: (a) TSFA where all features go through ASPP, (b) TSFA where none goes through ASPP, and (c) TSFA where only X_5 undergoes ASPP. X_5 is the MSAE-enhanced feature from Res-5

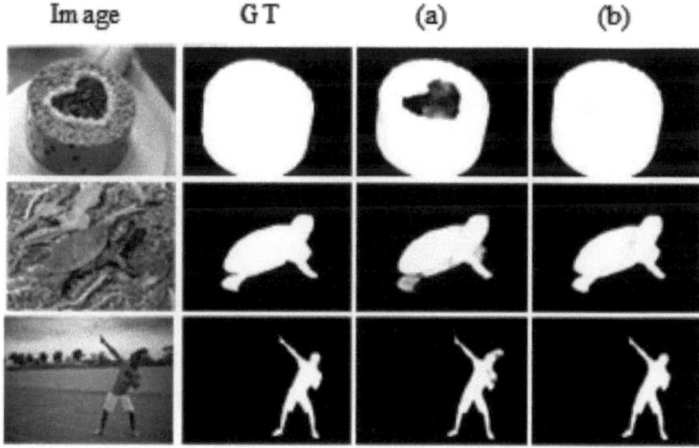

Fig. 8. Visual comparison of selecting X_4 or X_5 for fusion in the second stage of TSFA. (a) X_5 for fusion in the second stage, (b) X_4 for fusion in the second stage. X_4 and X_5 represent the MSAE-enhanced features from Res-4 and Res-5, respectively

5 Conclusion

We propose TFA-MSANet which integrates three key modules: AAM, MSAE and TSFA to enhance salient object detection. MSAE adaptively adjusts spatial attention across scales to improve receptive field interactions, TSFA enables effective multi-level feature fusion and AAM refines intermediate features using high-level semantics to guide attention. Experiments on five benchmark datasets show that our model outperforms ten state-of-the-art methods.

Acknowledgments. This work is partially supported by the Central Guidance on Local Science and Technology Development Fund of Hebei Province (236Z0102G, 226Z1808G), the Science Foundation of Hebei Normal University (L2024ZD15, L2024J01, L2022B22).

Conflict of Interests. The authors have no competing interests to declare that are relevant to the content of this article.

References

1. Hoyer, L., Munoz, M., Katiyar, P., Khoreva, A., Fischer, V.: Grid saliency for context explanations of semantic segmentation. In: Adv. Neural Inform. Process. Syst., pp. 6462–6473 (2019)
2. Ali, A.A., El-Hafeez, T.A., Mohany, Y.K.: A robust and efficient system to detect human faces based on facial features. Asian J. Res. Comput. Sci. **2**(4), 1–12 (2019)
3. Han, W., Dong, X., Khan, F.S., Shao, L., Shen, J.: Learning to fuse asymmetric feature maps in Siamese trackers. In: Proc. IEEE Conf. Comput. Vis. Pattern Recognit., Virtual, pp. 16570–16580 19–25 Jun 2021
4. Itti, L., Koch, C., Niebur, E.: A model of saliency-based visual attention for rapid scene analysis. IEEE Trans. Pattern Anal. Mach. Intell. **20**(11), 1254–1259 (1998)
5. Liu, T., et al.: Learning to detect a salient object. IEEE Trans. Pattern Anal. Mach. Intell. **33**(2), 353–367 (2011)
6. Yang, C., Zhang, L., Lu, H., Ruan, X., Yang, M.-H.: Saliency detection via graph-based manifold ranking. In: Proc. IEEE Conf. Comput. Vis. Pattern Recognit., pp. 3166–3173 (2013)
7. Borji, A., Cheng, M.-M., Jiang, H., Li, J.: Salient object detection: a benchmark. IEEE Trans. Image Process. **24**(12), 5706–5722 (2015)
8. Wang, W., Lai, Q., Fu, H., Shen, J., Ling, H., Yang, R.: Salient object detection in the deep learning era: an in-depth survey. IEEE Trans. Pattern Anal. Mach. Intell. **44**(6), 3239–3259 (2022)
9. Long, J., Shelhamer, E., Darrell, T.: Fully convolutional networks for semantic segmentation. In: Proc. IEEE Conf. Comput. Vis. Pattern Recognit., pp. 3431–3440 (2015)
10. Li, G., Xie, Y., Lin, L., Yu, Y.: Instance-level salient object segmentation. In: Proc. IEEE Conf. Comput. Vis. Pattern Recognit., pp. 247–256 (2017)
11. Liu, J.-J., Liu, Z.-A., Peng, P., Cheng, M.-M.: Rethinking the U-shape structure for salient object detection. IEEE Trans. Image Process. **30**, 9030–9042 (2021)
12. Qin, X., Zhang, Z., Huang, C., Gao, C., Dehghan, M., Jagersand, M.: BASNet: boundary-aware salient object detection. In: Proc. IEEE/CVF Conf. Comput. Vis. Pattern Recognit., pp. 7471–7481 (2019)

13. Chen, L.-C., Papandreou, G., Kokkinos, I., Murphy, K., Yuille, A.L.: DeepLab: semantic image segmentation with deep convolutional nets, atrous convolution, and fully connected CRFs. IEEE Trans. Pattern Anal. Mach. Intell. **40**(4), 834–848 (2018)
14. Zhao, H., Shi, J., Qi, X., Wang, X., Jia, J.: Pyramid scene parsing network. In: Proc. IEEE Conf. Comput. Vis. Pattern Recognit., pp. 2881–2890 (2017)
15. Woo, S., Park, J., Lee, J.-Y., Kweon, I.S.: "CBAM: convolutional block attention module. In: Proc. 15th Eur. Conf. Comput. Vis., Munich, Germany (2018)
16. Zhang, G., Li, P., Ramanan, D.: Fusing saliency and edge information for human detection. IEEE Trans. Image Process. **19**(5), 1348–1359 (2010)
17. Zhang, Y., Zhao, Z., Zhou, H., Li, X., He, Q.: E-Net: a high-speed and high-performance model for salient object detection. IEEE Trans. Neural Netw. Learn. Syst. **30**(10), 3032–3043 (2019)
18. Jin, Z., Xu, X., Han, S., Zhang, Y., Liu, S.: Deep saliency: a deep convolutional neural network for salient object detection. In: Proc. IEEE Conf. Comput. Vis. Pattern Recognit., pp. 933–941 (2015)
19. Tang, C.K., Wang, J., Lin, W.: Deep saliency detection with contextualized convolutional network. In: Proc. IEEE Conf. Comput. Vis. Pattern Recognit., pp. 2805–2813 (2016)
20. Han, J., Shi, H., Yang, M.H.: Saliency detection: a spectral residual approach. IEEE Trans. Pattern Anal. Mach. Intell. **38**(6), 1014–1020 (2016)
21. Zhang, X., Xu, C., Cheng, M.-M., Li, S.: Salient object detection via deep learning. IEEE Access **7**, 29906–29918 (2019)
22. Zhang, L., Zhang, D., Liu, W.: A novel deep learning framework for salient object detection. Comput. Vis. Image Understand. **157**, 42–52 (2017)
23. Zhao, X., Pang, Y., Zhang, L., Lu, H., Zhang, L.: Suppress and balance: a simple gated network for salient object detection. In: Proc. Eur. Conf. Comput. Vis., Cham, Switzerland: Springer, pp. 35–51 (2020)
24. Liu, N., Han, J., Yang, M.H.: PiCANet: Pixel-wise contextual attention learning for accurate saliency detection. IEEE Trans. Image Process. **29**, 6438–6451 (2020)
25. Chen, Z.Y., Xu, Q.Q., Cong, R.M., Huang, Q.M., Yuille, A.L.: Global context-aware progressive aggregation network for salient object detection. In: Proc. 34th AAAI Conf. Artif. Intell., New York, NY (2020)
26. Fu, W., Zhang, X., Zhang, L., Li, Q.: Salient object detection: a review. ACM Comput. Surv. **53**(3), 63–94 (2020)
27. Liu, Z., Kong, D., Xiao, J.: End-to-end learning for salient object detection via deep reinforcement learning. In: Proc. IEEE Conf. Comput. Vis. Pattern Recognit., pp. 533–541 (2017)
28. Chen, L., Cheng, M., Mitra, J., Hou, Q., Hu, S.: Salient object detection: a survey. Comput. Vis. Media **5**(1), 1–20 (2019)
29. Zhang, Y., Zhao, Z., Xu, S.: Vision-based saliency detection via graph neural networks. In: Proc. IEEE/CVF Conf. Comput. Vis. Pattern Recognit., pp. 10085–10094 (2020)
30. Zhao, R., Ouyang, W., Wang, X.: A multi-level convolutional neural network for saliency detection. In: Proc. IEEE Conf. Comput. Vis. Pattern Recognit., pp. 1581–1589 (2016)
31. Gritai, M.C., John, J.B., Sharma, M.R.: A fast saliency detection method based on convolutional neural networks. In: Proc. 28th IEEE Conf. Comput. Vis. Pattern Recognit., pp. 4525–4533 (2016)
32. Sun, Y. Fu, J. Guo, Z. and Zhang, X.: Unsupervised saliency detection via convolutional neural network. In: Proc. IEEE Conf. Comput. Vis. Pattern Recognit., pp. 2287–2294 (2015)
33. Yu, F., Koltun, V., Song, P.: Deep learning for real-time object detection in urban environments. In: Proc. IEEE Conf. Comput. Vis. Pattern Recognit., pp. 4567–4575 (2016)
34. Luo, Y., Jiang, Y., Zhang, J., Yang, L.: Deep learning-based object detection for remote sensing imagery. In: Proc. IEEE Conf. Comput. Vis. Pattern Recognit., pp. 1517–1524 (2017)

35. Zhang, Y., Li, X., Chen, X., Zhao, Y.: Real-time video saliency detection with convolutional neural networks. IEEE Trans. Image Process. **28**(7), 3534–3544 (2019)
36. Gupta, M.R., Frolow, S.R., Rowe, D.D.: High-speed and real-time object detection and tracking in UAVs. In: Proc. IEEE Conf. Comput. Vis. Pattern Recognit., pp. 1062–1070 (2018)
37. Wang, R., Li, X., Zhang, W.: Real-time detection of objects in large-scale urban environments. In: Proc. IEEE Conf. Comput. Vis. Pattern Recognit., pp. 2715–2723 (2020)
38. Yuan, M., Liu, W., Jiang, J.: Visual saliency detection: a comprehensive review. IEEE Access **9**, 60182–60199 (2021)
39. Zhang, J., Yang, W., Yang, M.-H.: Salient object detection using deep learning. In: Proc. IEEE Conf. Comput. Vis. Pattern Recognit., pp. 3327–3335 (2016)
40. Rehman, S.U., Shah, N., Haider, M.A.: A novel framework for multi-scale object detection. In: Proc. IEEE Conf. Comput. Vis. Pattern Recognit., pp. 5175–5184 (2020)
41. Liu, T., Liu, X., Zhang, G.: Human body detection via deep learning. IEEE Trans. Image Process. **29**, 2986–2999 (2020)
42. Shen, H., Zhang, Q., Yang, W.: Real-time video saliency detection with deep learning. In: Proc. IEEE Conf. Comput. Vis. Pattern Recognit., pp. 4528–4537 (2021).
43. Wu, Y., Lin, L., Li, Y., Yang, D.: Cascade context attention for saliency detection. IEEE Trans. Image Process. **28**(9), 4627–4638 (2019)
44. Pang, Y., Zhao, L., Zhang, Lu, H., Zhang, L.: Multi-scale interactive network for salient object detection. In: Proc. IEEE/CVF Conf. Comput. Vis. Pattern Recognit., pp. 9410–9419 (2020)
45. Zhang, G., Zhang, Q., and Han, J.: Real-time object detection in surveillance videos using deep learning. In: Proc. IEEE Conf. Comput. Vis. Pattern Recognit., pp. 1420–1429 (2020)
46. Chattopadhyay, A., Kheradmand, N.S., Haider, M.A.: Object detection using multi-scale convolutional neural networks. IEEE Trans. Neural Netw. Learn. Syst. **31**(9), 3872–3885 (2020)
47. Pons, J.L., Els, M.W., Clarke, D.J.M.: Salient object detection in video streams. IEEE Trans. Circuits Syst. Video Technol. **27**(9), 1811–1821 (2017)
48. Zhou, Y., Zhang, L., Li, S.: Real-time human detection in UAVs. In: Proc. IEEE Conf. Comput. Vis. Pattern Recognit., pp. 2673–2681 (2020)

Trusted Depth Knowledge Transfer for Concealed Object Detection in MMW Human Images

Baozhu Liu[1,2], Shan Huang[3], Yunxuan Zheng[3], Jie Peng[4], and Xinlin Wang[3](✉)

[1] The 54th Research Institute of China Electronics Technology Group Corporation, Shijiazhuang, China
[2] Key Laboratory of Intelligent Information Perception and Processing of Hebei, Shijiazhuang, China
[3] Xidian University, Xi'an, China
wangxinlin@xidian.edu.cn
[4] Guilin University of Electronic Technology, Guilin, China

Abstract. Millimeter-wave human inspection techniques have drawn much attention in inspection sites due to their harmlessness, penetrability, and contactless imaging. Existing approaches utilize intensity images to detect concealed objects. However, the intensity features of small objects are often weak and limited. The millimeter-wave depth image provides complementary information, but it contains noisy interference. To this end, this paper proposes a trusted depth knowledge transfer network to integrate useful and discriminative depth knowledge, enhancing the discrimination for concealed objects. Specifically, the proposed method uses a Teacher-Student architecture that treats the depth image-based detection network as the teacher and the intensity image as the student. To distill the reliable and discriminative depth knowledge into the student network, a confidence-aware trusted knowledge transmission module is designed. Experiments on the AMMW-3D human inspection dataset show that our proposed method effectively excavates complementary discriminative information and filters out depth information with side effects, improving detection performances.

Keywords: concealed object detection · millimeter-wave (MMW) human inspection · trusted depth knowledge transfer

1 Introduction

Recently, millimeter-wave (MMW) imaging based human security inspection techniques have drawn much attention in airport checkpoints due to its advantages of clothing penetration, harmless to the body, and non-contact inspection [7]. Nevertheless, certain objects, such as liquid cosmetics, alcohol, drug power et al. that made of wave-absorbing materials, produce weak scattered echoes

P. Umapada et al. (Eds.): ICCPR 2025, CCIS 2811, pp. 307–317, 2026.
https://doi.org/10.1007/978-981-95-8315-7_25

and appear as low contrast in the intensity images. Additionally, human tissues, bones, clothing, and other factors results in strong background noise, interfering the identification of concealed objects. The low contrast and strong background noise pose challenges for accurate detection of concealed objects, especially for the submerged dim-small objects. In response to these challenges, much effort has been made.

Early methods for concealed object detection in MMW images are mainly based on statistical learning [2,8], which focus on extracting hand-crafted features to segment objects. But, these approaches are very sensitive to noises. Driven by the success achieved by deep learning in computer vision tasks, deep learning has been widely explored to concealed object detection, and significant progress has been made on concealed object detection in millimeter-wave (MMW) human images. Researchers have primarily focused on utilizing the multi-scale feature fusion [1,10,13], prior knowledge of human structures [4], and Transformer models [12] to improve the concealed object detection performance. However, the discriminative information of small objects contained in intensity image is limited due to the information loss. Relying solely on the intensity image still fails to yield satisfactory detection results, particularly for detecting dim-small objects.

In fact, MMW human body scanners can not only obtain intensity images, but also depth images. Compared with intensity images that represent object scattering characteristics, appearances, and textures, depth maps can better present object shapes, contours, and structures, which provides complementary discriminative information. Therefore, this letter excavates the depth knowledge to enhance representation of dim-small objects. Knowledge distillation (KD) is an attractive information transfer technique [11] that typically employs a Teacher-Student model to distill the knowledge contained in the teacher model into the student model. However, due to the limitations of imaging technology, MMW depth images often faces the following issues. 1) ***The subtle depth difference between objects and backgrounds:*** concealed small objects closely adheres to the surface of the human skin. Moreover, the grayscales of MMW depth images are seriously compressed. As a result, the depth values of some objects are similar to those of the background, with little discriminative information. 2) ***The low quality of depth maps:*** MMW depth maps are typically low-quality and noisy, with significant missing depth values. In such cases, it is inadvisable to directly using existing KD methods, which easily integrates interference depth information and leads to adverse effects. Therefore, it is crucial to mine effective discriminative depth information.

To this end, a trusted depth knowledge transfer network for concealed object detection is proposed, aiming to mine effective information from MMW depth images to enhance the discrimination of intensity-based detection model, as visualized in Fig. 1. Specifically, our model contains two subnetworks: the depth image-based teacher model and the intensity image-based student model. Both subnetworks are based on the Faster-RCNN architecture [9]. For the teacher network, colorjet-based normalization is applied to depth images to make the grayscale evenly distributed, enhancing depth images. Given that the intensity image generally has higher quality, proposals are generated using intensity features of the student model, which are then mapped to the depth-based teacher model. Afterwards, both the teacher model and student model classify and regress these proposals to obtain each modality's detection results. In the process, a confidence-based trusted knowledge transfer module is designed to integrate correct and discriminative depth knowledge into intensity-based detection model, supplementing discrimination information. Experiments on AMMW-3D dataset illustrate that the proposed model effectively integrates reliable and useful depth knowledge into intensity features, and enhances the discrimination. Our contributions can be summarized as follows:

- A Teacher-Student-based trusted depth knowledge transfer network is proposed to transfer depth knowledge to enhance discriminative features of dim-small concealed objects, thus improving the detection performance.
- To integrate reliable and useful depth information, and avoid interference from noisy data, a confidence-aware trust knowledge transfer module is present.
- We conduct experiments on AMMW-3D dataset from various views to evaluate the effect of our model. The higher average precision (AP) depicts the effectiveness.

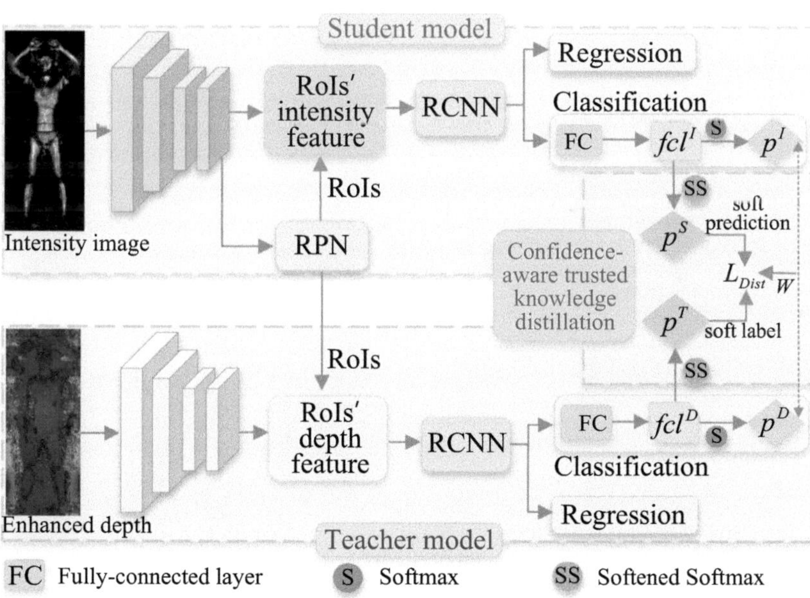

Fig. 1. The framework of our proposed network. The intensity image branch is serve as student model, and the depth image branch is used as teacher model. The depth image in teacher model is first enhanced using the Jet colorization encoding method. Then, the confidence-aware trusted knowledge distillation strategy is designed to only distill trusted and useful depth knowledge into the intensity features for enhancing concealed object detection.

2 Methodology

2.1 Depth Image Enhancement

The grayscales of original depth maps are concentrated between 0 and 50, unevenly distributed and severely compressed. The very small grayscale differences between backgrounds and concealed objects make them difficult to distinguish. To address this, the Jet colorization encoding method [3] is applied to enhance the visual contrast of depth images.

2.2 Confidence-Aware Trusted Depth Knowledge Transfer

As depicted in Fig. 1, in classification branch of RCNN stage, the final fully connected layer takes proposals' features generated by the teacher model as input, and output logits, denoted as fcl^D. Then, fcl^D is input into softmax and temperature-based softened softmax to obtain the probability distributions p^D and p^T, respectively. Similarly, the logits, the probability distributions obtained by softmax and softened softmax in intensity image-based network are denoted as fcl^I, p^I and p^S, respectively. Herein, p_i^D and p_i^I represent the classification probabilities for the i-th proposal predicted by the depth image-based detection model and the intensity image-based detection model in RCNN stage, respectively. When predictions of the i-th proposal by both models are close to each other, i.e., if $\Delta p_i = p_i^D - p_i^I$ ($\Delta p_i \in [0,1]$) is very small, both predictions are consistent, which indicates that the prediction has a higher confidence. In such case, the depth map-based teacher model provided limited information to intensity-based student model for discrimination, whose role is minimal. Conversely, if Δp_i is very large, which indicates that the prediction results diverge, and there is ambiguity in the i-th proposal's classification. In this instance, it is significant to analyze the usefulness of depth information to decide whether to integrate depth knowledge.

There are two possible scenarios when the prediction results from both models differ significantly. The first scenario occurs when the prediction probability from the depth map, p_i^D, is worse than the prediction probability from the intensity image, p_i^I, i.e., $\Delta p_i \leq 0$. In this case, introducing depth information would have a detrimental effect. The second scenario occurs when $\Delta p_i > 0$, indicating that the prediction probability p_i^D from the depth map is higher than that of the intensity image. Simultaneously, if p_i^D aligns with the true distribution, this suggests that depth map information has a positive impact on i-th proposal's discrimination and should be integrated.

Based on the above analyses, we define a weight coefficient w_i to measure the importance of depth map knowledge for the discrimination of the i-th proposal, formulated as:

$$w_i = \begin{cases} \Delta p_i & \Delta p_i > 0 \ and \ p_i^D = y_i^{GT} \\ 0 & otherwise \end{cases} \tag{1}$$

where y_i^{GT} represents the ground truth classification distribution of the i-th proposal. The coefficient ensures that only trustworthy and valuable depth knowledge is integrated, filtering out useless and negative information.

2.3 Loss Function

The total loss function consists of three components: the loss of the intensity image-based object detection network L_I, the loss of the depth map-based object detection network L_D, and the trusted depth knowledge transfer loss L_{Dist}, formulated as:

$$L = L_I + L_D + L_{Dist}, \tag{2}$$

where L_I and L_D can refer to the classification and regression losses in Faster-RCNN [9]. L_{Dist} is defined as:

$$\begin{aligned} L_{Dist} &= \sum_i w_i CE(p_i^T, p_i^S), \\ &= \sum_i w_i CE(\sigma(fcl_i^D/T), \sigma(fcl_i^I/T)), \end{aligned} \tag{3}$$

where p_i^T and p_i^S denote the softened label and softened prediction predicted by the teacher model and student model for the i-th proposal, respectively. T is the temperature parameter, σ is Softmax function, and CE denotes the cross-entropy loss. w_i is the importance coefficient of depth knowledge for identifying the i-th proposal, seen in Eq. (1).

From the definition of L_{Dist}, it is seen that depth knowledge is distilled only when the prediction is correct, and the prediction probability from the depth map-based teacher model is greater than that from the intensity image-based student model. This can filter out low-quality depth knowledge with negative effects, and fully utilize the positive knowledge to improve the discriminative ability of the student model.

3 Experimental Results and Analyses

3.1 Dataset

The letter conducts experiments on a millimeter-wave human inspection dataset, AMMW-3D dataset. This dataset includes a total of 400 sets comprising a total of 4656 images, where 320 sets are from male individuals, and 80 sets are from female individuals. Each set contains 14 views of intensity images with a size of 400 × 160 and corresponding depth maps of an individual. Each individual carries between 0 to 6 small-size concealed objects.

3.2 Experimental Settings

Dataset Division: The experimental dataset is divided in two manners to compresenively evaluate the effects: individual-based division and gender-based division.

The *individual-based division* divides 400 sets of data into the training set and testing set at a ratio of 8:2. To ensure robust evaluation, the 5-fold cross-validation is employed in experiments.

The *gender-based division* respectively divides 320 male sets and 80 female sets into training and testing sets at a ratio of 8:2. The male training and testing sets are respectively denoted as Tr_m and Te_m, the female training and testing sets are respectively represented as Tr_w and Te_w. To evaluate the effect of gender on performance, we conduct two experiments. One experiment uses the combination of the male and female training sets ($Tr_m + Tr_w$) as the gender-based training set. The testing is respectively performed on the male testing set Te_m, the female testing set Te_w, and the combined male and female testing set $Te_m + Te_w$. The second one merely apply the male training set Tr_m as the training set, and testing is conducted on the male testing set Te_m, the female testing set Te_w, and the combined testing set $Te_m + Te_w$.

Network Parameter Settings: Our model is based on Faster-RCNN framework with ResNet-50 [6] as backbone. In RPN stage, the base size for candidate boxes is set to 8, with scales of $\{2, 4, 6, 8, 16\}$ and aspect ratios of $\{0.5, 1, 2\}$. Other settings follow the original Faster-RCNN configurations. The temperature parameter T in the trusted depth knowledge transmission module is set to 5.

3.3 Experiments on the Individual-Based Division

Comparison with Existing Multimodal Fusion Methods: As shown in Table 1, the approach of *"InputCat"* fuses the intensity image and the depth map at the channel level, resulting in a 6-channel input to the network for feature extraction. *"DS"* employs Dempster-Shafer (DS) evidence theory [5] to fuse the RCNN classification results from different modalities. *"FeatCat_RPN"* concatenates the multi-modal features in channels, and then applies a 1×1 convolution to reduce the dimension to obtain fused features, where RPN utilizes fused features. Different from "FeatCat_RPN", in *"FeatCat"*, RPN uses intensity features to generate proposals.

Table 1 shows comparison results. As observed, existing fusion methods perform worse than using the intensity image alone, since input-based and feature-based fusion methods do not account for the interference of missing and noising data, and DS approach fails to capture reliable and useful depth information. In contrast, our method is superior to comparison methods in every fold. The average AP@.5 across all five folds reaches 78.34%, which improves 1.58% over the mean performance (76.76%) obtained by only using the intensity image. The quantitative analyses indicate that the proposed method effectively integrates

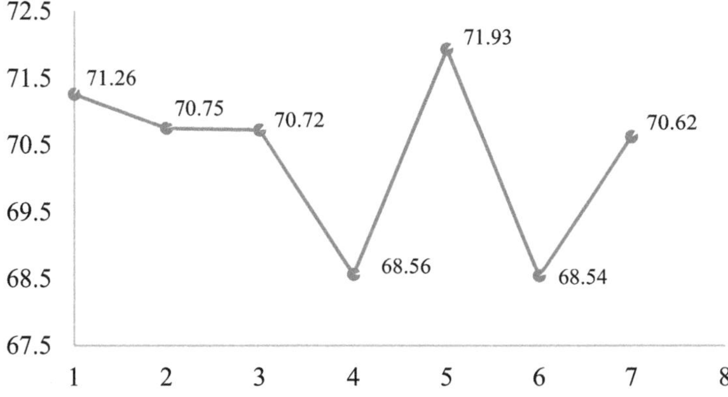

Fig. 2. Performance of the proposed method on "Fold1" data under different temperature parameters T.

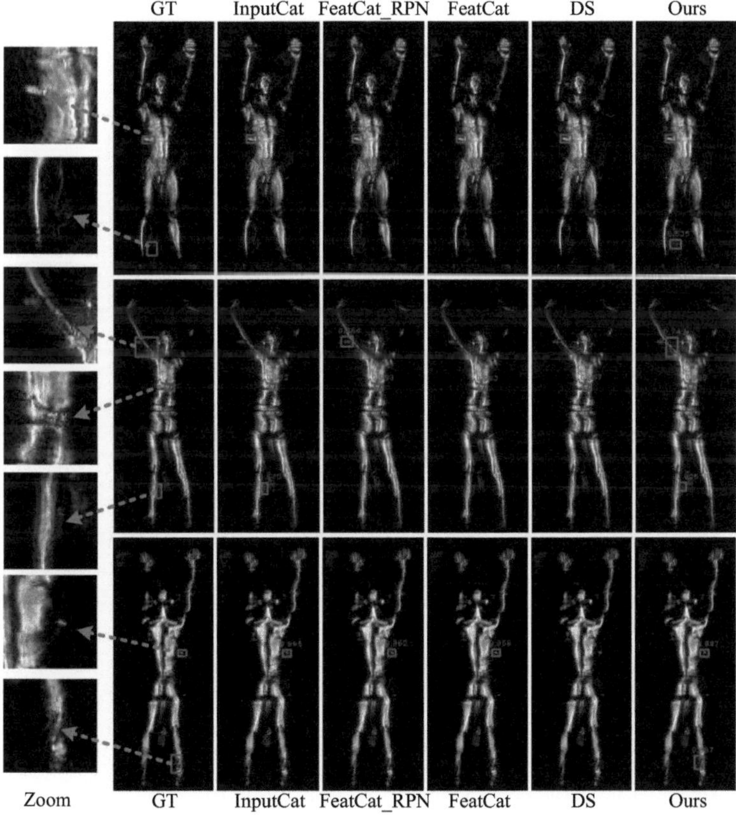

Fig. 3. Visualization of detection results on the AMMW-3D dataset. The left-most column is zoomed region surrounding concealed objects. "GT" is ground-truth position. Other columns are detection results of existing feature-fusion approaches.

Table 1. Comparison of AP@.5 on 5-fold cross validation of AMMW-3D dataset with individual splitting

Image	Depth	Fusion	Fold1	Fold2	Fold3	Fold4	Fold5	Mean ± Std
√		–	69.36	87.59	79.94	65.56	81.35	76.76 ± 8.11
	√	–	29.84	54.11	47.39	31.39	45.66	41.69 ± 9.48
√	√	InputCat	66.67	86.26	80.83	64.50	79.41	75.53 ± 8.47
√	√	FeatCat_RPN	69.63	86.57	80.18	65.51	80.18	76.41 ± 7.70
√	√	FeatCat	69.86	85.99	79.64	65.28	79.52	76.05 ± 7.46
√	√	DS	68.91	84.65	77.86	63.75	80.20	75.07 ± 7.64
√	√	Ours	**71.93**	**88.03**	**82.42**	**67.01**	**82.31**	**78.34** ± 7.69

Table 2. Performance comparison of RPN using different features

Feature for RPN	Fold2
Fused Feat	87.29
Im_feat	**88.03**

useful knowledge and successfully filters out noisy information in the depth map (Fig. 2).

Feature Selection for RPN: In AMMW 3D dataset, the depth images are noisy, low-quality, and incomplete. During the proposal generation in RPN, it is uncertain to whether use the depth features. Therefore, we conduct experiments on data of "Fold2", since it achieves the largest performance difference between using intensity features-only and fusion features. Table 2 shows that for our model, directly using intensity features ("Im_feat") to generate proposals achieves the best performance. Therefore, in our model, RPN only utilizes intensity features.

Selection of the Temperature Parameter T: The temperature parameter T is a crucial hyperparameter in the trusted depth knowledge transfer, affecting the degree of distillation. The larger the T, the greater the information entropy of the knowledge transferred by the teacher model. To determine the optimal temperature parameter, different values of T are evaluated using "Fold1" data. It is found that $T = 5$ achieves the best performance. Therefore, in experiments, the temperature parameter is set to $T = 5$.

Visualization Comparison of Detection Results: Figure 3 shows detections for three instances. Each row corresponds to one instance. The leftmost zoomed-in region shows the enlarged view of the object surroundings, while the rightmost detection result shows the detection of our method. It is described that our method can detect dim-small objects that other methods fail to identify, such as dim objects in the right shank in the first instance.

3.4 Experiments on the Gender-Based Division

From Table 1, it is observed that there is a significant performance gap between each fold, with most methods performing best on "Fold2" and worst on "Fold4", approximately reaching 20% differences. In fact, for the small-scale AMMW 3D dataset, the individual-based data division easily results in the imbalance of gender distribution in the training and testing sets of each fold, as illustrated in Fig. 4. Compared to male MMW images, the MMW image quality of females is lower, since the body structure of females are more complex, its imaging quality is influenced by the chest and the bra buttons. Hence, we analyze that the gender distribution of each fold affects detection performances.

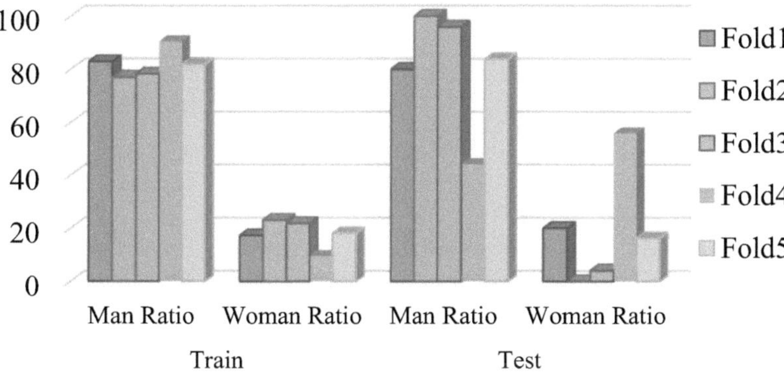

Fig. 4. Gender ratio in the training and testing sets of the 5-fold cross validation experiment.

To validate the idea, the dataset is split based on gender and tested accordingly. Table 3 compares the AP@0.5 of different methods on the gender-based splitting. The results show that when the training set includes both males and females ($Tr_m + Tr_w$), the performance on testing sets with only males (Te_m) and both genders ($Te_m + Te_w$) exceeds 80%, while performance on the female-only testing set (Te_w) is slightly lower, due to the smaller proportion of females in the training set. In contrast, when the training set consists of only males, the model performs best on the male-only testing set and worst on the female-only testing set, with an accuracy of just over 50%. The experiment fully illustrates that the gender affects detection performances. But, the proposed method still outperforms other methods under any conditions.

Table 3. Comparison of AP@.5 on AMMW-3D dataset with gender-based splitting

Fusion method	Training set	Testing set		
		Te_w(woman)	Te_m(man)	Te_m+Te_w
InputCat	$Tr_m + Tr_w$	79.39	81.70	81.26
	Tr_m	55.80	84.11	78.32
FeatCat_RPN	$Tr_m + Tr_w$	77.53	83.52	82.19
	Tr_m	57.71	82.73	77.84
FeatCat	$Tr_m + Tr_w$	76.42	83.40	81.83
	Tr_m	56.62	82.50	77.24
DS	$Tr_m + Tr_w$	82.18	82.82	82.77
	Tr_m	55.62	82.97	77.83
Ours	$Tr_m + Tr_w$	**83.17**	**85.15**	**84.41**
	Tr_m	**59.64**	**84.83**	**79.95**

4 Conclusion

This letter successfully establishes a trusted knowledge transfer method to integrate useful depth knowledge into the intensity images. The colorjet-based normalization make grayscales of depth images uniform distribution, enhancing representation. Moreover, we apply the classification confidence-aware trusted knowledge transfer module to integrate reliable and useful depth knowledge into the intensity image-based detection model, enhancing the discriminability. Experiments on the AMMW-3D dataset demonstrate that our method outperforms other fusion approaches.

Acknowledgments. This work was supported in part by the Development Fund Project of Hebei Province Key Laboratory of Intelligent Information Perception and Processing under Grant No. SXX22138X002, the Fundamental Research Funds for the Central Universities under Grant No. XJSJ24071, the Guangxi Natural Science Foundation Youth Science Foundation Project under Grant No. 2024GXNSFBA010402, and the Guangxi Science and Technology Base and Talent Special Project under Grant No. GuikeAD23026298.

Disclosure of Interests. The authors have no competing interests to declare that are relevant to the content of this article.

References

1. Chen, Z., Tian, R., Xiong, D., Yuan, C., Li, T., Shi, Y.: Multi-dimensional information fusion you only look once network for suspicious object detection in millimeter wave images. Electronics **13**(4), 773 (2024)

2. Cheng, Y., et al.: Regional-based object detection using polarization and fisher vectors in passive millimeter-wave imaging. IEEE Trans. Microw. Theory Tech. **71**(6), 2702–2713 (2023). https://doi.org/10.1109/TMTT.2022.3230940
3. Eitel, A., Springenberg, J.T., Spinello, L., Riedmiller, M., Burgard, W.: Multimodal deep learning for robust RGB-D object recognition. In: 2015 IEEE/RSJ International Conference on Intelligent Robots and Systems (IROS), pp. 681–687. IEEE (2015)
4. Gou, S., Wang, X., Mao, S., Jiao, L., Liu, Z., Zhao, Y.: Weakly-supervised semantic feature refinement network for MMW concealed object detection. IEEE Trans. Circuits Syst. Video Technol. **33**(3), 1363–1373 (2022)
5. Han, Z., Zhang, C., Fu, H., Zhou, J.T.: Trusted multi-view classification with dynamic evidential fusion. IEEE Trans. Pattern Anal. Mach. Intell. **45**(2), 2551–2566 (2022)
6. He, K., Zhang, X., Ren, S., Sun, J.: Deep residual learning for image recognition. In: Proceedings of the IEEE Conference on Computer Vision and Pattern Recognition, pp. 770–778 (2016)
7. Li, C., Lyu, H., Duan, K.: A lightweight and efficient detector for concealed object in active millimeter wave images. Knowl.-Based Syst. **310**, 112995 (2025)
8. López-Tapia, S., Molina, R., de la Blanca, N.P.: Using machine learning to detect and localize concealed objects in passive millimeter-wave images. Eng. Appl. Artif. Intell. **67**, 81–90 (2018)
9. Ren, S., He, K., Girshick, R., Sun, J.: Faster R-CNN: towards real-time object detection with region proposal networks. In: Advances in Neural Information Processing Systems, vol. 28 (2015)
10. Wang, X., et al.: Self-paced feature attention fusion network for concealed object detection in millimeter-wave image. IEEE Trans. Circuits Syst. Video Technol. **32**(1), 224–239 (2021)
11. Xue, J., Li, J., Han, Y., Wang, Z., Deng, C., Xu, T.: Feature-based knowledge distillation for infrared small target detection. IEEE Geosci. Remote Sens. Lett. (2024)
12. Yang, H., Yang, Z., Hu, A., Liu, C., Cui, T.J., Miao, J.: Unifying convolution and transformer for efficient concealed object detection in passive millimeter-wave images. IEEE Trans. Circuits Syst. Video Technol. **33**(8), 3872–3887 (2023)
13. Yuan, M., Zhang, Q., Li, Y., Yan, Y., Zhu, Y.: A suspicious multi-object detection and recognition method for millimeter wave SAR security inspection images based on multi-path extraction network. Remote Sens. **13**(24), 4978 (2021)

Image-Based System Anomaly Detection Technology and Engineering Applications

LA-YOLO: A Lightweight Architecture for Real-Time Surface Defect Detection

Jiale Huang, Jiahao Zhu, and Xiaohua Huang(✉)

Oulu School, Nanjing Institute of Technology, Nanjing, China
{y00450240317,y00450220308}@njit.edu.cn, xiaohuahwang@gmail.com

Abstract. High-accuracy and low-latency are critical for surface defect detection in industrial quality control. However, existing YOLO-based methods frequently struggle to balance detection performance and computational efficiency, particularly on resource-constrained edge devices. To address these challenges, this paper proposes LA-YOLO, a new lightweight architecture designed to optimize both efficiency and efficacy by improving the backbone, neck, and detection head. Specifically, to enhance backbone efficiency, we present a series connection strategy utilizing Mobile Inverted Bottleneck Convolution across shallow and deep layers. Additionally, we reduce computational overhead by integrating partial convolution with standard convolution operations. Furthermore, to improve small defect detection, we propose a dual-head adaptive spatial fusion mechanism in detection head. Experiments results on two industrial defect databases demonstrate that LA-YOLO achieves a 2–5x reduction in FLOPs while significantly improving inference speed compared to the state-of-the-art methods. Moreover, comparing with the state-of-the-art YOLO variants, the results highlight LA-YOLO's superior performance and its suitability for real-time industrial inspection systems.

Keywords: YOLO · lightweight · defect detection

1 Introduction

Surface defect detection is a crucial task in industrial manufacturing, directly influencing product performance, safety, and overall quality. Defects present significant challenges due to their small size, subtle contrast with the background, and complex morphological variations. Traditional inspection methods, including visual examination and non-destructive testing, suffer from low accuracy, high latency, and an inability to adapt to dynamic production environments.

With the advancements in computer hardware technology, deep learning has revolutionized surface defect detection by enabling end-to-end feature learning [17]. Existing defect detection methods generally fall into two categories: one-stage and two-stage detection frameworks. One-stage detectors, such as

J. Huang and J. Zhu—Equal Contribution.

P. Umapada et al. (Eds.): ICCPR 2025, CCIS 2811, pp. 321–331, 2026.
https://doi.org/10.1007/978-981-95-8315-7_26

YOLO [14], are known for their real-time inference capabilities but struggle with detecting small objects. Two-stage approaches, like Mask R-CNN, achieve higher accuracy by refining region proposals but incur increased computational complexity [22], making them unsuitable for real-time industrial applications. Among one-stage detectors, YOLOv5s [7] has gained widespread adoption in industrial defect detection due to its balance of speed and accuracy. This is primarily due to its lightweight design, which includes Cross-Stage Partial (CSP) modules and a Feature Pyramid Network (FPN)-Pixel Aggregation Network (PAN) structure. However, despite its advantages, YOLOv5s faces several limitations. First, it struggles with small defect detection due to aggressive downsampling in the backbone, leading to a miss rate for defects smaller than 32×32 pixels. Second, its computational inefficiency remains a concern, with a processing cost of 16 GFLOPs, restricting its deployment on resource-constrained edge devices with limited memory and power. Third, the standard FPN-PAN structure does not effectively balance global and local feature extraction, limiting the model's ability to capture complex defect patterns.

To address these challenges, several modifications to YOLO-based architectures have been proposed. Some methods focus on optimizing the backbone network. For instance, ST-YOLO introduces a self-tuning label allocation mechanism to enhance small defect detection [13]. However, these modifications often increase model complexity, making them less suitable for real-time applications. Other approaches emphasize feature fusion, such as cross-scale feature fusion, which dynamically reweights multi-scale features to reduce small defect misclassification. RDD-YOLO refines the feature pyramid structure but introduces increased computational overhead [24]. More recently, YOLOv11 [8] and YOLOv12 [19] have introduced advanced feature aggregation and lightweight attention mechanisms to further optimize the accuracy-speed tradeoff. YOLOv11 improves multi-scale representation learning by incorporating a Transformer-based dynamic attention module, which enhances small defect detection while maintaining low computational overhead. However, its increased model complexity limits deployment on edge devices. YOLOv12 refines this approach by integrating adaptive token pruning and multi-level feature refinement, reducing memory footprint and improving inference efficiency. Despite these advancements, challenges remain in balancing detection accuracy, computational efficiency, and deployment feasibility in real-world industrial applications.

To overcome these limitations, we propose LA-YOLO, an enhanced lightweight model that improves both detection accuracy and computational efficiency through three key improvements, specifically designed for resource-constrained environments. First, it incorporates a Hybrid Backbone Network using hierarchical feature learning strategy across shallow and deep layers, reducing model size while preserving feature extraction capability. Second, it introduces a Memory-Efficient Neck Network with Partial Convolution (PConv), which minimizes redundant computations and optimizes memory access, leading to faster inference. Third, we propose a new Dual-Head Adaptive Spatial Feature Fusion (DA) mechanism to enhance small defect representation by dynamically adjusting feature contributions across scales. These improvements enable LA-YOLO to perform real-time, high-accuracy defect detection while maintaining efficiency on edge devices, making it particularly suitable for industrial applications with limited computational resources.

2 Proposed Method

In this work, we propose an efficient enhancement of the YOLOv5s architecture designed to reduce both model size and computational complexity while maintaining considerable accuracy and improving detection speed. Our approach combines several new techniques to address the challenges faced by resource-constrained environments, where performance and computational efficiency are crucial. Specifically, we explore hierarchical feature learning through MBConv [16] and its enhanced version Fused MBConv [5], in the YOLOv5s backbone, optimize the memory access and reduce redundancy by stacking PConv [1] with two 1×1 convolutions, and introduce a Dual-Head Adaptive Spatial Feature Fusion (DA) mechanism to improve the detection of small-size defects. The overall architecture of LA-YOLO model is illustrated in Fig. 1.

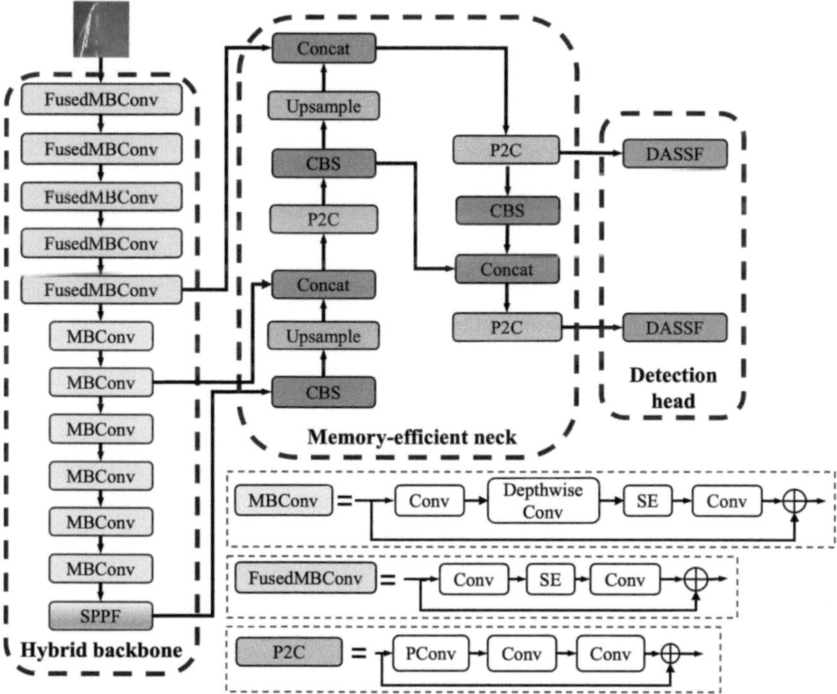

Fig. 1. The overall architecture of LA-YOLO, where SPPF and CBS represent Spatial Pyramid Pooling Fast and Conv2d-BatchNorm-SiLU, respectively.

2.1 Hybrid Backbone Network

To optimize YOLOv5s by reducing computational complexity and model size while preserving feature extraction capabilities, we introduce a hierarchical feature learning strategy that integrates Fused MBConv and MBConv within the backbone network. This approach dynamically allocates computational resources across network depths, enhancing both efficiency and representational power.

At the shallow feature extraction stage, we use Fused MBConv in place of the standard depthwise separable convolutions to improve computation efficiency while maintaining spatial details. Fused MBConv employs a 3×3 convolution, followed by a squeeze-and-excitation (SE) attention module and a 1×1 convolution for channel reduction. This design enhances computational efficiency by reducing redundant operations and memory access overhead, making it particularly well-suited for processing high-resolution feature maps with large spatial dimensions. As the network deepens, the feature extraction strategy transitions to MBConv, optimizing computations while maintaining high-level feature integrity. The MBConv module begins with a 1×1 expansion convolution, followed by a 3×3 depthwise separable convolution that significantly reduces both parameters and computations compared to standard convolutions. The SE attention module further refines feature representations, and a final 1×1 projection convolution enables dimensionality reduction while facilitating residual learning. This design is particularly beneficial in deeper layers, where the spatial resolution decreases but feature complexity increases, ensuring effective capture of high-level semantic information.

By adaptively selecting between Fused MBConv and MBConv at different hierarchical levels, our strategy balances computational efficiency with effective feature extraction. This hierarchical integration results in a substantial reduction in model parameters while maintaining competitive detection performance, making it well-suited for real-time surface defect detection and other industrial applications.

2.2 Memory-Efficient Neck Network

In the CSP module of YOLOv5s, traditional convolutions involve extensive memory access for each pixel, leading to redundant calculations and memory bottlenecks, particularly in deeper layers where the spatial resolution decreases. As noted by Chen *et al.* [1], Partial Convolution (PConv) mitigates this issue by performing calculations only on relevant regions of the input feature map, reducing unnecessary computations and improving processing speed.

Mathematically, the computational cost of PConv is given by: $K_w \times K_h \times C_{pi} \times W_o \times H_o \times C_{po}$, where C_{pi} and C_{po} represent the input and output channels, respectively, and K_w and K_h are the kernel dimensions. By setting the partial convolution ratio to $r = \frac{1}{4}$, the computational cost becomes $\frac{1}{16}$ of a standard convolution, significantly lowering both the number of calculations and memory usage. In the LA-YOLO model, we replace the CSP module with the P2C module, which incorporates PConv followed by two 1×1 convolutions. This design improves both computational efficiency and memory usage by addressing the limitations of traditional convolution operations. Additionally, a shortcut connection is included to reuse the input features, further enhancing the overall efficiency of the model.

2.3 A Dual-Head Adaptive Spatial Feature Fusion for Detection Head

Standard YOLOv5s employs multi-scale feature fusion via a Feature Pyramid Network (FPN) to handle features of varying sizes. However, fusing features across layers can introduce conflicting information, especially when detecting small targets in industrial applications. To mitigate this issue, we propose removing the output from the final layer of the network, which is primarily tuned for large-scale objects. This modification reduces computational complexity and prevents the fusion of irrelevant features, enabling a more focused approach for small defect detection. We introduce Dual-Head Adaptive Spatial Feature Fusion (DA) which refines the feature fusion process to focus on small-target features more effectively. In this method, the detection head is modified by eliminating the last layer's output. The updated feature representation is formulated as $z_{ij}^k = \alpha_{ij}^k \cdot x_{ij}^{1\to k} + \beta_{ij}^k \cdot x_{ij}^{2\to k}$, where z_{ij}^k represents the output vector of the kth feature level at position (i,j), and $x_{ij}^{1\to k}$ and $x_{ij}^{2\to k}$ denote the adjusted feature vectors from the first and second feature levels, respectively. The parameters α_{ij}^k and β_{ij}^k are learned weights that control the contributions from these two feature levels.

This streamlined feature fusion approach reduces model complexity, leading to faster inference times while significantly improving small-target detection performance. By enhancing the fusion of features, DA ensures that small defects, which are commonly encountered in industrial material inspection tasks, are detected efficiently and accurately.

3 Experiment

We evaluate the efficacy of LA-YOLO using two datasets: NEU-DET steel surface defect dataset [6] and PKU-Market-PCB dataset [12]. The NEU-DET dataset contains 1,800 grayscale images (200×200 pixels) across six defect types: crazing (Cr), inclusion (In), patch (Pa), pitted surface (Ps), rolled-in scale (Rs), and scratch (Sc), with 300 samples per category. The PKU-Market-PCB dataset includes 693 PCB defect images across six defect categories: 115 Missing Hole (Mh), 115 Mouse Bite (Mb), 116 Open Circuit (Oc), 116 Short Circuit (Sh), 115 Spur (Sp), and 116 Spurious Copper (Sc) defects. Both datasets are split into training, validation, and test sets in an 8:1:1 ratio. We trained and evaluated LA-YOLO on a Nvidia GeForce RTX 2080Ti GPU. The model was optimized with the SGD, using a batch size of 8 and an initial learning rate of 0.01 for 300 epochs. Four metrics were used for performance evaluation on validation set: mAP at an IoU threshold of 0.5, FPS, model size, and FLOPs.

3.1 Performance Comparison

NEU-DET Dataset. Table 1 shows that our proposed model achieves an excellent balance between detection accuracy, computational efficiency, and inference speed compared to existing YOLO-based methods. With a highest mAP of 0.873, our model outperforms all other YOLO variants, demonstrating superior defect detection precision. It surpasses YOLOv5s and EC-YOLO, further validating its effectiveness in defect detection. The performance gains are particularly pronounced across defect classes, where our model consistently delivers competitive accuracy.

Beyond accuracy, our model excels with a lowest FLOPs count of 3.1G, significantly outperforming YOLOv3, YOLOv5s, and YOLOv8s, reducing computational complexity for faster image processing and lower resource consumption. Furthermore, the model achieves an FPS of 92.57, ranking highly in comparison—slightly below YOLOv8s and EC-YOLO, but well above YOLOv3 and YOLOv5s. This ensures a balance of accuracy and speed, making our model highly suitable for real-time deployment in industrial applications where rapid processing is critical.

Therefore, the experimental results demonstrate that our model not only outperforms existing YOLO models in terms of accuracy but also maintains a lightweight structure with minimal computational demands. These attributes make it a promising solution for real-time defect detection in industrial environments, where both precision and efficiency are paramount.

PKU-Market-PCB Dataset. Table 2 reports a detailed performance comparison of LA-YOLO with various YOLO series models and state-of-the-art models on the PKU-Market-PCB dataset in terms of FPS and mAP. With a mAP of 0.951, our model surpasses all other methods while maintaining a high inference speed of 77.25 FPS, making it both highly accurate and efficient. This combination of accuracy and speed is particularly advantageous for real-time defect detection applications.

When compared to YOLO-based models, our approach consistently outperforms even the latest versions. For example, YOLOv12 achieves an mAP of 0.891, YOLOv11 reaches 0.877, and YOLO-World records 0.851, all of which fall short of our model's accuracy. Notably, YOLOv11 achieves the highest FPS at 78.12, slightly exceeding ours, but at the cost of lower detection accuracy. Additionally, our model demonstrates strong robustness across all defect classes, with particularly high performance in the Mouse bite (Mb) class and Spur (Sp) class. In comparison to transformer-based and two-stage models, our method remains highly competitive. Although Mask DINO with SwinL achieves a relatively high mAP of 0.942, its significantly lower inference speed of 12.6 FPS makes it impractical for real-time applications. Similarly, ResNet-50+FPN and Rev-RetinaNet reach mAPs of 0.915 and 0.897, respectively, but lack the necessary efficiency for rapid defect detection.

The experiment results on two datasets demonstrate that our model effectively balances accuracy and efficiency, which is crucial for real-time industrial

Table 1. The comparison of model accuracy and efficiency with other YOLO series models on NEU-DET database. Best result is in bold. NP, M, and G mean the number of parameters, megabyte, and giga, respectively. We re-implement the YOLO algorithms by using the released code on our own computation platform.

Method	NP (M)	FLOPs (G)	FPS	mAP	Class					
					Cr	In	Pa	Ps	Rs	Sc
YOLOv3 [15]	61.55	155.3	22.84	0.827	0.711	0.817	0.919	0.897	0.785	0.835
YOLOv5n	1.77	4.2	92.68	0.76	0.394	0.876	0.929	0.741	0.76	0.861
YOLOv5s	7.04	16	88.37	0.862	0.845	0.875	0.967	0.661	0.944	0.881
YOLOv8s	11.17	28.6	**106.38**	0.763	0.387	0.874	0.954	0.864	0.626	0.874
YOLOv9 [21]	7.2	26.7	72.79	0.728	0.371	0.784	0.925	0.78	0.635	0.874
YOLOv10 [20]	8.07	24.8	78.28	0.773	0.486	0.827	0.923	0.815	0.636	0.954
YOLOv11 [8]	9.42	21.3	98.04	0.806	0.495	0.935	0.981	0.707	0.757	0.962
YOLOv12 [19]	9.26	21.5	67.45	0.826	0.486	0.958	0.980	0.754	0.824	0.955
YOLO-World [3]	13.38	38.4	51.37	0.769	0.472	0.815	0.967	0.831	0.569	0.959
RDD-YOLO [24]	–	–	–	0.811	0.59	0.859	0.944	0.862	0.707	0.966
MD-YOLO [25]	9.0	14.1	59.1	0.782	0.467	0.814	0.913	0.851	0.726	0.920
DsP-YOLO [23]	11.13	28.5	86.9	0.804	0.545	0.840	0.950	0.821	0.727	0.941
EC-YOLO [4]	–	–	91.74	0.83	0.57	0.87	0.93	0.90	0.77	0.97
Ours	7.60	**3.1**	90.57	**0.873**	0.641	0.903	0.951	0.947	0.803	0.901

defect detection. While other high-accuracy models like Trans-YOLO achieve competitive results, their inference speeds are insufficient for practical real-time deployment. As such, our model outperforms existing state-of-the-art methods in both detection accuracy and inference speed, making it an ideal solution for real-world deployment.

3.2 Ablation Study

Table 3 reports the ablation study on three key networks including hybird backbone, memory-efficient neck, and detection head in improving the performance of the proposed model. It is observed that each module plays a distinct role in enhancing the model's accuracy and efficiency, with DA standing out as the most impact component.

Hybrid Backbone. It is a lightweight convolutional block designed to reduce the model's parameter count and computational cost. By replacing standard convolutions with efficient Mobile Inverted Bottleneck Convolutions, this module helps reduce the model size and FLOPs significantly, making it computationally efficient. However, its contribution to accuracy is limited, as seen in Case 2, where the model with only Fused MBConv suffers a decrease in mAP from 0.862

Table 2. Performance comparison of our model with other state-of-the-art models on PKU-Market-PCB dataset. The best results are in bold. The asterisks indicate our re-implementation based on the open source code.

Method	FPS	mAP	Class					
			Mh	Mb	Oc	Sh	Sp	Sc
YOLOv3 [15]*	54.51	0.893	0.985	0.895	0.867	0.925	0.790	0.901
YOLOv5s*	73.71	0.798	0.981	0.823	0.723	0.857	0.618	0.785
YOLOv8s*	74.36	0.813	0.980	0.834	0.719	0.875	0.670	0.798
YOLOv9 [21]*	68.28	0.808	0.968	0.816	0.736	0.838	0.674	0.776
YOLOv10 [20]*	71.4	0.819	0.981	0.795	0.768	0.854	0.716	0.799
YOLOv11 [8]*	**78.12**	0.877	0.968	0.895	0.811	0.944	0.780	0.793
YOLOv12 [19]*	62.52	0.891	0.995	0.930	0.878	0.944	0.746	0.855
YOLO-World [3]*	52.86	0.851	0.967	0.852	0.786	0.883	0.755	0.867
Trans-YOLO [2]	21	0.944	–	–	–	–	–	–
Rev-RetinaNet [18]	–	0.897	0.994	0.849	0.877	0.904	0.864	0.893
ResNet-50+FPN [9]	3.12	0.915	0.942	0.908	0.894	0.956	0.913	0.916
KD-LightNet [11]	35	0.888	–	–	–	–	–	–
Mask DINO with ResNet50 [10]*	5.7	0.926	0.988	0.935	0.948	0.892	0.886	0.909
Mask DINO with SwinL [10]*	12.6	0.942	0.988	0.965	0.955	0.908	0.880	0.954
Ours	77.25	**0.951**	0.975	0.988	0.905	0.933	0.883	0.927

to 0.823. This suggests that while it offers efficiency benefits, it alone cannot effectively capture the complex features required for accurate defect detection.

Memory-Efficient Neck. It addresses challenges related to incomplete or occluded input features, which is common in defect detection tasks. By learning to mask out invalid regions during convolution, it allows the model to focus on valid information, improving feature extraction. In Case 3, the introduction of PConv results in an mAP of 0.861, which is slightly higher than Fused MBConv alone but still falls short of the best-performing models. While PConv helps improve feature extraction, it does not fully exploit the potential of adaptive feature fusion, which limits its contribution in terms of detection accuracy.

DA Module. It is a key innovation in our model, significantly improving detection accuracy. It enables adaptive fusion of spatial features from different branches, allowing the network to focus on relevant regions and improving global context awareness. In Case 4, the model incorporating the DA module achieves the highest mAP of 0.912, showcasing its ability to enhance the model's feature fusion capacity, especially for surface defect detection. By enabling the model to better capture complex defect features, which are often spread across different spatial regions, the DA module leads to a notable accuracy increase.

Table 3. Ablation Study Results on NEU-DET dataset, where HB and ME are hybird backbone and memory-efficient neck, respectively.

Case	Module			Params (M)	FLOPs (G)	mAP	FPS
	HB	ME	DA				
1				7.04	16.6	0.862	88.37
2	✓			4.27	5.5	0.823	90.84
3		✓		6.66	15.4	0.861	80.11
4			✓	10.77	15.1	0.912	102.96
5	✓	✓		2.16	3.1	0.840	86.84
6	✓		✓	7.72	5.5	0.881	92.21
7	✓	✓	✓	7.60	3.1	0.873	92.57

Moreover, the DA mechanism maintains computational efficiency, as evidenced by its low FLOPs and high FPS in Case 4. When combined with other modules, as in Case 7, the model strikes a strong balance between accuracy and efficiency, achieving 0.873 mAP and 92.57 FPS. The DA module's ability to improve accuracy while minimizing computational cost contributes to an optimized solution for real-time defect detection.

4 Conclusion

In this paper, we presented LA YOLO, a lightweight architecture specifically designed to balance accuracy and computational efficiency for surface defect detection in industrial quality control. By replacing the conventional Conv2d-BatchNorm-SiLU structure with a series of connections leveraging Mobile Inverted Bottleneck Convolution, we improved the backbone's efficiency while enabling hierarchical feature learning. To further reduce computational cost, we integrated Partial Convolution with standard convolution operations, leading to a significant reduction in FLOPs without compromising detection performance. Additionally, we introduced a Dual-Head Adaptive Spatial Fusion mechanism in the detection head, which enhances feature aggregation and defect localization, particularly for small-sized defects. Extensive experiments on two industrial defect datasets demonstrated the superiority of LA-YOLO. LA-YOLO achieved a significant reduction in FLOPs and notable improvements in inference speed, making it particularly well-suited for deployment on resource-constrained edge devices.

Acknowledgement. This research was in part supported by the Basic Science (Natural Science) Research Project of Higher Education Institutions in Jiangsu Province under Grant 24KJA520003, in part by the 333 high-level talents in Jiangsu Province (2024), in part by the National Natural Science Foundation of China under Grant No. 62076122.

References

1. Chen, J., et al.: Run, don't walk: chasing higher flops for faster neural networks. In: Proceedings of CVPR, pp. 12021–12031 (2023)
2. Chen, W., Huang, Z., Mu, Q., Sun, Y.: PCB defect detection method based on transformer-yolo. IEEE Access **10**, 129480–129489 (2022)
3. Cheng, T., Song, L., Ge, Y., Liu, W., Wang, X., Shan, Y.: YOLO-world: real-time open-vocabulary object detection. arXiv preprint: arXiv:2401.17270 (2024)
4. Cheng, Z., Gao, L., Wang, Y., Deng, Z., Tao, Y.: EC-YOLO: effectual detection model for steel strip surface defects based on YOLO-v5. IEEE Access (2024)
5. Gupta, S., Akin, B.: Accelerator-aware neural network design using AutoML. arXiv preprint: arXiv:2003.02838 (2020)
6. He, Y., Song, K., Meng, Q., Yan, Y.: An end-to-end steel surface defect detection approach via fusing multiple hierarchical features. IEEE Trans. Instrum. Meas. **69**(4), 1493–1504 (2019)
7. Jocher, G.: YOLOv5 by ultralytics (2020). https://doi.org/10.5281/zenodo.3908559, https://github.com/ultralytics/yolov5
8. Khanam, R., Hussain, M.: YOLOv11: an overview of the key architectural enhancements. arXiv preprint: arXiv:2410.17725 (2024)
9. Li, C.J., Qu, Z., Wang, S.Y., Bao, K.H., Wang, S.Y.: A method of defect detection for focal hard samples PCB based on extended FPN model. IEEE Trans. Comp. Pack. Man. **12**(2), 217–227 (2021)
10. Li, F., et al.: Mask DINO: towards a unified transformer-based framework for object detection and segmentation. In: Proceedings of CVPR, pp. 3041–3050 (2023)
11. Liu, J., Li, H., Zuo, F., Zhao, Z., Lu, S.: KD-LightNet: a lightweight network based on knowledge distillation for industrial defect detection. IEEE Trans. Instrum. Meas. **72**, 1–13 (2023)
12. Liu, Q., Huang, X., Shao, X., Hao, F.: Industrial cylinder liner defect detection using a transformer with a block division and mask mechanism. Sci. Rep. **12**(1), 10689 (2022)
13. Liu, S., Jia, M.: An adaptive shunt model for steel defect detection based on YOLOx. In: Proceeding of ITNEC, vol. 6, pp. 950–954. IEEE (2023)
14. Redmon, J., Divvala, S., Girshick, R., Farhadi, A.: You only look once: unified, real-time object detection. In: Proceedings of CVPR, pp. 779–788 (2016)
15. Redmon, J., Farhadi, A.: YOLOv3: an incremental improvement. arXiv preprint: arXiv:1804.02767 (2018)
16. Sandler, M., Howard, A., Zhu, M., Zhmoginov, A., Chen, L.C.: MobileNetV2: inverted residuals and linear bottlenecks. In: Proceedings of CVPR, pp. 4510–4520 (2018)
17. Singh, S.A., Desai, K.A.: Automated surface defect detection framework using machine vision and convolutional neural networks. J. Intell. Manuf. **34**(4), 1995–2011 (2023)
18. Tang, J., Zhao, Y., Bai, D., Liu, Q.: Rev-RetinaNet: PCB defect detection algorithm based on improved RetinaNet. In: Proceedings of International Conference on Electrical Engineering, Big Data and Algorithms, pp. 653–658 (2023)
19. Tian, Y., Ye, Q., Doermann, D.: YOLOv12: attention-centric real-time object detectors. arXiv preprint: arXiv:2502.12524 (2025)
20. Wang, A., et al.: YOLOv10: real-time end-to-end object detection. arXiv preprint: arXiv:2405.14458 (2024)

21. Wang, C.Y., Yeh, I.H., Liao, H.Y.M.: YOLOv9: learning what you want to learn using programmable gradient information. arXiv preprint: arXiv:2402.13616 (2024)
22. Weng, Y., Xiao, J., Xia, Y.: Strip steel surface defect detection based on improved mask R-CNN algorithm. Comput. Eng. Appl. **57**, 235–242 (2021)
23. Zhang, Y., Zhang, H., Huang, Q., Han, Y., Zhao, M.: DsP-YOLO: an anchor-free network with DsPAN for small object detection of multiscale defects. Expert Syst. Appl. **241**, 122669 (2024)
24. Zhao, C., Shu, X., Yan, X., Zuo, X., Zhu, F.: RDD-YOLO: a modified YOLO for detection of steel surface defects. Measurement **214**, 112776 (2023)
25. Zheng, H., Chen, X., Cheng, H., Du, Y., Jiang, Z.: MD-YOLO: surface defect detector for industrial complex environments. Opt. Lasers Eng. **178**, 108170 (2024)

An Intelligent Detection Method for Factory Security Risks by Fusing Gradient Flow Features

Fuan Dong[1], Xiaoning Wu[1], Dong Gao[1], Ruixin Li[1], Li Sun[1], Pengyu Liu[2(✉)], and Lele Yuan[2]

[1] North Automatic Control Technology Institute, Beijing, China
[2] School of Information Science and Technology, Beijing University of Technology, Beijing, China
liupengyu@bjut.edu.cn

Abstract. To tackle challenges like small distant targets and occlusion in factory security detection, this paper proposes a YOLOX-based algorithm combining a gradient flow module and coordinate attention. The method enhances fine-grained feature extraction and global context modeling. A BiFPN structure is used for effective multi-scale feature fusion. Experiments in real factory scenarios show that the proposed method achieves 94.78% accuracy and outperforms mainstream detection algorithms in performance and robustness.

Keywords: Security risk detection · Coordinate attention mechanism · Feature extraction · BiFPN

1 Introduction

Enhancing safety risk detection in facilities through informatization and intelligent technologies is essential for modern safety supervision. The daily operation of large motors, transformers, and other equipment poses safety risks, making helmet detection, personnel intrusion detection, and pyrotechnic detection in key areas crucial. Manual inspections are inefficient and lack real-time capabilities, underscoring the urgent need for automated safety risk detection technologies to ensure smooth factory operation.

Current research on security risk detection in industrial park environments mainly includes traditional methods and deep learning approaches. Traditional techniques such as edge detection, feature association, and SVM have been applied to target detection tasks. For example, Rubaiya and Silva et al. used gradient histograms and frequency domain features with circular Hough transform to detect helmet use [1], while Dalal and Triggs proposed the HOG-SVM method for pedestrian detection [2]. However, due to their reliance on hand-crafted features, these methods often struggle in complex factory environments.

With the rise of deep learning, more advanced methods have emerged. Sun et al. proposed the HWD-YOLO model with multi-scale context aggregation and attention

P. Umapada et al. (Eds.): ICCPR 2025, CCIS 2811, pp. 332–343, 2026.
https://doi.org/10.1007/978-981-95-8315-7_27

mechanisms, improving mAP by 3.4% and speed from 17 fps to 33 fps based on YOLOv5X [3]. Jia et al. introduced a lightweight detection model optimized for edge devices [4]. Further improvements based on YOLOv8 combined with edge computing achieve real-time detection [5]. For fire and smoke detection, Wang et al. utilized multi-source data fusion and deep features [6], while Lee et al. designed a small-target enhancement module for better flame detection [7]. In personnel intrusion scenarios, Wang et al. applied a CNN-LSTM hybrid model to improve accuracy and timeliness [8]. These approaches significantly enhance detection performance and robustness in complex industrial environments.

To address limited feature extraction and environmental interference in factory security risk detection, this paper proposes a YOLOX-based algorithm enhanced with a gradient flow module and coordinate attention mechanism. A BiFPN structure is adopted for multi-scale feature fusion. Experiments show that the proposed method outperforms mainstream detection algorithms in factory scenarios, demonstrating superior performance and robustness.

2 Security Risk Detection Strategies for Industrial Facilities

This paper constructs a six-category safety risk detection dataset, enhanced with data augmentation for better generalization. The backbone network is improved using a C2F_CA module and coordinate attention to strengthen feature extraction, while replacing the original SPP module with SimSPPF to reduce computational complexity. A BiFPN structure in the feature fusion network efficiently integrates multi-scale information. The decoupled prediction head of YOLOX combines classification loss and CIOU regression loss for improved accuracy and localization. The proposed method balances accuracy and real-time performance, making it suitable for complex industrial and safety monitoring applications.

2.1 Security Risk Detection Dataset

The dataset comprises 3 500 field images captured under factory surveillance. To ensure robustness across real-world variations, we deliberately sampled under three lighting conditions (daylight, indoor LED, low-illuminance night) and from five camera angles (0°, ± 20°, ± 45°). Targets appear at scales ranging from 15 pixels (distant smoke) to 600 pixels (close helmet) and include partial occlusions (workers behind machinery) and inter-object closure (flames overlapping with smoke). Data augmentation further enlarges the diversity: random horizontal/vertical flips, rotations, scaling, translations, and mosaic cropping. All images are annotated with LabelImg and split 8: 1: 1, resized and padded to a uniform 640 × 640 input.

Key challenge scenarios are exemplified in Fig. 1: (a) Equipment obstruction - worker behind machine; (b) personnel obstruction - overlapping helmets; (c) lighting interference - LED glare on helmet; (d) small targets at long distances - 15-pixel helmet detection.

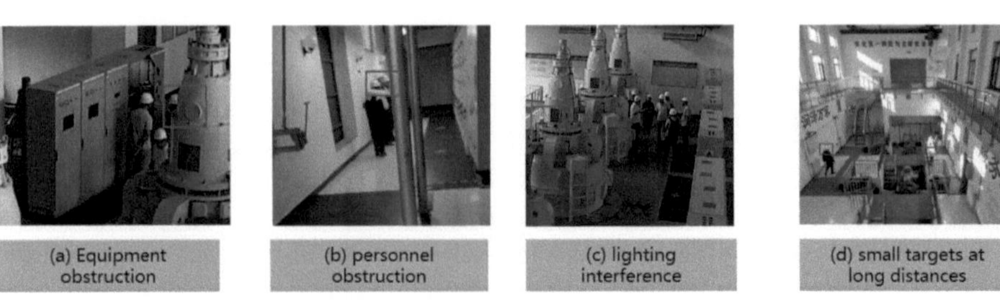

Fig. 1. Key challenge scenarios

2.2 Improved Security Risk Detection Network

The target detection network proposed in this paper mainly consists of four parts: input, backbone network, feature fusion network and prediction network, which are responsible for the preprocessing of the input image, feature extraction, feature fusion, and prediction of the target location and category, respectively. The gradient flow feature extraction module and coordinate attention mechanism are introduced into the feature extraction network, and the BiFPN structure is used in the feature fusion network to improve the fusion effect of target features of different sizes, and the specific network structure is shown in Fig. 2.

Improved Backbone Network

The backbone network is responsible for the feature extraction of the input image in the target detection model, and the phenomena of misdetection and omission frequently occurring in the security risk detection task are directly related to the feature extraction capability of the feature extraction network [7]. The improved model replaces the Res feature extraction module in the original YOLOX model with the C2F_CA module, which has stronger feature extraction capability, and the structure is shown in Fig. 3.

The original Res module integrates CSPNet's streaming concept with residual structures. It consists of two branches: the main branch applies CBS followed by multiple stacked Bottlenecks to downsample feature maps and reduce parameters, while the other branch undergoes CBS and is concatenated with the main branch output. However, the Res module in [9] only retains the last Bottleneck's output, ignoring intermediate feature layers during dimensionality reduction.

This paper proposes the C2F_CA module (Fig. 3b) inspired by ELAN. By removing sub-branches, it stacks Bottlenecks in parallel after CBS/Split operations and concatenates outputs to preserve gradient flow. Compared to Res, C2F_CA enhances feature extraction without extra computation [5–7].

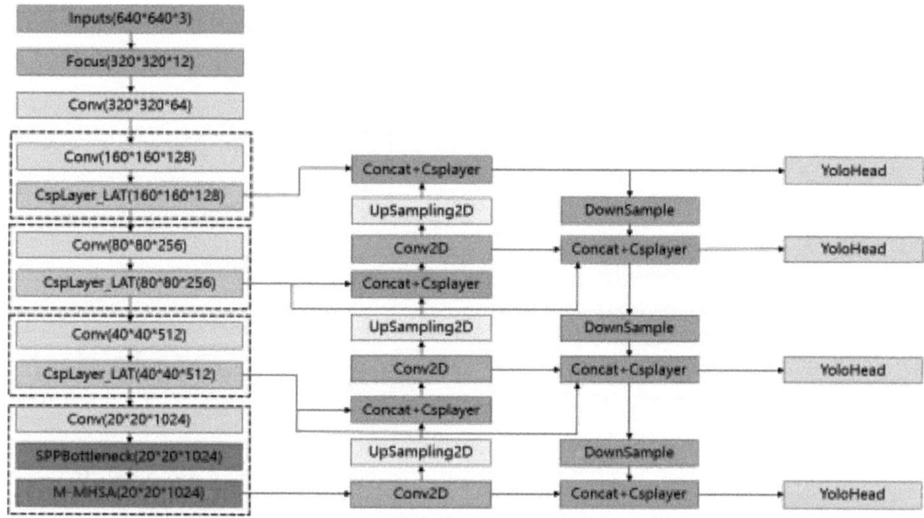

Fig. 2. Structure diagram of improved security risk detection network

In deep networks, channels encode diverse semantic information; modeling inter-channel relationships is vital for detection accuracy. We adopt the Coordinate Attention (CA) mechanism, which combines channel attention with spatial information via global average pooling along two spatial dimensions, capturing long-range dependencies and generating direction- and location-sensitive attention maps [9]. CA has been proven effective in lightweight detection models, enhancing performance without significant computational overhead [10–12]. Its structure is illustrated in Fig. 4.

Feature map X (C' × W' × H') is convolutionally mapped to U (C × W × H) for enhanced spatial receptive field [9]. U splits into two branches: one performs squeeze-excitation via global pooling to capture channel relationships; the other recalibrates features by channel-wise multiplication with attention vectors [10]. Compared to traditional pooling, CA prioritizes strong semantic channels, improving feature extraction.

In order to improve the detection speed of the factory security risk detection algorithm, this paper chooses to use the faster computational SimSPPF module to replace the SPP module in the original YOLOX-s [13], and the specific structure of the two modules is shown in Fig. 5.

The SPP module uses parallel max pooling (kernels: 5, 9, 13) plus identity branch, concatenating results for feature fusion [13]. SimSPPF decomposes large kernels (9/13) into sequential 5 × 5 max poolings, reducing operations while maintaining accuracy. Replacing SiLU with ReLU further lowers computation, easing mobile deployment [14].

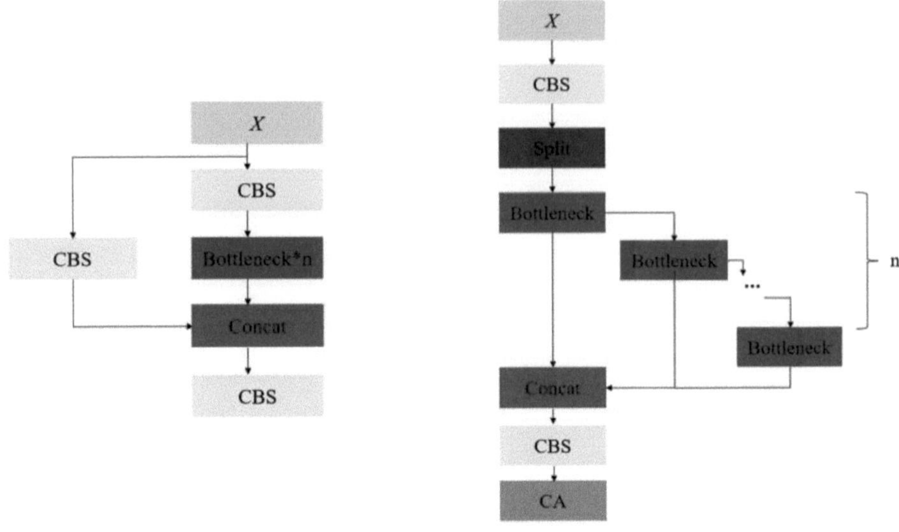

Fig. 3. Structure comparison diagram of feature extraction module

After ablating four lightweight variants, C2F_CA + SimSPPF achieved + 1.8% mAP over SE/CBAM with only + 0.3M parameters, as parallel Bottlenecks preserve gradients. Coordinate Attention's direction-aware encoding boosted small-object recall by + 2.4%. SimSPPF decomposed max-pooling into three 5 × 5 serial ops, improving FPS by 18% on Jetson Xavier NX versus ASFF, with unchanged accuracy. All modules are plug-and-play, hyperparameter-free, and deployment-friendly.

Improved Neck Network

The feature fusion network of the original YOLOX-s algorithm utilizes a FPN and PANet are combined to construct the feature pyramid structure, the specific structure is shown in Fig. 6(a). FPN transfers the strong semantic information from the deep layer to the shallow features, [14] while PANet transfers the strong location information from the shallow features to the deep features, and the combination of FPN and PANet realizes the parameter aggregation of the detection layers of different sizes, and ultimately realizes the feature fusion between different layers. The input of PA-Net is the feature information processed by FPN and lacks the original feature information extracted by the backbone network, which may lead to learning bias and feature information loss in feature fusion. Therefore, the G-YOLOX algorithm proposed in this paper adopts the BiFPN structure [13, 15] to improve the feature pyramid, and the specific structure is shown in Fig. 6(b).

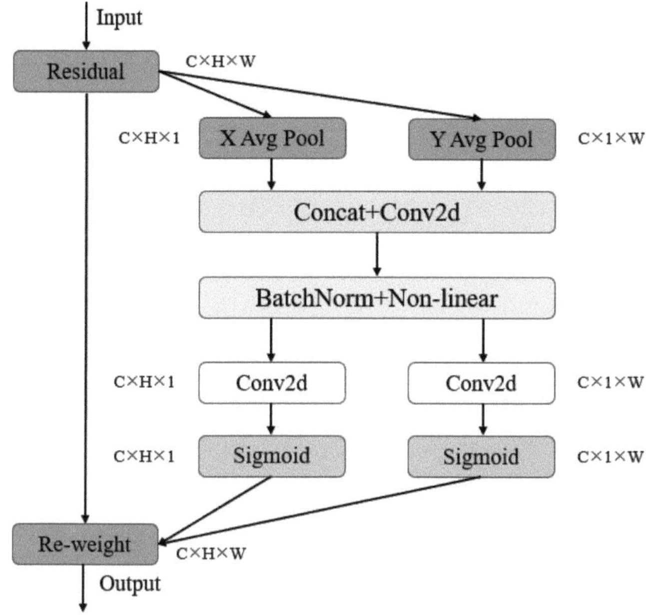

Fig. 4. Structure diagram of the structure of coordinate attention mechanism

BiFPN was chosen over FPN + PANet, NAS-FPN, and ASFF for: backbone + FPN feature reuse recovering PANet-lost small objects (+2.1% mAP); learnable weights reducing NAS-FPN parameters by 34% and latency by 3.7 ms; and direct YOLOX neck integration without search/tuning.

2.3 Predictive Network Optimization

The prediction header of H-YOLOX still follows the decoupled header structure of YOLOX-s, and the loss function is mainly divided into two parts: classification loss and regression loss, the classification loss adopts BCELoss, and the regression loss consists of CIOU_ Loss, and the final loss is obtained by the weighted calculation of the above losses. The final loss is weighted by the above losses. The calculation method of BCELoss is shown in the following formula:

$$\text{BCELoss} = \begin{cases} -\log p', & y = 1 \\ -\log(1 - p'), & y = 0 \end{cases} \tag{1}$$

where, P′-corresponds to the predicted value; y-determines whether it is the label of the real category or not, y = 1 means that the target predicted by the model is the target of the corresponding real category, and y = 0 means that the target predicted by the model is not the corresponding real target. The calculation of the CIOU_ Loss is shown as follows:

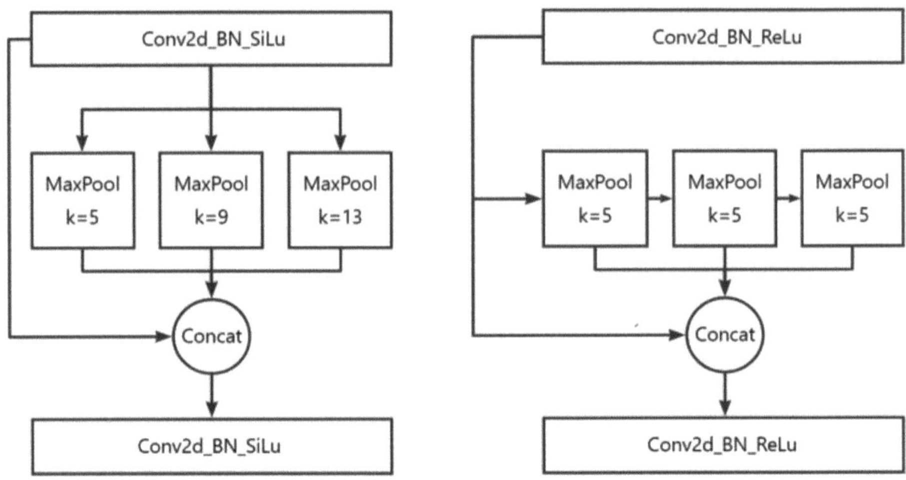

Fig. 5. Structure comparison between SPP module and SimSPPF module

$$L_{CIOU} = 1 - IOU + \frac{\rho^2(b, b^{gt})}{c^2} + \alpha v \tag{2}$$

where, ρ-Euclidean distance; b, b^{gt} -centers of the prediction frame, real frame; c-diagonal distance between the prediction frame and real frame; α-weighting coefficient; v-measured aspect ratio.

$$IOU = \frac{A \cap B}{A \cup B} \tag{3}$$

where, A-true frame; B-predicted frame; IOU-criteria for determining the degree of overlap between the true and predicted frames.

3 Experimental Results and Analysis

In this paper, experiments are conducted on Ubuntu 18.04 with an Intel i7-10875 CPU and an NVIDIA GeForce RTX 3080 GPU. The model is trained using pre-trained weights and a transfer learning strategy to accelerate convergence. The training parameters are set as follows: learning rate of 0.0005, batch size of 32, 100 epochs, and the Adam optimizer.

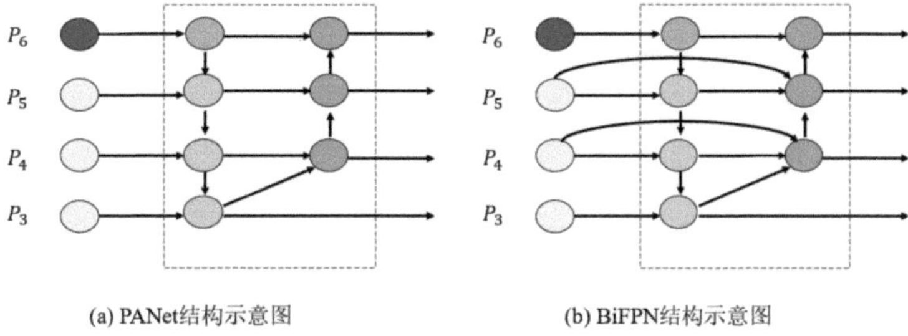

Fig. 6. Comparison of feature pyramid structure

3.1 Evaluation Indicators

For evaluation, we adopt Precision, Recall, and mean average precision (mAP), which are commonly used in object detection. Precision measures the proportion of correctly identified positive samples among all predicted positives, while Recall measures the proportion of correctly identified positive samples among all actual positives [14]. The specific calculations are shown in Eqs. (4)–(5).

$$Precision = \frac{TP}{TP + FP} \tag{4}$$

$$Recall = \frac{TP}{TP + FN} \tag{5}$$

where TP-model predicts positive cases and true labeling is positive; FP-model predicts positive cases and true labeling is negative; FN-model predicts negative cases and true labeling is positive.AP value is used to assess the prediction effect of a certain type of target, [14] is generally Precision value is in the interval.The AP value is used to evaluate the prediction effect of a certain type of target, [14] is generally the value obtained by integrating Precision's value in the interval (0, 1) to Recall, and mAP is the average value of AP for multiple types of targets, which is calculated as shown in Eqs. (6)–(7).

$$AP = \int_0^1 \frac{TP}{TP + FP} \lceil (Recall) \tag{6}$$

$$mAP = \frac{1}{n} \sum_{i=1}^{n} AP_i \tag{7}$$

where, n-number of target categories; i-serial number of current category.

3.2 Ablation Experiments

To validate the effectiveness of the proposed method, the original YOLOX is used as a baseline, and the modules developed in this paper are added incrementally.

Table ?? shows the detailed results. The original YOLOX achieves 89.21% precision, 84.17% recall, and an mAP of 86.91% on this dataset. Incorporating the modules proposed in this paper further improves these metrics, with the final algorithm reaching an mAP of 93.5%, demonstrating the feasibility of the proposed method for safety risk detection in the pumping station scenario.

Model	Precision/%	Recall/%	mAP/%
YOLOX	89.21	84.17	86.91
+ C2F_CA	92.37	85.18	88.07
+ SimSPPF	92.01	86.33	88.49
+ BiFPN	91.24	84.97	87.12
+ C2F_CA + SimSPPF	94.12	89.41	91.63
+ C2F_CA + SimSPPF + BiFPN	96.01	90.13	93.50

To comprehensively evaluate the effectiveness and robustness of the proposed H-YOLOX algorithm in complex factory scenarios, both qualitative and quantitative comparison experiments are conducted. The performance of H-YOLOX is compared against mainstream object detection models including YOLOv5, YOLOX, and YOLOv7 under consistent experimental settings. Evaluation metrics such as visual detection results and mean Average Precision (mAP) are used to demonstrate the superiority of the proposed method in safety risk detection tasks.

Qualitative Inorganic Analysis

To validate the detection effectiveness of the proposed H-YOLOX algorithm, it is qualitatively compared with three commonly used object detection models—YOLOv5, YOLOX, and YOLOv7—under the same dataset, experimental environment, and hyperparameter settings. All models are trained for 100 epochs using the security risk detection dataset constructed in this paper. As shown in Fig. 7, the proposed algorithm demonstrates superior performance in helmet detection, unauthorized personnel intrusion, and pyrotechnic detection. In contrast, other algorithms exhibit missed detections under the same conditions, indicating that H-YOLOX achieves more accurate and reliable detection in complex factory scenarios.

Quantitative Analysis

To further assess the performance of the proposed method, we use mean Average Precision (mAP) as the primary evaluation metric. The proposed H-YOLOX model achieves an mAP of 93.5%, and the change in loss across training epochs is illustrated in Fig. 8. Compared with YOLOv5, YOLOX, and YOLOv7, the proposed model shows notable improvements in detection accuracy. The performance comparison results are summarized in Table 1 further confirming the effectiveness of H-YOLOX in addressing safety risk detection challenges in complex environments.

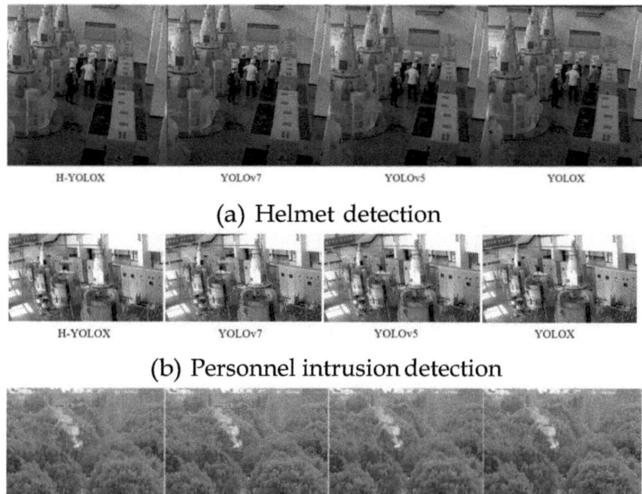

(c) Pyrotechnic detection

Fig. 7. Comparison of effects of different algorithms

Table 1. Comparison with mainstream object detection models

Model	mAP (%)	FPS
YOLOv5	85.69	29.2
YOLOX	86.91	27.9
YOLOv7	90.81	31.4
Proposed Method	94.78	35.1

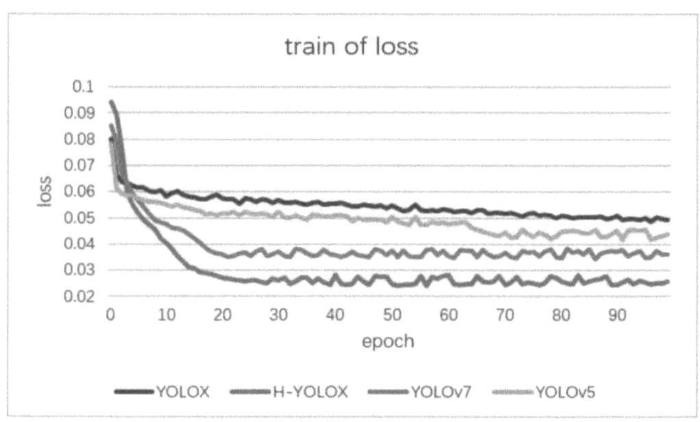

Fig. 8. Training loss line chart

4 Conclusion

To resolve low detection accuracy from small, distant targets and occlusion in factory environments, we propose a YOLOX-based algorithm integrating gradient flow feature extraction and coordinate attention for enhanced feature capture,and employs a BiFPN structure for improved multi-scale feature fusion.Experimental results demonstrate that the proposed method significantly enhances detection accuracy in factory environments, achieving a mAP of 94.78%. But limitations persist:

1. Missed detections for targets < 8px under heavy occlusion
2. Model size incompatible with < 4GB RAM edge devices
3. Untested in low-light/night conditions

Future work will:

(i) Deformable convs & transformers for occlusion
(ii) Compress to < 1 MB via distillation/pruning
(iii) Add night/fog datasets for robustness

References

1. Rubaiyat, A.H.M., et al.: Automatic detection of helmet uses for construction safety. In: Proceedings of the IEEE/WIC/ACM International Conference on Web Intelligence Workshops (WIW), pp. 135–142 (2016)
2. Dalal, N., Triggs, B.: Histograms of oriented gradients for human detection. In: Proceedings of the IEEE Conference on Computer Vision and Pattern Recognition (CVPR), vol. 1, pp. 886–893 (2005)
3. Sun, L., Li, H., Wang, L.: HWD-YOLO: a new vision-based helmetwearing detection method. Comput. Mater. Continua **80**(3) (2024)
4. Jia, X., et al.: High-precision and lightweight model for rapid safety helmet detection. Sensors **24**(21), 6985 (2024)
5. Thanh, H.D.N., Thien, P.N.N., Duc, N.P., Tuan, A.N.P., Nguyen, L.D.: Enhancing helmet violation detection and license plate recognition through optimization of YOLOv8 models with edge computing integration. In: Proceedings of the 9th International Conference on Intelligent Information Technology, pp. 82–85 (2024)
6. Wang, S., et al.: An advanced multi-source data fusion method utilizing deep learning techniques for fire detection. Eng. Appl. Artif. Intell. **142**, 109902 (2025)
7. Lee, J.-H., Cheng, C.-B.: Smoke emission detection by deep learning. In: Proceedings of the International Conference on Soft Methods in Probability and Statistics, pp. 241–248 (2024)
8. Wang, J., Si, C., Wang, Z., Fu, Q.: A new industrial intrusion detection method based on CNN-BiLSTM. Comput. Mater. Continua **79**(3) (2024)
9. Jiang, M., Luo, J., Zhao, H., Huang, X.: Improved dense residual network with the coordinate and pixel attention mechanisms for helmet detection. Int. J. Mach. Learn. Cybern. **15**(11), 5015–5031 (2024)
10. Huang, D., Yating, T., Zhang, Z., Ye, Z.: A lightweight vehicle detection method fusing GSConv and coordinate attention mechanism. Sensors **24**(8), 2394 (2024)
11. Jun, W., Dong, J., Nie, W., Ye, Z.: A lightweight YOLOv5 optimization of coordinate attention. Appl. Sci. **13**(3), 1746 (2023)

12. Doherty, J., Gardiner, B., Kerr, E., Siddique, N.: BiFPN-YOLO: one-stage object detection integrating bi-directional feature pyramid networks. Pattern Recogn. **160**, 111209 (2025)
13. Liu, S., Shao, F., Chu, W., Dai, J., Zhang, H.: An improved YOLOv8-based lightweight attention mechanism for cross-scale feature fusion. Remote Sens. **17**(6), 1044 (2025)
14. Chen, C., Li, B.: A transform module to enhance lightweight attention by expanding receptive field. Available at SSRN 4329707 (2024)

A CAN Bus Anomaly Detection Method Based on Time Series Prediction

Weiping Yao(✉), Qingbao Li, Zhifeng Chen, and Bocheng Xu

Information Engineering University, Zhengzhou, China
weiping129@163.com

Abstract. To address the high computational cost of feature extraction from CAN bus voltage signals of anomaly detection, the paper introduces a method that generates node-specific hardware fingerprints by extracting distinctive voltage subsequences, accurately capture the unique hardware characteristics of nodes. Furthermore, an anomaly detection model using a Long Short-Term Memory (LSTM) network for time series prediction is proposed, which effectively identifies the source nodes of CAN frames, detects anomalous frames, and discriminates attacks initiated by malicious or unauthorized nodes. Evaluations demonstrated that the proposed approach achieves a source identification accuracy of 99.10% in actual vehicles and 99.88% under low sampling rates in the prototype. Comparative results indicate that the method not only outperforms existing advanced techniques but also offers benefits such as low computational overhead and a reduced false positive rate, highlighting its strong practical applicability and potential for widespread adoption.

Keywords: Controller Area Network (CAN) · time series · anomaly detection · voltage fingerprint · Long Short-Term Memory (LSTM)

1 Introduction

With the development of artificial intelligence and communication technologies, modern vehicles are becoming increasingly intelligent and interconnected. Automobiles are gradually evolving into Software Defined Vehicles (SDVs). While intelligence and connectivity bring convenience and comfort, they also introduce more attack surfaces, making vehicles more vulnerable to external cyber threats. In 2023, as many as 295 attack incidents were disclosed, 95% of which were initiated remotely [46]. The software in modern vehicles runs on 70 to 100 microprocessor-based electronic control units (ECUs), which are interconnected via in-vehicle networks [9]. Given that CAN remains the de facto standard in the automotive industry and is the mainstream in-vehicle network for current commercial vehicles, most research on in-vehicle network security focuses on CAN. since CAN communicates via broadcast and lacks encryption and authentication mechanisms, it is susceptible to cyber attacks. Various mechanisms have been proposed to enhance safety and protect against threats. Key measures include

P. Umapada et al. (Eds.): ICCPR 2025, CCIS 2811, pp. 344–365, 2026.
https://doi.org/10.1007/978-981-95-8315-7_28

encryption [13,39], authentication [41,45,55], and intrusion detection [47,56,57]. Due to the limited computational resources of ECUs, implementing encryption and authentication technologies presents challenges related to compatibility and performance. For practical reasons, including operability and cost-effectiveness, intrusion detection and similar measures are more feasible for in-vehicle networks [30].Many studies leverage the CAN bus protocol, collecting physical layer or data link layer data such as bus voltage, message transmission frequency, inter-message time intervals, ID sequences, and message payloads. Features are extracted through manual statistics or autoencoders, and anomalies are detected using statistical or machine learning methods [11,14,15,28,35,42,50,58].

Many studies have performed anomaly detection based on the physical characteristics of bus voltage. Due to minor manufacturing variations in the transceivers within ECUs, each transceiver exhibits slight differences in internal resistance and capacitance. Research has shown that these characteristics are highly useful for high-precision ECU fingerprinting [18] and can serve as hardware-based fingerprints for ECUs. Research on anomaly detection based on bus voltage primarily involves two stages: feature extraction and anomaly detection.

Feature extraction can be categorized into three approaches: Simple statistical features involve manual extraction of bus voltage distribution characteristics. In 2014, Murvay et al. first proposed a CAN bus voltage fingerprint extraction method [29]. They measured voltage sequences of CAN messages with different IDs and applying low-pass filtering, used the filtered data as fingerprints. Shabbir et al. proposed a scheme calculating the voltage ratio between two points as a fingerprint [3] in 2021. Signal processing-based feature extraction involves deriving time domain and frequency domain features using signal processing techniques. In 2016, Choi et al. introduced signal processing methods for feature extraction in CAN bus voltage anomaly detection [10]. They extracted 40 time domain and frequency domain features, selecting 17 features as node fingerprints. Subsequent studies built upon this approach [6,11,15,25,26]. In 2022, Xun et al. proposed the VehicleEIDS method [50], extracting 14 time domain features from differential bus signals as fingerprints. In 2023, Xun et al. further divided differential voltage signals of CAN-H and CAN-L into rising edge, steady state, and falling edge segments, extracting 8 time domain and 8 frequency domain features per segment, totaling 48 features as fingerprints [51]. Long et al. also proposed a similar feature extraction method [18,47,54].Automated feature extraction involves using deep learning techniques such as autoencoders to automatically extract features from bus voltage waveforms or reconstruct waveforms for anomaly detection [8,53].

In anomaly detection, early researchers detected anomalies via threshold-based methods. Such as mean squared error (MSE) [29], Mahalanobis distance [15], Euclidean distance [33]. With advances in artificial intelligence, numerous machine learning and deep learning classification methods have been employed to detect anomalies, including SVM [4], OCSVM [11], KNN [36,40], CNN [2,7,24,51], DNN [1,49], LSTM [12,37,43,53], AE [8,19,48], and GAN [17]. Among existing studies, Xun et al. proposed an anomaly detection method that extracts the most features from raw voltage signals to date [51]. By dividing each dominant bit into three segments and extracting 16 time domain and frequency domain features per segment, they used 48 features for training and prediction. Combined with a hybrid method of FeatureBagging and CNN, their approach achieved a detection accuracy of 98.85%.

Current voltage-based anomaly detection methods face two main limitations: First, they often ignore CAN protocol characteristics, relying heavily on signal processing to extract high-dimensional features (e.g., 48 or up to 60 dimensions), which increases computational cost. Second, detection accuracy still needs to be further improved.In real vehicle networks, bus traffic is very high; even a 1% false alarm rate can generate numerous false positives, reducing system practicality.

The paper contributes the following three aspects:

(1) A voltage fingerprint extraction method is proposed. Through analysis and theoretical calculation, specific voltage subsequences are extracted as node fingerprints, effectively characterizing hardware features of the nodes.
(2) An anomaly detection model based on time series prediction is proposed. According to the characteristics of the fingerprint data, an LSTM-based anomaly detection model is designed to effectively identify the source of bus voltage transmissions and detect anomalies.
(3) Tests are conducted on a CAN prototype and actual vehicles, with comparative experiments against state-of-the-art methods. The results show that the proposed method achieves a identification accuracy of up to 99.10% and functions normally at low sampling rates. All data is open-sourced [52].

The paper is structured as follows: Sect. 1 discusses related work. Section 2 briefly describes the basic principles of the CAN bus; Sect. 3 outlines the threat model; Sect. 4 proposes the detection framework, details the fingerprint extraction method and anomaly detection model; Sect. 5 presents experimental evaluations; Sect. 6 offers prospects for future research on CAN bus voltage anomaly detection.

2 CAN Principles

2.1 CAN Protocol

The CAN bus protocol can be divided into three layers: the physical layer, data link layer, and application layer. Relevant standards and specifications have been established by organizations such as ISO [21–23].

According to the CAN physical layer protocol, CAN uses differential signaling, where the voltage difference between CAN-H and CAN-L transmits data. A differential voltage of approximately 2V represents a dominant bit ("0"), while a voltage near 0V represents a recessive bit ("1"). Differential signaling effectively eliminates electromagnetic interference and reduces noise. The use of twisted-pair cables further suppresses commonmode interference.

At the data link layer, the CAN protocol specifies non-return-to-zero (NRZ) bit encoding and employs a bit stuffing mechanism: after five consecutive identical bits, a bit of opposite polarity is inserted. The receiver automatically removes this stuffed bit. A CAN data frame includes fields such as SOF, ID, RTR, IDE, r0, DLC, DATA, CRC, ACK, and EOF, As depicted in Fig. 1. Since the SOF, arbitration, and ACK fields may involve signals from multiple nodes, the measured differential voltage in these fields is often higher than in other fields.

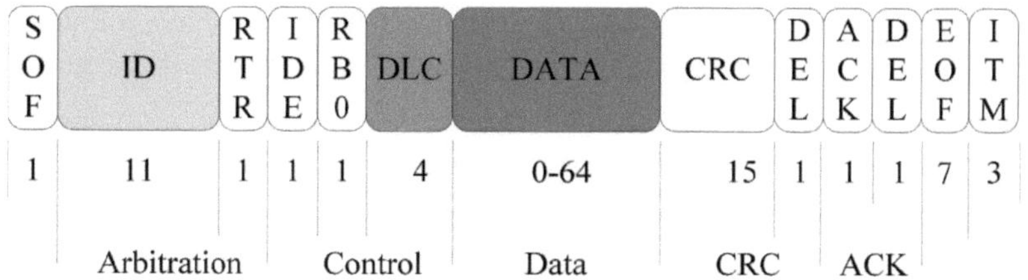

Fig. 1. CAN data frame format

2.2 Signal Rise Time

The ideal dominant bit voltage waveform is a square wave. However, due to factors such as transceiver drive capability, internal resistance, supply voltage, and bus impedance, the actual transmitted signal often deviates from a perfect square wave. Transitions between high and low levels require finite rise and fall times, leading to waveform distortion with rising and falling edges, and somctimes ringing.

In digital circuits, signal rise time refers to the time taken for a signal to transition from a low to a high level. It is a measure of the signal's transition rate, and precise control of rise time is key to achieving high-speed signal transmission. Industry standards typically define the low reference level as 10% of the voltage swing and the high reference level as 90%. The signal rising edge corresponds to the voltage sampling sequence during the rise time. Signal fall time and falling edge are defined similarly.

As illustrated in Fig. 2, the dominant bit width is 1, with the middle 500 points representing the bit waveform. The red and green dashed lines correspond to the high reference level (1.5218V) and low reference level (0.2931V) of the dominant bit. The rise time is approximately 4.1891 sampling points, or about 16.76 ns at a 4 ns sampling interval. The voltage rise slope is approximately 73.3 V/μs.

Different transceivers typically exhibit different voltage rise and fall slopes. For example, Infineon's TLE9251V High Speed CAN transceiver has a maximum voltage rise slope of 70 V/μs [44], while TI's SN65HVD230 transceiver has a rise slope of approximately 36 V/μs [20] in high-speed mode. Even transceivers of the same model from the same manufacturer may have slight variations due to production tolerances in internal resistance and circuit-specific differences, leading to variations in differential voltage edge slopes. As illustrated in Fig. 3, the equivalent input-output schematic of the SN65HVD230 transceiver illustrates that when the Rs pin is grounded, the transceiver operates in high-speed mode, enabling fast switching without limiting voltage rise/fall slopes. With a 10kΩ resistor between Rs and ground, the output differential voltage rise time is approximately 120 ns, with a voltage rise slope of about 15 V/μs. With a 100kΩ resistor, the rise time increases to approximately 800 ns, and the slope decreases to about 2 V/μs. Even resistors with the same nominal value have manufacturing tolerances, introducing variations in transceiver differential voltage rise slopes.

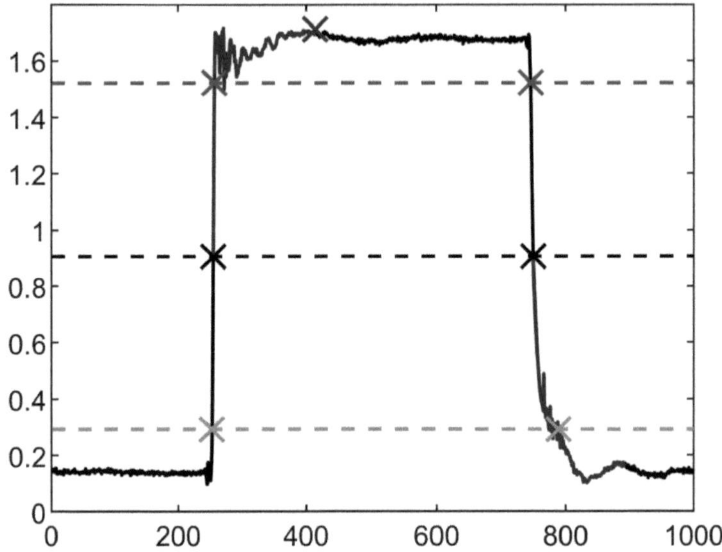

Fig. 2. dominant bit of CAN frame

2.3 Signal Settling Time

According to signal transmission theory, signals reflect when encountering impedance discontinuities. As illustrated in Fig. 4, assume the impedance of region 1 is Z_1 , the impedance of region 2 is Z_2 , the incident signal voltage is V_{inc}, the transmitted signal voltage after further transmission is V_{trans}, the reflected signal voltage is V_{ref}, and the calculation formulas for the reflected signal and the transmitted signal voltage after further transmission are as follows:

$$V_{ref}=\frac{Z_2-Z_1}{Z_2+Z_1}*V_{inc} \tag{1}$$

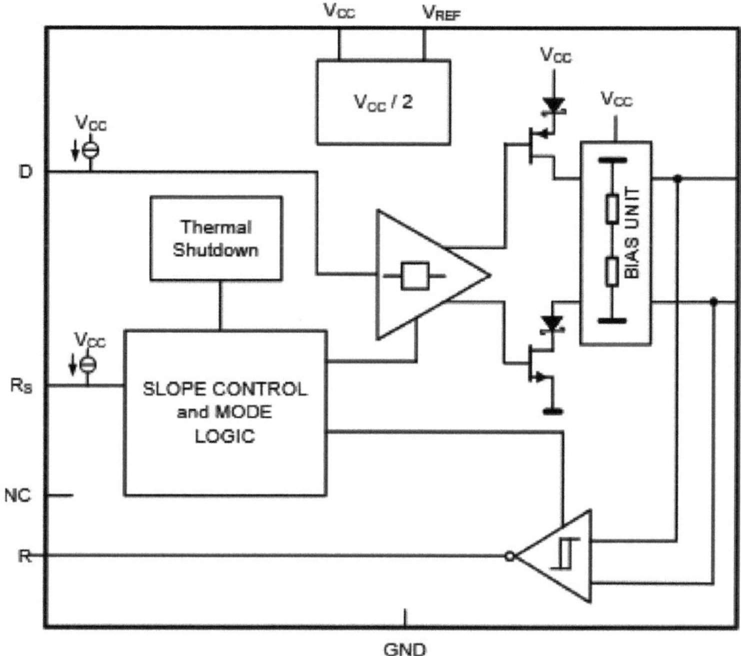

Fig. 3. equivalent input-output schematic diagram of SN65HVD230 [20]

$$V_{trans} = \frac{2 * Z_2}{Z_2 + Z_1} * V_{inc} \tag{2}$$

At bus cables, stub cables, connection points, and terminal nodes, impedance is often discontinuous. According to formula (1), if impedance suddenly increases, a positive voltage is reflected, causing overshoot. If impedance decreases, a negative voltage is reflected, causing undershoot. Multiple reflections between the output end and points of impedance discontinuity can cause ringing, as shown between the red and blue crosses on the left side of Fig. 2. Excessive ringing amplitude can affect communication quality.

After ringing, the bus voltage stabilizes. Signal settling time is the time required for an oscillating signal to enter a specified stable range. For a square wave, the settling time start is typically defined as when the signal crosses the mid-level (left black intersection in Fig. 2), and the end when it enters the stable range (blue inter-section in Fig. 2). The stable range is defined as the high-level voltage $\pm\ \alpha$, with α typically set to 2%.

2.4 Information Entropy

Information entropy describes the uncertainty and disorder of information. Shannon introduced the concept in the 1940s to measure the uncertainty of an information source [38]. For discrete sampling sequences, information entropy is calculated as follows:

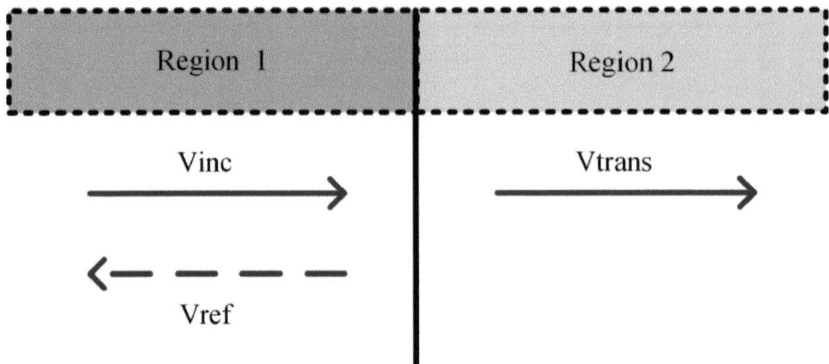

Fig. 4. signal reflection diagram

$$H(X) = -\sum P(x) * \log_2 P(X) \quad (3)$$

The upper part of Fig. 5, shows the dominant bit voltage sampling sequence, and the lower part shows the corresponding information entropy. When calculating entropy, the voltage sequence is processed using a sliding window of size 50. The black dashed line represents the average information entropy. The rising edge and ringing, as well as the falling edge and ringing, exhibit higher information entropy, indicating that these segments contain more information.

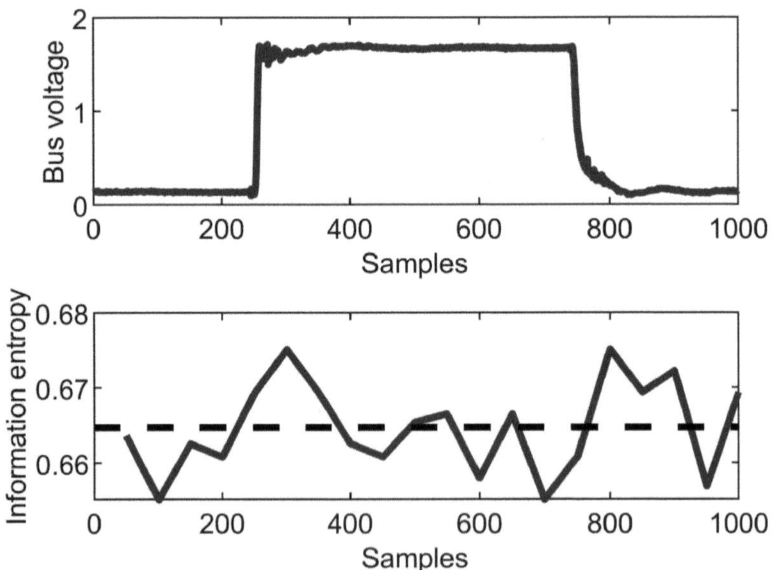

Fig. 5. information entropy distribution of the dominant bit

3 Threat Model

The CAN bus may face two types of attacks: (1) Remote attacks, where attackers gain access to controllable nodes remotely and inject spoofed messages onto the bus. (2) Physical attacks, where attackers replace existing nodes with malicious ones or add unauthorized nodes, using them to send messages masquerading as legitimate nodes. In automobiles, vulnerable points include OBD interfaces or easily accessible CAN bus locations [27].

This paper adds a monitoring unit to the CAN bus to sample and detect bus voltage signals at a specific rate.

4 System Framework

As illustrated in Fig. 6, the system framework comprises four main parts: voltage sampling, data preprocessing, fingerprint extraction, and training/prediction.

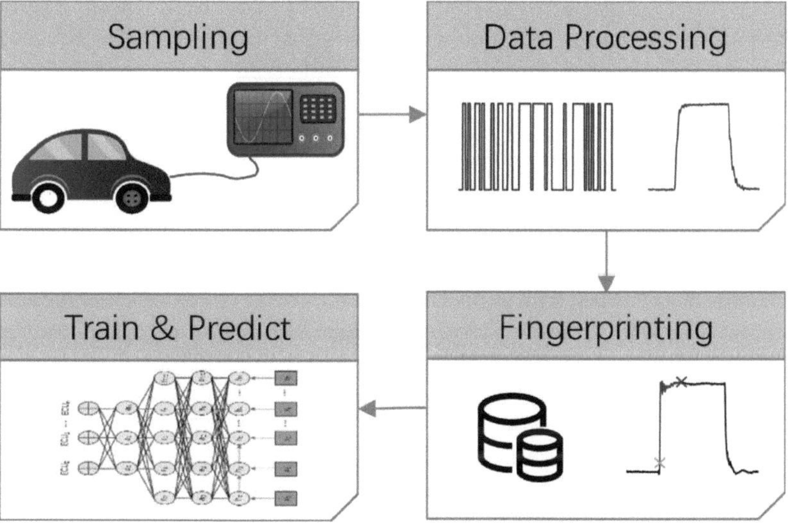

Fig. 6. system framework

4.1 Sampling

Compared to time-based or payload-based analysis, the proposed method relies solely on CAN bus voltage, which is difficult to forge. Since individual CAN-H and CAN-L voltages are susceptible to electromagnetic interference, extracting features from the differential bus voltage signal is more reliable. To reduce data volume and noise, the differential voltage is directly sampled and stored as a voltage sequence. This typically requires specialized sampling equipment, such as high-precision oscilloscopes or analog-to-digital converters (ADCs).

4.2 Data Preprocessing

Data preprocessing involves frame formation, bit separation, decoding, and bit filtering.

Frame Formation: Studies show that real vehicle CAN bus load ranges from 14% to 65%, mostly below 35% [16]. Since the bus is often idle, useful data frames are first separated to reduce analysis overhead. According to the CAN protocol, the end of a data frame includes a 1-bit recessive ACK delimiter and 7 recessive EOF bits, followed by a 3-bit interframe space. Thus, data between two consecutive sequences of 11 or more recessive bits may constitute a complete data or remote frame.

Bit Separation: Due to bit stuffing, frames contain variable-length dominant bits (1 to 5 bits). For each complete data frame, dominant bit separation is performed to obtain dominant bit voltage sequences.

Decoding and Bit Filtering: During anomaly detection, the CAN data frame ID field is used as a indirect label mapped to true ECU label. Frames are destuffed and decoded to obtain corresponding IDs. Since the SOF, ID, and ACK fields involve voltage contributions from multiple transceivers, to eliminate their influence and improve accuracy, dominant bits corresponding to these fields are excluded. Other dominant bit signals are selected to obtain the dominant bit waveform and corresponding ECU label dataset.

4.3 Fingerprinting

Existing works often use signal processing to extract up to 48-dimensional time domain and frequency domain features as fingerprints, requiring substantial computational resources. This paper theoretically analyzes and directly extracts specific dominant bit voltage subsequences as node voltage fingerprints.

From earlier analysis, when a node transceiver sends a dominant bit "0", the bus voltage transitions through six stages: from recessive bit stable voltage, through rising, ringing, entering dominant bit stable voltage, falling, ringing, and returning to recessive bit stable voltage.

During the rising stage, the transmitting transceiver charges the bus capacitance, acting as a power source with specific internal resistance and output voltage. The capacitive impedance of other nodes and cables is constant. The rising edge slope and subsequent ringing are closely related to the transmitting transceiver output voltage, internal resistance, and impedance of other transceivers and cables. Different transceivers yield different rising edge slopes and ringing. After ringing, the voltage enters a stable dominant state.

During the falling stage, residual charge on the bus discharges through the whole network. The residual charge on the bus primarily originates from the charging process of the bus parasitic capacitance. The transmitting transceiver drives this capacitive load through its output voltage, while the bus resistance determines the charging speed (i.e., the time constant).In other words,the charge

Q is the product of the output voltage V and the capacitance C. The resistance R determines the time required to reach this amount of charge.

$$Q = C \times V \tag{4}$$

where: Q is the total charge on the bus (unit: Coulomb, C). C is the parasitic capacitance to ground of the bus (unit: Farad, F), includes the capacitance of the bus itself, stub lines, PCB traces, connectors, and the input capacitance of all transceivers connected to the bus. V is the change in output voltage from the transmitting node (unit: Volt, V), specifically the difference between the recessive steady-state voltage and the dominant steady-state voltage.

Voltage falls, ringing occurs, and eventually stabilizes at the recessive level. During this process, the transmitting transceiver, like others, has its transistors turned off, acting as a resistor. The network impedance is determined by all nodes internal resistances, cable impedance, and terminal resistance. The falling edge slope and ringing depend on residual bus charge and overall impedance. Since overall impedance is constant, residual charge is proportional to the transceiver's dominant stable voltage. Minor differences in stable voltage across transceivers result in smaller differences in falling edge slope and ringing.

Thus, for a stable CAN bus topology, the rising edge slope and ringing are closely related to the transceiver's output voltage and internal resistance, while the falling edge slope is only related to output voltage. Compared to the falling edge, the rising edge better characterizes the transceiver. This paper selects voltage subsequences corresponding to the rising edge and ringing as node fingerprints. The subsequence starts at the rising edge onset and ends at the settling time conclusion. In practical CAN networks, the length of extracted voltage subsequences varies. The average length is used as a standard for extraction to facilitate training and prediction.

4.4 Training and Prediction

As described above, the extracted voltage subsequences exhibit temporal dependencies: during the rising edge, according to formula 5, capacitor charging causes each sampling point's voltage to depend on the previous point, transceiver output voltage, RC time constant, and sampling interval; during ringing, according to formula 1, signal reflection introduces additional temporal relationships. Thus, the subsequences constitute time series data with fixed sampled interval time.

$$V_t = V_{t0} + (V_{t1} - V_{t0}) * \left(1 - e^{-\frac{t}{RC}}\right) \tag{5}$$

As illustrated in Fig. 7, the anomaly detection model uses a Long Short-Term Memory (LSTM) network to handle temporal dependencies. The model includes an input sequence layer, LSTM layer, fully connected layers, etc. The input to the LSTM layer is voltage subsequence, and only the hidden state at the last time step is used to produce a single output. This approach works because the hidden state of the LSTM is designed to carry and integrate all relevant information from the beginning of the sequence up to the current time step. Theoretically, the final hidden state contains condensed information of the entire input sequence. During training, the cross-entropy loss computes the error between the predicted values and the true labels. This error is then backpropagated through the entire LSTM network, passing through the last hidden state and the output layer weights, back to the first time step, thereby updating all parameters. If the sequence is very long and critical information lies early in the sequence, the LSTM may struggle to retain such early information effectively. However, this issue is mitigated in this study by extracting relatively short subsequences sampled from sampled data of different lengths (1 to 5 bits).Two hidden layers are used, each with a size equal to 50% of the input layer size. A ReLU activation layer is added for training speed and robustness, and the output layer uses softmax. Dataset labels are ECU node numbers derived from message IDs via signal similarity mapping, as detailed in other works.

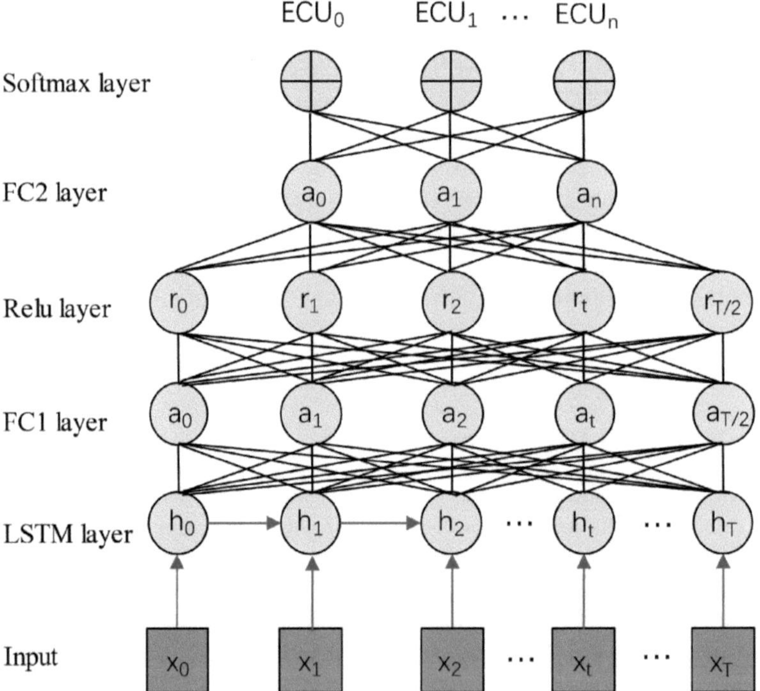

Fig. 7. anomaly detection model architecture

The dataset is split into training, validation, and test sets in a 7:1:2 ratio. The training set trains the model, the validation set tunes hyperparameters, and the test set evaluates generalization.

5 Evaluation

This section describes the experimental environment, evaluates the proposed method on a CAN prototype and a actual vehicle, and further assesses performance under attack and low-speed sampling scenarios.Finally, the comparison test was completed.

5.1 Environment

CAN Prototype. As illustrated in Fig. 8, a prototype with 5 CAN nodes was built to simulate a limited-node CAN bus network. Each node consisted of an Arduino UNO R4 Minima board [5] and a CAN transceiver. The Arduino board used a Renesas RA4M1 processor with an integrated CAN controller [34]. To introduce voltage diversity, two types of transceiver were used: NXP TJA1040 and TI SN65HVD230 [20,32]. The transceivers were connected via AWG26 twisted-pair cables, with 120Ω termination resistors at the first and fifth transceivers. Communication speed was set to 500 kbps.

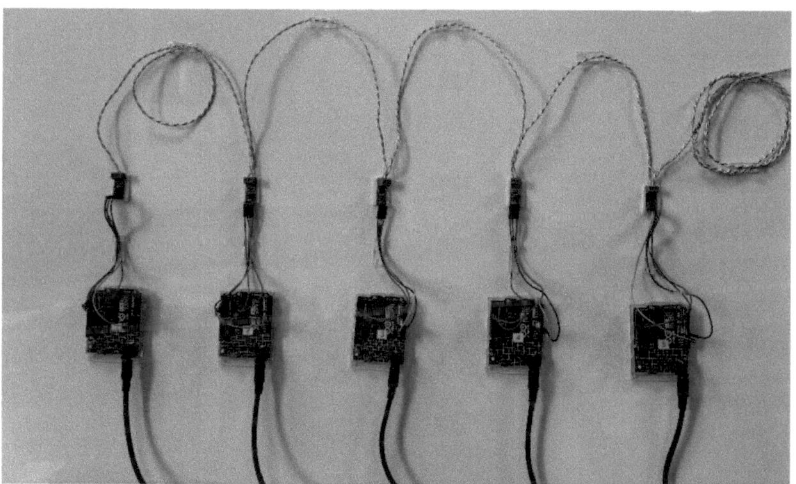

Fig. 8. CAN prototype

Actual Vehicle. A 2017 Nissan X-Trail was selected for actual vehicle evaluation. According to the electronic service manual (ESM) [31], the vehicle has 13 ECUs divided into two subnets: CAN1 and CAN2. CAN1 includes 11 ECUs, the OBD-II interface connects to CAN1. CAN2 includes two ECUs. BCM acts as a gateway between CAN1 and CAN2. The airbag sensor (A-BAG) only sends messages during collisions, so under normal conditions, up to 10 ECUs on CAN1 send messages.

Hardware. A Tektronix MSO54B mixed-signal oscilloscope sampled CAN bus voltage signals. Data preprocessing, training, and analysis were performed on a desktop with an Intel i7-10700K CPU, 16 GB RAM, and an NVIDIA GeForce RTX 2070 GPU.

5.2 Evaluation Metrics

To evaluate model performance under different scenarios, sampling rates, and attack conditions. Standard metrics including *Precision*, *Recall*, and $F1_{score}$ were used. For multi-class classification, macro-averaging was applied as follows (Fig. 9):

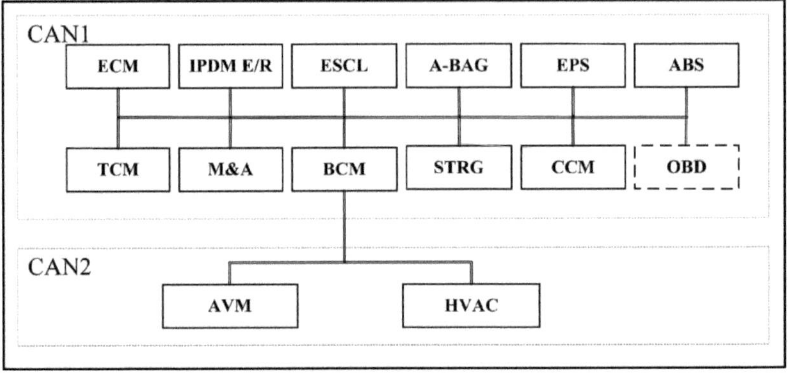

Fig. 9. Actual vehicle CAN

$$\text{Macro}_P = \frac{1}{n}\sum_{i=1}^{n}\text{Precision}_i \tag{6}$$

$$\text{Macro}_R = \frac{1}{n}\sum_{i=1}^{n}\text{Recall}_i \tag{7}$$

$$\text{Macro}_{F1} = \frac{1}{n}\sum_{i=1}^{n}\text{F1}_i \tag{8}$$

5.3 Result

CAN Prototype Evaluation. Data description: At 25 °C, using rising edge trigger mode, the bus voltage was sampled at 250 Msps with 12-bit precision. Preprocessing yielded 7220 CAN data frames with 13 unique IDs. From these, 271,160 dominant bit waveforms were isolated. After excluding SOF, ID, and ACK fields, 243,128 undisturbed dominant bit waveforms remained. The average number of sampling points from rising edge to steady state was 128. Thus,

128-point voltage subsequences were extracted as node fingerprints, resulting in 243,128 samples.

Test results: The dataset was split 7:1:2 (training: 170,180; validation: 24,294; test: 48,654). Labels were ECU numbers. The model was trained and tested, achieving 100% accuracy on the test set. The confusion matrix is shown in Fig. 10. Precision, recall, and F1 were all 1. It is plausible that the perfect accuracy rate stems from the controlled and simplified laboratory conditions of the CAN prototype. Its performance would likely fall below 1.0 when deployed in real-world environments.

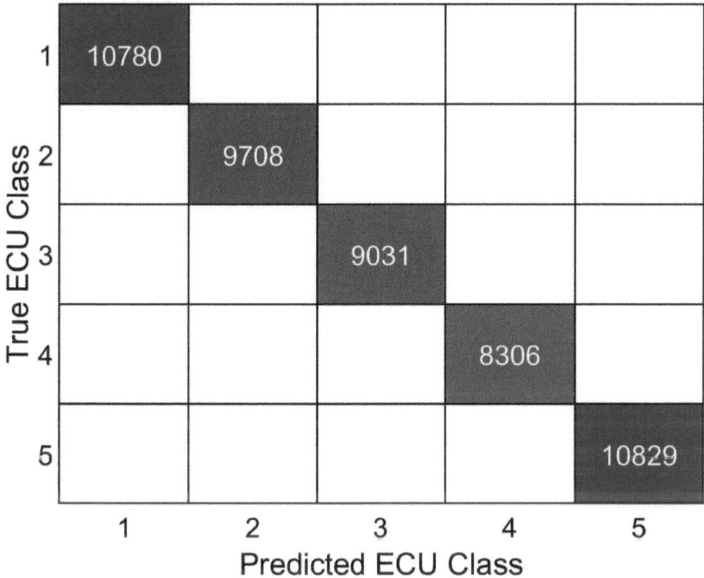

Fig. 10. Confusion Matrix of CAN protype

Actual Vehicle Evaluation. Data description: At 12 °C, the OBD interface of the parked vehicle was sampled at 250 Msps with 12 bits precision for 2 s. Preprocessing yielded 4113 CAN frames with 52 unique IDs. From these, 89,965 dominant bit waveforms were isolated. After exclusions, 70,906 undisturbed waveforms remained. Due to more complex topology and noise, signal ringing was significant. The average number of points from rising edge to steady state was 110. Thus, 110-point subsequences were extracted, yielding 70,906 samples. The dataset was split (training: 49,674; validation: 7,082; test: 14,150).

Test results: The model was trained and tested, achieving 100% accuracy on the test set. The confusion matrix is shown in Fig. 2. Precision was 0.9910, recall was 0.9983, and F1 was 0.9945.

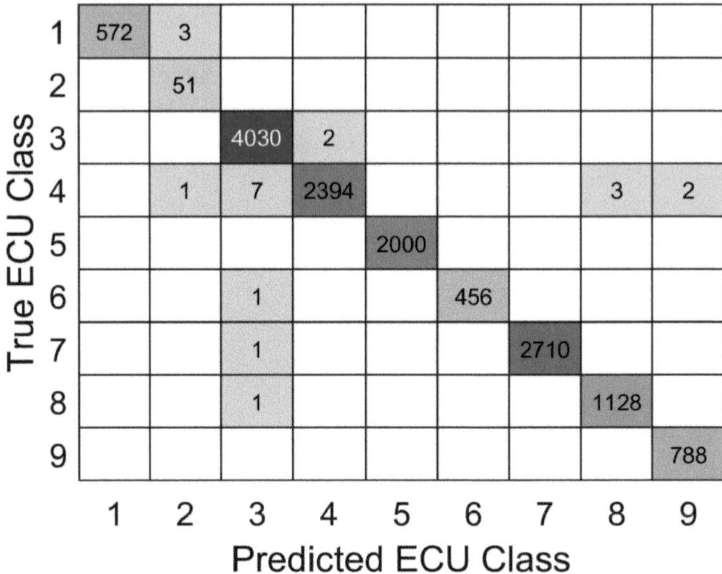

Fig. 11. Confusion Matrix of actual vehicle

Testing Under Attack from Additional Node. Data description: An attack node was added to the prototype, transmitting messages with ID 0x11 with slightly altered data, so as to distinguish the normal message from the attack message during the evaluation stage. Sampling yielded 6969 frames with 13 unique IDs. After preprocessing, 221,686 undisturbed dominant bit waveforms were obtained. 128-point subsequences were extracted as fingerprints (Fig. 11).

Test results: The test was conducted in two steps. First, a dataset without ID 0x11 (180,648 samples) was split 7:1:2 (training: 126,453; validation: 18,065; test: 36,130). The model achieved 100% accuracy (Fig. 12). Then, 22,075 normal and 18,963 attack samples with ID 0x11 were added to the test set (total 77,168). The model correctly predicted all normal samples and flagged all attack samples as anomalous (Fig. 13). Precision, recall, and F1 were all 1. Using signal similarity, attack messages showed only 0.0693 similarity to ECU4's normal messages (threshold 0.95), confirming they originated from an additional node.

Testing Under Low-Speed Sampling Speed. Data description: The prototype dataset was downsampled by 5× and 10×, reducing sampling rates to 50 Msps and 25 Msps. Subsequence lengths reduced to 26 and 13 points, respectively.

Test results: At 50 Msps, the confusion matrix is the same as Fig. 10, performance remained perfect (accuracy=1). At 25 Msps, the confusion matrix (Fig. 14) showed a slight increase in errors, but accuracy, recall, and F1 were all 0.9988, indicating robust performance even at low sampling rates.

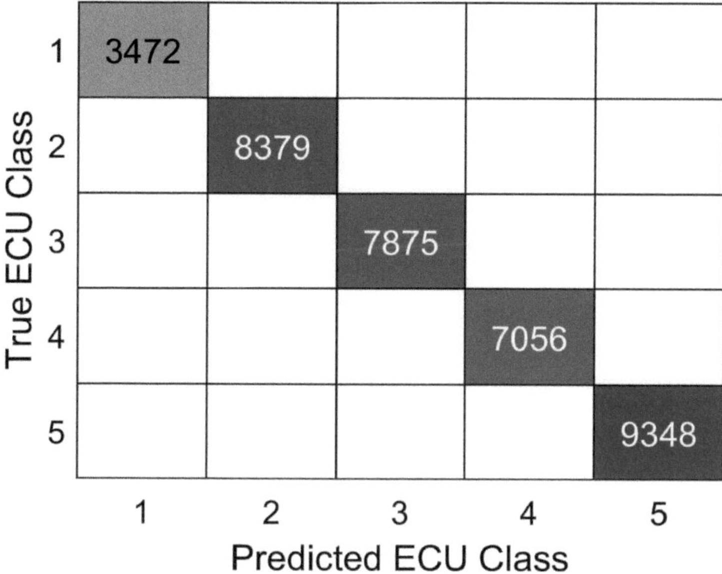

Fig. 12. Confusion Matrix I under attackN

5.4 Comparison Evaluation

The identification method proposed by Xun et al. [51], divides the dominant bit into rising edge, steady state, and falling edge three segments, and extracts up to 48-dimensional features in both the time and frequency domains. These features are then converted into images and detected using a CNN-based identification algorithm. To the best of our knowledge, this method currently employs the most extensive set of time-domain and frequency-domain features for identification, achieving an accuracy rate of 98.95%.

To ensure fairness, the prototype dataset along with its ground truth was used. Some adjustments were made to their method:

(1) Rising/falling edge start threshold changed from 0.08V to 0.2V (original too low).
(2) Rising edge end threshold set to 0.95× dominant high level (undefined originally).
(3) Sampling rate increased to 250 Msps (original 12.5 Msps insufficient for frequency features). Given a sampling speed of 12.5 Msps, which corresponds to an 80 ns interval between samples, a minimum sampling duration of 240 ns is required to obtain at least three data points for extracting the signal's frequency domain features. Statistical results show that the duration of the dominant bit's rising edge in the CAN prototype is significantly shorter than 240 ns, with an average of only 57.33 ns. Therefore, the sampling speed must be increased to capture sufficient data during the rising and falling edges for subsequent frequency domain feature extraction.

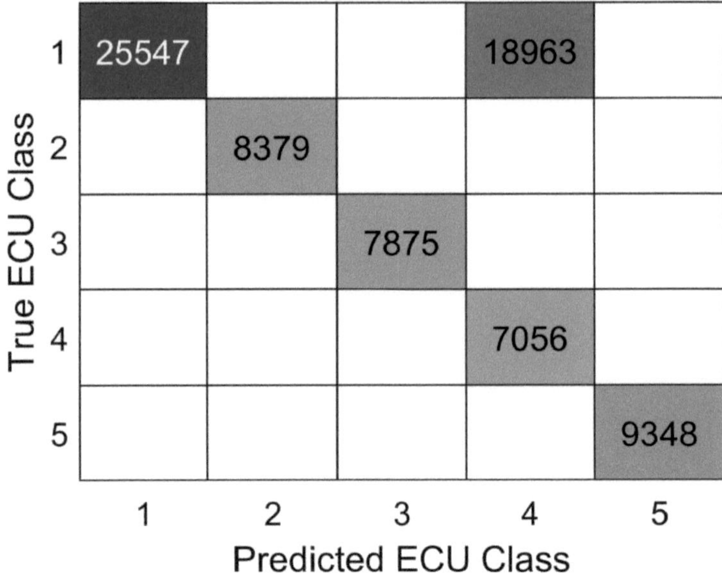

Fig. 13. Confusion Matrix II under attack

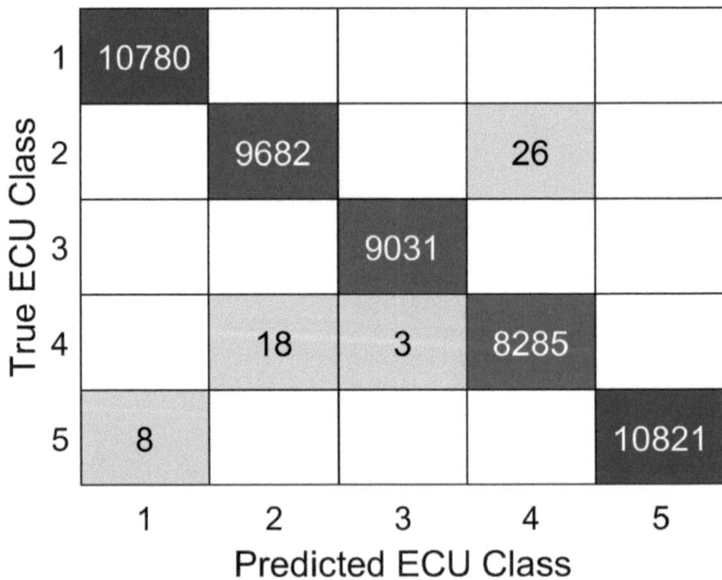

Fig. 14. Confusion Matrix under 25Msps sampling speed

Test results: Employing the approach proposed by Xun et al., feature extraction failed on 15 of 243,128 waveforms. The remaining 243,113 samples were split (training: 170,179; validation: 24,311; test: 48,623). The confusion

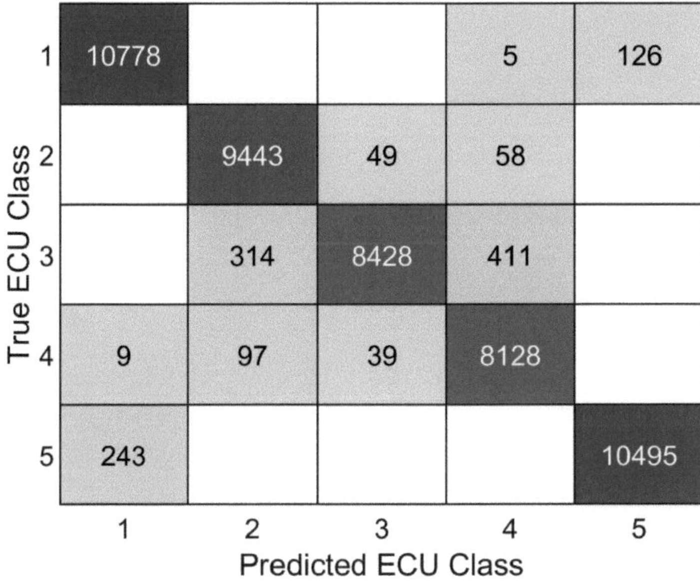

Fig. 15. Confusion Matrix with Xun method

matrix (Fig. 15) showed accuracy=0.9716, recall=0.9715, F1=0.9712. The proposed method outperformed this approach with higher accuracy and lower computational cost.

6 Conclusion

The paper focused on CAN bus voltage anomaly detection, proposing a novel scheme that extracts specific voltage subsequences as fingerprints and uses LSTM-based time series prediction. This approach avoids heavy feature computation and effectively identifies spoofing and additional node attacks. Specifically, by analyzing CAN signal transmission characteristics, voltage subsequences from rising edge start to steady state are extracted as node fingerprints. An LSTM-based model detects anomalies, identifies message sources, and discriminates attack types. Experiments showed 100% source identification accuracy in the prototype, 99.10% in actual vehicles, and maintained performance at low sampling rates (25 Msps, accuracy 99.88%). The method outperformed state-of-the-art techniques with lower false alarm rates and higher practicality.

Future work should address two issues: (1) identifying hidden nodes that do not send data frames, and (2) evaluating system overhead on embedded platforms to enhance practicality in actual vehicles and systems.

Acknowledgments. This research was supported by the National Key Research and Development Program of China(2021YFB3101804), Research and Practice Project on Higher Education Teaching Reform in Henan Province(2024SJGLX0214).

Disclosure of Interests. The authors declare that they have no Conflict of interest.

References

1. Metwaly, A., Elhenawy, I.: Sustainable intrusion detection in vehicular controller area networks using machine intelligence paradigm. Sustain. Mach. Intell. J. **4** (2023). https://doi.org/10.61185/SMIJ.2023.44104
2. Agbo, O., Hefeida, M., El-Wakeel, A.S.: Hybrid-CNN intrusion detection framework for CAN networks in connected and autonomous vehicles. IEEE Internet Things J. **12**(17), 36353–36369 (2025). https://doi.org/10.1109/JIOT.2025.3582118
3. Ahmed, S., Juliato, M., Gutierrez, C., Sastry, M.: Two-Point Voltage Fingerprinting: Increasing Detectability of ECU Masquerading Attacks (2021). https://doi.org/10.48550/arXiv.2102.10128
4. Alshammari, A., Zohdy, M., Debnath, D., Corser, G.: Classification approach for intrusion detection in vehicle systems. Wirel. Eng. Technol. **09**(04), 79–94 (2018). https://doi.org/10.4236/wet.2018.94007
5. Arduino Team: UNO R4 Minima. https://docs.arduino.cc/hardware/uno-r4-minima/
6. Avatefipour, O., Hafeez, A., Tayyab, M., Malik, H.: Linking received packet to the transmitter through physical-fingerprinting of controller area network. In: 2017 IEEE Workshop on Information Forensics and Security (WIFS), pp. 1–6 (2017). https://doi.org/10.1109/WIFS.2017.8267643
7. Baldini, G.: Voltage based electronic control unit (ECU) identification with convolutional neural networks and walsh-hadamard transform. Electronics **12**(1), 199 (2022). https://doi.org/10.3390/electronics12010199
8. Battaglia, A., Canino, N., Dini, P., Lombardo, G., Longo, F., Rossi, D.: Autoencoder-based detection of physical-layer anomalies in automotive CAN networks. In: 2025 IEEE 31st International Symposium on On-Line Testing and Robust System Design (IOLTS), pp. 1–4 (2025). https://doi.org/10.1109/IOLTS65288.2025.11116870
9. Charette, R.N.: This Car Runs on Code. https://spectrum.ieee.org/this-car-runs-on-code
10. Choi, W., Jo, H.J., Woo, S., Chun, J., Park, J., Lee, D.H.: Identifying ECUs using inimitable characteristics of signals in controller area networks. IEEE Trans. Veh. Technol. **67**(6), 4757–4770 (2016). https://doi.org/10.1109/TVT.2018.2810232
11. Choi, W., Joo, K., Jo, H.J., Park, M.C., Lee, D.H.: VoltageIDS: low-level communication characteristics for automotive intrusion detection system. IEEE Trans. Inf. Forensics Secur. **13**(8), 2114–2129 (2018). https://doi.org/10.1109/TIFS.2018.2812149
12. Chougule, A., Kulkarni, I., Alladi, T., Chamola, V., Yu, F.: HybridSecNet: in-vehicle security on controller area networks through a hybrid two-step LSTM-CNN model. IEEE Trans. Veh. Technol. **73**(10), 14580–14591 (2024). https://doi.org/10.1109/TVT.2024.3413849
13. Cui, J.: Lightweight encryption and authentication for controller area network of autonomous vehicles. IEEE Trans. Veh. Technol. **11**, 14756–14770 (2023). https://doi.org/10.1109/TVT.2023.3281276
14. Deng, Z., Liu, J., Xun, Y., Qin, J.: IdentifierIDS: a practical voltage-based intrusion detection system for real in-vehicle networks. IEEE Trans. Inf. Forensics Secur. **19**, 661–676 (2024). https://doi.org/10.1109/TIFS.2023.3327026

15. Foruhandeh, M., Man, Y., Gerdes, R., Li, M., Chantem, T.: SIMPLE: single-frame based physical layer identification for intrusion detection and prevention on in-vehicle networks. In: Proceedings of the 35th Annual Computer Security Applications Conference, pp. 229–244. ACM (2019). https://doi.org/10.1145/3359789.3359834
16. Gay, C., Matsumoto, T.: A model for CAN message timestamp fluctuations to accurately estimate transmitter clock skews. Int. J. Autom. Eng. **15**(1), 10–18 (2024). https://doi.org/10.20485/jsaeijae.15.1_10
17. Im, H., Kim, D., Lee, S.: Listening to CAN: anomaly detection to enhance in-vehicle network security. In: 2023 IEEE 13th International Conference on Consumer Electronics - Berlin (ICCE-Berlin), pp. 172–175. IEEE (2023). https://doi.org/10.1109/ICCE-Berlin58801.2023.10375627
18. Hafeez, A., Topolovec, K., Awad, S.: ECU fingerprinting through parametric signal modeling and artificial neural networks for in-vehicle security against spoofing attacks. In: 2019 15th International Computer Engineering Conference (ICENCO), pp. 29–38 (2019). https://doi.org/10.1109/ICENCO48310.2019.9027298
19. Hoang, T., Kim, D.: Detecting in-vehicle intrusion via semi-supervised learning-based convolutional adversarial autoencoders. Veh. Commun. **38** (2022). https://doi.org/10.1016/j.vehcom.2022.100520
20. Instruments, T.: SN65HVD23x 3.3-V CAN Bus Transceivers Datasheet. https://www.ti.com.cn/cn/lit/ds/symlink/sn65hvd230.pdf
21. ISO: Road vehicles-Controller area network (CAN) Part 3: Low-speed, fault-tolerant, medium-dependent interface (2003). https://www.iso.org/standard/36055.html, iSO 11898-3:2006
22. ISO: Road vehicles-Controller area network(CAN)-Part 1 Data link layer and physical signalling (2003). https://www.iso.org/standard/33422.html, iSO 11898-1:2003
23. ISO: Road vehicles-Controller area network (CAN)-Part 2 High-speed medium access unit (2003). https://www.iso.org/standard/33423.html, iSO 11898-2:2003
24. Javed, A., Rehman, S., Khan, M., Alazab, M., Reddy, G.: CANintelliIDS: detecting in-vehicle intrusion attacks on a controller area network using CNN and attention-based GRU. IEEE Trans. Netw. Sci. Eng. **8**(2), 1456–1466 (2021). https://doi.org/10.1109/TNSE.2021.3059881
25. Kneib, M., Huth, C.: Scission: signal characteristic-based sender identification and intrusion detection in automotive networks. In: Proceedings of the 2018 ACM SIGSAC Conference on Computer and Communications Security, pp. 787–800. ACM (2018). https://doi.org/10.1145/3243734.3243751
26. Kneib, M., Schell, O., Huth, C.: EASI: edge-based sender identification on resource-constrained platforms for automotive networks. In: Proceedings 2020 Network and Distributed System Security Symposium (2020). https://doi.org/10.14722/ndss.2020.24025
27. Labs, A.P.: CAN Injection Attacks: A New Form of Vehicle Theft - Auxilium Pentest Labs Blog (2024). https://blog.auxilium-labs.com/2024/02/can-injection-attacks-a-new-form-of-vehicle-theft/
28. Lokman, S.F., Othman, A.T., Abu-Bakar, M.H.: Intrusion detection system for automotive Controller Area Network (CAN) bus system: a review. EURASIP J. Wirel. Commun. Netw. **2019**, (2019). https://doi.org/10.1186/s13638-019-1484-3
29. Murvay, P.S., Groza, B.: Source identification using signal characteristics in controller area networks. IEEE Signal Process. Lett. **21**(4), 395–399 (2014). https://doi.org/10.1109/LSP.2014.2304139

30. Narayanan, S., Mittal, S., Joshi, A.: OBD_SecureAlert: an anomaly detection system for vehicles. In: 2016 IEEE International Conference on Smart Computing (SMARTCOMP), pp. 1–6 (2016). https://doi.org/10.1109/SMARTCOMP.2016.7501710
31. Nissan Motor Corporation: Electronic Service Manual, Nissian T32 LAN Section
32. NXP Semiconductors: TJA1040 High speed CAN transceiver. https://www.nxp.com.cn/docs/en/data-sheet/TJA1040.pdf
33. Popa, L., Groza, B., Jichici, C., Murvay, P.S.: ECUPrint–physical fingerprinting electronic control units on CAN buses inside cars and SAE J1939 compliant vehicles. IEEE Trans. Inf. Forensics Secur. **17**, 1185–1200 (2022). https://doi.org/10.1109/TIFS.2022.3158055
34. Renesas Electronics Corporation: RA4M1 Group Datasheet. https://www.renesas.com/en/document/dst/ra4m1-group-datasheet?r=1054146
35. Rogers, M., Weigand, P., Happa, J., Rasmussen, K.: Detecting CAN attacks on J1939 and NMEA 2000 networks. IEEE Trans. Dependable Secure Comput. **20**(3), 2406–2420 (2023). https://doi.org/10.1109/TDSC.2022.3182481
36. Qiu, S., Fei, J., Yang, H., Xiao, Y., Zhang, X.: A physical fingerprint-based intrusion detection and localization in fieldbus network. In: 2023 4th International Seminar on Artificial Intelligence, Networking and Information Technology (AINIT), pp. 294–302. IEEE (2023). https://doi.org/10.1109/AINIT59027.2023.10212629
37. Wu, S., Li, S., Sun, W.: ConvIDS: a convolutional LSTM based intrusion detection model for in-vehicle CAN bus. In: 2023 IEEE 6th International Conference on Automation, Electronics and Electrical Engineering (AUTEEE), pp. 285–290. IEEE (2023). https://doi.org/10.1109/AUTEEE60196.2023.10408040
38. Shannon, C.E.: A mathematical theory of communication. Bell Syst. Tech. J. **27**(3), 379–423 (1948). https://doi.org/10.1002/j.1538-7305.1948.tb01338.x
39. Soderi, S., Colelli, R., Turrin, F., Pascucci, F., Conti, M.: SENECAN: secure KEy DistributioN OvEr CAN through watermarking and jamming. IEEE Trans. Dependable Secure Comput. **20**(3), 2274–2288 (2023). https://doi.org/10.1109/TDSC.2022.3179562
40. Sugunaraj, N., Ranganathan, P.: Electronic control unit (ECU) identification for controller area networks (CAN) using machine learning. In: 2022 IEEE International Conference on Electro Information Technology (eIT), pp. 1–7. IEEE (2022). https://doi.org/10.1109/eIT53891.2022.9813928
41. Sun, H., et al.: MTDCAP: moving target defense-based CAN authentication protocol. IEEE Trans. Intell. Transp. Syst. **25**(9), 12800–12817 (2024). https://doi.org/10.1109/TITS.2024.3384054
42. Taylor, A., Japkowicz, N., Leblanc, S.: Frequency-based anomaly detection for the automotive CAN bus. In: 2015 World Congress on Industrial Control Systems Security (WCICSS), pp. 45–49. IEEE (2015). https://doi.org/10.1109/WCICSS.2015.7420322
43. Taylor, A., Leblanc, S.P., Japkowicz, N.: Anomaly detection in automobile control network data with long short-term memory networks. In: 2016 IEEE International Conference on Data Science and Advanced Analytics (DSAA), pp. 130–139 (2016). https://doi.org/10.1109/DSAA.2016.20
44. Technologies, I.: TLE9251VLE - Automotive CAN transceivers. https://www.infineon.com/part/TLE9251VLE
45. Umar, M., et al.: Physical layer authentication in the internet of vehicles through multiple vehicle-based physical attributes prediction. Ad Hoc Netw. **152** (2024). https://doi.org/10.1016/j.adhoc.2023.103303

46. Upstream Security: 2024 Global Automotive Cybersecurity Report (2024)
47. Wei, Y., Cheng, C., Xie, G.: OFIDS MISC learning-enabled and fingerprint-based intrusion detection system in controller area. IEEE Trans. Dependable Secure Comput. **20**(6), 4607–4620 (2023). https://doi.org/10.1109/TDSC.2022.3230501
48. Xing, L., Wang, K., Wu, H., Ma, H., Zhang, X.: FL-MAAE: an intrusion detection method for the internet of vehicles based on federated learning and memory-augmented autoencoder. Electronics **12**(10), 2284 (2023). https://doi.org/10.3390/electronics12102284
49. Xu, C., Shen, J., Du, X.: A method of few-shot network intrusion detection based on meta-learning framework. IEEE Trans. Inf. Forensics Secur. **15**, 3540–3552 (2020)
50. Xun, Y., Zhao, Y., Liu, J.: VehicleEIDS: a novel external intrusion detection system based on vehicle voltage signals. IEEE Internet Things J. **9**(3), 2124–2133 (2022). https://doi.org/10.1109/JIOT.2021.3090397
51. Xun, Y., Deng, Z., Liu, J., Zhao, Y.: Side channel analysis: a novel intrusion detection system based on vehicle voltage signals. IEEE Trans. Veh. Technol. **72**(6), 7240–7250 (2023). https://doi.org/10.1109/TVT.2023.3236820
52. Yao, W.: CAN Bus Voltage Data Set. https://doi.org/10.57760/sciencedb.27336
53. Yin, L., Xu, J., Chai, H., Wang, C.: A manipulated overlapped voltage attack detection mechanism for voltage-based vehicle intrusion detection system. In: Ahene, E., Li, F. (eds.) Frontiers in Cyber Security, vol. 1726, pp. 413–428. Springer, Singapore (2022)
54. Yin, L., Xu, J., Wang, C., Wang, Q., Zhou, F.: Detecting CAN overlapped voltage attacks with an improved voltage-based in-vehicle intrusion detection system. J. Syst. Architect. **143**, 102957 (2023). https://doi.org/10.1016/j.sysarc.2023.102957
55. Zhang, G., Shen, J., Li, J., Qin, W., Li, Y.: Payload processor : message authentication for in-vehicle CAN bus using data compression and tag filling. Comput. Netw. **259**, 111061 (2025). https://doi.org/10.1016/j.comnet.2025.111061
56. Zhao, Y., Xun, Y., Liu, J.: VehicleCIDS: an efficient vehicle intrusion detection system based on clock behavior. In: 2021 IEEE Global Communications Conference (GLOBECOM), pp. 1–6. IEEE (2021). https://doi.org/10.1109/GLOBECOM46510.2021.9685130
57. Zhao, Y., Xun, Y., Liu, J.: ClockIDS: a real-time vehicle intrusion detection system based on clock skew. IEEE Internet Things J. **9**(17), 15593–15606 (2022). https://doi.org/10.1109/JIOT.2022.3151377
58. Zhou, X., Jiang, R., Tian, M., Qu, H., Zhang, H.: Temperature-sensitive fingerprinting on ECU clock offset for CAN intrusion detection and source identification. In: Proceedings of the ACM Turing Celebration Conference - China, pp. 89–94. ACM (2020). https://doi.org/10.1145/3393527.3393543

Detection and Early Warning Methods for Unsafe Behaviors in the Emergency Response Process of Coal-Fired Power Enterprises

Songyu Zou[1,2] and Hongfu Gao[1,2](✉)

[1] Huayuan Power Plant, State Grid Energy Hami Coal and Power Co., Ltd., Hami 839000, Xinjiang, China
{20008333,17041588}@ceic.com

[2] State Grid Energy Hami Coal and Power Co., Ltd., Hami 839000, Xinjiang, China

Abstract. In the emergency response process of coal-fired power enterprises, working environments are often characterized by high dust concentration, insufficient illumination, and frequent occlusion, which severely restrict the performance of unsafe behavior detection models and make it difficult to meet real-time early warning requirements. To address this challenge, this study optimizes the YOLOv8 framework: on the one hand, a local feature filtering mechanism is introduced to effectively enhance the recognition of small and occluded targets; on the other hand, a triple attention module is integrated with the C2f_DWR module to achieve collaborative optimization of feature extraction and representation. Experimental results show that when the triple attention module and the C2f_DWR module are used independently, the detection speed decreases by approximately 6 FPS and 9 FPS, respectively, while both improve accuracy by more than 3%. When the two modules are combined, the detection speed decreases by about 13 FPS, while the mean Average Precision (mAP) significantly increases from 38.5% to 46%, achieving a substantial improvement in detection accuracy while maintaining acceptable detection speed. The findings demonstrate that the proposed method can effectively enhance unsafe behavior detection and early warning in complex industrial environments, providing a practical technical reference for safety monitoring systems in coal-fired power enterprises during emergency response.

Keywords: coal-fired power enterprise · emergency response · unsafe behavior detection · YOLOv8 · attention mechanism

1 Introduction

With the widespread application of artificial intelligence and deep learning technologies, the working environments of coal-fired power enterprises are gradually advancing toward intelligent and automated management. Although the level

P. Umapada et al. (Eds.): ICCPR 2025, CCIS 2811, pp. 366–377, 2026.
https://doi.org/10.1007/978-981-95-8315-7_29

of safety technology has been continuously improving in recent years, accidents caused by unsafe behaviors of operators still occur frequently. Statistics show that more than 70% of coal industry accidents are triggered by unsafe behaviors, making them a critical factor restricting workplace safety. Therefore, developing efficient and accurate unsafe behavior detection methods is of great importance for enhancing safety management in such environments.

Traditional manual inspection methods face numerous limitations in industrial workplaces, such as confined operational spaces, complex illumination conditions, and severe dust occlusion, which lead to low recognition efficiency and susceptibility to subjective bias. In recent years, with the rapid development of computer vision technology, image-based object detection methods have been widely applied to intelligent safety monitoring. However, existing methods still face challenges in complex environments, including weak robustness, difficulty in detecting small targets, and missed detections of occluded objects.

To address these challenges, this paper proposes an improved unsafe behavior detection method based on the YOLOv8 network. While retaining the efficiency of the YOLO architecture, two key improvements are introduced: (1) a triple attention module integrated into the backbone network to achieve feature interaction across spatial, channel, and directional dimensions, thereby enhancing the model's ability to extract critical behavioral features; and (2) a local feature filtering module designed for fine-grained regional modeling, which strengthens the model's focus on occluded targets and subtle abnormal behaviors.

Experimental results demonstrate that the proposed method achieves a significant improvement of 7.5% in mean Average Precision (mAP) while maintaining reasonable detection speed. This approach provides a deployable model structure and optimization strategy for intelligent monitoring of unsafe behaviors in complex industrial environments, and it has the potential to further advance the safety management capabilities of coal-fired power enterprises.

The main contributions of this paper are as follows:

(1) A detection method for unsafe behaviors in complex work environments is proposed, which integrates triple attention and local feature filtering to enhance the representation of multi-scale targets.
(2) A local feature weighting mechanism is introduced to address occlusion and illumination interference, significantly improving localization accuracy and discriminative ability.
(3) Extensive experiments on the DsLMF+ dataset validate the effectiveness and deployability of the proposed model, providing technical support for intelligent safety management in the coal industry.

2 Related

The YOLO series algorithms have been widely applied in energy and industrial hazardous object detection due to their efficiency and real-time performance, with improvements focusing on model efficiency, real-time deployment, and complex environment adaptation—key demands for emergency scenarios in coal-fired power enterprises. For model lightweight and real-time optimization, FL-YOLO [1] uses depthwise separable convolutions and inverted residual blocks for fast edge video analysis, providing a basis for mobile emergency monitoring in power plants; CAP-YOLO [2] enhances coal-energy enterprise surveillance with channel attention and adaptive image enhancement, which can be extended to smoke/dust interference reduction in power plant emergency detection; YOLO-BS [3] targets abnormal coal block detection in scraper conveyors, offering a reference for hazardous bulk material monitoring in power plant coal handling systems; CM-YOLOv8 [4] employs adaptive anchor boxes and pruning for lightweight coal-related scene detection, meeting real-time requirements of hazardous object detection in power plant workshops.

Image enhancement and backbone modification technologies are critical to addressing harsh conditions in coal-fired power emergencies. Histogram-based enhancement with YOLOv8 [5] strengthens feature expression of thermal imaging targets, adapting to equipment overheating emergency detection; MobileNet integration with dropout and augmentation [6] improves model robustness in complex backgrounds, suitable for multi-source interference scenarios in power plants; AYOLO [7] combines unsupervised learning with Vision Transformers, providing a new path for intelligent adaptive detection of unknown hazards; PP-YOLO [8] applies histogram and wavelet preprocessing, which can be used for preprocessing of fuzzy images caused by smoke in power plant emergencies; DSP-YOLO [9] focuses on occluded target detection, solving the problem of hazardous object occlusion by equipment in power plant workshops; LMFI-YOLO [10] enhances regression and attention mechanisms for lightweight pedestrian detection, supporting personnel hazard prevention near power plant hazardous areas; low-light feature interaction methods [11] improve low-illumination detection accuracy, adapting to emergency detection in power plant auxiliary workshops with insufficient lighting; CDD-YOLO [12] optimizes lightweight low-light detection, reducing resource consumption for edge deployment in emergency scenarios.

For scenario-specific optimization related to coal-energy and industrial hazardous detection, dual-backbone YOLOv8 [13] enhances feature fusion for road defect detection, providing inspiration for hazardous structural detection of power plant pipelines; YOLO-sea [14] improves undersea target detection, which can be extended to hazard detection in high-humidity environments of power plants; Ucm-YOLOv5 [15] optimizes underground coal mine target detection, verifying the adaptability of improved YOLOv5 to confined spaces of power plants; YOLOv7-SE [16] strengthens small-target detection with squeeze-and-excitation modules, solving the problem of small hazardous object recognition in power equipment; SD-YOLOv5s-4L [17] realizes multi-object detection of underground unmanned locomotives, offering a reference for multi-hazard simultaneous detection in power plant transportation systems; improved YOLOv7-Tiny [18] enhances underground track foreign object detection, applicable to foreign object hazard monitoring in power plant transmission channels; YOLOv5 optimized by pruning [19] improves safety helmet detection efficiency, supporting personnel protection supervision in power plant emergency operations; improved YOLOv8n [20] enhances underground personnel behavior detection, providing a basis for abnormal behavior-induced hazard prevention in power plants.

These studies highlight the application value of YOLO series algorithms in coal-energy and industrial hazardous object detection, but also expose challenges in coal-fired power emergencies: insufficient adaptability to smoke/high-temperature interference during equipment failures, and difficulty balancing real-time performance and precision in mobile emergency monitoring. This motivates the proposed hazardous object detection method tailored for coal-fired power emergencies, which combines multi-scale feature fusion and adaptive preprocessing to achieve robust real-time detection.

3 Underground Behavior Detection Method Based on YOLOv8

3.1 Triple-Attention Mechanism

The triple-attention module allocates attention weights by computing feature correlations, thereby enhancing the capture of target features and suppressing irrelevant background noise. It models feature interactions in three dimensions: channel–height, channel–width, and height–width. The overall structure is shown in Fig. 1, consisting of three parallel branches.

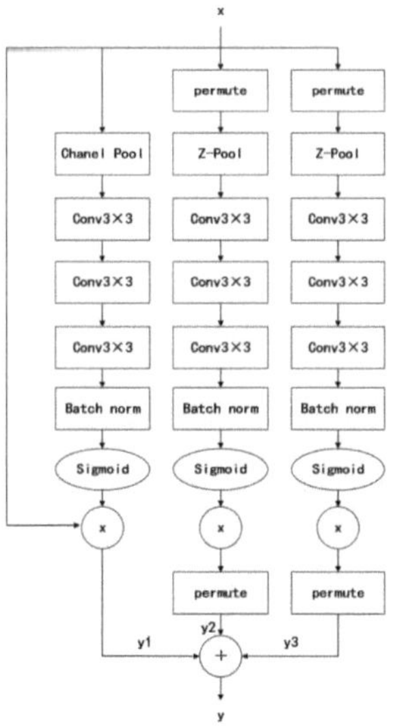

Fig. 1. Structure Diagram of the Triple Attention Module

The first branch focuses on spatial interaction (H, W). Given an input tensor $x \in \mathbb{R}^{C\times H\times W}$, channel pooling compresses the channel dimension, and the output x_1 is processed by three 3×3 convolutions, normalized by Layer Norm, and activated with a Sigmoid function to generate an attention weight map. This weight map is applied element-wise to x, enabling adaptive feature weighting:

$$\boldsymbol{x}_l = \text{channelpool}(\boldsymbol{x}) \tag{1}$$

$$\hat{x}_l = x_l \otimes \delta\left(\text{LN}\left(\text{Conv3} \times 3\left(\text{Conv3} \times 3\left(\text{Conv3} \times 3(x_l)\right)\right)\right)\right) \tag{2}$$

Here, *channelpool* reduces C to 2, LN denotes normalization, and Conv$k \times k$ is a convolution with kernel size $k \times k$.

The second branch models channel–width interactions. The input x is rotated along the W axis, followed by a Z-pool operation compressing H to 2. After three 3×3 convolutions, normalization, and Sigmoid activation, the attention map is applied, and the result is rotated back to restore the original shape:

$$\mathbf{x}_2 = \text{Per}(\text{Z-pool}(\mathbf{x})), \tag{3}$$

$$\hat{\mathbf{x}}_2 = \mathbf{x}_2 \otimes \delta\left(\text{LN}\left(\text{Conv3} \times 3\left(\text{Conv3} \times 3\left(\text{Conv3} \times 3(\mathbf{x}_2)\right)\right)\right)\right) \tag{4}$$

In Formula (3), Per denotes permutation, Z-pool compresses H to 2, δ is the Sigmoid function, and LN is normalization.

The third branch captures channel–height dependencies in a manner similar to the second branch, using transposition, pooling, convolution, normalization, and activation, then restoring the original tensor shape. Finally, the outputs of the three branches are combined:

$$\mathbf{y} = \frac{1}{3}(\mathbf{y}_1 \oplus \mathbf{y}_2 \oplus \mathbf{y}_3) \tag{5}$$

This cross-dimensional interaction mechanism improves global feature representation, strengthens focus on unsafe behaviors, and reduces false alarms in complex coal-fired power enterprise environments.

3.2 Local Attention Feature Filtering Module

To complement the triple-attention mechanism, a local attention feature filtering module is introduced to refine edge details of unsafe behavior features. As shown in Fig. 2, the module applies local average pooling and max pooling on the input feature y, extracting contextual information. Using normalization, ReLU, and Sigmoid, average and maximum attention weights (w_{avg}, w_{max}) are obtained. A 3×3 convolution is then applied to y, followed by weighted multiplication with w_{avg} and w_{max}, and the results are added to produce the final output z:

$$W_{\text{avg}} = \delta\left(\text{Relu}\left(\text{BN}\left(\text{Avg3} \times 3(y)\right)\right)\right) \tag{6}$$

$$W_{\text{max}} = \delta\left(\text{Relu}\left(\text{BN}\left(\text{Max3} \times 3(y)\right)\right)\right) \tag{7}$$

$$z = \left(W_{\text{avg}} \circledast \text{Conv3} \times 3(y)\right) \oplus \left(W_{\text{max}} \circledast \text{Conv3} \times 3(y)\right) \tag{8}$$

This design enhances the model's ability to highlight fine-grained unsafe behavior cues in complex backgrounds, reducing false detections caused by weak feature representations.

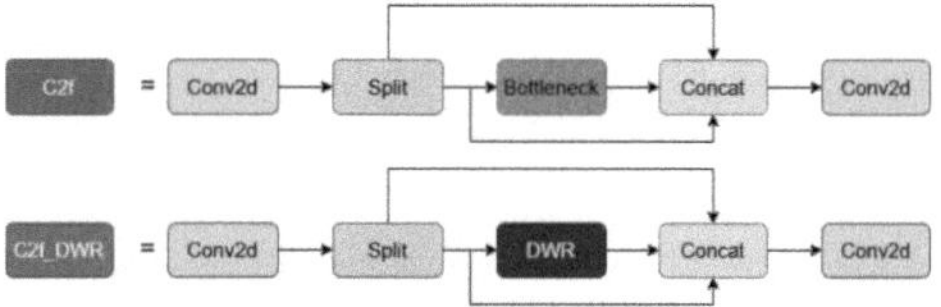

Fig. 2. Structure Diagram of C2f_DWR

To further strengthen multi-scale feature extraction, a C2f_DWR structure is proposed by replacing the Bottleneck unit in the original C2f module with a Dilation-wise Residual (DWR) module. As shown in Fig. 2, this design leverages dilated convolutions with different dilation rates to enlarge the receptive field, enabling the capture of richer contextual information. Residual connections are applied to fuse the original input with processed features, generating a stronger representation and improving the detection of small or subtle unsafe behaviors.

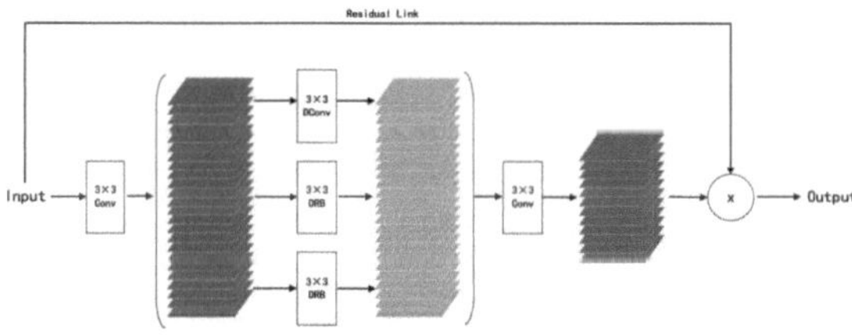

Fig. 3. Structure Diagram of Residual Block

The DWR module Fig. 3 works in two stages. First, 3×3 convolutions, BN, and ReLU expand the channel size from $2c$ to $3c$, establishing a basis for multi-scale extraction. Second, depthwise convolutions with different dilation rates (e.g., rate 2 for 5×5, rate 3 for 7×7) filter features at multiple scales. Outputs are concatenated to $4c$ channels, compressed back to $2c$ via BN and 1×1 convolution, and added to the original input through a residual connection. This produces enriched feature maps that retain original detail while embedding multi-scale semantics.

To balance accuracy and efficiency, the DWR module is introduced into the C2f module before the SPPF layer of the backbone, avoiding excessive computational overhead while preserving high-resolution feature detail. This enhances the model's capability to detect unsafe behaviors under dense, cluttered, and small-target conditions common in coal-fired power enterprise environments.

3.3 Module Integration and Improved Architecture

The two proposed modules were integrated into different stages of the YOLOv8 architecture to construct the complete improved model. Specifically, the triple-attention mechanism was incorporated into the backbone feature extraction stage, enhancing the semantic representation of mid- and high-level features. This integration enables the model to better capture the global characteristics of unsafe behaviors, thereby improving recognition accuracy in complex coal-fired power enterprise environments.

Meanwhile, the local attention feature filtering module was embedded within the neck feature fusion stage. By strengthening the representation of shallow feature maps, this module improves the model's ability to capture fine-grained cues such as edges, textures, and small-scale structural patterns often associated with unsafe actions. This design ensures that both global semantic information and local details are effectively preserved and utilized.

Importantly, the introduction of these modules only extends the original YOLOv8 architecture in a lightweight manner, without significantly increasing computational complexity. Thus, the model maintains high inference efficiency while achieving enhanced detection precision. The overall network data flow is illustrated in Fig. 4.

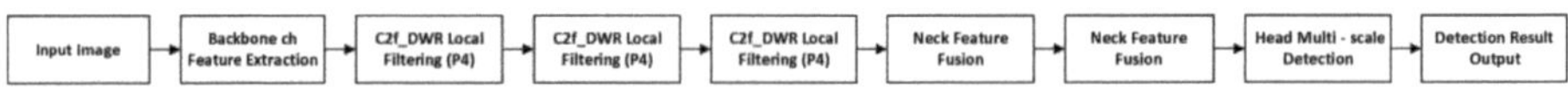

Fig. 4. Network Data Flow Diagram

4 Experimental Results and Analysis

4.1 Experimental Setup

The DsLMF+ dataset, constructed by Xi'an University of Science and Technology, contains 138,004 annotated images of underground coal mine scenes. Six types of targets are labeled: coal mine workers, hydraulic support guard plate postures, large coal blocks, shearer faults, miners' unsafe behaviors, and safety helmet wearing. Labels are provided in YOLO and COCO formats. The dataset covers multiple workplaces, lighting conditions, and complex backgrounds, simulating realistic underground scenarios.

Benchmark Method and Parameter Settings. The original YOLOv8n is used as the benchmark. Comparisons are made with:

1) YOLOv8 + Triple Attention module,
2) YOLOv8 + C2f_DWR module,
3) YOLOv8 + both modules.

The experimental parameters are summarized in Table 1. The model is initialized with COCO-pretrained YOLOv8n weights. Data augmentation includes random scaling, flipping, rotation, and Mosaic for the first 270 epochs. SGD is used with an initial learning rate of 0.01, momentum 0.937, weight decay 0.0005, total epochs 300, batch size 32, and input image size 640×640. Evaluation metrics include Precision, Recall, and mAP@0.5, with mAP under different IoU thresholds (0.3, 0.5, 0.7) for comprehensive assessment.

Table 1. Experimental Parameter Settings

Parameter	Value	Description
epochs	300	Total training epochs
batch	32	Batch size
imgsz	640	Input image size
workers	8	Data loading threads
lr0	0.01	Initial learning rate
lrf	0.01	Final learning rate
weight_decay	0.0005	Prevent overfitting
conf_thres	0.25	Confidence threshold

4.2 Model Performance Analysis

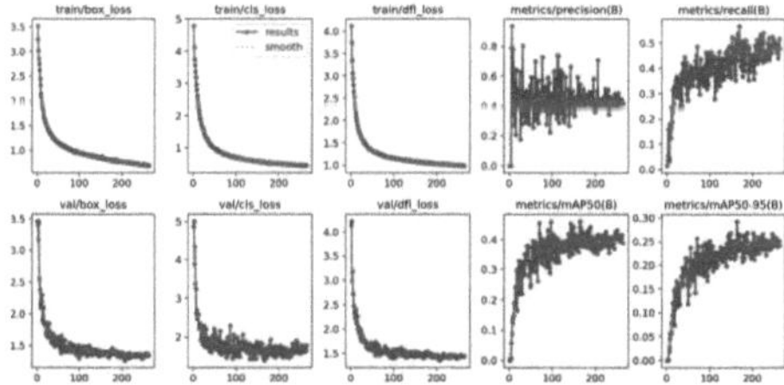

Fig. 5. Training Loss Curves and Performance Metrics

Training Convergence. Figure 5 shows the change of training losses. The model converges well, with train/box_loss dropping from 3.5 to 0.5 by epoch 100. Precision, Recall, and mAP indicators steadily increase, demonstrating good generalization and accurate localization.

Detection Visualization. Figure 6 illustrates detection results in real underground environments. Despite low illumination, occlusion, and irregular target shapes, the model accurately identifies key targets like hydraulic supports and shearers.

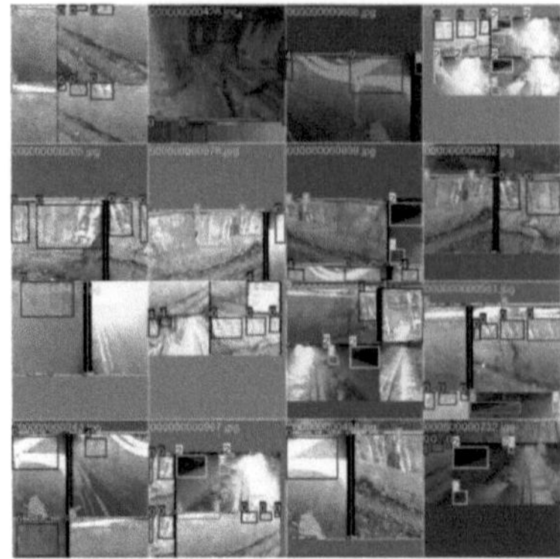

Fig. 6. Example Diagram of Object Detection Visualization Results.

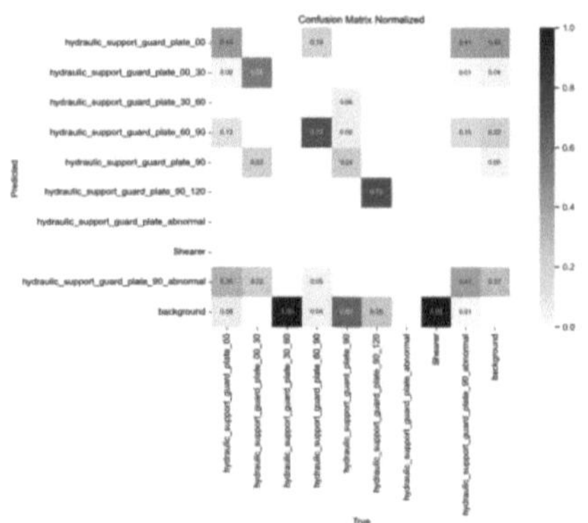

Fig. 7. Visualization of Detection Results

Confusion Matrix. Figure 7 presents the normalized confusion matrix. Categories with sufficient samples show high classification accuracy (0.72–0.75), while small-sample or deformed targets have some misclassification, indicating areas for further improvement.

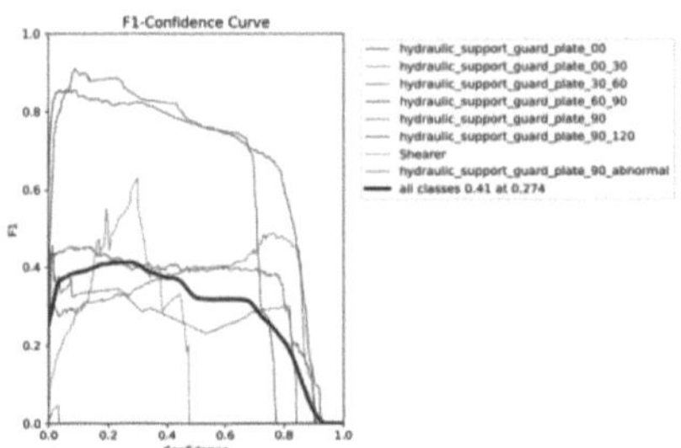

Fig. 8. Normalized Confusion Matrix

F1-Confidence Curve. Figure 8 shows the F1 score across different confidence thresholds. Well-performing categories maintain stable F1 scores, while challenging categories fluctuate, highlighting targets that require further optimization.

4.3 Module Ablation Study

Four model configurations are evaluated:

1) **Baseline:** Original YOLOv8n; 2) **Model 1:** C2f + Triple Attention

3) **Model 2:** C2f_DWR; 4) **Model 3:** C2f_DWR + Triple Attention (Fig. 9).

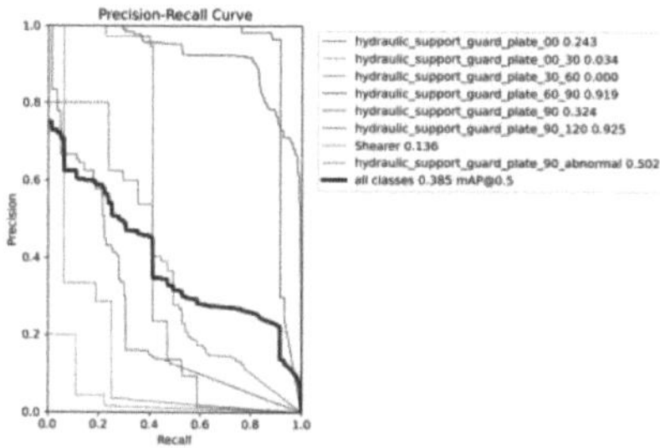

Fig. 9. Caption for first image

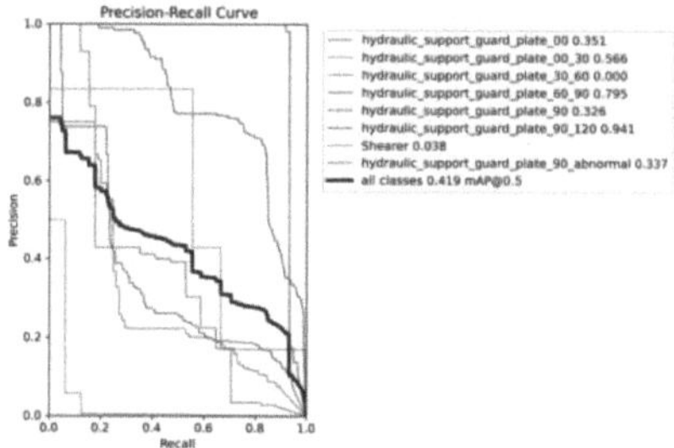

Fig. 10. Caption for second image

Triple Attention Module. As shown in Fig. 10, introducing Triplet Attention improves mAP@0.5 from 0.385 to 0.419 (+3.4%), especially in small-sample categories.

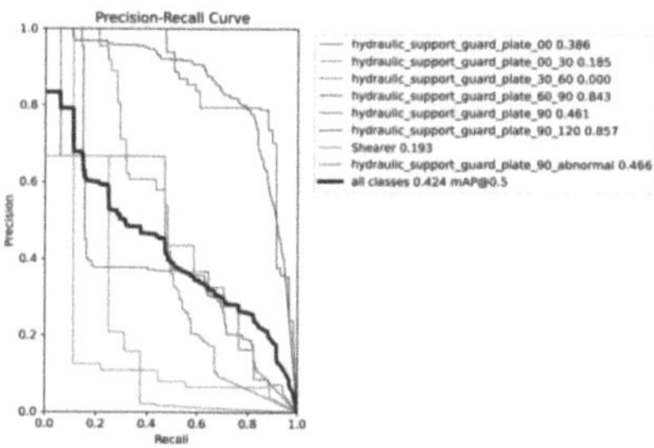

Fig. 11. Caption for first image

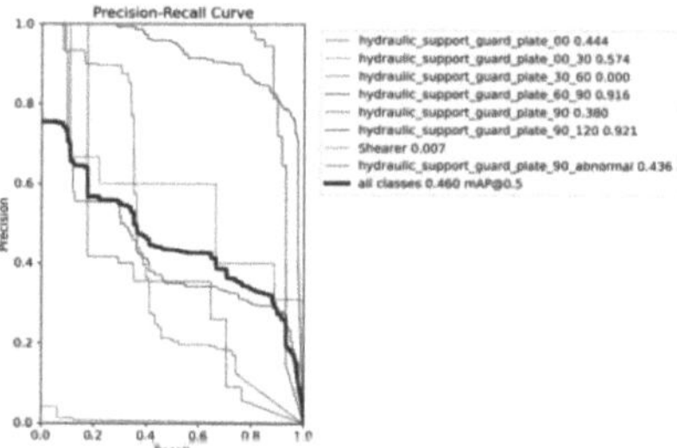

Fig. 12. Caption for second image

C2f_DWR Module. Figure 11 demonstrates that replacing the Bottleneck with C2f_DWR increases mAP@0.5 to 0.424 (+3.9%), improving multi-scale feature extraction and edge detection.

Combined Module. The dual-module scheme (Fig. 12) achieves mAP@0.5 of 0.460 (+7.5%), showing synergistic benefits. Small-sample and difficult targets see significant improvements.

Efficiency and Deployability. Introducing attention and DWR modules slightly reduces inference speed (up to 13 FPS), but the accuracy gain outweighs the cost, making the model deployable in industrial underground monitoring.

4.4 Summary

The experiments show (Table 2):

Table 2. Improved Model Performance Comparison

Model	mAP@0.5	Improvement
Baseline YOLOv8n	0.385	—
C2f + Triplet Attention	0.419	+3.4%
C2f_DWR	0.424	+3.9%
Combined Module	0.460	+7.5%

1) Single modules enhance performance in specific aspects: Triplet Attention improves global feature representation; C2f_DWR enhances small-target recognition and edge detection.
2) The combined module achieves the best overall performance, particularly for small-sample and high-difficulty targets.
3) The model maintains reasonable computational efficiency for practical deployment.

Overall, the improved YOLOv8 model effectively detects unsafe behaviors in complex underground coal mine environments, verifying the practicality of the proposed method.

5 Conclusion and Limitations

As coal-fired power enterprises expand, traditional manual inspection-based safety methods have become inadequate in terms of timeliness and coverage. This paper proposes an improved YOLOv8 model with a Triple Attention module and a local attention feature filtering module for automatic detection and early warning of unsafe behaviors in complex underground environments. A dedicated dataset and ablation experiments verify the model's effectiveness, showing a 7.5% improvement in mAP@0.5 while maintaining real-time performance at 105 FPS. The method significantly reduces missed detections of small targets and dynamic behaviors, providing a solid foundation for intelligent safety monitoring. However, challenges remain, including the need for more robust perception in extreme underground conditions like dust and low illumination, which could be addressed through multi-modal data fusion. Additionally, temporal behavior prediction should be explored for proactive risk assessment. Practical deployment will also require lightweight optimization and integration with existing safety platforms. Future work will focus on developing a closed-loop "perception-analysis-response" early warning system to support intelligent, unmanned safety production in coal-fired power enterprises.

References

1. Xu, Z., Li, J., Zhang, M.: A surveillance video real-time analysis system based on edge-cloud and FL-YOLO cooperation in coal mine. IEEE Access **9**, 68482–68497 (2021)
2. Xu, Z., Li, J., Meng, Y., Zhang, X.: Cap-YOLO: channel attention based pruning YOLO for coal mine real-time intelligent monitoring. Sensors **22**(12), 4331 (2022)
3. Wang, Y., Guo, W., Zhao, S., Xue, B., Zhang, W., Xing, Z.: A big coal block alarm detection method for scraper conveyor based on YOLO-BS. Sensors **22**(23), 9052 (2022)
4. Fan, Y., Mao, S., Li, M., Wu, Z., Kang, J.: CM-YOLOv8: lightweight YOLO for coal mine fully mechanized mining face. Sensors **24**(6), 1866 (2024)
5. Mudavath, T., Niranjan, V.: Person detection in thermal images using kurtosis based histogram enhancement and YOLOv8. Earth Sci. Inf. **18**(2), 381 (2025)
6. Pasupuleti, S., Ramalakshmi, K., Gunasekaran, H., Arokiaraj, R.M., Debnath, S., Jebaseeli, T.J.: An enhancement of object detection using YOLO v8 and mobile net in challenging conditions. SN Comput. Sci. **6**(4), 321 (2025)
7. Yılmaz, A., Yurtay, Y., Yurtay, N.: AYOLO: development of a real-time object detection model for the detection of secretly cultivated plants. Appl. Sci. **15**(5), 2718 (2025)
8. Akdoğan, C., Özer, T., Oğuz, Y.: PP-YOLO: deep learning based detection model to detect apple and cherry trees in orchard based on histogram and wavelet preprocessing techniques. Comput. Electron. Agric. **232**, 110052 (2025)
9. Fan, H., Yan, X., Cao, X.: An intelligent detection algorithm for unsafe personnel states on coal mine belt conveyors based on DSP-YOLO. Coal Sci. Technol. (2025)
10. Yuan, T.: LMFI-YOLO: lightweight pedestrian detection algorithm in complex scenes. Comput. Eng. Appl. **61**(15), 111–123 (2025)
11. Qin, J.Q.: Low illumination image object detection method based on ICFIE-YOLO. Acta Electron. Sin. **53**(2), 514–526 (2025)
12. Shi, L.: Lightweight low-light object detection algorithm based on CDD-YOLO. Comput. Eng. Appl. **61**(6), 106–117 (2025)
13. Ye, F.: DB-YOLOdual backbone YOLOv8 model with feature enhancement fusion for road defect detection. Comput. Eng. Appl. **60**(24), 260–269 (2024)
14. Li, R.: YOLO-sea: improved complex undersea target detection algorithm for YOLOv7-tiny. Comput. Eng. Appl. **61**(2), 247–258 (2025)
15. Kou, F.: Research on target detection in underground coal mines based on improved YOLOv5. J. Electron. Inf. Technol. **45**(220725), 2642 (2023)
16. Cao, S., Dong, L., Deng, F., Gao, F.: A small object detection method for coal mine underground scene based on YOLOv7-SE. J. Mine Autom. **50**(3), 35–41 (2024)
17. Zhao, W., Wang, S., Zhao, D.: Multi object detection of underground unmanned electric locomotives in coal mines based on SD-YOLOv5s-4L. J. Mine Autom. **49**(11), 121–128 (2023)
18. Xue, X., Wang, X., Li, F., Zhu, W.: Underground track foreign object detection method in coal mines based on improved YOLOv7-tiny. J. Optoelectronics·Laser (2025)
19. Ru, H., Liang, Y., Wang, G.: Coal mine underground safety helmet detection method based on YOLOv5 improved by pruning algorithm. J. Heilongjiang Univ. Sci. Technol. **34**(3), 452–456 (2024)
20. Wang, Q., Yang, Y., Chen, L.: Underground personnel behavior detection study based on improved YOLOv8n algorithm. Coal Mine Mach. **46**(5), 195–200 (2025)

Inner-Hole Anomaly Detection with Split-Attention Patch Distribution Model

Fan Yang[1], Jiazheng Xu[2(✉)], Zhenshen Qu[2], and Jinyan Xue[3]

[1] China Electronics Technology Group Corporation 27th Research Institute, Zhengzhou 450047, China
3337477194@qq.com

[2] Harbin Institute of Technology, Harbin 150080, China
190678185@qq.com, miraland@hit.edu.cn

[3] Guochuang Robot Innovation Center (Harbin) Technology Co., Ltd., Harbin 150080, China
xuejinyan@hit.edu.cn

Abstract. Surface defect detection for workpieces with inner holes is a critical task in manufacturing quality control. However, current methods often struggle to satisfy industrial standards due to the scarcity of defective samples and the subtle, indistinct nature of surface anomalies. To address these challenges, we propose a novel anomaly detection framework based on Split-Attention Patch Distribution Modeling. By integrating the split-attention mechanism, our approach significantly enhances feature extraction capabilities, enabling more robust discrimination between normal and anomalous patterns even amidst complex backgrounds. Specifically, we employ multivariate Gaussian distribution modeling to accurately capture intra-feature correlations and utilize the Mahalanobis distance for precise anomaly scoring. Extensive experiments demonstrate that our model outperforms state-of-the-art algorithms in both AUROC and precision. Notably, the proposed method exhibits exceptional robustness in the challenging task of tiny defect detection, marking a significant improvement over existing baselines.

Keywords: Inner hole defect detection · Anomaly detection · Patch distribution modeling · Split-attention

1 Introduction

Inner-hole workpieces are ubiquitous components in mechanical design and manufacturing, widely used in aerospace, automotive, machinery, and defense industries—for instance, automobile/aircraft engine parts and steam turbines [1]. During processing and operation, inner-hole surface integrity is vulnerable to damage, leading to defects like blowholes, scratches, pores, and cracks. These flaws reduce component sealing performance, increase leakage risks of high-pressure fluids/gases, and compromise equipment lifespan and stability. Thus, timely detection, localization, and handling of inner-hole defects are critical to ensuring operational safety, lowering failure rates, and protecting lives and property.

P. Umapada et al. (Eds.): ICCPR 2025, CCIS 2811, pp. 378–390, 2026.
https://doi.org/10.1007/978-981-95-8315-7_30

Traditional manual visual inspection for inner-hole defects is inefficient and error-prone, driving research into machine vision-based automated detection (framed as object detection). Yin et al. combined Otsu's algorithm with the maximum entropy principle for brake oil pipe defect detection [2]. Li proposed a method integrating morphological processing and difference image technology to identify bamboo strip wormholes [3]. Wei developed specialized algorithms: one for damage/inclusion detection via brightness morphology and adaptive thresholding, and another for color variation defects using channel difference analysis [4].

Recently, deep learning has become mainstream in defect detection. Satrajit et al. noted its superior generalization over traditional image processing [5]. By incorporating Inception modules and residual mechanisms, they improved neural network performance in classification tasks, achieving high accuracy and efficiency in commutator inner-hole defect detection [6]. However, these methods rely on supervised learning, which requires large sample sizes and precise defect labeling, hard to satisfy in practical production.

Anomaly detection is an effective defect detection approach, aiming to identify anomalous images and localize anomalies accurately. Compared with conventional learning-based methods, deep learning-based unsupervised anomaly detection requires fewer samples while maintaining high accuracy, thus attracting wide attention in recent years [7]. It is mainly categorized into three paradigms: image reconstruction-based, generative model-based, and deep feature embedding-based methods.

Image reconstruction-based methods train models to only reconstruct normal images; anomalies are identified by comparing original and reconstructed images. Normal samples yield high-quality reconstructions, while anomalous ones do not. Baur et al. first applied autoencoders (AEs) to brain MRI anomaly segmentation [8]; Youkachen et al. used convolutional AEs (CAEs) for hot-rolled strip surface defect segmentation [9]; Nakanishi et al. proposed a weighted frequency domain loss to improve reconstruction clarity and localization accuracy [10]. However, CAE performance is limited by lacking prior information, relying heavily on the latent layer's ability to represent normal features.

To address reconstruction flaws, generative models are introduced to approximate real data distributions. Anomalies correspond to discrepancies between generated and input data, which can be captured in latent/feature spaces (unlike AEs focusing only on final reconstruction). Matsubara et al. applied Variational AEs (VAEs) to industrial anomaly detection [11]; Akcay et al. proposed GANomaly with an "encode-decode-encode" structure [12]; Rudolph et al. developed DifferNet using normalizing flow-based density estimation [13]. However, DifferNet prioritizes image-level classification (not pixel-level localization) and prone to false detections due to poor normal region generation.

The two aforementioned methods often have erroneous reconstructions (caused by insufficient feature-level discrimination) leading to false/missed detections. Deep feature embedding-based methods, by contrast, perform better, consisting of feature extraction and anomaly estimation (generating pixel-level anomaly maps via feature comparison). Defard et al.'s PaDiM uses pre-trained CNN features and multivariate Gaussian models (only for aligned datasets) [14]; Roth et al.'s PatchCore achieves over 99% AUROC on MVTec AD but has poor localization [15].

To address the limitations of existing approaches, this paper proposes an anomaly detection algorithm based on Split-attention Patch Distribution Modeling (SAPaDiM). By incorporating an attention mechanism, the model enhances feature extraction capabilities, strengthens the discriminability between normal and anomalous samples, and improves the detection accuracy for tiny defects and those in complex backgrounds. Experimental results indicate that compared with state-of-the-art algorithms, the proposed method achieves superior performance in terms of AUROC (Area Under the Receiver Operating Characteristic Curve) and precision—with particular advantages demonstrated in the challenging task of tiny defect detection.

2 Split-attention Patch Distribution Model

The feature extraction network generates image feature vectors, which underpin all subsequent operations of the anomaly detection algorithm. Thus, feature extraction effectiveness directly dictates the final accuracy of anomaly detection, classification, and localization. This paper incorporates a split-attention architecture into the anomaly detection network: by integrating a branch attention mechanism, it optimizes cross-channel information fusion. Leveraging the split-attention mechanism, the network accurately assesses the importance of each channel, enabling targeted extraction of region-of-interest features.

The overall architecture of the SAPaDiM is illustrated in Fig. 1, comprising three modules: feature extraction and embedding vector generation, multivariate Gaussian distribution modeling, and Mahalanobis distance scoring. During training, a pre-trained split-attention network performs feature extraction, while multi-level features from the CNN are used to exploit inter-semantic-level correlations. A multivariate Gaussian distribution is then employed to model the probabilistic representation of normal classes. For detection, the Mahalanobis distance between the test image's embedding vector (generated by the feature extraction network) and the pre-constructed Gaussian distribution is computed to identify anomalous regions in inner-hole images.

Figure 2 illustrates the main architecture of the split-attention block. Through continuous splitting and fusion of feature maps, the network achieves sensitive perception and effective integration of multi-channel information. Specifically, it first subdivides the input feature map twice: the first subdivision splits it into t base blocks, each further divided into N independent slices in the second subdivision. These slices are then processed by 1×1 and 3×3 convolutional kernels, respectively, before being fed into the split-attention block for in-depth fusion. After processing, the slice information is concatenated for integration, followed by a 1×1 convolution to restore the feature map to its original channel count.

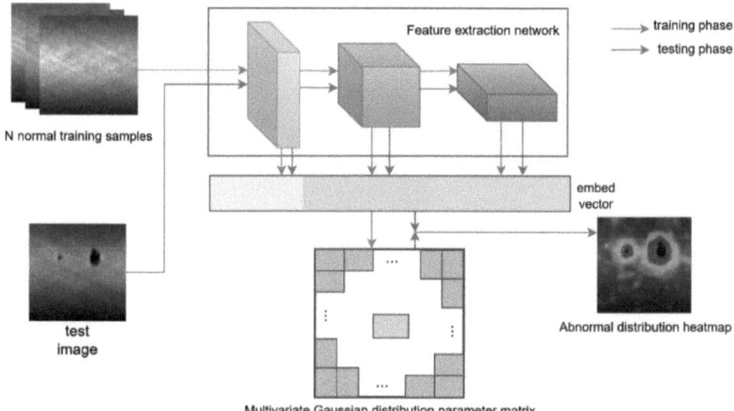

Fig. 1. SAPaDiM Network Architecture

The resulting feature map not only integrates cross-channel information but also retains the same dimensional specifications as the input. This design enables it to function as a modular component, easily embeddable into more complex neural network architectures.

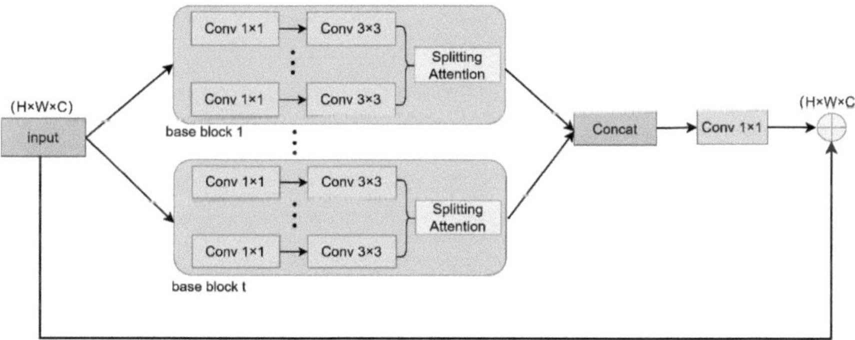

Fig. 2. Architecture of the split-attention block

Figure 3 details the structure of the split-attention block, which processes the N slices (from Fig. 2) after 3×3 convolution. The module first aggregates these slices into a single feature map through summation. It then employs an average pooling layer to capture global contextual information, which is fed into fully connected layers and normalization layers, followed by a ReLU activation function to generate attention weights for each slice. These weights are multiplied with and fused into their corresponding slices, enabling both cross-channel information integration and explicit prioritization of channel importance.

This design allows ResNeSt to model interrelationships between cross-channel features, enabling the network to learn key features more efficiently without increasing parameter count—thereby significantly enhancing recognition accuracy.

To facilitate the explanation of the model principle, let Φ denote the ResNeSt feature extraction module employed. The training set is denoted as $X_N = \{x_k | k = 1, 2, ..., N\}$, which contains N normal sample images. Here, x_k represents the k-th sample image in the training set, with a height of h, width of w, and c channels; the test set is denoted as $Q_M = \{y_q | k = 1, 2, ..., M\}$.

During training, images from the training set are input into the split-attention network for feature extraction, as formulated below:

$$F_k^l = \Phi(x_k) \tag{1}$$

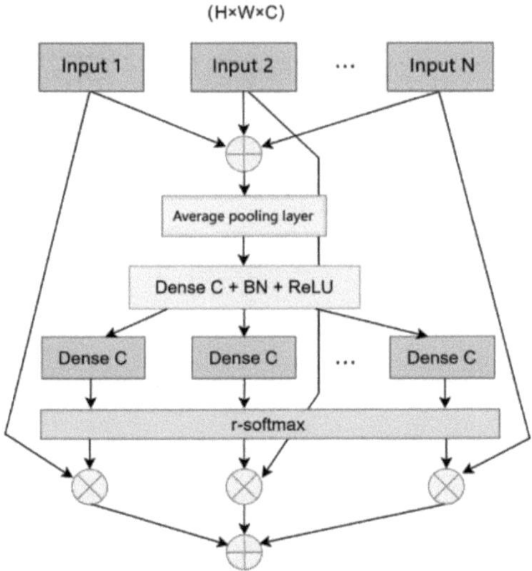

Fig. 3. Detailed structure of split-attention block

Let F_k^l denote the l-th layer feature obtained by extracting features from image x_k. In the experiment, $l = 1, 2, 3$ is selected. The features of each layer are then concatenated, as formulated below:

$$e_k = F_k^0 \oplus \text{upsample}\left(F_k^1\right) \oplus \text{upsample}\left(F_k^2\right) \tag{2}$$

In the equation, e_k denotes the embedding vector formed by concatenating the hierarchical features extracted from image x_k; $\oplus$ represents feature concatenation; and "upsample" refers to the upsampling of features with bilinear interpolation adopted.

3 Distribution Modeling and Distance Scoring

We derive embedded vectors e_k from training images x_k with feature extraction and multi-level feature concatenation. To learn the normal image features at position (i, j), a set of embedded vectors corresponding to this position is calculated from (i, j) normal images.

$$E^{ij} = \left\{e_k^{ij}, k \in [1, N]\right\} \tag{3}$$

Information carried by the set is aggregated and represented by E_{ij}, which is generated from a multivariate Gaussian distribution $N\left(\mu_{ij}, \Sigma_{ij}\right)$. In this distribution, μ_{ij} denotes the sample mean and Σ_{ij} the sample covariance, with their estimation formulas given as follows:

$$\Sigma_{ij} = \frac{1}{N-1}\sum_{k=1}^{N}\left(e_k^{ij} - \mu^{ij}\right)(e_k^{ij} - \mu^{ij})^T + \varepsilon I \tag{4}$$

For each embedding vector of images corresponding to position (i, j), the Gaussian parameters $\left(\mu_{ij}, \Sigma_{ij}\right)$ are trained from the embedding vector set $\left\{e_k^{ij}, k \in [1, N]\right\}$ consisting of N images. A regularization term εI ensures the sample covariance matrix Σ_{ij} is full-rank and invertible, where I denotes the identity matrix and ε is a constant smaller than 0.01.

Finally, Gaussian parameter matrices connect each embedding vector position to a multivariate Gaussian distribution. Consequently, each estimated multivariate Gaussian distribution $N\left(\mu_{ij}, \Sigma_{ij}\right)$ captures the similarity of the same coordinate point across different feature layers.

Anomaly region detection in images is performed by calculating the distance between multi-level features of the test image and their corresponding Gaussian distributions. The Mahalanobis distance is used to compute anomaly scores, where $M\left(e_q^{ij}\right)$ represents the anomaly score of the feature embedding vector at position (i, j) for test image y_q. This score is obtained by calculating the Mahalanobis distance between the embedding feature vector of test image y_q and the learned distribution $N\left(\mu_{ij}, \Sigma_{ij}\right)$. The formula for $M\left(e_q^{ij}\right)$ is as follows:

$$M\left(e_q^{ij}\right) = \sqrt{(e_q^{ij} - \mu^{ij})^T \sum_{ij}^{-1}\left(e_q^{ij} - \mu^{ij}\right)} \tag{5}$$

Thus, the Mahalanobis distance matrix that constitutes the anomaly map can be computed:

$$M = (M\left(e_q^{ij}\right))_{1<i<w, 1<j<h, 1<q<M} \tag{6}$$

In image-level anomaly detection tasks, a test image is classified as anomalous if any of its regions exhibits abnormal features. The algorithm proposed uses the maximum value in the model-predicted anomaly map as the metric to assess the overall anomaly level of test image y_q. The image-level anomaly detection score s is defined as follows:

$$s = \max(M) \quad (7)$$

4 Experiments and Analysis

To evaluate and verify the performance of the proposed method, we constructed the inner-hole anomaly detection dataset. First, the model evaluation metrics and relevant experimental environment configurations are introduced. Subsequently, comparative experiments between the proposed algorithm and other anomaly detection algorithms were conducted on the dataset, with the experimental results and their visualizations presented and evaluated.

4.1 Evaluation Metrics

For the evaluation of detection and segmentation performance, and in line with industrial requirements, AUROC, Accuracy, False Negative Rate (FNR), and False Positive Rate (FPR) are selected as evaluation metrics. Part of the metrics are derived from the confusion matrix, whose basic form is presented in Table 1.

Table 1. Confusion matrix

	Prediction-Positive(P)	Prediction-Negative(N)
Actual-True(T)	TP	FN
Actual-False(F)	FP	TN

Table 2. System requirements

Metric	Requirements
Image-level AUC	≥0.95
Pixel-level AUC	≥0.90
Accuracy	≥0.90
MDR	≤0.05
FDR	≤0.10

- Actual-True (T) refers to actual positive samples, which represent actual defective samples in inner-hole anomaly detection;

- Actual-False (F) refers to actual negative samples, which represent actual non-defective samples in inner-hole anomaly detection;
- Prediction-Positive (P) refers to predicted positive samples, which represent model-predicted defective samples in inner-hole anomaly detection;
- Prediction-Negative (N) refers to predicted negative samples, which represent model-predicted non-defective samples in inner-hole anomaly detection.

Accuracy reflects the accuracy level of a model in prediction. This metric is calculated as the ratio of the number of samples correctly identified by the model to the total number of samples. Generally, the higher the accuracy of a model, the more reliable its prediction effect and the better its performance. The specific calculation formula for Accuracy is as follows:

$$Accuracy = \frac{TP + TN}{TP + FN + FP + TN} \times 100\% \tag{8}$$

To better align with the requirements of actual industrial production processes, this study selected the Missed Detection Rate (MDR) and False Detection Rate (FDR) as key metrics for evaluating model performance. The MDR refers to the proportion of undetected defective products to the total number of actual defective products, while the FDR refers to the proportion of qualified products incorrectly classified as defective products to the total number of actual qualified products.

The required metrics for the practical system are listed as follows:

5 Experimental Environment

The configuration of the computation unit is show in Table 3.

Table 3. Configuration of the computation unit

Configuration	Information
CPU	Intel(R) Core(TM) i7-6300HQ
Memory	12G
GPU	GTX950
Operating system	Windows11 64-bitt
Programming Language	Python3.8
Deep Learning framework	Pytorch

6 Experimental Results

The detection model is trained on the dataset using the Distracted Attention Patch Distribution Model. Partial results of the model's anomaly detection on the test set are shown in Fig. 4, which includes six test images. From left to right, these images are four anomaly-containing images and two normal images.

The anomaly heatmap represents the anomaly score of each pixel—higher scores indicate a greater probability that the corresponding pixel contains an anomaly. The predicted mask map and segmentation results are the outcomes of the model's segmentation and localization of anomaly regions.

From the visualization results, we can see the model can accurately identify scratches, blowholes, and some small-scale defects. It can classify anomaly-containing images and normal images; besides, it can also localize the anomalies. For defects with distinct discriminative features (such as blowholes), the model achieves excellent recognition and segmentation performance, with high-quality segmentation results. Even for more challenging defect recognition and segmentation tasks—such as the small defect located at the edge of the third image—the model can still achieve good recognition and localization. For light-colored scratches (which have high similarity to the background), the model can detect such defects; however, from the anomaly heatmap and final segmentation results, the localization effect is only observed in a portion of the scratch defect. For normal images, the model exhibits good recognition and localization performance. Nevertheless, the anomaly heatmap shows some areas with relatively high values, which are attributed to factors such as uneven illumination.

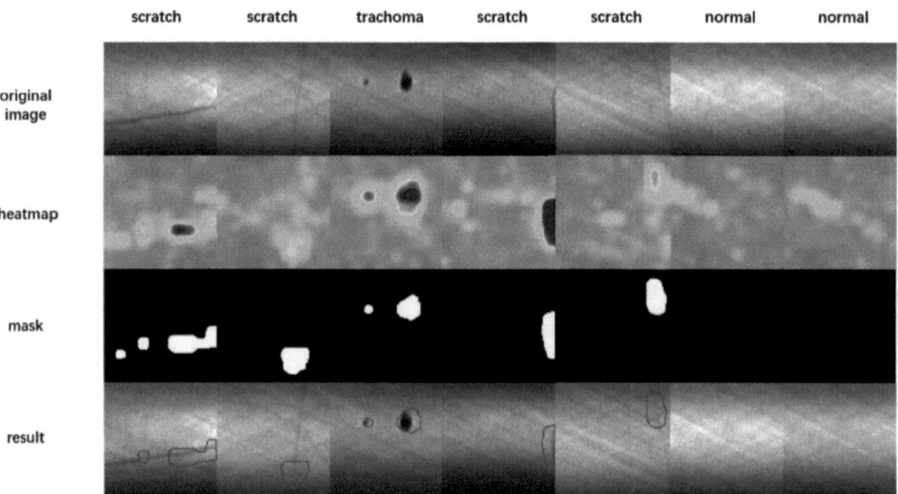

Fig. 4. Visualization of model detection results

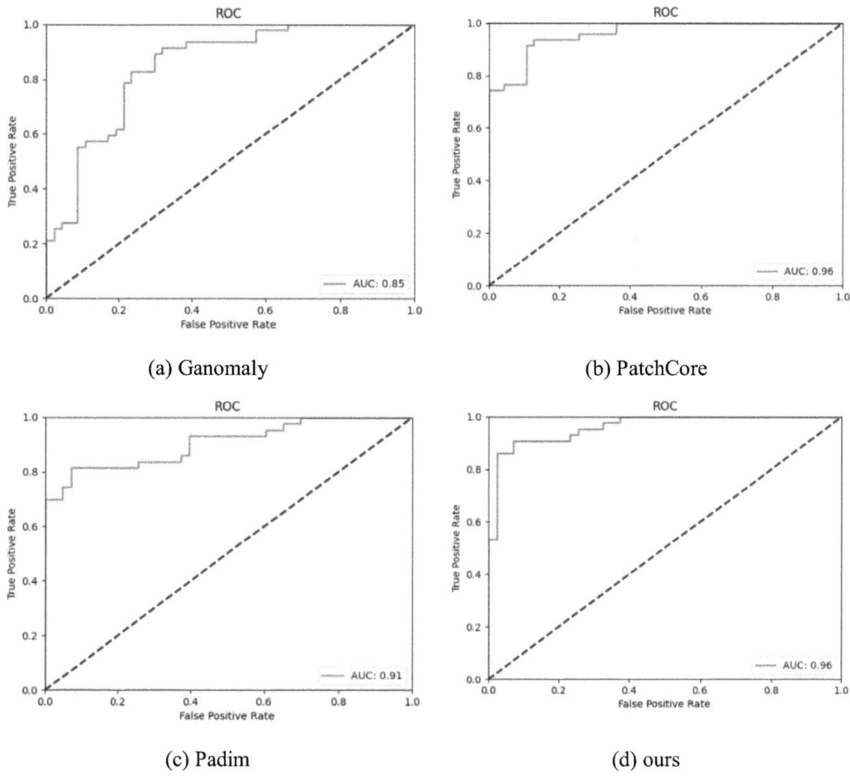

(a) Ganomaly (b) PatchCore

(c) Padim (d) ours

Fig. 5. Image-level AUC result comparison

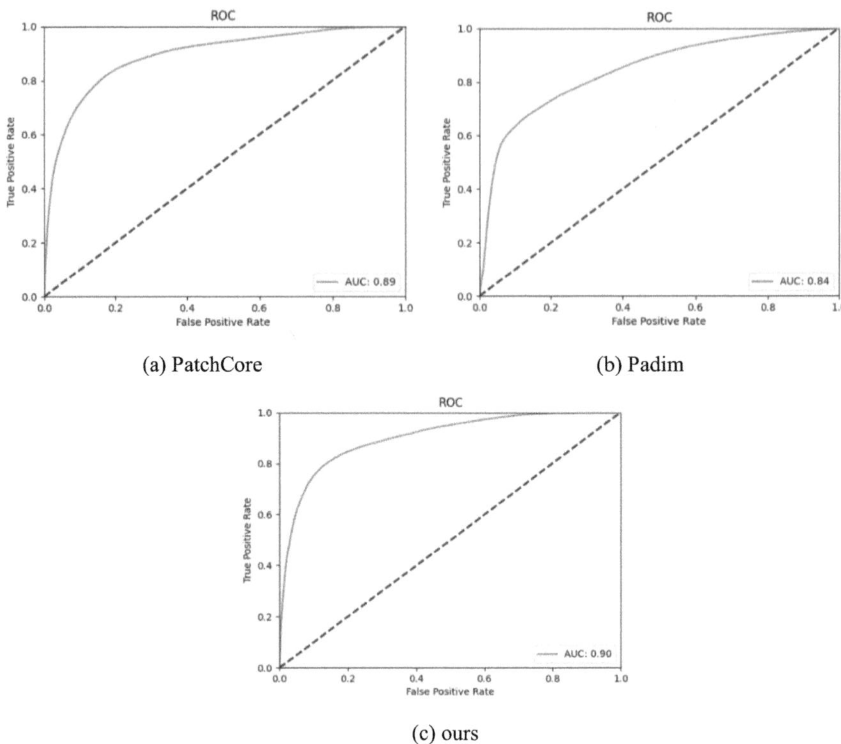

(a) PatchCore

(b) Padim

(c) ours

Fig. 6. Pixel-level AUC result comparison

Meanwhile, the proposed algorithm was compared with existing anomaly detection algorithms in terms of performance metrics on the inner-hole anomaly detection dataset. The image-level AUC curves for each algorithm are presented in Fig. 5, the pixel-level AUC curves in Fig. 6. The mean values of detection results for each category are detailed in Table 4.

Table 4. The average value of test results for each class

Algorithm	Image-level AUC	Pixel-level AUC	Accuracy	MDR	FDR
Ganomaly	0.85	N/A	0.70	0.27	0.38
PathCore	0.96	0.89	0.85	**0.09**	0.21
PaDiM	0.91	0.84	0.79	0.17	0.19
Ours	**0.96**	**0.90**	**0.89**	0.11	**0.11**

By comparing the detection results, the proposed method demonstrates better classification capability, superior localization performance, and higher accuracy than other models for inner-hole anomaly detection, while also attaining the lowest FDR. Among all compared models, the PatchCore model achieves the lowest MDR; however, its FDR is relatively high—indicating a strong bias toward classifying samples as anomalous.

7 Conclusions

The paper proposes a novel anomaly detection method for inner-hole defects. By integrating the attention mechanism, the proposed model can more effectively extract features, distinguish between normal and abnormal samples, and improve the recognition accuracy of tiny defects as well as defects in complex backgrounds. Experimental results demonstrate that compared with other existing algorithms, the proposed method exhibits superior performance in terms of AUROC score and precision—particularly showing its advantages in addressing challenging tiny defect detection tasks. Meanwhile, through comparative analysis, our method also achieves a better balance between the MDR and FDR.

Acknowledgments. This study was funded by the Basic Scientific Research Program (Grant No. JCKY2024210C006).

References

1. Zeng, B.: Research on defect detection system and key technologies for inner hole sidewall. Master's thesis. Hunan University (2016)
2. Yin, L.: Research on defect detection system of brake oil pipe heading based on machine vision. Master's thesis, Harbin Institute of Technology (2018)
3. Li, J.: Design of bamboo strip surface defect detection and color classification system based on machine vision. Master's thesis, Guangxi Normal University (2017)
4. Wei, F.: Research on online detection technology of inner hole defects of commutator based on machine vision. Ph.D. thesis, Huazhong University of Science and Technology (2019)
5. Chatterjee, S., Zielinski, P.: On the generalization mystery in deep learning (2022). arXiv preprint arXiv:2203.10036
6. Zhang, Z.: Research on visual detection method of inner hole defects of motor commutator based on deep learning. Master's thesis, Huazhong University of Science and Technology (2022)
7. He, H., et al.: A diffusion-based framework for multi-class anomaly detection. In: Proceedings of the AAAI Conference on Artificial Intelligence (AAAI 2024), vol. 38, no. 8, pp. 8472–8480. AAAI Press (2024)
8. Baur, C., Wiestler, B., Albarqouni, S., Navab, N.: Deep autoencoding models for unsupervised anomaly segmentation in brain MR images. In: Crimi, A., Bakas, S., Kuijf, H., Menze, B., Reyes, M. (eds.) BrainLes 2018, LNCS, vol. 11384, pp. 161–169. Springer, Cham (2019)
9. Youkachen, S., Ruchanurucks, M., Phatrapomnant, T., Kaneko, H.: Defect segmentation of hot-rolled steel strip surface by using convolutional auto-encoder and conventional image processing. In: 2019 10th International Conference on Information and Communication Technology for Embedded Systems (IC-ICTES), pp. 1–5. IEEE, Piscataway (2019)
10. Nakanishi, M., Sato, K., Terada, H.: Anomaly detection by autoencoder based on weighted frequency domain loss (2021). arXiv preprint arXiv:2105.10214
11. Matsubara, T., Sato, K., Hama, K., Tachibana, R., Uehara, K.: Deep generative model using unregularized score for anomaly detection with heterogeneous complexity. IEEE Trans. Cybern. **52**(6), 5161–5173 (2020)
12. Akcay, S., Atapour-Abarghouei, A., Breckon, T.P.: GANomaly: semi-supervised anomaly detection via adversarial training. In: Asian Conference on Computer Vision. LNCS, vol. 11363, pp. 622–637. Springer, Cham (2018)

13. Rudolph, M., Wandt, B., Rosenhahn, B.: Same same but DifferNet: semi-supervised defect detection with normalizing flows. In: IEEE/CVF Winter Conference on Applications of Computer Vision (WACV), pp. 1907–1916. IEEE, Piscataway (2021)
14. Defard, T., Setkov, A., Loesch, A., Audigier, R.: PaDiM: a patch distribution modeling framework for anomaly detection and localization. In: International Conference on Pattern Recognition (ICPR). LNCS, vol. 12664, pp. 475–489. Springer, Cham (2021)
15. Roth, K., Pemula, L., Zepeda, J., Schölkopf, B., Brox, T., Gehler, P.: Towards total recall in industrial anomaly detection. In: IEEE/CVF Conference on Computer Vision and Pattern Recognition (CVPR), pp. 14318–14328. IEEE, Piscataway (2022)

Key Technologies of Computer Vision and Image Processing Based on Multimodal Fusion

MAG-VCNet: Multi-Scale Attention-Guided Visual Concept Classifier for Pulmonary Nodule Assessment

Tianwen Wang[1], Guangxu Li[2], and Tohru Kamiya[1](✉)

[1] Kyushu Institute of Technology, Fukuoka City, Japan
kamiya@cntl.kyutech.ac.jp
[2] Tiangong University, Tianjin, China

Abstract. Accurate classification of pulmonary nodules is critical due to their strong association to lung cancer. However, despite advances in CT imaging, the inherent variability and morphological ambiguity of nodules remain significant barriers to reliable differentiation between benign and malignant forms. To overcome this challenge, we propose a novel multi-scale classification framework for pulmonary nodules based on structure-aware learning. The model introduces directionally-constrained von Mises-Fisher (vMF) clustering and spatial assignment coefficients to learn discriminative structural prototypes in an unsupervised manner. These prototypes effectively capture stable and interpretable patterns within pulmonary nodules. Moreover, the model integrates multi-scale features from the first four ResNet-18 stages via a Feature Pyramid Network (FPN), enabling the construction of a unified high-dimensional representation that encodes both fine-grained details and contextual semantics. CBAM attention modules are incorporated at each backbone stage to enhance structural responsiveness. The final classification is performed using a multilayer perceptron (MLP) based on a probabilistic visual concept embedding derived from the fused features. Evaluated on the LIDC-IDRI dataset, MAG-VCNet achieves a test AUC of 90.57%, outperforming several state-of-the-art methods. These results demonstrate the effectiveness of combining structure-aware learning, vMF-based visual concepts, and multi-scale fusion for robust and interpretable pulmonary nodule classification.

Keywords: Pulmonary nodule classification · Structure-aware learning · multi-scale feature fusion · Visual concept embedding · von Mises-Fisher clustering

1 Introduction

Lung cancer, with its high incidence and mortality, represents a foremost threat to human health [1, 2]. Its asymptomatic early progression significantly undermines survival and complicates timely clinical intervention. Computed tomography (CT) [3, 4], with its high spatial resolution and multi-slice imaging capabilities, significantly enhances the

P. Umapada et al. (Eds.): ICCPR 2025, CCIS 2811, pp. 393–402, 2026.
https://doi.org/10.1007/978-981-95-8315-7_31

efficiency and accuracy of pulmonary abnormality detection. It plays a critical clinical role in the early detection and management of high-mortality diseases such as lung cancer, effectively improving treatment success rates and patient survival outcomes.

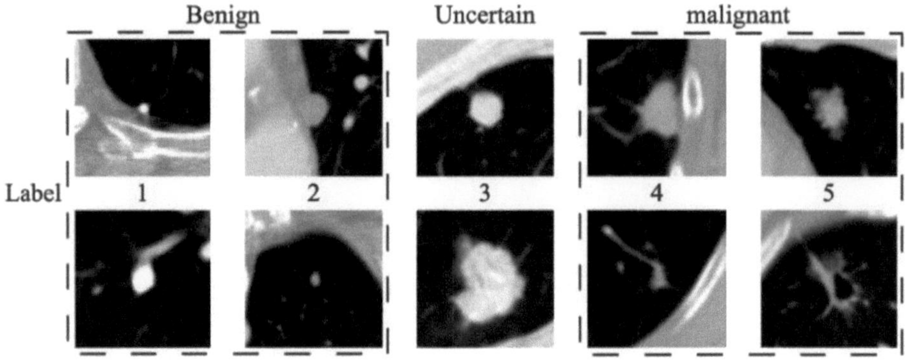

Fig. 1. Visualization of nodules used for classification.

Due to their small size, indistinct margins, and morphological variability, pulmonary nodules are challenging to detect and evaluate on CT scans. Accurate diagnosis remains highly dependent on radiologists' expertise and experience [5]. Against this backdrop, numerous lung nodule classification methods have been proposed to improve diagnostic accuracy and efficiency, while also alleviating the workload of radiologists. In pulmonary nodule classification tasks, nodules are typically categorized into five types based on their imaging characteristics, as illustrated in Fig. 1. Categories 1 and 2 are generally considered benign, while categories 4 and 5 are classified as malignant. Category 3, due to its ambiguous features and uncertain diagnostic outcomes, is often excluded during model training to enhance classification accuracy and stability. Broadly, these approaches can be categorized into two main types: traditional classification methods based on handcrafted features, and deep learning-based methods that automatically learn features through neural networks.

Traditional image recognition methods predominantly rely on mathematical theory and rule-based feature extraction techniques. Handcrafted features are constructed based on geometric structures, texture patterns, or statistical properties of the image, and are subsequently classified using conventional classifiers such as support vector machines (SVM), k-nearest neighbors (k-NN), or Bayesian classifiers. Kawata et al. [6] proposed a hybrid classification approach that integrates k-means clustering with a linear discriminant analysis, leveraging curvature and CT density features to differentiate between benign and malignant pulmonary nodules. Srija et al. [7] proposed a KNN and logistic regression-based classification approach, incorporating image feature extraction, 10-fold cross-validation, and statistical analysis to model and assess the classification performance between pulmonary nodule patients and healthy subjects. Li et al. [8] improved the random forest algorithm by integrating grey-level co-occurrence matrix (GLCM) and Gabor filter techniques to extract texture and edge features of pulmonary nodules and further optimized the classifier structure for intelligent classification of benign and malignant nodules in CT images.

Traditional image segmentation methods, while not reliant on large, annotated datasets, are limited by complex design, low efficiency, and lack of end-to-end capability, making them less suitable for modern data-intensive and high-efficiency applications. Deep learning–based methods, with their end-to-end modeling capability and powerful feature learning capacity, have increasingly become a research hotspot and a prevailing trend in the field of pulmonary nodule detection and classification [9]. Huang et al. [10] proposed a 3D convolutional neural network framework (SSTL-DA) that integrates adaptive slice selection (ASS), self-supervised representation learning, and domain-adaptive transfer learning, effectively enhancing the accuracy of benign-malignant classification of lung nodules using limited annotated CT data. Sun et al. [11] proposed the Nodule-CLIP model, which integrates 3D U-Net segmentation, contrastive learning between image and attribute features, and multi-feature alignment mechanisms to effectively capture the underlying associations between CT images and diagnostic labels in pulmonary nodule classification. This approach significantly enhances both classification performance and model interpretability. Wang et al. [12] proposed a multi-scale residual neural network (MResNet) that integrates a ResNet backbone with a pyramid pooling module (PPM), enabling efficient extraction and fusion of multi-scale features of pulmonary nodules. This approach significantly improves model performance and enhances its clinical applicability in benign–malignant classification tasks. Kaushik et al. [13] proposed the Unsupervised Generative Transition (UGT) method, which enhances robustness in image classification under occlusion and complex out-of-distribution (OOD) conditions by unsupervised learning of a transitional von Mises-Fisher (vMF) feature dictionary between source and target domains, combined with spatial structure modelling derived from the source domain.

Inspired by the above two articles, we propose the MAG-VCNet model, a structure-aware classification framework tailored for pulmonary nodule analysis. The model first processes three-view CT images using a modified ResNet-18 [14] backbone, in which the max-pooling layer is removed to preserve spatial resolution. Each stage of the backbone is enhanced using the Convolutional Block Attention Module (CBAM) [15] to enhance structural responsiveness. To better model the fine-grained structural characteristics of pulmonary nodules, multi-scale features derived solely from the first four hierarchical stages are integrated via a top-down Feature Pyramid Network (FPN) [16], yielding a unified 128-channel high-dimensional representation that effectively captures both local detail and contextual semantics. Subsequently, a visual concept dictionary is constructed using von Mises-Fisher (vMF) [17] clustering, which transforms the fused features into a probabilistic structural representation p that models the directional distribution of critical visual patterns. Finally, a multilayer perceptron (MLP) [18] classifier is employed to perform the final classification, enabling robust modelling and precise discrimination of subtle structural variations in pulmonary nodules. The main contributions of this work are as follows:

1. We propose a structure-aware multi-scale classification framework that integrates traditional structural modelling concepts with modern deep learning techniques.

2. To capture the core discriminative structures of pulmonary nodules, directionally-constrained von Mises-Fisher clustering is applied to construct structural prototypes. Additionally, group-wise depth wise convolutions are utilized to encode local context, enhancing structural awareness while maintaining parameter efficiency and interpretability.
3. Drop Block regularization is introduced to simulate potential local occlusions and information loss in medical images, effectively mitigating overfitting and significantly enhancing the model's robustness and generalization capability.

2 Method

2.1 A Subsection Overall Framework

In this chapter, we provide a detailed explanation of the proposed model, MAG-VCNet. The architecture of the model is illustrated in Fig. 2. Centered on the pulmonary nodule, 64×64 2D slices are extracted from the axial, coronal, and sagittal views and concatenated channel-wise to form a three-channel image, which serves as the model input to capture spatial structural information from multiple views.

The proposed model employs a pre-trained ResNet-18 as the backbone network, with structural adaptations made to accommodate the relatively small input image size. Specifically, the original 7×7 convolutional kernel is replaced with a 3×3 kernel, and the initial max pooling layer is removed to prevent early down sampling and associated information loss, thereby enhancing the model's ability to retain fine-grained structural features of pulmonary nodules. To enhance the network's ability to model critical features, the Convolutional Block Attention Module (CBAM) is integrated after each stage of the ResNet-18 backbone. By sequentially applying channel and spatial attention mechanisms, CBAM enables the model to automatically focus on the most discriminative regions of pulmonary nodules, thereby improving feature representation and enhancing classification performance.

To more effectively model the structural characteristics of pulmonary nodules, the proposed framework employs the first four stages of a ResNet-18 backbone enhanced with CBAM to extract multi-scale, attention-refined feature maps. These features are then integrated via a Feature Pyramid Network (FPN) to produce a unified $128 \times 64 \times 64$ feature map that preserves both local detail and global semantic context. This fused representation is subsequently fed into a von Mises-Fisher (vMF) clustering module to construct a visual dictionary that captures prototypical structural patterns of nodules. By projecting each sample into the space defined by these cluster centers, the model achieves feature compression and enhances the compactness of the representation.

This structured abstraction facilitates more accurate differentiation of nodule morphologies in downstream classification tasks, while simultaneously improving model robustness and interpretability. The resulting cluster centers serve as stable and discriminative morphological prototypes in the structural feature space. To further model the spatial heterogeneity of structural responses, a spatial coefficient module is introduced to perform position-aware weighting across the feature map, yielding a spatially sensitive structural representation. This representation is subsequently flattened and passed through a Multi-Layer Perceptron (MLP) [18] classifier to produce the final prediction of nodule malignancy.

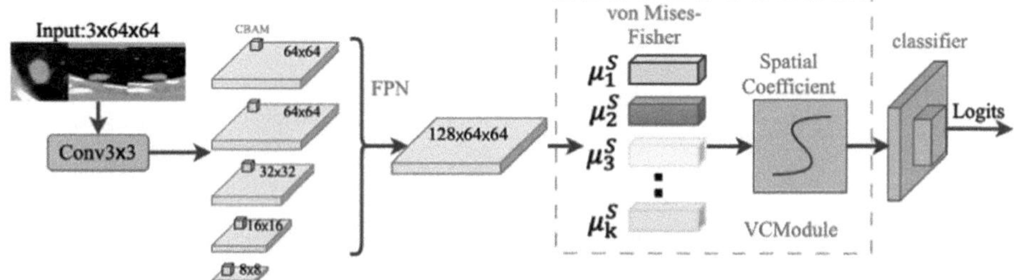

Fig. 2. Architecture of MAG-VCNet.

2.2 Attention-Feature Pyramid Network

For small pulmonary nodules, their diminutive size, irregular morphology, and the frequent presence of complex surrounding structures—such as blurred boundaries and low contrast—collectively result in a limited set of extractable discriminative features. To address this challenge, this study incorporates an attention mechanism module into the network architecture, enabling the model to adaptively allocate greater attention weights to critical feature regions, thereby effectively suppressing irrelevant features and noise and enhancing the robustness and expressiveness of feature extraction. Furthermore, in the process of feature propagation within the network, convolutional operations and down sampling procedures, such as pooling, inevitably lead to information loss. To mitigate this issue, this study further integrates a cross-level feature fusion strategy, allowing the network to comprehensively aggregate multi-level and multi-scale semantic information. This approach partially compensates for the feature information loss caused by increased network depth and down sampling, thereby effectively alleviating information degradation during feature propagation and enhancing the completeness and robustness of feature representations.

The introduction of the lightweight Convolutional Block Attention Module (CBAM) has enhanced the network's capability for feature selection: the spatial attention module assists the network in identifying "where to focus," while the channel attention module directs the network to determine "what to focus on" [19].

By effectively integrating local fine-grained details with global contextual information, this strategy improves the model's ability to accurately identify small pulmonary nodules.

FPN is a top-down feature fusion architecture featuring lateral connections [20], as depicted in Fig. 3. The feature maps produced by the first four layers of the ResNet-18 network, after being recalibrated by the CBAM attention module, are subsequently fed into the FPN. Initially, FPN employs 1×1 convolutions to align the channel dimensions of each feature map, thereby ensuring dimensional consistency. Starting from the highest-level feature map, FPN then progressively upsamples these feature maps to achieve spatial resolution matching with the adjacent lower-level feature maps. At each fusion stage, lateral connections are leveraged to perform element-wise addition between the upsampled features and the corresponding lower-level features of the same resolution, thereby ensuring the effective integration of deep semantic information with

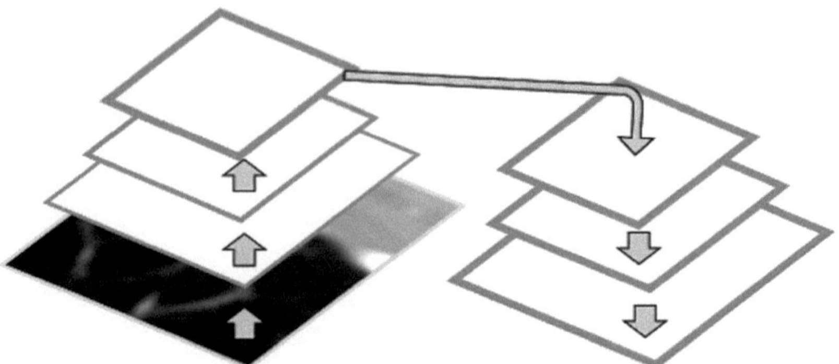

Fig. 3. Architecture of MAG-VCNet.

shallow, fine-grained details. This iterative fusion process proceeds in a top-down manner, culminating in the generation of multi-scale feature maps that are well-suited for downstream tasks.

As the network depth increases, the detailed features of pulmonary nodules gradually diminish during the feature propagation process. To maximize the extraction and utilization of these detailed features, this paper performs feature fusion only on the output feature maps of the first four layers of the network. Prior to fusion, an attention mechanism is introduced to enhance the discriminative power of feature weights, thereby improving the quality of the fused feature representations. Additionally, the attention mechanism's ability to filter and prioritize features effectively reduces redundant information, resulting in more compact and efficient fused features. This approach ultimately helps to better preserve and enhance the local details of pulmonary nodules.

2.3 Attention-Feature VCModule

The multi-scale feature map generated by the Feature Pyramid Network (FPN) is fed into the VCModule, enabling the unsupervised extraction of visual concepts based on the von Mises-Fisher (vMF) dictionary. Although the dataset includes label supervision during training, the learning process of the structural prototypes (i.e., feature centers mu) is inherently unsupervised. These prototypes are not directly optimized using ground truth labels but rather emerge through similarity-based soft clustering driven by structural consistency within the fused features. The choice of the number of clusters is crucial as it directly impacts the model's accuracy and interpretability. Too few prototypes may prevent the model from capturing the complexity of the data, leading to underfitting, while too many prototypes can introduce irrelevant features, affecting the model's learning and generalization. Based on the complexity of the data, task requirements, and experimental results, we carefully selected the optimal number of clusters, balancing the model's expressiveness and generalization ability to ensure its effectiveness and stability. The vMF-based representation thus allows the network to discover stable and interpretable structural patterns in a self-organizing manner, without explicit label guidance. The input feature map is first processed by a structure-aware context enhancement module, which strengthens the modeling of local spatial structures within each channel by extracting spatial patterns independently. Additionally, a DropBlock operation is employed during training to randomly occlude local regions, simulating partial structural loss or occlusion. This mechanism effectively improves the model's robustness to missing local information, blurred boundaries, and incomplete structures, encouraging the network to focus on stable and discriminative structural patterns.

The structural probability map p is derived by computing the normalized cosine similarity between the input feature map and the vMF structural prototype dictionary mu, followed by a SoftMax operation. This map encodes the probabilistic assignment of each spatial location to the learned structural prototypes, effectively capturing the distributional tendency of each pixel within the structural semantic space. The structural probability map p undergoes global average pooling to produce an image-level embedding, which is classified by a multilayer perceptron (MLP) under label supervision to learn discriminative semantics from structure-aware representations.

3 Method

We evaluate the classification performance of the proposed model using the LIDC-IDRI dataset [21]. CT images and corresponding nodule annotations are extracted, preprocessed, and saved using the PyLIDC toolkit. A total of 3,634 images of size 64x64 are obtained after preprocessing, including 2,027 benign nodules and 1,528 malignant nodules. The dataset is then split into training and testing sets using an 8:2 ratio. T The model uses the Adam optimizer and was trained for 10 epochs with a batch size of 4 to optimize its classification performance. During the training process, the learning rate and other hyperparameters were appropriately adjusted to ensure stable convergence and efficient learning. The final learning rate was set to 1e-4. Due to the smaller batch size, which allows for more frequent gradient updates, the model was able to adapt to the data more quickly, resulting in a faster convergence process. The use of small batches also introduces a moderate level of randomness, helping to avoid overfitting and improving the model's generalization ability. To comprehensively assess the classification performance, we adopt five commonly used evaluation metrics: Accuracy, Area Under the Receiver Operating Characteristic Curve (AUC), Precision, Recall, and F1-score. Table 1 separately presents the training and testing results of the experiments.

Table 1. Table captions should be placed above the tables.

	Training	Testing
AUC (%)	91.27	90.57
Acc (%)	81.78	84.21
Precision (%)	82.08	81.74
Recall (%)	87.18	83.74
F1 (%)	84.56	82.73

Table 2. Comparison between models with and without DropBlock regularization.

	AUC (%)	Acc (%)	Precision (%)	Recall (%)	F1 (%)
DropBlock	90.57	84.21	81.74	83.74	82.73
Without DB	88.89	80.75	73.79	88.96	80.67

Table 3. Performance of the proposed method with other methods

	AUC (%)	Acc (%)	Precision (%)	Recall (%)	F1 (%)
CNN [21]	88.56	79.64	82.08	70.25	75.70
ImprovedCNN [22]	89.24	81.85	84.84	72.09	77.94
DAC-Xception [23]	87.24	79.64	79.34	74.23	76.70
Nodule-CLIP [24]	88.17	79.50	74.32	83.44	78.61
Swim-Transformer [25]	85.44	77.95	78.57	72.09	73.78
MAG-VCNet	90.57	84.21	81.74	83.74	82.73

The model demonstrates stable performance during both the training and testing phases, indicating strong generalization ability and effective classification capability for pulmonary nodules.

Table 2 presents the classification results on the testing set, comparing models with and without the DropBlock regularization. The results show that introducing DropBlock leads to consistent improvements across all evaluation metrics, with the most notable gain observed in Precision, which increases from 73.79% to 81.74%. This indicates that DropBlock helps reduce false positives. In addition, AUC and Accuracy improve by approximately 1.68% and 3.46%, respectively, demonstrating enhanced generalization capability. Although Recall slightly decreases (from 88.96% to 83.74%), the overall F1-score still improves by 2.06%, suggesting a better balance between precision and recall.

Table 3 presents a performance comparison between our proposed model and other existing methods. ImprovedCNN achieves the highest precision, but its recall is relatively low (72.09%), indicating that it may miss some actual malignant nodules. In contrast, Nodule-CLIP demonstrates strong sensitivity in detecting malignant nodules but suffers from a higher false positive rate. In summary, our proposed model outperforms others in all evaluation metrics except precision, highlighting its advantages in overall classification capability, model robustness, and sensitivity.

4 Result

Accurate identification and proactive classification of pulmonary nodules are of great significance for safeguarding patient health and enhancing clinical diagnostic efficiency. Thereby, this paper proposes a structure-aware multi-scale classification framework,

MAG-VCNet. MAG-VCNet, built on a ResNet backbone and integrated with a von Mises–Fisher (vMF) clustering mechanism, enables structure-aware, multi-scale feature extraction for robust lung nodule classification. Evaluated on the LIDC-IDRI dataset, the model achieves superior overall performance, demonstrating strong generalization and reliability, particularly in high-recall clinical screening scenarios.

Acknowledgments. This work was supported by China Scholarship Council. (CSC No. 202209347003).

Disclosure of Interests. Authors have no competing interests.

References

1. Schluger, N.W., Koppaka, R.: Lung disease in a global context: a call for public health action. Ann. Am. Thorac. Soc. **11**(3), 407–416 (2014). https://doi.org/10.1513/AnnalsATS.201312-420PS
2. Siegel, R.L., Miller, K.D., Wagle, N.S., Jemal, A.: Cancer statistics, 2023. CA Cancer J. Clin. **73**(1), 17–48 (2023). https://doi.org/10.3322/caac.21763
3. Karagür, E.R.: Advances in lung cancer diagnosis. Curr. Res. Health Sci. **III**, 81 (2023)
4. Wang, T., Nelson, R.A., Bogardus, A., Grannis, F.W., Jr.: Five-year lung cancer survival: which advanced stage nonsmall cell lung cancer patients attain long-term survival? J. Thorac. Oncol. **5**(10), 1226–1232 (2010). https://doi.org/10.1097/JTO.0b013e3181e8b3f4
5. Winkel, D.J., et al.: Others: a novel deep learning-based computer-aided diagnosis system improves the accuracy and efficiency of radiologists in reading biparametric magnetic resonance images of the prostate. Invest. Radiol. **56**(10), 605–613 (2021). https://doi.org/10.1097/RLI.0000000000000780
6. Kawata, Y., et al.: Hybrid classification approach of malignant and benign pulmonary nodules based on topological and histogram features. In: MICCAI 2000. LNCS, vol. 1935, pp. 297–306. Springer, Berlin, Heidelberg (2000)
7. Srija, G.C., Thirunavukkarasu, U.: Classification of normal and nodule lung images from LIDC-IDRI datasets using K-NN and logistic classifiers. In: ICETMCCT 2021. AIP Publishing (2023)
8. Li, X.X., Li, B., Tian, L.F.: Others: automatic benign and malignant classification of pulmonary nodules in thoracic computed tomography based on RF algorithm. IET Image Process. **12**(7), 1253–1264 (2018)
9. Cai, G., et al.: Medical AI for early detection of lung cancer: a survey (2024). arXiv:2410.14769
10. Huang, H., Wu, R., Li, Y., Peng, C.: Self-supervised transfer learning based on domain adaptation for benign-malignant lung nodule classification on thoracic CT. IEEE J. Biomed. Health Inform. **26**(8), 3860–3871 (2022). https://doi.org/10.1109/JBHI.2022.3161533
11. Sun, L., Zhang, M., Lu, Y., Zhu, W., Yi, Y., Yan, F.: Nodule-CLIP: lung nodule classification based on multi-modal contrastive learning. Comput. Biol. Med. **175**, 108505 (2024). https://doi.org/10.1016/j.compbiomed.2024.108505
12. Wang, H., Zhu, H., Ding, L.: Others: a diagnostic classification of lung nodules using multiple-scale residual network. Sci. Rep. **13**(1), 11322 (2023)
13. Kaushik, P., Kortylewski, A., Yuille, A.: A Bayesian approach to OOD robustness in image classification. In: CVPR 2024, pp. 22988–22997 (2024)

14. Wang, T., Li, G., Tang, C., Kamiya, T.: Res BCDU-SCB Net: a lung CT image segmentation based on SC-BLSTM. In: ICBSIP 2023, pp. 42–46 (2023)
15. Woo, S., Park, J., Lee, J.-Y., Kweon, I.S.: CBAM: convolutional block attention module. In: ECCV 2018, pp. 3–19 (2018)
16. Kirillov, A., Girshick, R., He, K., Dollár, P.: Panoptic feature pyramid networks. In: CVPR 2019, pp. 6399–6408 (2019)
17. Gopal, S., Yang, Y.: Von Mises-Fisher clustering models. In: ICML 2014, pp. 154–162. PMLR (2014)
18. Desai, M., Shah, M.: An anatomization on breast cancer detection and diagnosis employing multi-layer perceptron neural network (MLP) and convolutional neural network (CNN). Clin. eHealth **4**, 1–11 (2021). https://doi.org/10.1016/j.ceh.2021.01.001
19. Liu, Z., et al.: Swin transformer: hierarchical vision transformer using shifted windows. In: ICCV 2021, pp. 10012–10022. IEEE (2021)
20. Lin, T.Y., Dollár, P., Girshick, R., He, K., Hariharan, B., Belongie, S.: Feature pyramid networks for object detection. In: CVPR 2017, pp. 2117–2125 (2017)
21. Suji, R.J., Bhadouria, S.S., Dhar, J., Godfrey, W.W.: Optical flow methods for lung nodule segmentation on LIDC-IDRI images. J. Digit. Imaging **33**(5), 1306–1324 (2020). https://doi.org/10.1007/s10278-020-00365-3
22. Yang, X., Yu, S., Xu, W.: Enhanced convolutional neural networks for improved image classification (2025). arXiv:2502.00663
23. Sun, L., Zhang, M., Lu, Y., others: Improved Xception with dual attention mechanism and feature fusion for face forgery detection (2024). arXiv:2001.00001
24. Sun, L., Zhang, M., Lu, Y.: Others: Nodule-CLIP: lung nodule classification based on multi-modal contrastive learning. Comput. Biol. Med. **175**, 108505 (2024). https://doi.org/10.1016/j.compbiomed.2024.108505
25. Liu, Z., Lin, Y., Cao, Y., others: Swin Transformer: hierarchical vision transformer using shifted windows. In: ICCV, pp. 10012–10022 (2021). https://doi.org/10.1109/ICCV48922.2021.00982

Consistency-Induced Incomplete Multi-view Clustering

Jinghai Chen and Hong Jia(✉)

Guangdong Provincial Key Laboratory of Intelligent Information Processing, College of Electronics and Information Engineering, Shenzhen University, Shenzhen, China
2310433040@email.szu.edu.cn, hongjia1102@szu.edu.cn

Abstract. Multi-view data commonly existed in different practical applications, such as disease diagnosis, multimedia analysis and recommendation systems. However, sometimes the collected data may have samples with missing information in some views. Incomplete multi-view clustering (IMVC) is an important unsupervised analysis method for this kind of data. However, most of the existing IMVC methods focus on the consistency of different views while ignoring the potential inconsistencies between them. To solve this problem, we propose a novel graph learning based multi-view clustering method by taking into account both of consistent similarity and inconsistent similarity between samples under different views. Subsequently, a missing information completion method based on multi-view consistency is proposed for incomplete data analysis. Extensive experiments show the promising performance of the proposed method in comparison with several state-of-the-art algorithms.

Keywords: Multi-view clustering · Incomplete data · Graph learning

1 Introduction

Multi-view learning has become one of the research hotspots recently, including multi-view clustering. Most multi-view clustering methods are based on the common assumption that all view information is fully captured. But in practical applications, not all sample views are available in many cases, which leads to the failure of traditional multi-view clustering methods. Clustering on such incomplete multi-view data is called incomplete multi-view clustering (IMVC).

Generally, there are two simplest methodsto treat incomplete data: (1) Delete the samples with missing view data, and then cluster the remaining samples with full view. (2) Set the missing data to zero or the average values [17]. The first method defeats the purpose of clustering to divide all data points into their own clusters. In the second method, the filled missing data will bring extra noise in the clustering process, especially in the case of high missing rate. Therefore, we need a better method to deal with the missing data.

From the perspective of the main learning models adopted, we can divide the existing IMVC methods into four categories [13]: (1) Matrix factorization(MF)

P. Umapada et al. (Eds.): ICCPR 2025, CCIS 2811, pp. 403–414, 2026.
https://doi.org/10.1007/978-981-95-8315-7_32

based IMVC, such as DAIMC [4] and SEMI [11]. These methods use matrix decomposition to learn the consistent feature representation of incomplete data for clustering. (2) Kernel based IMVC, such as MKKMIK [9] and EE-R-IMVC [8]. These methods learn kernels from every view to explore high-dimensional relationships between samples for subsequent missing data repair and clustering. (3) Graph based IMVC, such as TCIMC [15] and NGSP-CGL [14]. These methods learn a consensus graph from single view graphs. (4) Deep learning based IMVC, such as IMC-DSM [18] and CPSPAN [5]. These methods perform feature representation and clustering via neural networks. Among them, graph learning is a promising multi-view clustering technique that has attracted a lot of attention. However, most of the existing graph based IMVC methods focus on the consistency of the multi-view, while ignoring the potential multi-view inconsistencies, which often bring extra noise in the learning of the consensus graph.

To solve this problem, we propose a new multi-view Clustering method based on graph learning, called Consistency induced incomplete multi-view Clustering. This method divides the similarity graph into two parts. One is called consistent graph, which represents the consistent information between different views. The other one is called inconsistent graph, which represents the inconsistent information contained by each view. By learning these two kind of graphs, the multi-view consistency and inconsistency can be well explored. At the same time, we propose a missing information completion method based on multi-view consistency, which extends the model to incomplete multi-view clustering.

2 The Proposed Method

2.1 Consistency and Inconsistency Award Clustering

Define complete multi-view data $\mathbf{X} = \left\{\mathbf{X}^1, \mathbf{X}^2, \cdots, \mathbf{X}^V\right\}$, where $\mathbf{X}^v \in \mathbb{R}^{n \times d_v}$ represents the data for the v-th view, n represents the number of samples, d_v represents the feature dimension of v-th view. The similarity graph of the v-th view is defined as $\mathbf{S}^v \in \mathbb{R}^{n \times n}$, $\mathbf{S}^v \geq 0$. In order to represent consistent information and inconsistent information between multiple views, we divide $\mathbf{S}^v$ into two parts: consistent embedding $\mathbf{S}_a^v$ and inconsistent embedding $\mathbf{S}_b^v$, let $\mathbf{S}^v = \mathbf{S}_a^v + \mathbf{S}_b^v$. It is more reasonable to learn the uniform similarity graph $\mathbf{S}$ from the consistent embedding $\mathbf{S}_a^v$, so we have:

$$\min_{\mathbf{S}} \sum_{v=1}^{V} \|\mathbf{S} - \mathbf{S}_a^v\|_F^2 \quad , \quad s.t. \mathbf{S} \geq 0 \tag{1}$$

Inconsistency can be seen as a concept broader than noise. Due to the diversity of different views, it is intuitive that the inconsistent parts should be inconsistent with each other, otherwise it contradicts the inconsistent definition. To ensure that inconsistencies between different views are sparse, we should make the sum of the products of the inconsistencies small. In addition, we usually do not expect the inconsistency to be too large, so we have:

$$\min_{S_b^V} \sum_{v,w=1}^{V} \mathbf{B}_{vw} \operatorname{Tr}\left(\boldsymbol{S}_b^v \cdot \boldsymbol{S}_b^{wT}\right) \tag{2}$$

where $\mathbf{B}$ is a parameter matrix composed of hyperparameters γ and β, whose diagonal elements are γ and the remaining elements are β.

2.2 Local Structure

To explore the local structure, we introduce the grouping effect (also known as smooth regularization) [3]: if two instances x_i and x_j are similar, then their subspace representations s_i and s_j should also be close to each other.

In our method, we learn the consistent embedding $\mathbf{S}_a^v$ from $\mathbf{S}^v$ for the subsequent clustering, which can be guided by the grouping effect. In multi-view clustering, we expect that the local information in all views can directly affect the learning of the unified similarity graph. For the v-th view, if x_i^v and x_j^v are similar, that is, the similarity $\mathbf{S}_{ij}^v$ is large, then $\mathbf{S}_{i:}$ and $\mathbf{S}_{j:}$ should also be close to each other, which can be achieved by optimizing the following:

$$\min_{\mathbf{S}_a^v} \frac{1}{2} \sum_{i,j=1}^{n} \mathbf{S}_{ij}^v \left\|\mathbf{S}_{i:} - \mathbf{S}_{j:}\right\|_2^2 \tag{3}$$

Let the Laplacian matrix of $\mathbf{S}^v$ be $\mathbf{L}_S^v = \mathbf{D}_S^v - \mathbf{S}^v$, where $\mathbf{D}_S^v$ is the degree matrix of $\mathbf{S}^v$, $\mathbf{D}_{S.ii}^v = \sum \mathbf{S}_{i:}^v$, then Eq. (3) can be equivalent to:

$$\min_{\mathbf{S}} \sum_{v=1}^{V} \operatorname{Tr}\left(\boldsymbol{S} \cdot \boldsymbol{L}_S^v \cdot \boldsymbol{S}^T\right) \tag{4}$$

2.3 General Clustering Model

Finally, we apply a low-rank constraint on $\mathbf{S}$ with nuclear norm $\|\cdot\|_*$ to reduce the redundancy of the matrix and enhance the clustering structure. And for different views, adaptive weight coefficient α is introduced. The general clustering model:

$$\begin{aligned}
\min_{\mathbf{S},\{\mathbf{S}_a^v\}_{v=1}^V,\{\mathbf{S}_b^v\}_{v=1}^V,\boldsymbol{\alpha}} & \sum_{v=1}^{V} \left[\|\mathbf{S} - \alpha^v \mathbf{S}_a^v\|_F^2 + \lambda_1 \alpha^v \operatorname{Tr}\left(\mathbf{S} \cdot \mathbf{L}_\mathbf{S}^v \cdot \mathbf{S}^T\right)\right] \\
& + \lambda_2 \|\mathbf{S}\|_* + \sum_{v,w=1}^{V} \alpha^v \alpha^w \mathbf{B}_{vw} \operatorname{Tr}\left(\mathbf{S}_b^v \cdot \mathbf{S}_b^{wT}\right) \\
s.t.\mathbf{S}^v = & \mathbf{S}_a^v + \mathbf{S}_b^v, \mathbf{S} \geq 0, \mathbf{S}_a^v \geq 0, \mathbf{S}_b^v \geq 0, \boldsymbol{\alpha} > 0, \sum \boldsymbol{\alpha} = 1
\end{aligned} \tag{5}$$

2.4 Incomplete Clustering Model

For incomplete multi-view clustering, we adopt a missing information recovery strategy based on the consistent graphs $\{\mathbf{S}_a^v\}_{v=1}^V$. Therefore, the optimization of incomplete clustering model is divided into two stages. First, we learn a series of graphs $\{\mathbf{S}_a^v\}_{v=1}^V$ that are structurally reliable. Then, repair the missing data based on $\{\mathbf{S}_a^v\}_{v=1}^V$ and perform clustering.

To represent the missing data on each view and align incomplete multi-view data, we define a series of matrices $\{\mathbf{G}^v\}_{v=1}^V$, $\mathbf{G}^v \in \mathbb{R}^{n \times n_v}$:

$$\mathbf{G}_{ij}^v = \begin{cases} 1, & \text{if } x_i \text{ is the j-th sample in v-th view} \\ 0, & \text{otherwise} \end{cases}$$

Then the incomplete multi-view data can be represented as $\hat{\mathbf{X}} = \{\mathbf{X}^1\mathbf{G}^1, \mathbf{X}^2\mathbf{G}^2, \cdots, \mathbf{X}^V\mathbf{G}^V\}$. The similarity graph we get from each single view is $\mathbf{S}^v \in \mathbb{R}^{n_v \times n_v}$. n_v represents the number of unmissed samples in v-th view.

Define $\widetilde{\mathbf{S}}_b^v$ and $\widetilde{\mathbf{S}}^v$ as matrices after the missing part of $\mathbf{S}_b^v$ and $\mathbf{S}^v$ is repaired, $\widetilde{\mathbf{S}}_b^v, \widetilde{\mathbf{S}}^v \in \mathbb{R}^{n \times n}$. We can rewrite Eq. (4) as:

$$\min_{\mathbf{S}} \sum_{v=1}^{V} \operatorname{Tr}\left(\boldsymbol{S} \cdot \boldsymbol{L}_{\tilde{S}}^v \cdot \boldsymbol{S}^T\right) \tag{6}$$

where $\mathbf{L}_{\tilde{S}}^v$ is the Laplacian matrix of $\tilde{\mathbf{S}}^v$.

Then we can further rewrite Eq. (5) as:

$$\begin{aligned} \min_{\mathbf{S}, \mathbf{S}_a^v, \mathbf{S}_b^v, \alpha} & \sum_{v=1}^{V} \left[\left\| \mathbf{G}^{vT} \mathbf{S} \mathbf{G}^v - \alpha^v \mathbf{S}_a^v \right\|_F^2 + \lambda_1 \alpha^v \operatorname{Tr}\left(\mathbf{S} \cdot \mathbf{L}_{\tilde{\mathbf{S}}}^v \cdot \mathbf{S}^T\right) \right] \\ & + \lambda_2 \|\mathbf{S}\|_* + \sum_{v,w=1}^{V} \alpha^v \alpha^w \mathbf{B}_{vw} \operatorname{Tr}\left(\widetilde{\mathbf{S}_b^v} \cdot \widetilde{\mathbf{S}_b^w}^T\right) \\ s.t. \mathbf{S}^v = & \mathbf{S}_a^v + \mathbf{S}_b^v, \mathbf{S} \geq 0, \mathbf{S}_a^v \geq 0, \mathbf{S}_b^v \geq 0, \boldsymbol{\alpha} > 0, \sum \boldsymbol{\alpha} = 1 \end{aligned} \tag{7}$$

By initializing $\widetilde{\mathbf{S}^v}$ with the mean value and optimizing Eq. (5) to convergence, we can learn the reliable consistent structure, so that we can reliably fix missing information. At this time, in order to make full use of the recovered data, the objective function can be modified as follows:

$$\begin{aligned} \min_{\mathbf{S}, \mathbf{S}_a^v, \mathbf{S}_b^v, \alpha} & \sum_{v=1}^{V} \left[\left\| \mathbf{S} - \alpha^v \widetilde{\mathbf{S}^v} \right\|_F^2 + \lambda_1 \alpha^v \operatorname{Tr}\left(\mathbf{S} \cdot \mathbf{L}_{\tilde{\mathbf{S}}}^v \cdot \mathbf{S}^T\right) \right] \\ & + \lambda_2 \|\mathbf{S}\|_* + \sum_{v,w=1}^{V} \alpha^v \alpha^w \mathbf{B}_{vw} \operatorname{Tr}\left(\widetilde{\mathbf{S}_b^v} \cdot \widetilde{\mathbf{S}_b^w}^T\right) \\ s.t. \mathbf{S}^v = & \mathbf{S}_a^v + \mathbf{S}_b^v, \mathbf{S} \geq 0, \mathbf{S}_a^v \geq 0, \mathbf{S}_b^v \geq 0, \boldsymbol{\alpha} > 0, \sum \boldsymbol{\alpha} = 1 \end{aligned} \tag{8}$$

By solving Eq. (8), we can learn a consensus graph $\mathbf{S}$ for subsequent clustering.

2.5 Incomplete Graph Recovery Strategy

Since the inconsistent embedding $\mathbf{S}_b^v$ is the part that needs to be removed, it is not necessary to recover missing information, so we can directly fill zeros to the missing part, setting $\widetilde{\mathbf{S}_b^v} = \mathbf{G}^v\mathbf{S}_b^v\mathbf{G}^{vT}$. As for $\mathbf{S}^v$, directly setting $\widetilde{\mathbf{S}^v} = \mathbf{G}^v\mathbf{S}^v\mathbf{G}^{vT}$ will lead to loss of information.

Therefore, we propose an incomplete graph recovery strategy. Specifically, we use the consistent graph $\mathbf{S}_a^v$ to find the k nearest neighbors of the missing samples, and then the weighted sum of the embedding in the missing view of these neighbors is used as the recovery embedding. We can define an indicator matrix $\mathbf{A}^{vw} \in \mathbb{R}^{n\times n}$ to represent the sample neighbor relationship between the v-th view and the w-th view, defining as:

$$\mathbf{A}_{ij}^{vw} = \begin{cases} 1, & \text{if } x_i \text{ is missing in v-th view and} \\ & x_i \text{ is neighbor of } x_j \text{ in w-th view} \\ 0, & \text{otherwise} \end{cases}$$

Note that by definition, $\mathbf{A}^{vw}$ is not equivalent to $\mathbf{A}^{wv}$. For $\mathbf{A}^{vw}$, only if the i-th sample is missing in the v-th view, neither the i-th sample nor the j-sample is missing in the w-view, and they are neighbors of each other by $\mathbf{S}_a^v$, $\mathbf{A}_{ij}^{vw}$ will be set as 1. Then we can recover $\widetilde{\mathbf{S}^v}$ by following:

$$\widetilde{\mathbf{S}_{i:}^v} = \frac{1}{V}\sum_{\substack{w=1 \\ w\neq v}}^{V} \frac{\sum_{j=1}^{n} \mathbf{A}_{ij}^{vw} \cdot \widetilde{\mathbf{S}_a^w}_{ij} \cdot \widetilde{\mathbf{S}_{j:}^v}}{\sum_{j=1}^{n} \mathbf{A}_{ij}^{vw} \cdot \widetilde{\mathbf{S}_a^w}_{ij}} \tag{9}$$

where $\widetilde{\mathbf{S}_a^w} = \mathbf{G}^w\mathbf{S}_a^w\mathbf{G}^{wT}$. The above equation recoveries the data in row vectors. We further let $\widetilde{\mathbf{S}_{:i}^v} = \widetilde{\mathbf{S}_{i:}^v}$ to make the similarity graph a symmetric matrix.

Algorithm 1: K-nearest neighbor incomplete graph recovery strategy

Input: Incomplete similarity graph $\mathbf{S}^v$, Indicator matrix $\mathbf{G}^v$, Consistent graph $\mathbf{S}_a^v$, number of neighbors k

Output: Similarity graph $\widetilde{\mathbf{S}^v}$

```
Initialize S̃^v, calculate A^vw by S_a^v and G^v;
for v = 1,2,...,V do
    Update S̃^v_i: by Eq(9);
    Let S̃^v_:i = S̃^v_i:;
end
```

3 Optimization

We optimize the objective function by using an alternating minimization algorithm. To make the problem separable, we introduce an auxiliary variable $\mathbf{C}$ for the common representation matrix $\mathbf{S}$, let $\mathbf{C} = \mathbf{S}$. Then the corresponding augmented Lagrangian matrix of Eq. (5) can be written as follows:

$$\begin{aligned}\min_{\mathbf{S},\mathbf{S}_a^v,\mathbf{S}_b^v,\boldsymbol{\alpha}} &\sum_{v=1}^{V}\left[\left\|\mathbf{G}^{vT}\mathbf{S}\mathbf{G}^v-\alpha^v\mathbf{S}_a^v\right\|_F^2+\lambda_1\alpha^v\operatorname{Tr}\left(\mathbf{S}\cdot\mathbf{L}_{\tilde{\mathbf{S}}}^v\cdot\mathbf{S}^T\right)\right]+\lambda_2\|\mathbf{C}\|_* \\ &+\sum_{v,w=1}^{V}\alpha^v\alpha^w\mathbf{B}_{vw}\operatorname{Tr}\left(\widetilde{\mathbf{S}_b^v}\cdot\widetilde{\mathbf{S}_b^w}^T\right)+\frac{\mu}{2}\left\|\mathbf{S}-\mathbf{C}+\frac{\mathbf{Y}}{\mu}\right\|_F^2\end{aligned} \tag{10}$$

By doing the same processing on Eq. (8), we can obtain:

$$\begin{aligned}\min_{\mathbf{S},\mathbf{S}_a^v,\mathbf{S}_b^v,\boldsymbol{\alpha}} &\sum_{v=1}^{V}\left[\left\|\mathbf{S}-\alpha^v\widetilde{\mathbf{S}^v}\right\|_F^2+\lambda_1\alpha^v\operatorname{Tr}\left(\mathbf{S}\cdot\mathbf{L}_{\tilde{\mathbf{S}}}^v\cdot\mathbf{S}^T\right)\right]+\lambda_2\|\mathbf{C}\|_* \\ &+\sum_{v,w=1}^{V}\alpha^v\alpha^w\mathbf{B}_{vw}\operatorname{Tr}\left(\widetilde{\mathbf{S}_b^v}\cdot\widetilde{\mathbf{S}_b^w}^T\right)+\frac{\mu}{2}\left\|\mathbf{S}-\mathbf{C}+\frac{\mathbf{Y}}{\mu}\right\|_F^2\end{aligned} \tag{11}$$

3.1 Update S

With $\mathbf{S}_b^v$ and $\mathbf{C}$ fixed, the variable $\mathbf{S}$ is updated. Specifically, Eq. (10) can be rewritten as follows:

$$\mathcal{L}_1(\mathbf{S})=\sum_{v=1}^{V}\left[\left\|\mathbf{G}^{vT}\mathbf{S}\mathbf{G}^v-\alpha^v\mathbf{S}_a^v\right\|_F^2+\lambda_1\alpha^v\operatorname{Tr}\left(\mathbf{S}\cdot\mathbf{L}_{\tilde{\mathbf{S}}}^v\cdot\mathbf{S}^T\right)\right]+\frac{\mu}{2}\left\|\mathbf{S}-\mathbf{C}+\frac{\mathbf{Y}}{\mu}\right\|_F^2 \tag{12}$$

Let $\frac{\partial\mathcal{L}_1(\mathbf{S})}{\partial\mathbf{S}}=0$, $\mathbf{E}=\sum_{v=1}^{V}2\alpha^v\mathbf{S}_a^v+\mathbf{G}^{vT}(\mu\mathbf{C}-\mathbf{Y})\mathbf{G}^v$, $\mathbf{D}^v=\left(2\mathbf{I}+\mu\mathbf{I}+2\lambda_1\alpha^v\mathbf{L}_{\tilde{S}}^v\right)$, $\mathbf{D}_i=[\sum_v\mathbf{W}_{i1}^v\mathbf{D}_{i1}^v,\sum_v\mathbf{W}_{i2}^v\mathbf{D}_{i2}^v,\cdots,\sum_v\mathbf{W}_{in}^v\mathbf{D}_{in}^v]$, then $\mathbf{S}$ can be updated as follows:

$$\mathbf{S}_{i:}\mathbf{D}_i=\mathbf{E}_{i:} \tag{13}$$

where $\mathbf{W}^v\in\mathbb{R}^{n\times n}$ is defined as follows:

$$\mathbf{W}_{ij}^v=\begin{cases}1, & \text{if } x_i \text{ and } x_j \text{ is both unmissed in v-th view}\\ 0, & \text{otherwise}\end{cases}$$

Similarly, Eq. (11) can be rewritten as follows:

$$\mathcal{L}_2(\mathbf{S})=\sum_{v=1}^{V}\left[\left\|\mathbf{S}-\alpha^v\widetilde{\mathbf{S}^v}\right\|_F^2+\lambda_1\alpha^v\operatorname{Tr}\left(\mathbf{S}\cdot\mathbf{L}_{\tilde{\mathbf{S}}}^v\cdot\mathbf{S}^T\right)\right]+\frac{\mu}{2}\left\|\mathbf{S}-\mathbf{C}+\frac{\mathbf{Y}}{\mu}\right\|_F^2 \tag{14}$$

Let $\frac{\partial\mathcal{L}_2(\mathbf{S})}{\partial\mathbf{S}}=0$, we have:

$$\mathbf{S}\cdot\left[2\sum_{v=1}^{V}\left(\mathbf{I}+\lambda_1\alpha^v\mathbf{L}_{\tilde{S}}^v\right)+\mu\mathbf{I}\right]=2\sum_{v=1}^{V}\alpha^v\widetilde{\mathbf{S}^v}+\mu\mathbf{C}-\mathbf{Y} \tag{15}$$

Then $\mathbf{S}$ can be updated by solving Eq. (15).

3.2 Update C, Y

With $\mathbf{S}_b^v$ and $\mathbf{S}$ fixed, the variable $\mathbf{S}$ is updated. We can update $\mathbf{C}$ via the singular value thresholding (SVT) operator:

$$\mathbf{C} = \mathbf{U}_C \delta_{\frac{\lambda_2}{\mu}} \left(\Sigma_C\right) \mathbf{V}_C^T \tag{16}$$

where $\mathbf{U}_C \Sigma_C \mathbf{V}_C^T$ denotes the singular value decomposition(SVD) of $\left(\mathbf{S} + \frac{\mathbf{Y}}{\mu}\right)$, $\delta_{\frac{\lambda_2}{\mu}}(\cdot)$ denotes the shrinkage operator, which can be defined as:

$$\delta_{\frac{\lambda_2}{\mu}} \left(\Sigma_C\right) = \max\left(0, \Sigma_C - \frac{\lambda_2}{\mu}\right) + \min\left(0, \Sigma_C + \frac{\lambda_2}{\mu}\right) \tag{17}$$

The multiplier $\mathbf{Y}$ can be updated by $\mathbf{Y} = \mathbf{Y} + \mu\left(\mathbf{S} - \mathbf{C}\right)$.

3.3 Update $\mathbf{S}_a^v, \mathbf{S}_b^v$

With $\mathbf{C}$ and $\mathbf{S}$ fixed, the variable $\mathbf{S}_b^v$ is updated. Specifically, as $\mathbf{S}_a^v = \mathbf{S}^v - \mathbf{S}_b^v$, Eq. (10) can be rewritten as follows:

$$\mathcal{L}\left(\widetilde{\mathbf{S}_b^v}\right) = \sum_{v=1}^{V} \left[\left\|\mathbf{G}^v \mathbf{G}^{vT} \mathbf{S} \mathbf{G}^v \mathbf{G}^{vT} - \alpha^v \left(\mathbf{S}^v - \mathbf{S}_b^v\right)\right\|_F^2\right] + \sum_{v,w=1}^{V} \alpha^v \alpha^w \mathbf{B}_{vw} \operatorname{Tr}\left(\widetilde{\mathbf{S}_b^v} \cdot \widetilde{\mathbf{S}_b^w}^T\right) \tag{18}$$

Let $\frac{\partial \mathcal{L}\left(\widetilde{\mathbf{S}_b^v}\right)}{\partial \widetilde{\mathbf{S}_b^v}} = 0$, we can obtain the following system of linear equations:

$$\mathbf{M} \cdot \begin{bmatrix} vec\left(\widetilde{\mathbf{S}_b^1}\right) \\ vec\left(\widetilde{\mathbf{S}_b^2}\right) \\ \vdots \\ vec\left(\widetilde{\mathbf{S}_b^V}\right) \end{bmatrix} = \begin{bmatrix} vec\left(\left(\alpha^1\right)^2 \widetilde{\mathbf{S}^1} - \alpha^1 \mathbf{S}\right) \\ vec\left(\left(\alpha^2\right)^2 \widetilde{\mathbf{S}^2} - \alpha^2 \mathbf{S}\right) \\ \vdots \\ vec\left(\left(\alpha^V\right)^2 \widetilde{\mathbf{S}^V} - \alpha^V \mathbf{S}\right) \end{bmatrix} \tag{19}$$

where $\mathbf{M}$ is defined as:

$$\mathbf{M} = \begin{bmatrix} \left(\alpha^1\right)^2 + \mathbf{B}_{11}\left(\alpha^1\right)^2 & \mathbf{B}_{12}\alpha^1\alpha^2 & \cdots & \mathbf{B}_{1V}\alpha^1\alpha^V \\ \mathbf{B}_{21}\alpha^2\alpha^1 & \left(\alpha^2\right)^2 + \mathbf{B}_{22}\left(\alpha^2\right)^2 & \cdots & \mathbf{B}_{2V}\alpha^2\alpha^V \\ \vdots & \vdots & \ddots & \vdots \\ \mathbf{B}_{V1}\alpha^V\alpha^1 & \mathbf{B}_{V2}\alpha^V\alpha^2 & \cdots & \left(\alpha^V\right)^2 + \mathbf{B}_{VV}\left(\alpha^V\right)^2 \end{bmatrix} \tag{20}$$

Then $\widetilde{\mathbf{S}_b^v}$ can be updated by solving Eq. (19). Meanwhile, we need to satisfy that $\mathbf{S}_a^v \geq 0, \mathbf{S}_b^v \geq 0$:

$$\mathbf{S}_{b\,ij}^v = \begin{cases} 0, & \text{if } \mathbf{S}_{b\,ij}^v < 0 \\ \mathbf{S}_{ij}^v, & \text{if } \mathbf{S}_{b\,ij}^v > \mathbf{S}_{ij}^v \\ \mathbf{S}_{b\,ij}^v, & \text{otherwise} \end{cases} \tag{21}$$

We summarize the overall algorithm of our method in Algorithm 2. The final clustering result is obtained by spectral clustering on $\mathbf{S}$.

Algorithm 2: Consistency Induced Incomplete Multi-View Clustering

Input: Incomplete similarity graph $\mathbf{S}^v$, Indicator matrix $\mathbf{G}^v$, number of views V, number of clusters c, number of neighbors k, parameter $\beta, \gamma, \lambda_1, \lambda_2$

Output: Similarity graph $\mathbf{S}$

Initialize $\widetilde{\mathbf{S}^v}$ with average value, initialize $\mathbf{S}_b^v$ with 0, initialize $\boldsymbol{\alpha}$ with $\frac{1}{V}$;

while *not convergent* **do**

  Update $\mathbf{S}$ by Eq(13);

  Update $\mathbf{C}, \mathbf{Y}$ by Eq(16);

  Update $\mathbf{S}_a^v$ and $\mathbf{S}_b^v$ by Eq(19) and Eq(21), update $\boldsymbol{\alpha}$;

end

while *not convergent* **do**

  Update $\widetilde{\mathbf{S}^v}$ by Algorithm1;

  Update $\mathbf{S}$ by Eq(15);

  Update $\mathbf{C}, \mathbf{Y}, \mathbf{S}_a^v, \mathbf{S}_b^v, \boldsymbol{\alpha}$;

end

4 Experiments

4.1 Experimental Settings

Detailed information of the utilized datesets is shown in Table 1. In the experiments, we compared the proposed method with three normal multi-view clustering methods, w.r.t. SC, JMCI [6], MVGL [16], as well as seven incomplete multi-view clustering methods, w.r.t. MIC [10], IMSCAGL [12], MKKM-IK [9], TCIMVC [15], PIMVC [1], SCSL [7], RMoGL [2].

To construct incomplete multi-view datasets, we randomly removed 10%, 20%, 30%, 40% and 50% of instances from every view. Note that the SCSL method needs to ensure that all samples contain at least one view information, otherwise its performance will collapse in the incomplete multi-view clustering task. Therefore, for this method, experiments on incomplete multi-view data with this special restriction are denoted as SCSL-w. As a comparison, experiments on the same incomplete data as the other methods are denoted as SCSL-o.

Table 1. Details of datasets

Dataset	view	class	instance	feature
3Sources	3	6	169	[3560,3631,3068]
MSRCv1	5	7	210	[24,576,512,256,254]
Yale	3	15	165	[4096,3304,6750]
ORL	3	40	400	[4096,3304,6750]
movies617	2	17	617	[1878,1398]
20NewsGroups	3	5	500	[2000,2000,2000]
Caltech101-7	6	7	1474	[48,40,254,1984,512,928]
COIL20	3	20	1440	[30,19,30]
100leaves	3	100	1600	[64,64,64]

4.2 Performance of Multi-view Clustering

Table 2. ACC(%) scores of multi-view clustering on different datasets.

Dataset	SP	JMCI	MVGL	MIC	IMSCAGL	MKKMIK	TCIMVC	PIMVC	SCSL	RMoGL	Ours
3Sources	50.12	48.28	39.64	45.27	70.41	57.40	74.96	**78.11**	62.54	65.68	<u>77.57</u>
MSRCv1	71.86	78.57	69.02	66.43	**85.71**	62.38	72.09	<u>85.71</u>	83.17	74.86	80.00
Yale	65.09	58.00	47.76	46.42	**79.39**	61.82	<u>67.28</u>	61.21	62.39	67.27	61.58
ORL	59.35	**78.35**	53.94	56.25	75.25	69.00	77.49	70.50	77.30	75.60	<u>77.53</u>
movies617	20.29	19.72	13.13	16.86	24.80	26.26	26.57	<u>32.25</u>	28.09	28.85	**32.41**
20NewsGroups	21.20	47.50	24.20	22.34	**99.00**	97.00	96.60	97.80	97.21	96.60	<u>98.80</u>
Caltech101-7	48.25	50.16	59.91	43.26	48.71	33.99	60.31	**68.59**	36.09	65.07	<u>67.62</u>
COIL20	100.00	100.00	100.00	86.90	97.85	92.78	100.00	100.00	100.00	100.00	**100.00**
100leaves	44.55	<u>89.83</u>	55.94	66.19	82.06	37.13	59.33	82.63	81.22	**91.44**	76.63
Avg-Score	53.41	63.38	51.50	49.99	73.69	59.75	70.51	**75.20**	69.78	73.93	<u>74.68</u>
Avg-Rank	7.44	5.56	8.44	9.78	4.44	8.33	4.44	<u>3.11</u>	4.67	3.67	**3.00**

Table 3. NMI(%) scores of multi-view clustering on different datasets.

Dataset	SP	JMCI	MVGL	MIC	IMSCAGL	MKKMIK	TCIMVC	PIMVC	SCSL	RMoGL	Ours
3Sources	37.92	42.85	12.64	26.38	64.82	57.83	**69.19**	67.30	65.45	62.20	<u>68.34</u>
MSRCv1	56.17	69.85	62.55	58.52	<u>76.98</u>	50.16	64.39	76.47	72.50	69.28	**78.27**
Yale	67.20	63.42	51.00	53.94	**78.06**	63.24	68.36	65.02	66.26	<u>68.73</u>	63.52
ORL	79.05	87.58	68.39	76.06	86.89	83.62	87.83	85.28	<u>89.10</u>	89.01	**91.78**
movies617	22.12	18.42	8.62	17.99	26.14	28.42	28.31	**32.61**	27.32	27.44	<u>31.34</u>
20NewsGroups	3.75	22.31	4.98	6.39	**96.69**	90.41	89.58	92.86	91.46	89.87	<u>96.23</u>
Caltech101-7	34.48	42.48	51.31	49.56	47.82	22.57	45.81	**58.35**	52.23	54.68	<u>54.69</u>
COIL20	100.00	100.00	100.00	96.57	98.54	97.60	100.00	100.00	100.00	100.00	**100.00**
100leaves	70.79	<u>95.37</u>	64.77	86.75	91.96	64.52	78.11	92.75	91.79	**95.75**	88.57
Avg-Score	52.39	60.25	47.14	52.46	74.21	62.05	70.18	<u>74.52</u>	72.90	73.00	**74.75**
Avg-Rank	7.89	6.11	8.67	9.11	4.67	8.33	4.78	<u>3.11</u>	3.89	3.67	**2.67**

The performance of different algorithms on complete multi-view data with regard to ACC and NMI are reported in Table 2 and Table 3, respectively. The best score on each dataset is marked in bold and the second best score is marked with an underline. It can be observed that the proposed method achieves the best or second best performance on 3Sources, 20NewsGroups, COIL20, ORL, movies617 and Caltech101-7 datasets in both ACC and NMI. On MSRCv1 dataset, it does not achieve the best in terms of ACC, but performs best in terms of NMI. Although our method does not achieve the best performance on datasets Yale and 100leaves, it is still competitive. Moreover, the average score and average rank for different methods have also be reported. It can be observed that the proposed method achieves the best average NMI score and the second best ACC score. Also, it gets the best rank in both ACC and NMI.

4.3 Performance of Incomplete Multi-View Clustering

The performance on incomplete multi-view data is reported by Fig1 and Fig2. As shown in these figures, the performance of SP, JMCI and MVGL deteriorates rapidly as the data missing rate increases, which indicates that a good learning strategy for missing data is important. On Caltech101-7, ORL and COIL20 datasets, our method can achieve the best or second best performance when the missing rate rises from 10% to 50%. On MSRCv1 dataset, our method does not achieve the best performance when the missing rate is 10%, but the stability of performance as the missing rate increases is pretty good. Therefore, it achieves the best performance when the missing rate comes to 50%. On 3Sources, 100leaves, Yale, movies617 and 20NewsGroups datasets, although our method can't achieve the best performance, the performance stability with the increase of data missing rate is pretty good. On 20NewsGroups dataset, although the result is not very good on ACC, it is much better on NMI, which indicates that the cluster structure obtained by the proposed method on this dataset is actually reasonable.

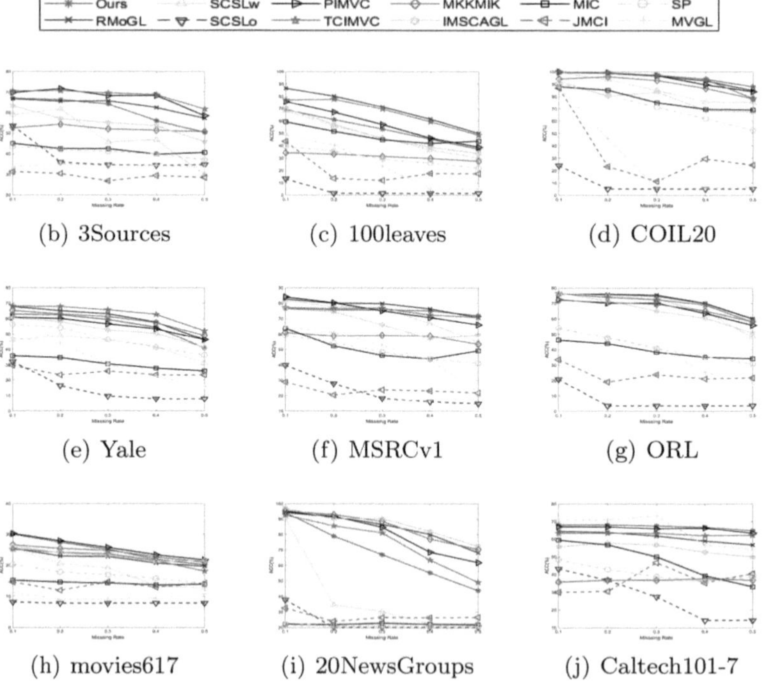

(b) 3Sources (c) 100leaves (d) COIL20

(e) Yale (f) MSRCv1 (g) ORL

(h) movies617 (i) 20NewsGroups (j) Caltech101-7

Fig. 1. ACC(%) scores of incomplete multi-view clustering on different datasets.

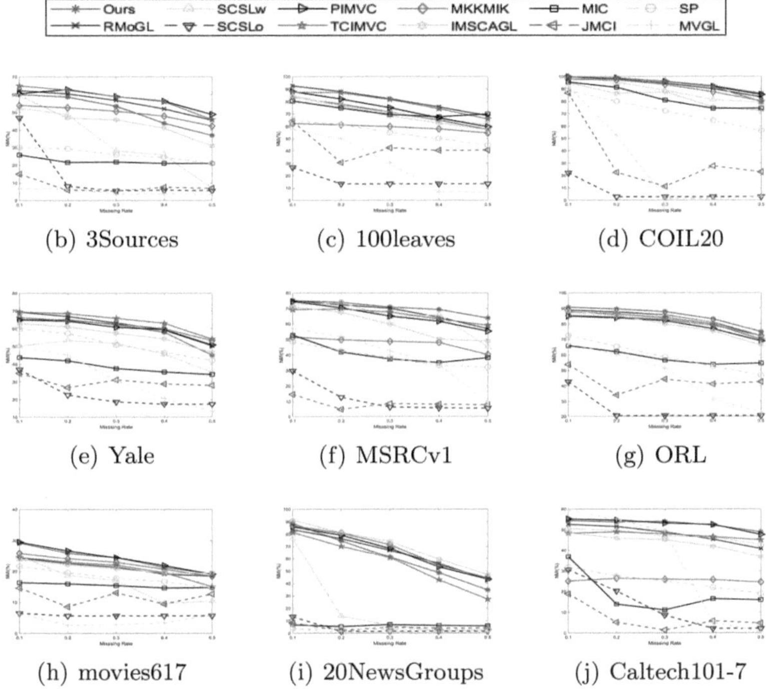

Fig. 2. NMI(%) scores of incomplete multi-view clustering on different datasets.

5 Conclusion

In this paper, we present a novel incomplete multi view clustering approach termed CI-IMVC. In the proposed approach, to simultaneously capture the cross-view commonness and inconsistencies, we decompose the subspace representation on each view into two matrices, i.e., the consistent matrix and the inconsistent matrix. The consistent matrix is further constrained by the view-consensus grouping effect, which exploits the multi-view local structures to promote the learning of the common representation. To strengthen the cluster structure and improve the clustering robustness, the low-rank representation is incorporated via the nuclear norm. Moreover, a novel incomplete graph recovery strategy is introduced to cope with the incomplete multi view clustering. Experimental results on a variety of incomplete multi-view datasets have demonstrated the superior of the proposed approach.

References

1. Deng, S., Wen, J., Liu, C., Yan, K., Xu, G., Xu, Y.: Projective incomplete multi-view clustering. IEEE Trans. Neural Netw. Learn. Syst. (2023)
2. Guo, W., Che, H., Leung, M.F., Jin, L., Wen, S.: Robust mixed-order graph learning for incomplete multi-view clustering. Inf. Fusion **115**, 102776 (2025)

3. Hu, H., Lin, Z., Feng, J., Zhou, J.: Smooth representation clustering. In: Proceedings of the IEEE conference on computer vision and pattern recognition, pp. 3834–3841 (2014)
4. Hu, M., Chen, S.: Doubly aligned incomplete multi-view clustering. arXiv preprint arXiv:1903.02785 (2019)
5. Jin, J., Wang, S., Dong, Z., Liu, X., Zhu, E.: Deep incomplete multi-view clustering with cross-view partial sample and prototype alignment. In: Proceedings of the IEEE/CVF conference on computer vision and pattern recognition, pp. 11600–11609 (2023)
6. Liang, Y., Huang, D., Wang, C.D., Yu, P.S.: Multi-view graph learning by joint modeling of consistency and inconsistency. IEEE Trans. Neural Netw. learn. Syst. **35**(2), 2848–2862 (2022)
7. Liu, S., et al.: Sample-level cross-view similarity learning for incomplete multi-view clustering. In: Proceedings of the AAAI conference on artificial intelligence. vol. 38, pp. 14017–14025 (2024)
8. Liu, X.: Efficient and effective regularized incomplete multi-view clustering. IEEE Trans. Pattern Anal. Mach. Intell. **43**(8), 2634–2646 (2020)
9. Liu, S.: Multiple kernel k k-means with incomplete kernels. IEEE Trans. Pattern Anal. Mach. Intell. **42**(5), 1191–1204 (2019)
10. Shao, W., He, L., Yu, P.S.: Multiple incomplete views clustering via weighted nonnegative matrix factorization with regularization. In: Appice, A., Rodrigues, P., Santos Costa, V., Soares, C., Gama, J., Jorge, A. (eds.) Joint European Conference on Machine Learning and Knowledge Discovery in Databases, pp. 318–334. Springer, Cham (2015). https://doi.org/10.1007/978-3-319-23528-8_20
11. Wang, Z., Li, L., Ning, X., Tan, W., Liu, Y., Song, H.: Incomplete multi-view clustering via structure exploration and missing-view inference. Inf. Fusion **103**, 102123 (2024)
12. Wen, J., Xu, Y., Liu, H.: Incomplete multiview spectral clustering with adaptive graph learning. IEEE Trans. Cybern. **50**(4), 1418–1429 (2018)
13. Wen, J., Zhang, Z., Fei, L., Zhang, B., Xu, Y., Zhang, Z., Li, J.: A survey on incomplete multiview clustering. IEEE Transactions on Systems, Man, and Cybernetics: Systems **53**(2), 1136–1149 (2022)
14. Wong, W.K., Liu, C., Deng, S., Fei, L., Li, L., Lu, Y., Wen, J.: Neighbor group structure preserving based consensus graph learning for incomplete multi-view clustering. Information Fusion **100**, 101917 (2023)
15. Xia, W., Gao, Q., Wang, Q., Gao, X.: Tensor completion-based incomplete multiview clustering. IEEE Transactions on Cybernetics **52**(12), 13635–13644 (2022)
16. Zhan, K., Zhang, C., Guan, J., Wang, J.: Graph learning for multiview clustering. IEEE transactions on cybernetics **48**(10), 2887–2895 (2017)
17. Zhao, H., Liu, H., Fu, Y.: Incomplete multi-modal visual data grouping. In: IJCAI. pp. 2392–2398 (2016)
18. Zhao, L., Chen, Z., Yang, Y., Wang, Z.J., Leung, V.C.: Incomplete multi-view clustering via deep semantic mapping. Neurocomputing **275**, 1053–1062 (2018)

Finger Vein Image Quality Assessment by Mated Comparison Score Prediction

Felix Garcia Funk[1], Olaf Henniger[2(✉)], and Arjan Kuijper[1,2]

[1] Department of Computer Science, Technical University of Darmstadt, Darmstadt, Germany
[2] Fraunhofer Institute for Computer Graphics Research IGD, Darmstadt, Germany
olaf.henniger@igd.fraunhofer.de

Abstract. Assessing the quality of finger vein images can help improve the biometric recognition performance. Current methods achieve quality assessment by training a machine learning model against predefined quality labels. The model approximates the target labels and is thereby limited by the labels in use. Therefore, the assignment of quality labels plays an important role for the performance of the quality assessment and provides an additional source of error. The absence of a commonly agreed definition for biometric sample quality makes the assignment of quality labels a research topic on its own. While many of the related studies focus on the machine-learning part, unsophisticated quality assignment practices such as the manual annotation of images are still in use. This paper presents an alternative to the explicit assignment of quality labels. The proposed method uses mated comparison scores as training targets. Without explicit quality labels, the model freely adapts to the underlying data and produces continuous quality scores.

Keywords: Biometrics · Finger vein pattern · Biometric sample quality

1 Introduction

Many types of biometric characteristics such as fingerprints, face, voice and iris have in common that they can be easily captured using standard equipment and then spoofed. Vascular patterns, mainly hand and finger vein patterns, provide a harder-to-capture alternative since they are only visible under near-infrared light, which passes through cellular tissue but is absorbed by blood [10].

Quality assessment systems can be used to filter out inadequate biometric samples that can cause recognition errors. Most quality assessment methods are based on machine learning and share a common structure: First, training, validation and testing images are labeled with quality targets (quality assignment), and then a quality prediction model is trained, validated and tested against these targets. The quality targets have a considerable influence on the system, and the quality assignment process needs to be well designed. With two sources of

P. Umapada et al. (Eds.): ICCPR 2025, CCIS 2811, pp. 415–427, 2026.
https://doi.org/10.1007/978-981-95-8315-7_33

inaccuracy, the chances of error increase, and additional attention is required to optimize both, quality assignment and quality prediction, against the underlying data. A single end-to-end system that does not require explicit quality labels for training but defines quality implicitly during training would be desirable.

The main contributions of this paper are

1. An end-to-end system that takes pairs of mated finger vein patterns as input and uses comparison scores as targets, thereby implicitly defining the quality of each sample during training. No explicit quality assignment is necessary.
2. A data augmentation scheme based on image generation by a conditional generative adversarial network to tackle the data imbalance (only a fraction of the available training images represents low quality).
3. An evaluation scheme that uses standardized metrics (error vs. discard characteristics) and utilizes publicly available datasets to form a basis for comparison for future work.

2 Related Work on Finger Vein Image Quality Assessment

Image quality was first defined with respect to the human visual system [30]. For a human observer, the quality of a vein image may be affected under certain lighting conditions, by finger displacement, loss of contrast or other clear visual artefacts. Early approaches to assessing the quality of vascular images focused on defining handcrafted features describing such irregularities [5,15,31]. Biometric sample quality, however, corresponds to the utility or usefulness of the sample for automated recognition [2]. Therefore, later methods tried to detect image features responsible for higher recognition error rates [4,20,21,25].

Most finger vein image quality assessment methods use some sort of machine learning. Before the advances in automatic image feature detection with convolutional neural networks (CNNs), handcrafted features formed the basis for machine learning models [16–18,23,27,32]. CNNs detect the necessary image descriptors themselves [9,19–21,25,29]. Lightweight CNNs decrease the computational costs [21,29].

Some related papers focus on quality prediction while paying less attention to quality assignment. Several approaches use manually labeled data as a basis for the training process [9,17,23,27,32]. Automated quality assignment based on similarity scores, however, improves the reproducibility of results [21].

Some approaches only distinguish between discrete classes of quality such as high and low quality [9,16,18,19,21,25]. High quality is ascribed to images participating in true matches and low quality to images participating in false non-matches. Binary quality scores, however, lack explainability. Users may want to know the degree of quality to make more informed decisions. Therefore, multiple quality score values are preferable. Usually, they range from 0 to 100 [2]. While this creates a more granular quality evaluation, predicting continuous scores instead of binary ones is a harder problem. Few approaches display quality granularly using multiple labels or even a continuous scale [32].

Some related work is evaluated only on proprietary datasets [16,27]. Use of one or more publicly available datasets, however, increases reproducibility and comparability of results. In some related papers [9,20,21,29], the SDUMLA dataset [28] is used for the evaluation; in some [9,17–20,32], the HKPUFID dataset [11] is used; in some [9,20,21,25,29], the MMCBNU_6000 dataset [14], and in some [9,19–21], the FV-USM dataset [3] is used.

Sometimes, the prediction performance is only evaluated against the labels of the test set [23,25,29]. High prediction performance on the test set is a key ingredient for a successful quality prediction, but only if the target labels do represent quality. Otherwise, predictions close to the assigned targets have little meaning. Therefore, the effect of discarding low-quality images on recognition performance should also be evaluated. Some related papers use the change of the equal error rate (EER) after discarding low-quality images as a performance metric [9,16–21,27,32]. Some related papers use the change of the prediction accuracy at a fixed false match rate (FMR) after discarding low-quality images as a performance metric [18–20].

The combination of datasets, comparison algorithms and performance metrics is close to unique for each proposed finger vein image quality assessment method, making benchmarking cumbersome.

3 Quality Assessment Methodology

3.1 Introduction

We define the set of mated comparison scores as the set $S_{\mathrm{m}} = \{s_{ij,ik} \mid j \neq k\}$ where i refers to the ith finger and j and k refer to the jth and k^{th} images of the ith finger, respectively. A comparison algorithm C produces a mated comparison score $s_{ij,ik} = C(f_{ij}, f_{ik})$ based on image features f_{ij} and f_{ik}. A feature extractor F creates the image features based on a vascular image I_{ij}. Therefore, the mated comparison score is defined as $s_{ij,ik} = C(F(I_{ij}), F(I_{ik}))$.

Grother and Tabassi proposed that the mated similarity score can be approximated through image quality [4]. In their discussion, a predictor T is used to predict the mated similarity scores $s_{ij,ik}$ based on quality scores q_{ij} and q_{ik}: $s_{ij,ik} = T(q_{ij}, q_{ik}) + \epsilon_{ij,ik} = T(Q(I_{ij}), Q(I_{ik})) + \epsilon_{ij,ik}$ where Q is a function that retrieves a quality score from an image, and $\epsilon_{ij,ik}$ is an error term. If $Q = F$ and $T = C$, ϵ becomes zero. They used discrete quality scores produced by NFIQ 1 [24] to test the approximation and chose linear and cubic predictors for T.

Beside the comparison algorithm in use, image quality is the main influence on mated similarity scores. Low quality results in lower similarity scores. A pair of quality scores (q_1, q_2) where one quality is high while the other one is low will lead to a low mated similarity score. On the other hand, if both quality scores are high, a high mated similarity score is expected. When both quality scores are low, a low mated similarity score is expected. When comparing non-mated samples, a low similarity score is expected regardless of the quality of the compared samples. Given these assumptions, a coarse mapping from the image qualities to the mated similarity score is defined as:

$$T(q_1, q_2) = \begin{cases} \text{high} & \text{if } q_1 = \text{high and } q_2 = \text{high} \\ \text{low} & \text{otherwise} \end{cases} \tag{1}$$

Given continuous scores q_1 and q_2 in the interval $[0, 1]$, the simplest predictor for mated similarity scores would be a linear mapping

$$T(q_1, q_2) = \frac{q_1 + q_2 - |q_2 - q_1|}{2} \tag{2}$$

3.2 System Structure

Based on the assumptions made before, a system M taking two mated vascular images as input and generating a similarity score as output is defined as

$$M(I_{ij}, I_{ik}) = T(Q(I_{ij}), Q(I_{ik})) = \hat{s}_{ij,ik} \tag{3}$$

If M is differentiable, the overall system or parts of it can be parameterized and trained through backpropagation to find the best fitting mapping between the quality scores of two mated images and their similarity score. The targets for the training are the mated similarity scores in S_{m}, and the quality prediction is implicitly defined through Q.

The system will converge to a representation of C and choose Q in such a way that a quality predictor relative to the underlying comparison algorithm C emerges. The process can be formalized as the minimization of a given loss function $l(M(\widetilde{I}; \phi), S)$ with respect to weights ϕ, where $\widetilde{I} = \{(I_{ij}, I_{ik}) \mid j \neq k\}$ denotes the set of mated image pairs. T and Q can be parameterized resulting in $\phi = \phi_T \cup \phi_Q$, but also an unparameterized mapping T can be used with $\phi = \phi_Q$. In this case, T is called "fixed".

After minimizing the loss function l through training, Q can be extracted and be used as a stand-alone predictor of image quality. This finalized version is denoted as $\hat{Q}$. The two quality generators share their parameters during training to ensure a single predictor of quality and prevent a biased adaptation towards certain image combination patterns. This helps the generalizability of the system.

3.3 Quality Output

By considering the mated comparison scores as input, the proposed system can learn the relation between mated comparison scores and sample quality during training. The pairwise relation between images is directly displayed in the two-dimensional T function. For any given image I_{ij} and its quality score q_{ij}, there exists a cross section along the axis representing the one-dimensional similarity score progression $W^{(1)}$ and $W^{(2)}$ associated with q_{ij}

$$W_{ij}^{(1)}(q) = T(q_{ij}, q) \mid q \in \mathbb{R} \tag{4}$$

$$W_{ij}^{(2)}(q) = T(q, q_{ij}) \mid q \in \mathbb{R} \tag{5}$$

T is symmetric if and only if $W_{ij}^{(1)} = W_{ij}^{(2)}$ for every $q \in \mathbb{R}$. If a pair of images produces the same similarity score independent of the order of the images, a valid T function would in theory always be symmetric. In practice, however, if T is parameterized, the inaccuracy of the training may produce an asymmetric T function (see Fig. 1). In this case, a function m_w (e.g., minimum, maximum or mean) is necessary to combine $W^{(1)}$ and $W^{(2)}$ into a single representation of similarity score progression,

$$\widetilde{W}_{ij}(q) = m_w(W_{ij}^{(1)}(q), W_{ij}^{(2)}(q)) \tag{6}$$

In Sect. 5, the maximum was used for m_w.

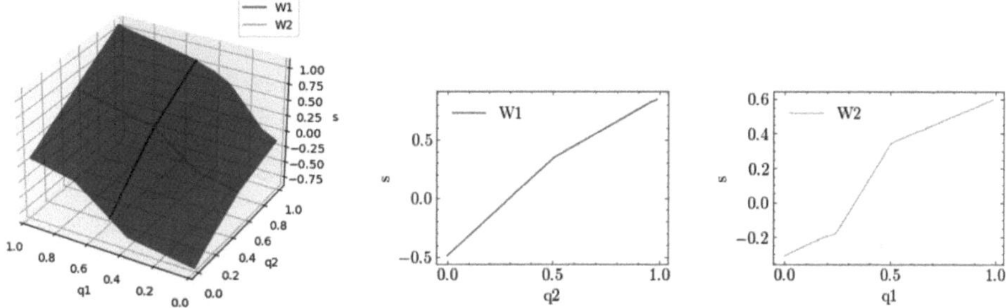

Fig. 1. Asymmetric T function and cross sections $W^{(1)}$ and $W^{(2)}$ at $q = 0.5$. s are similarity scores.

To normalize the quality scores into the required range (0 to 100 [2]), e.g., min-max normalization can be used as post-process function

$$\text{pp}(u_{ij}) = \frac{u_{ij} - \min(\Omega)}{\max(\Omega) - \min(\Omega)} \cdot L \tag{7}$$

where Ω is the set of all quality score values for the training images, and L is a constant such as 100. It can be beneficial to leave room for outliers or to use a sigmoid function with the range limits as asymptotes, instead.

3.4 Loss Function

Since the model predicts continuous scores, a regression loss function is utilized during training. As low-quality samples are underrepresented in most datasets, the mean squared error and mean absolute error are insufficient. Ren et al. proposed a balanced mean squared error (BMSE) loss capable of handling imbalanced data [22]. The proposed loss function has the following form:

$$\text{BMSE}(X, Y) = -\log(\mathcal{N}(y; x, \sigma_{\text{noise}}^2 I) + \log \int_Y \mathcal{N}(y'; x, \sigma_{\text{noise}}^2 I) \cdot p_{\text{train}}(y')dy' \tag{8}$$

As its calculation is non-trivial, they proposed practical ways to approximate it. We use their batch sampling approach [22] as it produced the best results.

4 Experimental Setup

4.1 Finger Vein Image Datasets

Multiple finger vein image datasets are publicly available for research purposes. We use the following datasets that were also used in related work:

- SDUMLA-HMT [28]: Along with face, gait, iris and fingerprint samples, this multimodal dataset contains 3 816 finger vein images from 106 subjects. The index, middle and ring fingers of the left and right hands were each captured six times. The image size is 320 × 240 pixels. The capture device was an in-house implementation.
- MMCBNU_6000 [14]: This dataset consists of 6 000 finger vein images from 100 volunteers, aged between 16 and 72 years and coming from 20 different countries. The index, middle and ring fingers of the left and right hands were each captured ten times. The image size is 640×480 pixels. The capture device was a portable, lab-made MMCBNU_6000 device. This dataset contains some low-quality samples due to rotation, scale, uneven illumination and finger pressure.
- FV-USM [3]: This dataset consists of 5 904 finger vein images, captured over two sessions (2 952 images per session) from 123 volunteers (83 males and 40 females), aged between 20 and 52 years. The index and middle fingers of the left and right hands were each captured six times. The image size is 640 × 480 pixels. The capture device was an in-house implementation.

4.2 Training, Validation and Testing

Let the set of image pairs and corresponding mated similarity scores be

$$D = \{(I_{ij}, I_{ik}, s_{ij,ik}) \mid j \neq k\} \tag{9}$$

where I_{ij},$I_{ik} \in \widetilde{I}$ and $s_{ij,ik} \in S_{\mathrm{m}}$. The data subsets for training D_{train}, validation D_{val} and testing D_{test} are created in such a way that all samples of a finger instance belong to either one of the subsets.

4.3 Data Augmentation

Only a fraction of the available finger vein images is of low quality. This makes it difficult to learn how to accurately predict low quality. As a remedy, we augmented the datasets with synthetic finger vein images of low quality. These images were generated using a conditional generative adversarial network, a Pix2Pix [7] model trained and executed on $D_{\mathrm{train}} \cup D_{\mathrm{val}}$. The number of synthetic images was chosen to be similar to the number of natural images given.

4.4 Preprocessing

All images were resized to 256 × 256 pixels, the input size of Q, and standardized regarding the overall mean $\overline{I}$ and standard deviation σ_I of each dataset:

$$I_{ij}^{\text{prep}} = \frac{I_{ij} - \overline{I}}{\sigma_I} \quad (10)$$

4.5 Comparison Algorithms

We use the following comparison algorithms to demonstrate the method's adaptability to different algorithms:

- ASAVE (anatomy structure analysis-based vein extraction) [26] as introduced by Yang et al. based on a structural analysis of the vascular pattern;
- SIFT (scale-invariant feature transform) [12] as proposed by Kim et al. for finger vein image comparison [8].

Both ASAVE and SIFT provide similarity scores (the higher, the more similar). For both, we use implementations from the open-source toolkit VeinPLUS+ [13].

4.6 Machine Learning

We utilize a classic CNN with an input size of 256 × 256 pixels as the quality prediction system Q. It is handcrafted and quite small with 11 321 parameters.

The parameterized T models are small, fully connected networks with 130 parameters each. To encourage T to diverge from the comparison mapping C and approach a function fulfilling Eq. (1) during training, its weights may be initialized so that T resembles the linear function presented in Eq. (2).

All models are trained for 150 epochs with a learning rate of 0.001 and a batch size of 16. Random stochastic descent is utilized as optimizer and the BMSE (see Sect. 3.4) as loss function.

5 Experimental Results

5.1 T Evaluation

This experiment compares the performance of the fixed linear T function proposed in Eq. (2), denoted as T_{fix}, against parameterized versions. The parameterized T functions are split into two groups. The first group is initialized such that it resembles the linear functions before the training starts. It is denoted as T_{free}. The second group is initialized randomly and is denoted as T_{freeni}.

Figures 2 and 3 show false non-match and false match error vs. discard characteristics (EDCs) with respect to different comparison algorithms and datasets. EDCs show the dependence of an error rate at a fixed decision threshold on the percentage of samples discarded based on lowest quality scores (discard

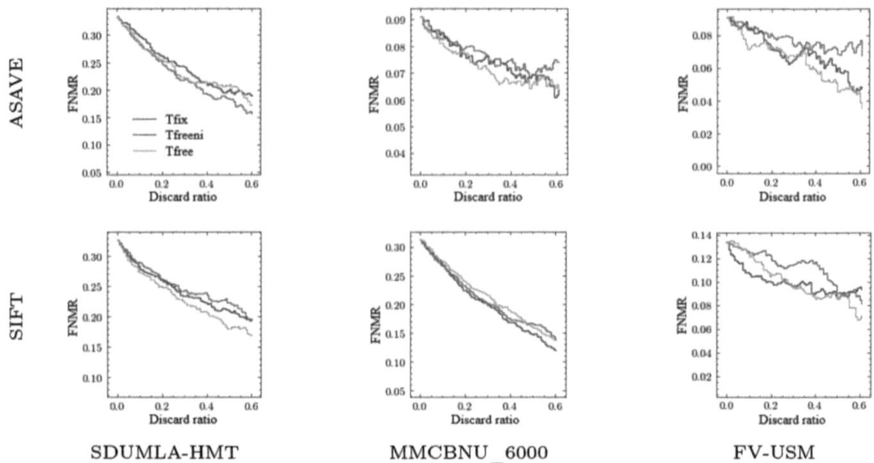

Fig. 2. False non-match EDCs regarding T_{fix}, T_{freeni} and T_{free} at a fixed decision threshold yielding an initial FMR of about 1%

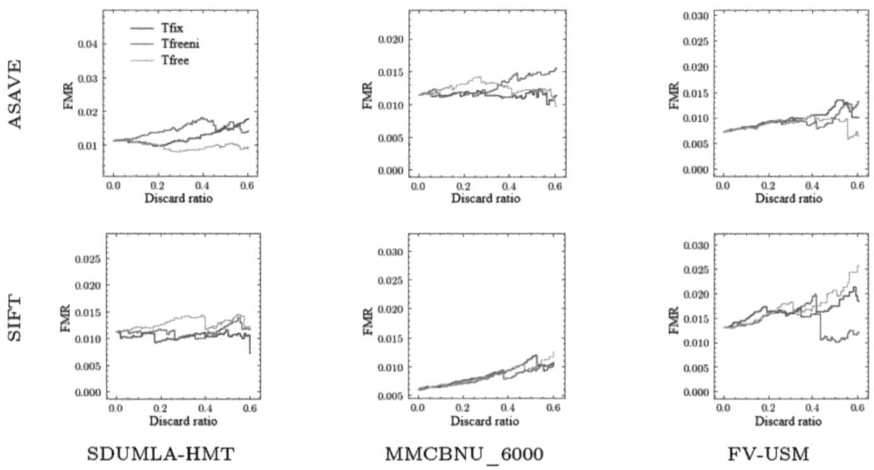

Fig. 3. False match EDCs regarding T_{fix}, T_{freeni} and T_{free} at a fixed decision threshold yielding an initial FMR of about 1%

ratio) [2,4]. Higher discard ratios can be omitted for lack of statistical reliability (according to the Rule of 30 [1], at least 30 errors must be observed to be 90% confident that the true error rate is within ±30% of the observed error rate).

To examine the full impact of the omission of samples on the recognition error rates, a false match EDC should be considered together with the corresponding false non-match EDC. A decrease in the false non-match rate (FNMR) may inadvertently come along with an increase in the false match rate (FMR) if inappropriate samples are assigned low quality scores. However, if the FMR values are much lower than the corresponding FNMR values, it is often considered safe to omit the false match EDCs. Because the EDCs depend on the decision threshold, it is common to plot EDCs for several threshold values. Due to space restrictions, we chose only one threshold value, without limiting generality.

The steeper the false non-match rate (FNMR) decreases with an increasing discard ratio and without significantly increasing the FMR, the better. Because T_{fix} cannot adapt to the underlying data, we suspected it to perform worse than T_{free} and T_{freeni}. However, T_{fix} achieves similar performance. This suggests that the linear translation is to some extent sufficient.

5.2 Quality Prediction Evaluation

For an evaluation of the quality prediction capabilities, our method needs to be compared with the result of another quality prediction system. To the best of our knowledge, only one related work [32] predicts continuous quality scores. The related work did not publish their code and did not report error vs. discard characteristics. This prevents direct benchmarking.

To evaluate whether our approach of defining quality implicitly is able to keep up with the common two-part structure (first quality assignment, then machine learning), another baseline is introduced: The trained quality prediction system $\hat{Q}$ is compared to a quality prediction system $\widetilde{Q}$ that is trained against target utility scores proposed by Henniger et al. [6]. The utility of a biometric sample I is measured as normalized arithmetic mean of I's mated similarity scores

$$u_I = \frac{\mu_{\text{m}I}}{\sigma_{\text{m}}} \tag{11}$$

where $\mu_{\text{m}I}$ is the arithmetic mean of the similarity scores for I and mated references and σ_{m} is the standard probe deviation of all mated similarity scores. The training of $\widetilde{Q}$ is formulated as a standard regression optimization and uses the BMSE loss (see Sect. 3.4) to counter target imbalance.

Figures 4 and 5 show the false non-match and false match EDCs for the quality scores calculated on testing data and the baseline. Our proposed method of quality assessment by mated comparison score prediction keeps up with or is ahead of the baseline. This implies that the system successfully transfers the relation between mated similarity scores and sample quality to the test environment. It also indicates that the pairwise image relations utilized during the training support the generalizability of the model.

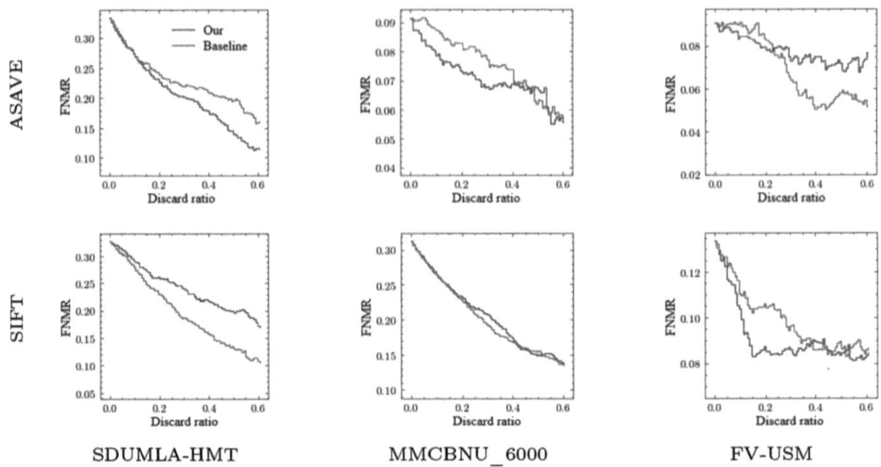

Fig. 4. False non-match EDCs at a fixed decision threshold yielding an initial FMR of about 1%

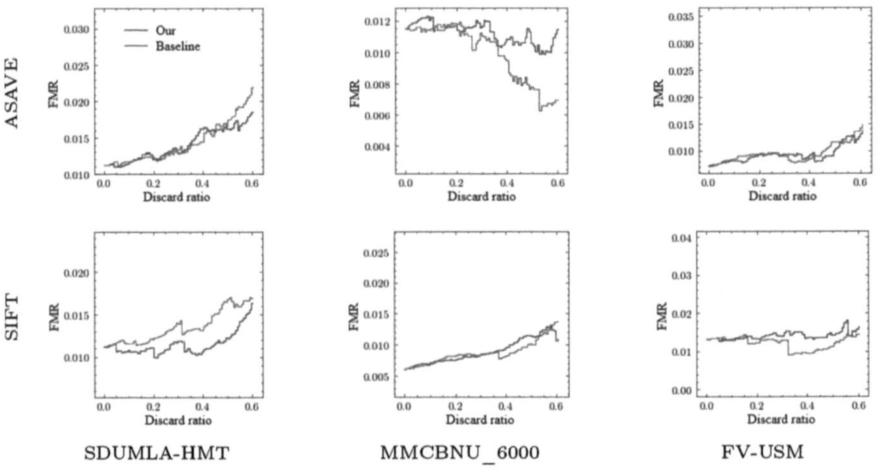

Fig. 5. False match EDCs at a fixed decision threshold yielding an initial FMR of about 1%

6 Conclusion

The proposed method of implicitly optimizing a quality assessment system while training against mated similarity scores has shown to be feasible and to produce reasonable quality scores. The results show that an explicit quality assignment step is not necessary and can be omitted.

The proposed finger vein image quality assessment method adapts to the decision patterns of the comparison algorithm at hand. This can result in better performance. In case that generalizability has a higher priority than performance, quality scores having predicitive value for multiple comparison algorithms may be more appropriate. For other comparison algorithms and for other capture devices, the quality assessment should be retrained to achieve better results.

Future investigations may evaluate different model architectures for Q and T or examine different parameter settings as well as different loss functions. The approach described is not limited to finger vein images. It may be transferred to other biometric recognition techniques where two samples are compared.

Acknowledgments. This research work has been funded by the German Federal Ministry of Education and Research and the Hessian Ministry of Higher Education, Research, Science and the Arts within their joint support of the National Research Center for Applied Cybersecurity ATHENE.

References

1. Information technology - Biometric performance testing and reporting - Part 1: Principles and framework. International Standard ISO/IEC 19795–1 (2021)
2. Information technology - Biometric sample quality - Part 1: Framework. International Standard ISO/IEC 29794–1 (2024)
3. Asaari, M.S.M., Suandi, S.A., Rosdi, B.A.: Fusion of band limited phase only correlation and width centroid contour distance for finger based biometrics. Expert Syst. Appl. **41**(7), 3367–3382 (2014). https://doi.org/10.1016/j.eswa.2013.11.033
4. Grother, P., Tabassi, E.: Performance of biometric quality measures. IEEE Trans. Pattern Anal. Mach. Intell. **29**(4), 531–543 (2007). https://doi.org/10.1109/TPAMI.2007.1019
5. Hartung, D., Martin, S., Busch, C.: Quality estimation for vascular pattern recognition. In: International Conference on Hand-Based Biometrics, pp. 1–6. IEEE (2011). https://doi.org/10.1109/ICHB.2011.6094332
6. Henniger, O., Fu, B., Kurz, A.: Utility-based performance evaluation of biometric sample quality measures. EURASIP J. Image Video Process. (2024). https://doi.org/10.1186/s13640-024-00644-1
7. Isola, P., Zhu, J.Y., Zhou, T., Efros, A.A.: Image-to-image translation with conditional adversarial networks. In: IEEE/CVF Conference on Computer Vision and Pattern Recognition (CVPR) (2017). https://doi.org/10.1109/CVPR.2017.632
8. Kim, H.G., Lee, E.J., Yoon, G.J., Yang, S.D., Lee, E.C., Yoon, S.M.: Illumination normalization for SIFT based finger vein authentication. In: 8th International Symposium on Advances in Visual Computing (ISVC), LNCS, pp. 21–30. Springer, Cham (2012). https://doi.org/10.1007/978-3-642-33191-6_3
9. Kirchgasser, S., Kauba, C., Wimmer, G., Uhl, A.: Advanced image quality assessment for hand- and fingervein biometrics. IET Biometrics (2025). https://doi.org/10.1049/bme2/8869140
10. Kono, M.: A new method for the identification of individuals by using of vein pattern matching of a finger. In: 5th Symposium on Pattern Measurement, pp. 9–12 (2000)
11. Kumar, A., Zhou, Y.: Human identification using finger images. IEEE Trans. Image Process. **21**(4), 2228–2244 (2012). https://doi.org/10.1109/TIP.2011.2171697
12. Lindeberg, T.: Scale invariant feature transform. Scholarpedia 7 (2012). https://doi.org/10.4249/scholarpedia.10491

13. Linortner, M., Uhl, A.: Veinplus+: a publicly available and free software framework for vein recognition. In: International Conference of the Biometrics Special Interest Group (BIOSIG) (2021). https://doi.org/10.1109/BIOSIG52210.2021.9548286
14. Lu, Y., Xie, S.J., Yoon, S., Wang, Z., Park, D.S.: An available database for the research of finger vein recognition. In: 6th Int. Congress on Image and Signal Processing (CISP), vol. 1, pp. 410–415. IEEE (2013). https://doi.org/10.1109/CISP.2013.6744030
15. Ma, H., Wang, K., Fan, L., Cui, F.: A finger vein image quality assessment method using object and human visual system index. In: 3rd Sino-Foreign-Interchange Workshop on Intelligent Science and Intelligent Data Engineering (IScIDE), pp. 498–506. Springer, Cham (2012). https://doi.org/10.1007/978-3-642-36669-7_61
16. Nguyen, D.T., Park, Y.H., Shin, K.Y., Park, K.R.: New finger-vein recognition method based on image quality assessment. KSII Trans. Internet Inf. Syst. **7**(2) (2013). https://doi.org/10.3837/tiis.2013.02.010
17. Peng, J., Li, Q., Niu, X.: A novel finger vein image quality evaluation method based on triangular norm. In: 10th International Conference on Intelligent Information Hiding and Multimedia Signal Processing, pp. 239–242. IEEE (2014). https://doi.org/10.1109/IIH-MSP.2014.66
18. Qin, H., Chen, Z., He, X.: Finger-vein image quality evaluation based on the representation of grayscale and binary image. Multimed. Tools Appl. **77**, 2505–2527 (2018). https://doi.org/10.1007/s11042-016-4317-y
19. Qin, H., El-Yacoubi, M.A.: Finger-vein quality assessment by representation learning from binary images. In: 22nd International Conference on Neural Information Processing (ICONIP), LNCS, pp. 421–431. Springer, Cham (2015). https://doi.org/10.1007/978-3-319-26532-2_46
20. Qin, H., El-Yacoubi, M.A.: Deep representation for finger-vein image-quality assessment. IEEE Trans. Circuits Syst. Video Technol. **28**(8), 1677–1693 (2017). https://doi.org/10.1109/TCSVT.2017.2684826
21. Ren, H., Sun, L., Guo, J., Han, C., Cao, Y.: A high compatibility finger vein image quality assessment system based on deep learning. Expert Syst. Appl. **196**, 116603 (2022). https://doi.org/10.1016/j.eswa.2022.116603
22. Ren, J., Zhang, M., Yu, C., Liu, Z.: Balanced MSE for imbalanced visual regression. In: IEEE/CVF Conference on Computer Vision and Pattern Recognition (CVPR) (2022). https://doi.org/10.1109/CVPR52688.2022.00777
23. Shaheed, K., Qureshi, I.: A hybrid proposed image quality assessment and enhancement framework for finger vein recognition. Multimed. Tools Appl. 1–26 (2022). https://doi.org/10.1007/s11042-021-11877-x
24. Tabassi, E., Wilson, C.L., Watson, C.I.: NIST fingerprint image quality. NISTIR 7151 (2004). https://doi.org/10.6028/NIST.IR.7151
25. Wang, Y., Fang, P.: A finger-vein image quality assessment algorithm combined with improved SMOTE and convolutional neural network. In: IEEE 11th International Conference on Software Engineering and Service Science (ICSESS), pp. 1–4 (2020). https://doi.org/10.1109/ICSESS49938.2020.9237657
26. Yang, L., Yang, G., Yin, Y., Xi, X.: Finger vein recognition with anatomy structure analysis. IEEE Trans. Circuits Syst. Video Technol. **28**(8), 1892–1905 (2017). https://doi.org/10.1109/TCSVT.2017.2684833
27. Yang, L., Yang, G., Yin, Y., Xiao, R.: Finger vein image quality evaluation using support vector machines. Opt. Eng. **52**(2), 027003–027003 (2013). https://doi.org/10.1117/1.OE.52.2.027003

28. Yin, Y., Liu, L., Sun, X.: SDUMLA-HMT: A multimodal biometric database. In: 6th Chinese Conference on Biometric Recognition (CCBR), LNCS, pp. 260–268. Springer, Cham (2011). https://doi.org/10.1007/978-3-642-25449-9_33
29. Zeng, J., Chen, Y., Qin, C.: Finger-vein image quality assessment based on light-CNN. In: 14th IEEE International Conference on Signal Processing (ICSP), pp. 768–773 (2018). https://doi.org/10.1109/ICSP.2018.8652381
30. Zhai, G., Min, X.: Perceptual image quality assessment: a survey. SCIENCE CHINA Inf. Sci. **63**, 1–52 (2020)
31. Zhao, Y., Sheng, M.Y.: Application and analysis on quantitative evaluation of hand vein image quality. In: International Conference on Multimedia Technology, pp. 5749–5751. IEEE (2011). https://doi.org/10.1109/ICMT.2011.6001948
32. Zhou, L., Yang, G., Yang, L., Yin, Y., Li, Y.: Finger vein image quality evaluation based on support vector regression. Int. J. Signal Process. Image Process. Pattern Recognit. **8**(8), 211–222 (2015). https://doi.org/10.14257/ijsip.2015.8.8.23

Deep Image Restoration Method for Low-Quality Visualization in Emergency Response Coordination for Coal-Fired Power Plants

Yue Zhu[1(✉)] and Yuanhui Gu[2]

[1] Production Technology Department, Garden Power Plant, Hami Coal and Power Co., Ltd., State Grid Energy Group, Xinjiang, China
20008328@ceic.com

[2] New Energy Branch, Hami Coal and Power Co., Ltd., State Grid Energy Group, Xinjiang, China
17008608@ceic.com

Abstract. The safe production of coal-fired power enterprises is an important foundation for ensuring the stability of energy supply and the safety of employees' lives and property. Real-time and precise monitoring of the working scenes in mines is an effective means to prevent the occurrence of safety accidents. However, factors such as non-uniform lighting and high-concentration dust that are widespread in the underground environment have led to severe degradation of monitoring images, restricting the accuracy and timeliness of risk identification. To solve the above problems, this paper proposes a model-data dual-driven total variation model of depth maps (DGTV), embedding the spectral graph theory into the interpretable network structure to achieve depth restoration of low-quality images in coal mine shafts. The experimental results show that the DGTV algorithm proposed in this paper improves the peak signal-to-noise ratio (PSNR) by 6.00 dB and 1.93 dB respectively compared with the traditional classic denoising algorithm BM3D and the DnCNN based on deep convolutional neural network residual learning under medium light conditions. This method integrates spectral graph theory with deep networks to construct an end-to-end, trainable and interpretable image depth restoration scheme, providing high-definition visual data for emergency linkage, improving the efficiency of coal mine early warning and collaborative response, and ensuring personnel safety and facility stability.

Keywords: Coal-fired power plant safety production · Model-data dual-driven approach · Spectral graph theory · Image depth restoration

1 Introduction

As intelligent mining and new power system construction deepen, coal-fired power enterprises—core entities in energy supply—face critical challenges where

P. Umapada et al. (Eds.): ICCPR 2025, CCIS 2811, pp. 428–439, 2026.
https://doi.org/10.1007/978-981-95-8315-7_34

emergency coordination directly impacts production safety and power stability [1,2]. Underground operations like mining, transportation, and maintenance rely on cameras for real-time imagery to enable emergency alerts, equipment monitoring, and safety protocols. However, extreme underground conditions—including low illumination, dust-laden environments, and moving objects—result in severe degradation of monitoring image quality [3,4], significantly hindering the accuracy and timeliness of risk identification, disaster assessment, and emergency command. Traditional algorithms struggle to effectively suppress noise while preserving details in images plagued by heavy noise and blurring [5].

Due to the complex and diverse nature of noise, classical model-based natural image denoising algorithms (e.g., BM3D, Non-Local Means) [6–8] perform poorly in underground coal mine environments. This underperformance stems from two primary reasons: First, these methods typically rely on strong image texture priors [9] and noise distribution assumptions [10] (e.g., BM3D's assumption of Gaussian noise and collaborative filtering based on similar blocks [11,12]), making them ill-suited to address mine-specific degradation factors like dust scattering and directional motion blur. Second, these methods often lack explicit constraints for segmented smoothing of edge structures [13], leading to edge jaggedness or artifacts in high-contrast underground scenes. Since Rudin et al. [8] introduced Total Variation (TV) regularization, gradient-norm-based denoising models have become a significant branch of image restoration. While TV models effectively preserve edges, they often cause excessive smoothing of fine textures. To mitigate this issue, researchers proposed sparse representation methods combining dictionary learning or wavelet transforms [14]. These utilize pre-trained overcomplete dictionaries to sparsely encode image blocks, suppressing noise while partially retaining texture information. However, such approaches typically incur high computational costs, limiting their practical application.

With the advancement of deep learning, methods like DnCNN [15] leverage large-scale training data and the powerful expressive capabilities of convolutional neural networks to excel in additive Gaussian noise removal tasks. However, when confronted with non-Gaussian, non-stationary noise in mining environments, pre-trained models often suffer from degraded generalization due to distribution mismatches between training and testing datasets. Subsequently, researchers proposed FFDNet [16], which introduces a noise intensity map at the input layer, enabling a single model to adapt to multiple noise levels. Variants like REDNet [17] and MemNet [18] further enhance the network's detail recovery capability through symmetric Encoder–Decoder structures or multi-scale feature fusion. While these networks demonstrated improved performance under synthetic noise conditions, they still faced challenges such as "insufficient generalization when training and testing noise are inconsistent" and "large network size, making embedded deployment difficult" [12]. In recent years, interpretable methods based on graph signal processing [19–22]—such as Graph Total Variation (GTV) [23–25]—have gained increasing attention. GTV treats image pixels as graph nodes, constructs a weighted graph based on pixel similarity, and employs a L1 norm prior to characterize image segmentation smoothing properties, achieving both edge preservation and noise robustness. However, directly optimizing the non-smooth L1 norm functional typically relies on iterative reweighting strategies, approximated as Graph Laplacian Regularization (GLR) [21], which suffers from low computational efficiency.

The DGTV algorithm introduces an algorithmic unfolding concept based on GTV, transforming the traditional iterative optimization process into a deep network architecture. The main contributions of this paper include:

1. Proposing a model-data dual-driven DGTV scheme tailored for mine image restoration, integrating graph signal priors with deep learning feature representation capabilities to adapt to complex degrading conditions underground.
2. Enhancing the traditional GTV model: Unfolding the iterative optimization process into a multi-layer cascaded differentiable network module. Utilizing a convolutional neural network CNN_F to extract pixel-level feature vectors, while employing CNN_μ to predict the regularization strength coefficient for each layer, achieving adaptive balance between smoothness and detail preservation. Simultaneously, implementing efficient and precise inversion of the sparse Laplacian matrix to enhance computational efficiency and restoration accuracy.
3. Establishing an end-to-end underground image restoration workflow to eliminate error accumulation inherent in traditional stepwise processing. This approach supports edge computing deployment, meeting coal-fired power enterprises' real-time demands for high-precision, low-latency visual data within emergency response systems.

2 Foundational Theory

2.1 Graph Construction

An 8-neighborhood graph $G = (V, E, W)$ is constructed using a node set V and an edge set E to represent an image block with N pixel values. Each image pixel i is represented by a node $i \in V$ in the graph. Let w_{ij} denote the edge weight of (i, j), representing the edge connecting nodes i and j. The graph edge weights are defined as a similarity matrix W, where $\mathrm{W_{ij}} = \mathrm{w_{ij}}$ is the weight of edge (i, j). Edge weights represent pixel similarity, a larger weight indicates stronger similarity between nodes i and j.

Given a graph G , define a diagonal matrix D as $D_{ii} = \sum_i w_{ij}$. The composite graph Laplacian matrix L [21] is defined as $L = D - W$. Assuming $\mathrm{w_{ij}} > 0$, for $(i, j) \in E$, then L is a positive semi-definite (PSD) matrix. Given L is symmetric and positive definite, it can be factorized as $L = V\Lambda V^T$, where V is the eigenvector matrix, Λ is the diagonal matrix with eigenvalues λ_k on its diagonal, and V^T is the matrix known as the graph Fourier transform (GFT) [26], which transforms the graph signal from the graph domain to the graph frequency domain.

2.2 GTV

A popular graph signal regularization method is GLR, defined as:

$$x^T L x = \sum_{(i,j)\in E} w_{ij} (x_i - x_j)^2 = \sum_{k=1}^{N} \lambda_k \alpha_k^2 \tag{1}$$

where λ_k is the eigenvalue of L and α_k is the graph Fourier coefficient. Minimizing Eq. (1) implies reducing the high-frequency components of the signal x in the graph frequency domain, i.e., low-pass filtering. Since L is the PSD, GLR is bounded from below by zero, i.e., $\forall x \in \mathbb{R}^N$. Another popular graph signal prior is GTV regularization, which preserves image edges and suppresses noise by measuring differences between adjacent pixels in the graph. It is defined as:

$$\| x \|_{GTV} = \sum_{(i,j)\in \mathcal{E}} w_{ij} |x_i - x_j| \tag{2}$$

GTV is also bounded from below by zero and only if $w_{ij} \geq 0$ holds for all $(i, j) \in \varepsilon$.

2.3 Deep Algorithm Unfolding Network

The model-data dual-driven network [25,27,28], proposed in recent years based on deep algorithmic unfolding, offers a solution that combines interpretability with adaptability. This approach constructs a network architecture with clear structure and semantic consistency between layers by explicitly mapping each iteration of traditional optimization algorithms to a layer within the neural network. Under this design, each layer not only receives updates from the preceding layer end-to-end but also allows critical hyperparameters (e.g., regularization coefficients, step size factors) to exist in a learnable form, enabling automatic training and optimization through backpropagation. This design, combining model-driven and data-driven concepts, not only preserves the physical interpretability of the original optimization algorithm but also significantly enhances the model's adaptability and expressive power through data-guided training.

3 Algorithm Development

3.1 Algorithm Framework

As shown in Fig. 1, the network architecture in this paper consists of two layers, each implementing one iteration of the algorithm. Each layer is constructed by stacking multiple modules (Blocks). During runtime, the first module takes the original image as input y, performs denoising, and outputs an image as the "original image" for the next module.

The DGTV architecture incorporates multiple GTV-Layers, each containing several submodules. During denoising, each GTV-Layer includes two CNN: one computes the feature vector F for each pixel, and the other calculates the weight parameter μ for each pixel. The parameters of are learned end-to-end using training data.

Overall, the DGTV architecture is composed of the GTV-Layer cascaded at the T layer. Through layer-by-layer refinement, it achieves the final image denoising effect. The CNN network and GTV iteration process within each layer work synergistically, making the image denoising process more precise and efficient.

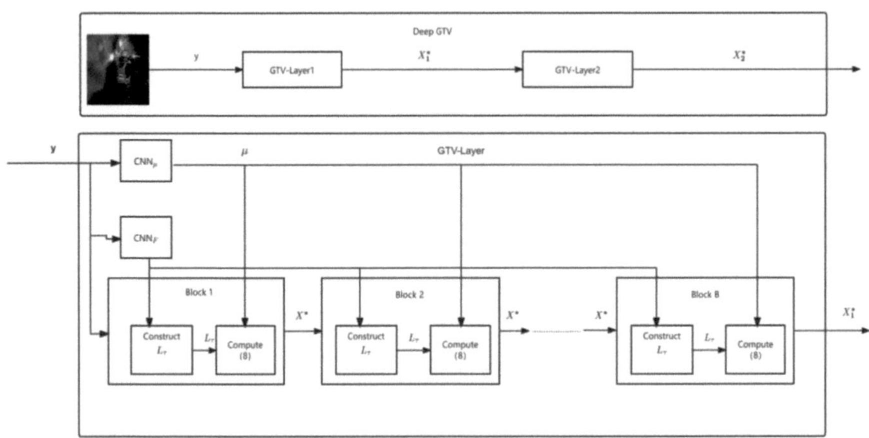

Fig. 1. DGTV Framework

3.2 GTV-Based Image Denoising

For a common image generation model where additive noise contaminates image blocks:

$$y = x + n \tag{3}$$

where $y \in \mathbb{R}^N$ is the contaminated observation, $x \in \mathbb{R}^N$ is the original image block, and $n \in \mathbb{R}^N$ is the noise term. The objective of this paper is to estimate x solely from y. Given a graph with edge weights w_{ij}, this paper constructs an optimization problem using GTV as the signal prior to reconstruct, as follows:

$$\min_{x} \| y - x \|_2^2 + \mu \sum_{(i,j)\in E} w_{ij} \left| x_i - x_j \right| \tag{4}$$

where $\mu > 0$ is the parameter balancing the fidelity term and the prior term. Equation (4) is convex and can be solved using the Proximal Gradient Method (PG) [29] iterative approach, though it lacks a closed-form solution. Furthermore, the resulting solution lacks spectral interpretability. In contrast, as described in [12], this paper transforms the GTV l_1-norm term in Eq. (4) into a quadratic term. The specific method is outlined below. First, a new adjacency matrix Γ is defined, with edge weights Γ_{ij} defined as follows given the signal estimate x° :

$$\Gamma_{ij} = \frac{w_{ij}}{\max\{|x_i^\circ - x_j^\circ|, \rho\}} \tag{5}$$

where $\rho > 0$ is a parameter chosen to avoid numerical instability, used when $|x_i^\circ - x_j^\circ| \approx 0$. Assuming $|x_i^\circ - x_j^\circ|$ is sufficiently large and the estimate x° is sufficiently close to the signal, we obtain:

$$\Gamma_{ij}|x_i - x_j|^2 \approx w_{ij}|x_i - x_j|^2 \tag{6}$$

This implies that by employing Γ, GTV can be expressed in quadratic form. Specifically, this paper defines a l_1-Laplace matrix L_Γ as follows:

$$L_\Gamma = \mathrm{diag}\left(\Gamma_1\right) - \Gamma \tag{7}$$

where 1 is a full-one vector of length N. The paper then uses L_Γ to transform Eq. (4) into a form with a quadratic regularization term:

$$x^* = \arg\min_{x} \parallel y - x \parallel_2^2 + \mu x^T L_\Gamma x \tag{8}$$

For a given estimate x° and the resulting Laplace matrix L_Γ, Eq. (8) has a closed-form solution:

$$x^* = (I + \mu L_\Gamma)^{-1} y \tag{9}$$

To solve Eq. (4) and Eq. (8), multiple iterations are required. In each iteration,$b+1$, the solution x_b^* obtained from the previous iteration b is used to update the edge weights Γ in Eq. (5). Within the DGTV architecture, each block solves Eq. (8) once. Each layer consists of blocks, and the cascade of layers forms the DGTV architecture presented herein, as shown in Fig. 1.

3.3 Feature and Weight Learning Based on CNN

In Eq. (4), assume a graph G exists with edge weights w_{ij}. As performed in [15,30], a CNN_F is employed at each layer to compute the appropriate feature vector $f_i \in \mathbb{R}^K$ for each pixel node, which is then used to construct the graph. Specifically, given the feature vectors f_i and f_j for pixel nodes i and j, respectively, a non-negative edge weight w_{ij} is computed using a Gaussian kernel, defined as:

$$w_{ij} = \exp\left(-\frac{\sum_{k=1}^{K}\left(f_i^k - f_j^k\right)^2}{\epsilon^2}\right) \tag{10}$$

where $\epsilon > 0$ is a parameter. To learn the parameters in the CNN, when calculating f_i, L_1 and L_Γ are constructed using training data. The gradient with respect to the mean squared error (MSE) between the reconstructed image x^*

and the true image x can then be computed, enabling parameter updates in the CNN via backpropagation. MSE is defined as:

$$\mathcal{L}_{\mathrm{MSE}} = \frac{1}{\mathrm{N}} \sum_{\mathrm{i}=1}^{\mathrm{N}} (\mathrm{x}_{\mathrm{i}}^{*} - \mathrm{x}_{\mathrm{i}})^2 \tag{11}$$

The computed edge weights w_{ij} form the adjacency matrix W. Another CNN is employed at each layer—using CNN_{μ} to compute the weight parameters μ for each noisy image block. Similar to CNN_F, μ appears in Eq. (9), enabling learning of CNN_{μ} through backpropagation. Each GTV layer in DGTV filters image blocks by solving Eq. (4), allowing training of all CNN in DGTV by supervising the error between the final reconstructed image x^* and the true image x.

4 Experiments

We conducted experiments on the proposed DGTV. We performed an in-depth analysis of the application effectiveness of three image denoising algorithms in the complex environment of coal mines. The experimental results demonstrate significant differences among the algorithms in terms of denoising quality, computational efficiency, and scene adaptability, providing important reference for selecting and optimizing image restoration techniques in coal mines.

4.1 Datasets

This paper utilizes the "Underground Drilling Site Target Detection Dataset for Coal Mines" [5], which comprises 70,948 color images at a resolution of 1280×720. For this study, two datasets were selected from the collection, corresponding to different lighting conditions and viewing angles: the Medium-Light group and the Low-Light group, each containing ten images. Additionally, to simulate extreme shooting environments, Gaussian noise with a standard deviation of 25 was added to these images. Both the noisy and clean images were stored together in project files.

4.2 Evaluation Metrics

Common objective metrics for evaluating denoising performance include PSNR, Structural Similarity(SSIM) and Peak Video Random Access Memory(Peak VRAM). PSNR measures the average error between the denoised image and the reference clean image , defined as:

$$\mathrm{PSNR} = 10 \cdot \log_{10}\left(\frac{\mathrm{L}^2}{\mathrm{MSE}}\right), \tag{12}$$

$$\mathrm{MSE} = \frac{1}{\mathrm{MN}} \sum_{\mathrm{i}=1}^{\mathrm{M}} \sum_{\mathrm{j}=1}^{\mathrm{N}} [\mathrm{x}(\mathrm{i},\mathrm{j}) - \hat{\mathrm{x}}(\mathrm{i},\mathrm{j})]^2 \tag{13}$$

where $M \times N$ is the image resolution and L is the maximum pixel value ($L = 255$ for an 8-bit image). A higher PSNR value indicates a smaller mean squared error between the reconstructed image and the original image. However, PSNR only reflects the pixel-level average difference and cannot accurately capture the human eye's subjective perception of structure and contrast.

SSIM provides a comprehensive measure of image similarity across luminance, contrast, and structural aspects. Its calculation formula is:

$$\mathrm{SSIM}(\mathrm{x}, \hat{\mathrm{x}}) = \frac{(2\mu_{\mathrm{x}}\mu_{\hat{\mathrm{x}}} + \mathrm{C}_1)(2\sigma_{\mathrm{x}\hat{\mathrm{x}}} + \mathrm{C}_2)}{(\mu_{\mathrm{x}}^2 + \mu_{\hat{\mathrm{x}}}^2 + \mathrm{C}_1)(\sigma_{\mathrm{x}}^2 + \sigma_{\hat{\mathrm{x}}}^2 + \mathrm{C}_2)} \tag{14}$$

where μ_x and $\mu_{\hat{x}}$ are the local means of x and $\hat{x}$ respectively, σ_x^2 and $\sigma_{\hat{x}}^2$ are the variances, $\sigma_{x\hat{x}}$ is the covariance, and C_1, C_2 are small constants to stabilize the division. The SSIM metric ranges from $[0, 1]$, values closer to 1 indicate better preservation of structure and contrast, reflecting higher visual quality.

Peak VRAM is a key performance metric for measuring the demand of GPU hardware memory resources for deep learning workloads. Peak VRAM directly determines the maximum Batch Size that the model can support. A larger Batch Size can usually increase the training speed and improve the stability of gradient estimation.

4.3 Image Restoration Results

This paper systematically evaluates three classical denoising methods: BM3D, DnCNN, and DGTV. DGTV employs two approaches to compute Eq. (9).

(1) Legacy Mode: During runtime, CNN_Fcomputes a suitable feature vector f_i for each pixel i in the N-limited patch. These feature vectors are used to construct the in Eq. (9). Additionally, CNN_μ calculates the weight parameter for each noisy image block to ensure optimal noise suppression and detail preservation.
(2) Non-Legacy Mode: Solves the sparse Laplacian matrix inverse explicitly for each patch, sacrificing slightly lower denoising quality for reduced memory consumption.

Table 1. Average Comparison of Three Methods on the 5-Image Test Set in the Medium-Light Group

Method	Average PSNR (dB)	Average SSIM	Peak VRAM (GB)
BM3D	23.96	0.350	10.5
DnCNN	28.03	0.501	12.5
DGTV (legacy)	29.96	0.710	23.6
DGTV	29.94	0.702	23.5

Table 2. Average Comparison of Three Methods on the 5-Image Test Set in the Low-Light Group

Method	Average PSNR (dB)	Average SSIM	Peak VRAM (GB)
BM3D	23.26	0.310	10.5
DnCNN	27.29	0.44	12.5
DGTV (legacy)	28.67	0.610	23.6
DGTV	28.21	0.500	23.5

We conducted experiments on BM3D, DnCNN, DGTV, and the Legacy mode of DGTV under both medium and low illumination conditions. The results are summarized in Table 1 and Table 2. DGTV demonstrated outstanding performance under both medium and low illumination conditions. Particularly in low-light conditions, DGTV achieved an average PSNR of 28.67 dB and SSIM of 0.610, demonstrating strong capabilities in both noise reduction and detail preservation. In contrast, DnCNN yielded an average PSNR of 27.29 dB and SSIM of 0.44 under low-light conditions, revealing a significant disadvantage in such challenging environments. Among the three algorithms, the PSNR and SSIM values in the high-light group were higher than those in the low-light group. However, DGTV achieved a higher absolute restoration level under both lighting conditions. Under medium-light conditions, DGTV achieved a PSNR of 29.96 dB and an SSIM of 0.710, significantly outperforming DnCNN (PSNR 28.03 dB, SSIM 0.501) and BM3D (PSNR 23.96 dB, SSIM 0.350). Generally,

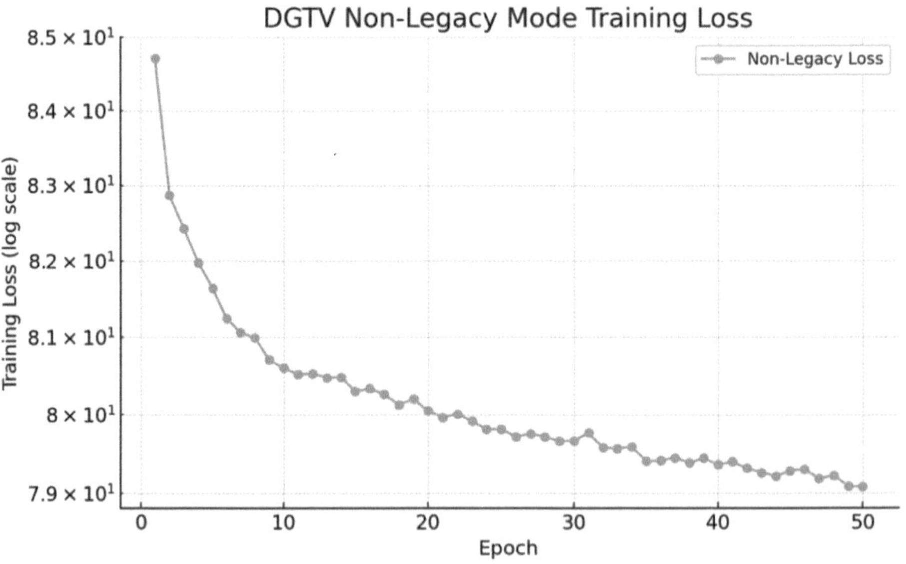

Fig. 2. DGTV training loss curve for the light group.

DGTV can achieve higher PSNR and SSIM values in the denoising task, compared to BM3D and DnCNN, particularly under complex lighting conditions. From the perspective of computational resources, Non-Legacy Mode occupies less memory than Legacy Mode.

In additon, by combining GTV priors with deep learning, DGTV can self-learn and optimize during iterations (see Fig. 2). Figure 3 provides the visual results of proposed DGTV method and comparison methods. From the enlarged regions, we can see that, compared to BM3D and DnCNN, the DGTV-restored image can offer smoother noise regions, indicating its clear advantage in noise reduction and detail preservation within the complex mine environment.

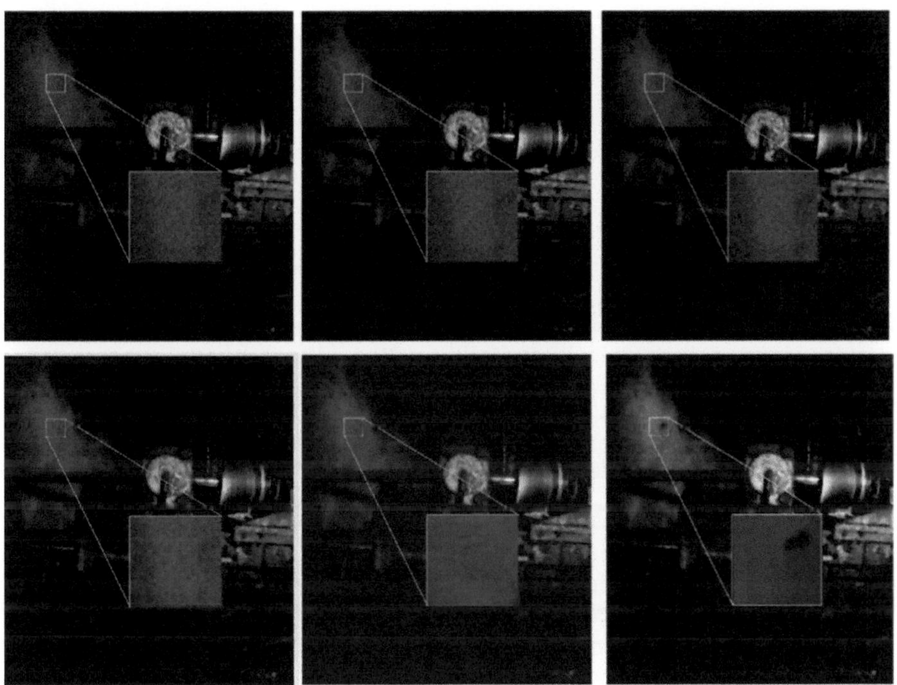

Fig. 3. Comparison of various denoising algorithms: The first column shows BM3D results, the second column shows DnCNN results, and the third column shows DGTV results. The top three images depict the original noise patterns, while the bottom three display the denoised results.

5 Summary

Addressing visualization demands for coal mining enterprises' safety production and emergency response coordination, effective image restoration in complex underground environments is crucial for enhancing risk identification and command decision-making capabilities. This paper tackles severe image degradation caused by non-uniform illumination, high dust concentration, and noise interference in coal mines by proposing DGTV—a method based on spectral graph theory and algorithmic unfolding. It achieves remarkable results in noise suppression and detail preservation.

Future work will explore multimodal information fusion and lightweight network design to enhance the algorithm's robustness under extreme degradation conditions and inference efficiency on embedded devices, thereby expanding its practical applications in intelligent mine monitoring and emergency command systems.

References

1. Xuefei, W., Li, H., Wang, B., Zhu, M.: Review on improvements to the safety level of coal mines by applying intelligent coal mining. Sustainability **14**(24), 16400 (2022)
2. Deqiang, C., et al.: Review on key technologies of ai recognition for videos in coal mine. Coal Sci. Technol. **51**(2), 349–365 (2023)
3. Cheng, J., et al.: Architecture and key technologies of coalmine underground vision computing. COAL Sci. Technol. **51**(9), 202–218 (2023)
4. Zhang, X.-H., et al.: Clearing research on fog and dust images in coalmine intelligent video surveillance. J. China Coal Soc. **39**(1), 198–204 (2014)
5. Zhou, W., et al.: A dataset of drilling_site object detection in underground coal mines. China Sci. Data **9**(2), 1–10 (2024)
6. Dabov, K., Foi, A., Katkovnik, V., Egiazarian, K.: Image denoising by sparse 3-d transform-domain collaborative filtering. IEEE Trans. Image Process. **16**(8), 2080–2095 (2007)
7. Elad, M., Aharon, M.: Image denoising via learned dictionaries and sparse representation. In: 2006 IEEE Computer Society Conference on Computer Vision and Pattern Recognition (CVPR'06), vol. 1, pp. 895–900. IEEE, 2006
8. Beck, A., Teboulle, M.: Fast gradient-based algorithms for constrained total variation image denoising and deblurring problems. IEEE Trans. Image Process. **18**(11), 2419–2434 (2009)
9. Muresan, D.D., Parks, T.W.,: Adaptive principal components and image denoising. In: Proceedings 2003 International Conference on Image Processing (Cat. No. 03CH37429), vol. 1, pp. I–101. IEEE, 2003
10. Foi, A., Katkovnik, V., Egiazarian, K.: Pointwise shape-adaptive dct for high-quality denoising and deblocking of grayscale and color images. IEEE Trans. Image Process. **16**(5), 1395–1411 (2007)
11. Dabov, K., Foi, A., Katkovnik, V., Egiazarian, K.: Image denoising with block-matching and 3d filtering. In: Image Processing: Algorithms and Systems, Neural Networks, and Machine Learning, vol. 6064, pp. 354–365. SPIE, 2006
12. Vemulapalli, R., Tuzel, O., Liu, M.Y.: Deep gaussian conditional random field network: a model-based deep network for discriminative denoising. In: Proceedings of the IEEE Conference on Computer Vision and Pattern Recognition, pp. 4801–4809, 2016

13. Kim, K.R., Kim, C.S.: Adaptive smoothness constraints for efficient stereo matching using texture and edge information. In: 2016 IEEE International Conference on Image Processing (ICIP), pp. 3429–3433. IEEE, 2016
14. Elad, M., Aharon, M.: Image denoising via sparse and redundant representations over learned dictionaries. IEEE Trans. Image Process. **15**(12), 3736–3745 (2006)
15. Zhang, K., Zuo, W., Chen, Y., Meng, D., Zhang, L.: Beyond a gaussian denoiser: residual learning of deep cnn for image denoising. IEEE Trans. Image Process. **26**(7), 3142–3155 (2017)
16. Zhang, K., Zuo, W., Zhang, L.: Ffdnet: Toward a fast and flexible solution for cnn-based image denoising. IEEE Trans. Image Process. **27**(9), 4608–4622 (2018)
17. Mao, X., Shen, C., Yang, Y.B.: Image restoration using very deep convolutional encoder-decoder networks with symmetric skip connections. Adv. Neural Inf. Process. Syst. **29** (2016)
18. ai, Y., Yang, J., Liu, X., Xu, C.: Memnet: a persistent memory network for image restoration. In: Proceedings of the IEEE International Conference on Computer Vision, pp. 4539–4547, 2017
19. Cheung, G., Magli, E., Tanaka, Y., Ng, M.K.: Graph spectral image processing. Proc. IEEE **106**(5), 907–930 (2018)
20. Su, X., Rizkallah, M., Maugey, T., Guillemot, C.: Graph-based light fields representation and coding using geometry information. In: 2017 IEEE International Conference on Image Processing (ICIP), pp. 4023–4027. IEEE, 2017
21. Pang, J., Cheung, G.: Graph laplacian regularization for image denoising: analysis in the continuous domain. IEEE Trans. Image Process. **26**(4), 1770–1785 (2017)
22. Ortega, A., Frossard, P., Kovačević, J., Moura, J.M., Vandergheynst, P.: Graph signal processing: overview, challenges, and applications. Proc. IEEE **106**(5), 808–828 (2018)
23. Couprie, C., Grady, L., Najman, L., Pesquet, J.-C., Talbot, H.: Dual constrained tv-based regularization on graphs. SIAM J. Imag. Sci. **6**(3), 1246–1273 (2013)
24. Berger, P., Hannak, G., Matz, G.: Graph signal recovery via primal-dual algorithms for total variation minimization. IEEE J. Sel. Top. Signal Process. **11**(6), 842–855 (2017)
25. Vu, H., Cheung, G., Eldar, Y.C.: Unrolling of deep graph total variation for image denoising. In: ICASSP 2021–2021 IEEE International Conference on Acoustics, Speech and Signal Processing (ICASSP), pp. 2050–2054. IEEE, 2021
26. Wei, H., Cheung, G., Ortega, A.: Intra-prediction and generalized graph fourier transform for image coding. IEEE Signal Process. Lett. **22**(11), 1913–1917 (2015)
27. Monga, V., Li, Y., Eldar, Y.C.: Algorithm unrolling: interpretable, efficient deep learning for signal and image processing. IEEE Signal Process. Mag. **38**(2), 18–44 (2021)
28. Chen, S., Eldar, Y.C., Zhao, L.: Graph unrolling networks: interpretable neural networks for graph signal denoising. IEEE Trans. Signal Process. **69**, 3699–3713 (2021)
29. Elmoataz, A., Lezoray, O., Bougleux, S.: Nonlocal discrete regularization on weighted graphs: a framework for image and manifold processing. IEEE Trans. Image Process. **17**(7), 1047–1060 (2008)
30. Su, W.T., Cheung, G., Wildes, R., Lin, C.W.: Graph neural net using analytical graph filters and topology optimization for image denoising. In: ICASSP 2020–2020 IEEE International Conference on Acoustics, Speech and Signal Processing (ICASSP), pp. 8464–8468. IEEE, 2020

Neuronal Biology-Inspired Multi-pooling CNNs for Skin Lesion Classification

Jiawei Sun[1,3], Youran Zhang[1,3], Emily Ding[2,3(✉)], and Robert Hou[2,3]

[1] Rangitoto College, 564 East Coast Road, Mairangi Bay, Auckland 0630, New Zealand
{183872,184217}@cloud.rangitoto.school.nz

[2] Vineyards AI Lab, Flat 1, 555 East Coast Road, Browns Bay, Auckland 0630, New Zealand
{emily.ding,robert.hou}@vineyardsailab.com

[3] Rangitoto and Vineyards AI Innovation Centre, 564 East Coast Road, Mairangi Bay, Auckland 0630, New Zealand

Abstract. Convolutional Neural Networks (CNNs) is a classic and powerful model widely used in computer vision. This paper presents a novel CNN architecture inspired by the information processing mechanisms of biological neurons, applied to skin cancer detection. Specifically, the new model is inspired by a creative analogy: applying the concepts of "convergence" and "divergence" from neural systems to CNNs. We observed that convolutional layers exhibit both convergent and divergent properties, while pooling layers mainly reflect convergence and lack a mechanism for divergence. Based on this, we propose an adaptive multi-pooling CNNs to enhance feature diversity. This study was grounded in the task of automatic classification of skin lesions, which aimed to provide a practical application for validating the effectiveness of the proposed model. Experimental results show that it achieves modest performance improvements over traditional CNNs. More importantly, this study aims to highlight that the value of scientific research lies not only in performance gains but also in its ability to inspire new ideas and meaningful innovation. This work displays a journey from AI users to learners and to creators, demonstrating the value of interdisciplinary thinking in AI research and education.

Keywords: Skin Lesion · Automatic Classification · Convergence and Divergence · CNNs · Adaptive Multi-Pooling

1 Introduction

Artificial Intelligence (AI) has evolved from a specialised academic term into a widely used technological tool [1,2]. As it becomes increasingly prevalent, understanding how to access and apply it effectively has become a topic of great interest. Many people aim not merely to be users of AI, but to understand its underlying principles, explore broader applications, and contribute to

P. Umapada et al. (Eds.): ICCPR 2025, CCIS 2811, pp. 440–451, 2026.
https://doi.org/10.1007/978-981-95-8315-7_35

its advancement. This paper presents a journey from AI users to learners and collaborative innovators.

The research started with Convolutional Neural Networks (CNNs) because it's the most classic and foundational models in AI, widely applied to visual tasks such as object detection, image generation, and image segmentation in computer vision [3,4]. In addition, CNNs and their variants, such as one-dimensional CNNs (1D-CNNs) and three-dimensional CNNs (3D-CNNs) [5,6], have been extended to numerous other fields, including natural language processing, speech recognition, time series analysis, video understanding, and medical imaging, demonstrating great versatility and potential. We draw an analogy between CNNs and neural processing mechanisms that involve both convergence and divergence [7]. While convolutional layers exhibit these dual properties, pooling layers lack the divergence aspect. To address this limitation, we propose an adaptive multi-pooling layer to compensate for this deficiency. To validate the effectiveness of the proposed model, we applied it to a skin lesion dataset [8]. Skin lesions are a significant global health concern, where early diagnosis is crucial [9–11]. Experimental results demonstrate that our model outperforms traditional CNNs in terms of performance.

Our key contributions are twofold:

1. Biological Interpretation and Model Innovation: We interpret CNN architecture through the lens of neural convergence-divergence theory and, inspired by this, introduce a multi-pooling module to augment feature representation in skin lesion classification.
2. Educational Impact and Youth Innovation: This project highlights how a team of high school students, guided by interdisciplinary insights, can contribute to AI algorithm development. It serves as a compelling example of the educational value of involving youth in real-world AI research.

2 The Pathway of Our Learning and Innovation

This section outlines our journey of learning and innovation, starting with a study of CNNs. We then explored the biological concepts of neuronal convergence and divergence, which inspired us to design a new model.

2.1 Convolutional Neural Networks

CNNs are a powerful type of artificial neural network which were designed for image and pattern recognition. Over the years, a series of increasingly sophisticated CNNs have been proposed, including VGG, GoogLeNet, ResNet, and DenseNet etc. [12–16]. In addition, CNNs has been widely used in medical image classification tasks and have shown promising performance [17,18]. Figure 1 displays the architecture of CNNs.

In Fig. 1, we can see that two of the most critical components are the convolutional layers and pooling layers, which work together to extract increasingly

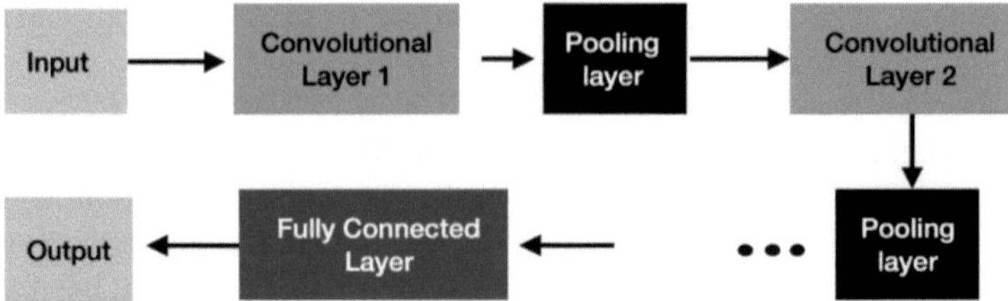

Fig. 1. Architecture of the Convolutional Neural Network (CNN)

abstract features from the input data. The convolutional layer applies multiple filters or kernel matrices of the same size, with each filter designed to detect a specific pattern in the input image. The kernel acts as a sliding window that moves across the matrix of pixels representing the image, performing convolution operations to extract relevant features such as edges, textures, or shapes. The pooling layer is designed to select the most significant features from the convoluted feature matrix. It reduces the dimension of the feature map by applying some aggregation operations such as max pooling, average pooling, etc.

2.2 Neural Convergence and Divergence in Biology

Neural convergence and divergence are opposing mechanisms of information processing in the nervous system. In visual processing, the concept of neural convergence refers to a neuron receiving multiple inputs from other neurons and integrating the signals into one [19]. The process is observed in early sensory cortices and higher-order association cortices [20]. Neurophysiological evidence in both cat and primate superior colliculus (SC) shows that afferents from different sensory modalities converge on individual neurons, enabling multisensory integration crucial for perception and cognition [21] (Fig. 2).

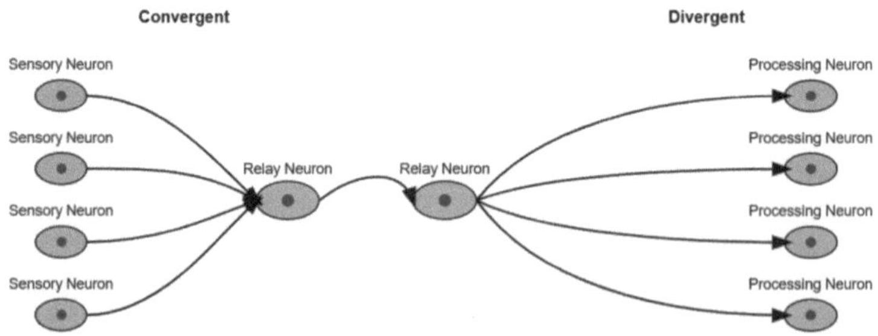

Fig. 2. The process of neural convergence and divergence in biology, which happens during signal transmission in the nervous system. These two mechanisms work together for neural communication.

In contrast, neural divergence amplifies and distributes information, enabling parallel processing and coordinated responses across different regions of the nervous system. For example, retinal ganglion cells (RGCs) in rodents and primates often project collaterally to numerous brain regions, including the lateral geniculate nucleus (LGN) and superior colliculus (SC), demonstrating anatomical divergence in visual pathways [22].

In summary, the convergence mechanism facilitates signal integration and abstraction, while the divergence mechanism supports signal amplification and parallel processing through multiple pathways. Together, these two mechanisms work in concert to build an efficient and robust neural information processing network.

2.3 Inspirations for Convergence and Divergence in CNNs

We analyse the convergence and divergence mechanisms in CNNs using the following figure as a representative example.

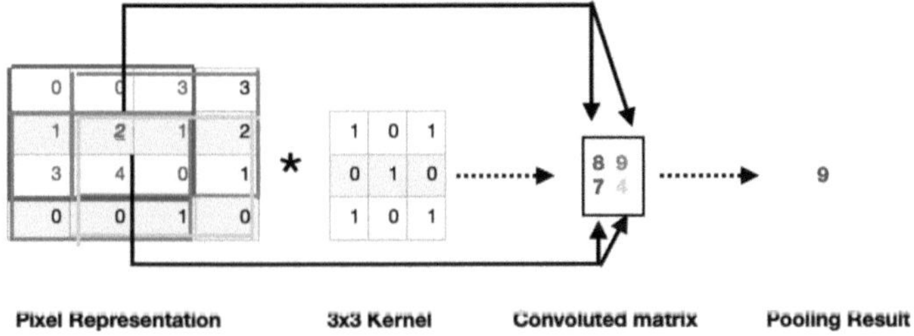

Fig. 3. Demonstration of Computations for Convolution

In image convolution, the input image undergoes element-wise multiplication with a kernel, and the resulting values are summed to produce a single output value. By sliding the window across the image with a fixed step size, known as the stride, the original 4 × 4 image is transformed into a 2 × 2 feature map. Pooling layers are then used to further reduce the spatial dimensions of the feature map. For example, using a 2 × 2 max pooling operation, only the maximum value 9 is retained, while the other values are discarded.

This layered transformation closely mirrors biological convergence in the nervous system. Convolutional and pooling layers progressively condense spatially distributed information into abstract representations. Convolution captures essential local features like edges or textures, while pooling highlights key responses and filters out redundancy, ultimately forming a compact, high-level encoding of the original input.

In addition, we can also observe that convolutional layers do not solely exhibit convergence mechanisms, but also embed a form of divergence. For instance, in Fig. 3, the red-marked input element '2' is involved in multiple local computations due to the sliding nature of the convolutional kernel. As a result, this single input contributes to several output values in the feature map, such as elements 8, 9, 7, and 4.

However, we observe that a single pooling layer lacks a divergent process. This limitation might lead to the loss of diverse contextual representations and restrict the network's ability to capture multi-scale features or subtle patterns. Inspired by this observation, we aim to design a multi-pooling CNN architecture and introduce it in the next part.

3 Adapt Multi-pooling CNNs for Skin Lesion Classification

3.1 The Architecture of Adapt Multi-Pooling CNNs

To enhance the divergence of information in the pooling layers, we designed a multi-pooling structure to replace the traditional single pooling method. Each pooling branch is assigned a learnable weight, which is automatically adjusted during training. This allows the network to adaptively balance the contribution of each pooling result. We refer to this architecture as an adaptive multi-pooling convolutional neural network (adaptive multi-pooling CNN), which was shown in Fig. 4.

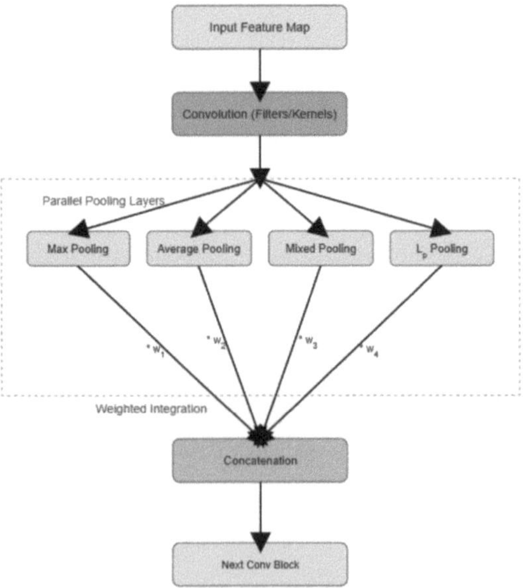

Fig. 4. Displayed above is an example of the solution with 4 parallel pooling methods implemented: maxing, average, mixed, and LP pooling.

The pooling layers include, but are not limited to max pooling, average pooling, and mixed pooling, LP pooling. The output of each pooling layer is then integrated together into a singular product, which is fed to the next convolution block, where the same process takes place.

However, given the flexibility of parallel processing and the consideration that different features may be more important or redundant at different stages of processing, an individual weight is assigned to the output of each pooling layer when integrating. The weight is then trained as well, so that at each different stage, the pooling layers applied carry different significance.

3.2 Mathematical Description

Each pooling layer should extract different aspects of the same feature map. For example, max pooling selects the most notable feature within the region examined; other lesser values, which may not be insignificant, are discarded. While the average pooling was used as a form of compensation, where the mean of all values is extracted, outputting a result which also weighs in the other values at the expense of compromising more outstanding features. Next, we'll list the pooling methods we'll be using.

The convoluted matrix can be written as:

$$\mathbf{Y}_k = \sigma(\mathbf{P}_{k-1} * \mathbf{F}_k + b_k) \tag{1}$$

where $\mathbf{Y}_k$ represents the kth layer's convoluted matrix, $\mathbf{P}_{k-1}$ is the output of the pooling layer from the previous layer, $\mathbf{F}_k$ denotes the convolutional kernel, b_k is the bias associated with the convolutional kernel, and $\sigma(\cdot)$ is the activation function (e.g., ReLU).

Max Pooling can be written as:

$$\mathbf{P}_{max} = max_{\mathbf{E}}(\mathbf{Y}_k(i,j)) \tag{2}$$

where $\mathbf{P}_k$ is the result of the pooling layer. max_E being the max pooling function, selecting the highest value within a kernel that is represented by $\mathbf{E}$ (which is $\mathbf{E}(m,n)$ where m and n are the height and width of the pooling kernel), and $\mathbf{Y}_k$ is the kth layer's convoluted matrix, and i,j is the height and width of the input feature map.

The other pooling functions used in the model can be written as below in sequential order: average, mixed, and L_p pooling.

$$\mathbf{P}_{avg} = ave_{\mathbf{E}}(\mathbf{Y}_k(i,j)) = \frac{1}{\mathbf{E}_m \cdot \mathbf{E}_n} \cdot \sum_{u=0}^{\mathbf{E}_{m-1}} \sum_{v=0}^{\mathbf{E}_{n-1}} \mathbf{Y}_k(s_i \cdot i + u, s_j \cdot j + v, c) \tag{3}$$

where $\mathbf{E}_m$ and $\mathbf{E}_n$ are the pooling kernel's height and width, respectively; and $\mathbf{Y}_k$ is the input feature map, and i and j are the height and width of the feature map, and c is the channel index.

$$\mathbf{P}_{mixed} = mixed_{\mathbf{E}}(\mathbf{Y}_k(i,j)) = \frac{1}{2}(\mathbf{P}_{avg} + \mathbf{P}_{max}) \tag{4}$$

$$\mathbf{P}_{L_p} = L_P(\mathbf{Y}_k(i,j)) = (\mathbf{P}^{\mathbf{p}}_{\mathbf{avg}} + \mathbf{P}^{\mathbf{p}}_{\mathbf{max}})^{\frac{1}{p}} \tag{5}$$

where $\mathbf{P}_{avg}$, $\mathbf{P}_{mixed}$, $\mathbf{P}_{L_p}$ are the outputs of average, mixed, and L_P pooling. $\mathbf{E}$ being the pooling kernel, and $\mathbf{Y}_k$ is the kth layer's convoluted matrix (input feature map), $i\&j$ are the dimensions of the pooling kernel. The $ave_{\mathbf{E}}(\mathbf{Y}_k)$ function takes the mean value of kernel $\mathbf{E}$, $mixed_{\mathbf{E}}(\mathbf{Y}_k)$ takes the mean of the sum of a max-pooled and an average-pooled value of the kernel $\mathbf{E}$, whereas $L_{p_{\mathbf{E}}}(\mathbf{Y}_k)$ takes the p-norm-based output of the max and the average pooled value of kernel $\mathbf{E}$.

The outputs of the four pooling layers $[\mathbf{P}_{max}, \mathbf{P}_{avg}, \mathbf{P}_{mixed}, \mathbf{P}_{L_p}]$ are concatenated into a single matrix: $\mathbf{P}_c = contact(\mathbf{P}_{max}, \mathbf{P}_{avg}, \mathbf{P}_{mixed}, \mathbf{P}_{L_p})$

Where $\mathbf{P}_c$ represents the concatenated tensor. Then it is passed through a dense layer with a softmax activation to compute or update the weights.

$$\mathbf{W} = softmax(\mathbf{P}_c \cdot \mathbf{w}_d + b_d) \tag{6}$$

where $\mathbf{W} = [w_{max}, w_{avg}, w_{mixed}, w_{L_p}]$ are the normalized weights for each pooling method. $\mathbf{w}_d$ and b_d are the trainable weights and biases of the dense layer. The softmax activation ensures that the weights $\mathbf{W}$ sum to 1.

Each pooling output is multiplied element-wise by its corresponding weight:

$$\mathbf{P}^k_{i-th(pooling-method)} = w^k_{i-th(pooling-method)} \mathbf{P}^k_{i-th(pooling-method)} \tag{7}$$

The weighted pooling outputs are summed to produce the final output:

$$\mathbf{P}^k = \sum_{i=1}^{N} \mathbf{P}^k_{i-th(pooling-method)} \tag{8}$$

where N is the number of pooling methods.

4 Experiments

This study aims not to outperform state-of-the-art models, but to demonstrate the effectiveness of our approach and offer inspiration. Thus, we compare it only with traditional CNNs using single pooling to highlight the benefits of multi-pooling.

4.1 Dataset Introduction

To validate our approach, we used the HAM10000 dataset, which contains 10,015 high-resolution dermatoscopic images across seven skin lesion categories: nv, mel, bkl, bcc, akiec, vasc, and df. Figure 5 shows the sample distribution, which is imbalanced but offers diverse categories for research. Due to computational constraints, we used a stratified subset of 5,000 samples for training and 1,500 for testing. Table 2 are sample distribution for testing and training subsets (Fig. 6).

To establish clinically meaningful severity levels for classification tasks, we referred to paper [9], which highlights the varying degrees of malignancy and urgency among different skin lesion types, as shown in Table 1.

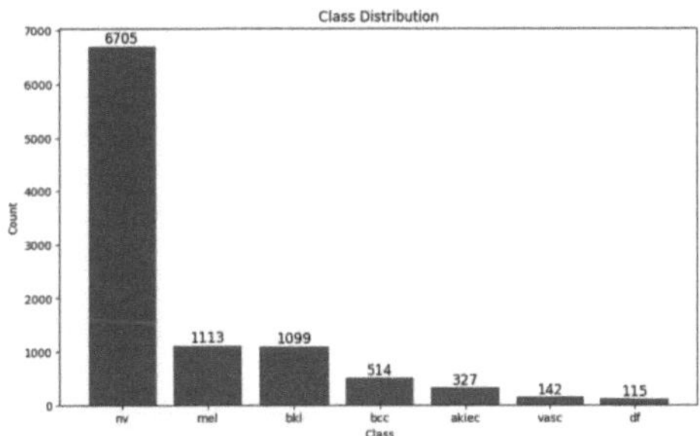

Fig. 5. Distribution of each sample

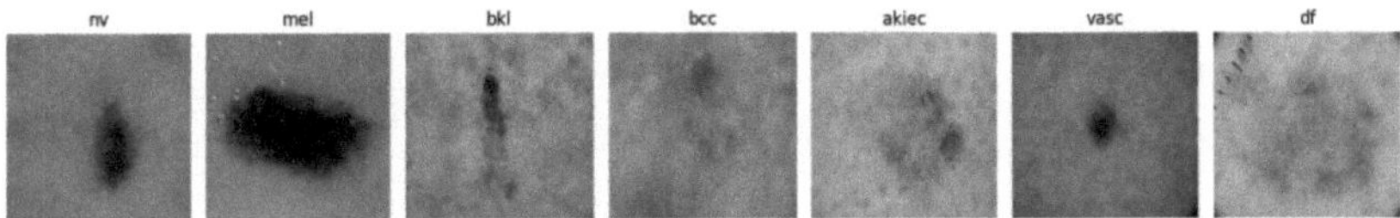

Fig. 6. Samples of each category

4.2 Experimental Results

The model was implemented in Colab utilizing the free GPU resources provided by Google. Table 3 shows the hyperparameters for the environment setup.

We compared our method against models using max, average, mixed, and L-p pooling to evaluate its effectiveness. The evaluation methods are based on accuracy score, confusion matrix, precision, sensitivity(recall), specificity, false positive rate (FPR), and false negative rate (FNR) [23,24].

To evaluate our multi-pooling strategy, we compared CNNs with 3–5 convolutional layers using different pooling methods. Results are shown in Table 4.

For the model with 3 convolutional layers, the highest accuracy among single-pooling methods was achieved by max pooling (76.20%), while others like average and mixed pooling fell below 73%. In contrast, our multi-pooling models achieved comparable or better performance, with combinations like Max & Mixed reaching 75.53% and a higher average accuracy of 74.66%, outperforming the single-pooling average of 73.41%.

For the model with 4 layers, the performance gap widened. The Max & L-p combination achieved the highest accuracy of 76.54%, surpassing all configurations of both single and multi-pooling in 3-layer models. The average accuracy

Table 1. Clinical Severity Levels of Skin Cancer Classification

Severity	Disease Category
Severe	Melanoma (Mel)
Moderate	Basal Cell Carcinoma (bbc), Actinic Keratoses and Intraepithelial Carcinoma (akiec)
Mild	Benign Keratosis-like lesions(bkl); Vascular lesions(vasc), Dermatofibroma(df), Melanocytic Nevi(nv)

Table 2. Sample distribution for testing

Total	nv	mel	bkl	bcc	akiec	vasc	df	dataset
1469	1004	166	164	76	48	21	17	Testing
4996	3347	555	548	256	163	70	57	Training

Table 3. Hyperparameters for the model

Hyperparameter	Description
Number of Convolutional Layers	3, (4, 5)
Number of filters	32, 64, 128, (256, 512)
Batch size	32
Optimizer	Adam
Classes	7
Image size	(128, 128, 3)
Activation function	Relu
Loss function	Sparse categorical cross-entropy
Epochs	30

for this layer rose to 75.45%, demonstrating the robustness and scalability of the multi-pooling design.

With 5 convolutional layers, our model again showed strong results. The Max & L-p setting achieved 75.27%, confirming that the integration of multiple pooling types continues to enhance feature representation and classification performance as the model complexity grows.

In summary, the results show that multi-pooling strategies inspired by neuronal convergence and divergence lead to more effective learning, particularly in complex classification tasks such as skin cancer detection.

In the analysis of experimental results, we shouldn't solely focus on the values of measurement metrics. It is necessary to consider factors such as the sever-

Table 4. Accuracy of various conditions

Invididual Pooling		max-pooling	average -pooling	mixed-pooling	L-p Pooling	-	Avg
3-layers	ACC	76.20%	72.19%	71.93%	73.33%	-	73.41%
	Best Iteration Time	36	30	30	27	-	34
Multi-Pooling		Ave&Max	Ave&Max&Mixed	Max&Mixed	Max&L_p	Max&Ave&Mixed&L_p	Avg
3-layers	ACC	75.13%	73.93%	75.53%	74.47%	74.26%	74.46%
	Best Iteration Time	27	27	26	29	26	27
4-layers	ACC	74.13%	74.53%	76.07%	76.54%	76.00%	**75.45%**
	Best Iteration Time	24	29	30	22	20	25
5-layers	ACC	73.03%	73.66%	74.40%	75.27%	72.14%	**75.45%**
	Best Iteration Time	27	28	28	25	21	26

ity of the category and the number of samples within that category. Table 5 presents the recall, specificity, false positive rate (FPR), and false negative rate (FNR) for each class. Measurements marked with 1 correspond to the baseline model using only max-pooling, while those marked with 2 refer to our proposed multi-pooling approach incorporating four pooling methods. Figure 7 shows the confusion matrix for our model.

Table 5. Recall, Specificity, FPR, and FNR across each class

	Class 0 bkl	Class 1 nv	Class 2 df	Class 3 mel	Class 4 vasc	Class 5 bcc	Class 6 akiec
Recall_1	50.61%	94.52%	5.88%	**27.71%**	71.43%	50%	**16.67%**
Recall_2	39.63%	91.93%	17.65%	**37.35%**	61.90%	30.26%	**41.67%**
specificity_1	95.35%	59.55%	1	96.69%	99.80%	97.25%	99.38%
specificity_2	95.65%	60.16%	99.66%	94.96%	99.93%	98.31%	97.51%
FPR_1	4.65%	40.45%	0	3.31%	0.20%	2.75%	0.62%
FPR_2	4.35%	39.84%	0.34%	5.04%	0.07%	1.69%	2.49%
FNR_1	49.39%	5.49%	94.12%	72.29%	28.57%	50%	83.33%
FNR_2	60.37%	8.07%	82.35%	62.65%	38.10%	69.74%	58.33%

Melanoma is a very severe disease, making early treatment crucial. Compared with traditional CNNs, the recall was only 27.71%, missing over 70% of cases. Our proposed method improved recall to 37.35%, reducing missed diagnoses. For akiec, recall increased from 16.67% to 41.67%, showing a significant improvement in recall too. Although the recall for bcc decreased from 50% to 30.26%, the model maintained high specificity, meaning it remained effective at correctly identifying non-bcc cases and avoiding false positives. For mild cases like nv and bkl, our model maintains high recall (nv: 91.93%) with minimal compromise in specificity. Although bkl recall dropped slightly, its FNR remains acceptable considering its lower risk.

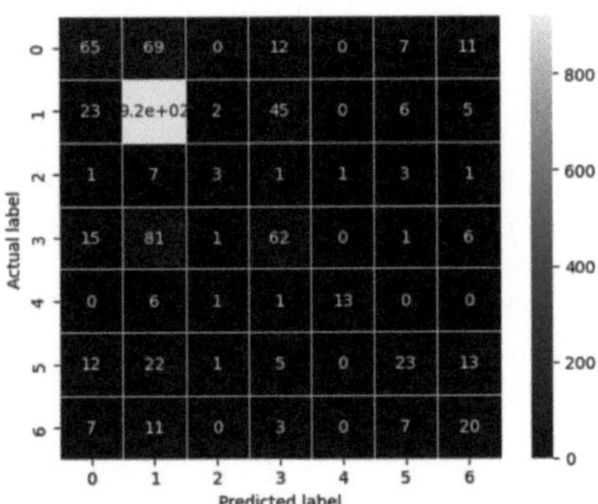

Fig. 7. Confusion Matrix

In addition, the model maintains or improves specificity for most classes. For example, the specificity of vasc improved from 99.80% to 99.93%, which means fewer false alarms. Therefore, our solution achieved better results with clinical priorities, reducing false negatives in high-risk cancers while preserving specificity in mild cases. However, the model shows lower recall for categories like melanoma due to class imbalance in the HAM10000 dataset, resulting in limited feature learning and more false negatives.

5 Conclusion

In this paper, we proposed a novel CNN-based method for skin lesion classification, inspired by the biological concepts of neuron convergence and divergence. Observing the lack of divergence in pooling layers, we introduced an adaptive multi-pooling strategy that showed slight performance gains. More importantly, this work serves as an example of how interdisciplinary insights can inspire AI model design, emphasising the value of aligning neural network architecture with biological principles. Beyond performance, our approach highlights the educational value for young learners. In the future, our work will focus on addressing data imbalance and enhancing feature representation. We also suggest exploring biological structures and systems biology for further model design insights.

References

1. Rashi, A.B., Kausik, M.A.K.: AI revolutionizing industries worldwide: a comprehensive overview of its diverse applications. Hybrid Adv. **7**, Art. no. 100277 (2024). https://doi.org/10.1016/j.hybadv.2024.100277
2. Wang, S., Wang, F., Zhu, Z., Wang, J., Tran, T., Du, Z.: Artificial intelligence in education: a systematic literature review. Expert Syst. Appl. **252**, Part A, Art. no. 124167 (2024). https://doi.org/10.1016/j.eswa.2024.124167
3. LeCun, Y., Bengio, Y., Hinton, G.: Deep learning. Nature **521**(7553), 436–444 (2015)
4. Krizhevsky, A., Sutskever, I., Hinton, G.E.: ImageNet classification with deep convolutional neural networks. Adv. Neural. Inf. Process. Syst. **25**, 1097–1105 (2012)
5. Wang, D., Yan, W., Oates, T.: Time series classification from scratch with deep neural networks: a strong baseline. In: Proceedings of the International Joint Conference on Neural Networks (IJCNN), pp. 1578–1585 (2017). https://doi.org/10.1109/IJCNN.2017.7966039
6. Chen, H., Qi, X., Yu, L., Heng, P.A.: DCAN: deep contour-aware networks for accurate gland segmentation. In: Proceedings of the IEEE Conference on Computer Vision and Pattern Recognition (CVPR), pp. 2487–2496 (2016)
7. Man, K., Kaplan, J., Damasio, H., Damasio, A.: Neural convergence and divergence in the mammalian cerebral cortex: from experimental neuroanatomy to functional neuroimaging. J. Comparative Neurol. **521**(18), 4097–4111 (2013). https://doi.org/10.1002/cne.23408
8. Tschandl, P., Rosendahl, C., Kittler, H.: The HAM10000 dataset, a large collection of multi-source dermatoscopic images of common pigmented skin lesions. Sci. Data **5**, Art. no. 180161 (2018). https://doi.org/10.1038/sdata.2018.161

9. Urban, K., Mehrmal, S., Uppal, P., Giesey, R.L., Delost, G.R.: The global burden of skin cancer: a longitudinal analysis from the global burden of disease study, 1990–2017. JAAD Int. **2**, 98–108 (2021). https://doi.org/10.1016/j.jdin.2020.10.013
10. World Cancer Research Fund, Skin cancer statistics (2025). https://www.wcrf.org/preventing-cancer/cancer-statistics/skin-cancer-statistics/. Accessed 15 Apr 2025
11. Dildar, M., et al.: Skin cancer detection: a review using deep learning techniques. Int. J. Environ. Res. Public Health **18**(10), Art. no. 5479 (2021)
12. Simonyan, K., Zisserman, A.: Very deep convolutional networks for large-scale image recognition, arXiv preprint arXiv:1409.1556 (2014)
13. Szegedy, C., et al.: Going deeper with convolutions. In: Proceedings of the IEEE Conference on Computer Vision and Pattern Recognition (CVPR), pp. 1–9 (2015)
14. He, K., Zhang, X., Ren, S., Sun, J.: Deep residual learning for image recognition. In: Proceedings of the IEEE Conference on Computer Vision and Pattern Recognition (CVPR), pp. 770–778 (2016)
15. Huang, G., Liu, Z., Van Der Maaten, L., Weinberger, K.Q.: Densely connected convolutional networks. In: Proceedings of the IEEE Conference on Computer Vision and Pattern Recognition (CVPR), pp. 4700–4708 (2017)
16. Zhang, J., Xie, Y., Xia, Y., Shen, C.: Attention residual learning for skin lesion classification. IEEE Trans. Med. Imaging **38**(9), 2092–2103 (2019)
17. Shen, D., Wu, G., Suk, H.: Deep learning in medical image analysis. Annu. Rev. Biomed. Eng. **19**, 221–248 (2017). https://doi.org/10.1146/annurev-bioeng-071516-044442
18. Chen, C., Isa, N.A.M., Liu, X.: A review of convolutional neural network based methods for medical image classification. Comput. Biol. Med. **185**, 109507 (2025). https://doi.org/10.1016/j.compbiomed.2024.109507
19. Stein, M.B., Stanford, T.R.: Multisensory integration: current issues from the perspective of the single neuron. Nat. Rev. Neurosci. **9**(4), 255–266 (2008). https://doi.org/10.1038/nrn2331
20. Man, K., Kaplan, J., Damasio, H., Damasio, A.: Neural convergence and divergence in the mammalian cerebral cortex: from experimental neuroanatomy to functional neuroimaging. J. Comp. Neurol. **521**(18), 4097–4111 (2013). https://doi.org/10.1002/cne.23408
21. Stein, B.E., Stanford, T.R.: Multisensory integration: current issues from the perspective of the single neuron. Nat. Rev. Neurosci. **9**(4), 255–266 (2008). https://doi.org/10.1038/nrn2331
22. Sterratt, D., Lyngholm, D., Willshaw, D., Thompson, I.D.: Principles of retinal projection patterns: implications for development, plasticity and repair. Prog. Retin. Eye Res. **32**, 88–121 (2013). https://doi.org/10.1016/j.preteyeres.2012.08.002
23. Haenssle, H.A., Fink, C., Schneiderbauer, R., Toberer, F., Buhl, T., Blum, A., et al.: Man against machine: diagnostic performance of a deep learning convolutional neural network for dermoscopic melanoma recognition in comparison to 58 dermatologists. Ann. Oncol. **29**(8), 1836–1842 (2018). https://doi.org/10.1093/annonc/mdy166
24. Classification: Accuracy, recall, precision, and related metrics, Google for Developers. https://developers.google.com/machine-learning/crash-course/classification/accuracy-precision-recall. Accessed 15 Apr 2025

An Image Defogging Algorithm Based on Enhanced Convolutional Block Attention Module and Dual-Scale Gated Feedforward Network

Zonghao Wang[1,2], Juntao Li[1,2(✉)], Ruiping Yuan[1,2], and Duojing Yun[1,2]

[1] Beijing Key Laboratory of Intelligent Logistics System, Beijing Wuzi University, Fuhe St. 321, Tongzhou 101149, Beijing, China
[2] School of Information, Beijing Wuzi University, Fuhe St. 321, Tongzhou 101149, Beijing, China
ljtletter@126.com

Abstract. In autonomous driving systems and related fields, foggy weather has a significant impact on the imaging capabilities of cameras and other sensors, often manifesting itself as obstruction in the near field and blurring in the far field. This paper proposes an image defogging algorithm (DCGNet) based on an enhanced convolutional block attention module and a dual-scale gated feedforward network. The algorithm adopts the U-Net architecture as the backbone network, combining the enhanced convolutional block attention module (Enhanced CBAM), the dual scale gated feedforward network (DGFF) and residual connections to effectively enhance the image defogging performance. The enhanced CBAM module integrates three channels' attention mechanisms, maximum pooling, average pooling, and variance pooling, and employs multiscale spatial attention to capture feature information at different scales. The DGFF achieves multiscale feature extraction through parallel deep convolution and dilated convolution, and adaptively fuses features of different scales via a gating mechanism. Experimental results demonstrate that the proposed DCGNet achieves outstanding defogging performance on the Haze4K dataset, outperforming existing mainstream defogging algorithms in evaluation metrics such as PSNR and SSIM.

Keywords: Image defogging · CBAM · Dual-scale feature extraction

1 Introduction

Image defogging is an important research topic in the field of computer vision, with the aim of restoring clear, fog-free images from foggy images. Haze and foggy weather severely impair image visibility and quality, posing significant challenges for applications such as autonomous driving, video surveillance, and remote sensing image analysis. Traditional image defogging methods rely primarily on atmospheric scattering models, such as the Dark Channel Prior (DCP) [1]

P. Umapada et al. (Eds.): ICCPR 2025, CCIS 2811, pp. 452–463, 2026.
https://doi.org/10.1007/978-981-95-8315-7_36

and Color Attenuation Prior (CAP) [2]. However, these methods often perform poorly when dealing with heavily foggy or hazy images.

In recent years, deep learning technology has made significant progress in the field of image defogging. Defogging algorithms based on convolutional neural networks (CNNs) can automatically learn the mapping relationship between hazy images and clear images through end-to-end learning. DehazeNet [3] was the first to introduce deep learning into image defogging, using a multilayer CNN structure to estimate transmittance. However, its network structure is relatively simple, limiting its performance in complex scenes. AODNet [4] proposed a lightweight dehazing network that directly learns the mapping relationship between hazy images and clear images using a reformulated atmospheric scattering model. Although it has high computational efficiency, its feature extraction capabilities are insufficient. FFANet [5] introduced a feature fusion attention mechanism, combining channel attention and pixel attention to achieve multi-scale feature fusion, significantly improving dehazing performance. However, its complex attention structure results in high computational overhead. GCANet [6] employs a global context aggregation module to enhance feature extraction capabilities through dense connections and non-local attention, but it tends to over-smooth details during processing. GridDehazeNet [7] proposes a multi-scale network based on attention, combining pre-processing and post-processing modules to improve defogging quality, but its complex architecture poses significant training challenges. MSBDN [8] employs a multi-scale dense connection network, effectively improving feature propagation efficiency through dense feature fusion, but the network has a large number of parameters, resulting in high deployment costs.

In the design of feedforward networks, traditional methods typically employ single-scale convolution operations, which struggle to effectively capture multi-scale feature information. Recently, Sun et al. [9] proposed the Dual-Scale Gated Feedforward Network (DGFF) in the Histogram Transformer, which achieves adaptive feature fusion through parallel convolution operations at different scales and gating mechanisms, demonstrating excellent performance in adverse weather image restoration tasks. However, existing attention mechanisms often focus solely on local features, failing to effectively utilize global contextual information, and struggle to strike a good balance between multi-scale feature extraction and computational efficiency.

While Transformer-based methods excel at modeling global dependencies, they often incur substantial computational overhead and require large-scale datasets to train effectively. In contrast, convolutional neural networks (CNNs) offer a more efficient inductive bias for processing local image features like textures and edges, which are crucial for dehazing. To harness the strengths of both architectures, our DCGNet adopts a hybrid design. It retains the U-Net backbone, which is efficient and effective for dense prediction tasks, and enhances it with an advanced attention mechanism and a gated multi-scale module. This approach allows us to achieve global feature awareness and multi-scale fusion without the excessive computational burden of standard Transformers, making it more suitable for practical applications where efficiency is paramount.

This paper proposes an image defogging algorithm (DCGNet) based on an enhanced attention mechanism and a dual-scale gated feedforward network. The main contributions of this algorithm include:

1. Enhanced attention mechanism with multi-pooling fusion. Innovatively introduces variance pooling into the CBAM [10] framework, forming a triple pooling fusion mechanism with max pooling and average pooling to more comprehensively capture channel feature statistical information. Additionally, multi-scale spatial attention (3×3, 5×5, and 7×7 convolution kernels) replaces the traditional single-scale design, significantly enhancing the ability to perceive fog features at different scales.
2. Improved Dual-Scale Gated Feedforward Network. Based on the DGFF concept from the Histogram Transformer, an improved version tailored for image defogging tasks is designed. By constructing dual-path feature extraction through parallel 5×5 depth convolutions and 3×3 dilation convolutions, and introducing a gated fusion mechanism using the Mish activation function [11], we achieve an optimal balance between computational efficiency and feature expression capability. Compared to the original DGFF, our improved version is optimized for the texture features of haze images.
3. Residual-enhanced U-Net architecture. A DCGAB module with dual residual connections is designed, with each module containing two residual connections (after ECBAM and DGFF, respectively), effectively alleviating the gradient vanishing problem in deep networks. At the same time, a jump connection design with instance normalization is adopted to enhance the feature propagation efficiency between the encoder and decoder.

2 Related Work

2.1 Traditional Image Defogging Methods

Traditional image defogging methods are primarily based on atmospheric scattering models:

$$I(x) = J(x)t(x) + A(1 - t(x)) \tag{1}$$

where I(x) is the foggy image, J(x) is the clear image, t(x) is the transmittance, and A is the atmospheric light value. Based on this model, researchers have proposed various prior knowledge, such as dark channel prior (DCP) and color attenuation prior (CAP). These methods perform well in handling light haze, but often fail to achieve satisfactory results in handling heavy haze or complex scenes.

2.2 Deep Learning-Based Defogging Methods

Deep learning-based defogging methods can be categorized into various types based on network architecture and training strategies. Among convolutional neural network (CNN) methods, DuRN [12]proposed a dual residual network, leveraging residual learning and dense connections to enhance feature propagation. Generative adversarial network (GAN) methods such as EPDN [13] adopt an enhanced-purified dual network architecture, improving defogging quality through adversarial training; FD-GAN [14] guides the generator to learn the true defogging mapping relationship through the discriminator. Methods based on Transformers, such as DehazeFormer [15], introduce self-attention mechanisms into the defogging task to model long-range dependencies. Multi-scale network methods, such as PMNet [16], propose a progressive multi-scale strategy that gradually restores image quality from coarse to fine.

2.3 Application of Attention Mechanisms in Image Restoration

Attention mechanisms play a crucial role in image restoration tasks, enabling the network to focus on key regions and features. In the field of image defogging, FFANet was the first to introduce attention mechanisms into defogging networks, combining feature fusion attention (FFA) modules with channel attention and pixel attention to effectively enhance multi-scale feature fusion capabilities. GCANet employs a gated context aggregation module to enhance the extraction of global contextual information using the attention mechanism. GridDehazeNet proposes an attention-based multi-scale network architecture, improving defogging quality through attention-guided feature selection.

In the field of image denoising, RIDNet [17] combines residual-dense blocks with attention mechanisms, re-calibrating feature responses through channel attention to significantly improve denoising performance. MPRNet [18] proposes a multi-stage progressive image restoration network, integrating a supervised attention module (SAM) at each stage to adaptively select useful features. Restormer [19] introduces the Transformer architecture into image restoration, modeling long-range dependencies through self-attention mechanisms, achieving excellent results in multiple restoration tasks such as dehazing and denoising.

In the field of adverse weather image restoration, TransWeather [20] adopts a Transformer encoder-decoder architecture, utilizing self-attention mechanisms to handle various weather degradations. Chen et al. [21] proposed a two-stage knowledge learning mechanism based on a multi-teacher-student architecture, capable of simultaneously handling multiple weather conditions such as rain, snow, and fog. Histogram Transformer [9] introduced a dual-scale gated feedforward network in adverse weather image restoration, achieving adaptive feature fusion through gated attention mechanisms.

In terms of hybrid attention mechanisms, CBAM combines channel and spatial attention and is widely used in image restoration tasks. Many defogging networks adopt CBAM or its variants to enhance feature representation capabilities. However, traditional CBAM only uses max pooling and average pooling, which are insufficient for extracting feature statistical information in complex haze scenes, providing room for improvement for the enhanced CBAM proposed in this paper.

3 Method

3.1 Network Architecture

The proposed DCGNet adopts an improved U-Net [22] architecture, as shown in Fig. 1(a). The network consists of two parts: an encoder and a decoder, which achieve multi-scale feature fusion through skip connections. The encoder contains four downsampling stages, each of which includes a DCGAB module and a downsampling operation, with the feature map size halved and the number of channels doubled in each stage. The decoder adopts a symmetric upsampling structure, restoring the feature map resolution through transposed convolution and fusing feature information from the corresponding encoder layers via skip connections.

The entire network comprises eight DCGAB modules, as shown in Fig. 1(b). Each DCGAB module consists of batch normalization (BN), ECBAM, DGFF, and residual connections. The input features first undergo batch normalization, then attention enhancement via ECBAM, followed by multi-scale feature extraction via DGFF, and finally residual connections to maintain gradient flow. This design ensures effective feature propagation while enhancing the network's ability to perceive haze features.

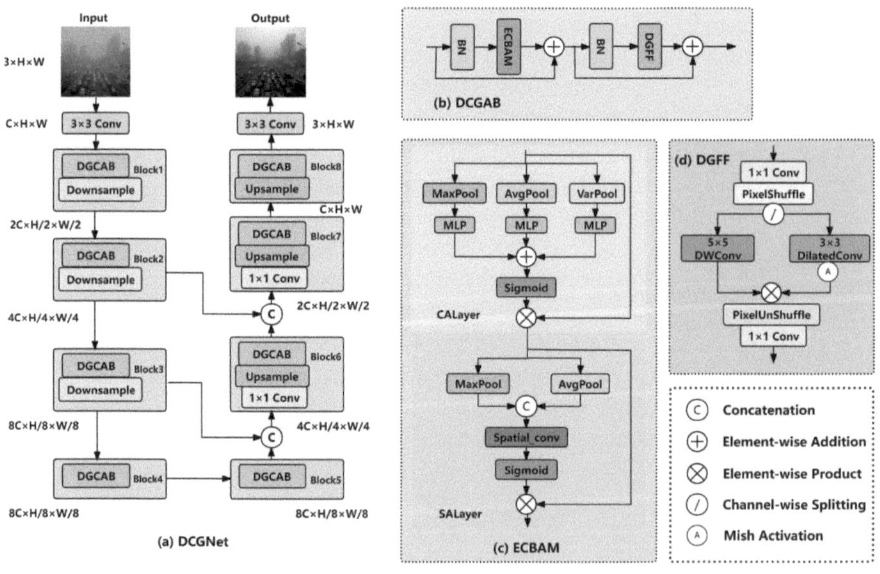

Fig. 1. DCGNet overall architecture diagram

3.2 Dual-Scale Gated Convolutional Attention Block

DCGAB is the core building block of DCGNet, as shown in Fig. 1(b). Each DCGAB module consists of batch normalization (BN), enhanced CBAM

(ECBAM), dual-scale gated feedforward network (DGFF), and residual connections. The mathematical expression of the module is:

$$F_{out1} = F_{ecbam}(F_{bn}(F_{input})) \oplus F_{input} \tag{2}$$

$$F_{out} = F_{dgff}(F_{bn}(F_{out1})) \oplus F_{out1} \tag{3}$$

where F_{input} is the input feature, $\oplus$ denotes element-wise addition, F_{bn} is the batch-normalized feature, F_{ecbam} is the feature after passing through the ECBAM module, F_{out1} is the feature after the first residual connection, F_{dgff} is the feature after passing through the DGFF module, and F_{out} is the final feature after the second residual connection. The DCGAB module employs a dual residual connection design, adding residual connections after both the ECBAM and DGFF modules. This effectively mitigates the vanishing gradient problem in deep networks while enhancing the expressive power of the features.

3.3 Enhanced Convolutional Block Attention Module

Traditional CBAM modules only use max pooling and average pooling for channel attention calculation, which is insufficient for extracting feature statistical information in complex haze scenes. The enhanced CBAM (ECBAM) module proposed in this paper is shown in Fig. 1(c), which integrates three pooling methods to capture channel feature statistical information more comprehensively.

Channel Attention Layer (CALayer): Combines max pooling, average pooling, and variance pooling, with the calculation formula as follows:

$$\begin{aligned} F_c = \sigma\big(&MLP(\text{MaxPool}(F)) \oplus MLP(\text{AvgPool}(F)) \\ &\oplus MLP(\text{VarPool}(F))\big) \odot F \end{aligned} \tag{4}$$

where F is the input feature, σ is the sigmoid activation function, $\odot$ denotes element-wise multiplication, and MLP is a multi-layer perceptron.

Spatial Attention Layer (SALayer): Its main operations are as follows:

$$F_{s_1} = Concat(MaxPool(F_c), AvgPool(F_c)) \tag{5}$$

$$Spatial\_conv = \sigma(Conv_{3\times3}(F_{s_1}) \oplus Conv_{5\times5}(F_{s_1}) \oplus Conv_{7\times7}(F_{s_1})) \tag{6}$$

$$F_s = \sigma(Spatial\_conv) \oplus F_c \tag{7}$$

where $Concat$ denotes channel concatenation, and $Spatial\_conv$ denotes multi-scale spatial convolution operations. By expanding the number of channels in F_{s_1} to three times its original size, multi-core multi-scale feature extraction is performed, thereby adaptively focusing on haze regions at different scales.

3.4 Dual-Scale Gated Forward Network

Traditional feedforward networks typically employ single-scale convolution operations, which struggle to effectively capture multi-scale feature information. The DGFF module designed in this paper, as shown in Fig. 1(d), achieves efficient multi-scale feature processing through parallel dual-scale feature extraction and a gated fusion mechanism.

Based on the DGFF structure shown in Fig. 1(d), this module uses PixelShuffle to separate channels, employs 5 × 5 depth convolutions and 3 × 3 dilated convolutions to process dual-scale features in parallel, and finally fuses them through the Mish gating mechanism:

$$F_{dgff} = Conv_{1\times1}\big(Pixel\_Unshuffle\big(Mish(DilatedConv(X))\big) \odot DWConv(X)\big) \tag{8}$$

$$X = Pixel\_Shuffle(Conv_1(F_{input})) \tag{9}$$

4 Experiment

4.1 Experimental Preparation

Dataset and Experimental Environment. The experiment used the Haze4K dataset for training and testing. This dataset contains 4,000 high-quality image pairs covering different scenes and different levels of haze concentration. The dataset is divided in a 3:1 ratio, with 3,000 image pairs used for training and 1,000 image pairs used for testing. Each image has a corresponding hazy version and a clear, haze-free version, providing reliable ground truth labels for supervised learning. All experiments were conducted on an NVIDIA RTX 4060 GPU using the PyTorch 1.12.0 deep learning framework. During training, a single GPU configuration was used with a batch size of 4, and image input dimensions were uniformly resized to 256 × 256 pixels.

Optimization Strategy and Loss Function. The network was trained using the AdamW optimizer, with an initial learning rate of 10^{-4} and a weight decay coefficient of 10^{-4}. The ReduceLROnPlateau learning rate scheduling strategy was adopted, where the learning rate automatically halves (factor 0.5) when the training loss stops decreasing, with a patience value set to 5 epochs. The total number of training epochs is set to 150, with a batch size of 4. Gradient clipping (maximum norm 1.0) is used during training to prevent gradient explosion. This paper proposes an adaptive hybrid loss function that dynamically combines L1 loss and SSIM loss:

$$L = \alpha \times L_{L_1} + (1-\alpha) \times L_{ssim} \tag{10}$$

where the weight α is adaptively adjusted based on the current SSIM value, ranging from [0.3, 0.7]. During the early stages of training when SSIM values are low, the focus is on structural similarity; during the later stages when SSIM values are high, the emphasis shifts to pixel accuracy. SSIM loss is computed using a Gaussian kernel with a window size of 11, with constants $C1 = 0.01^2$ and $C2 = 0.03^2$.

4.2 Experimental Results and Analysis

Table 1 and Fig. 2 present the quantitative and qualitative comparison results between DCGNet and mainstream defogging algorithms(FFANet, GCANet, DCP, AodNet). In terms of quantitative evaluation, DCGNet achieves the best performance in both PSNR (25.6576 dB) and SSIM (0.9581) metrics, outperforming the second-best FFANet by 0.0291 dB and 0.0065, respectively. It also demonstrates a significant advantage in computational efficiency, with FLOPs of 26.43 G, reducing FLOPs by approximately 58% compared to FFANet, with an inference speed of 31.38 FPS, capable of meeting real-time application requirements. In terms of qualitative analysis, visual comparisons reveal that DCGNet effectively removes haze while preserving image details and color accuracy. It demonstrates enhanced robustness in handling severe haze scenarios, avoiding artifacts caused by over-enhancement, and exhibits consistent de-hazing performance across diverse environments.

Fig. 2. Visual comparison of different defogging methods on the Haze4K dataset. The green box indicates the detail image to be enlarged, and the red box indicates the enlarged detail image.

Table 1. The SSIM, PSNR, computational efficiency, and model complexity statistics for various dehazing methods on the Haze4K dataset. Best results are highlighted in bold.

Method	PSNR↑	SSIM↑	Time (ms)↓	FPS↑	Params (M)↓	FLOPs (G)↓
FFANet	25.6285	0.9516	36.21	27.62	1.28	63.53
GCANet	25.3513	0.9472	14.64	68.28	0.71	14.18
ADONet	19.1336	0.8799	**3.86**	**259.30**	**0.002**	**0.08**
DCP	18.8132	0.8532	64.21	–	–	–
DCGNet (ours)	**25.6576**	**0.9581**	31.86	31.38	8.15	26.43

4.3 Ablation Study

To validate the effectiveness of each core component of DCGNet, a systematic ablation experiment was designed, with the results shown in Table II. Three comparison versions were set up: replacing the DGFF module with a 3×3 convolution (DGFF→3×3 conv), removing the residual connection in the DCGAB module (no residual), and using traditional CBAM to replace enhanced CBAM (ECBAM→CBAM). The experimental results show that the DGFF module achieves a 0.384 dB improvement in PSNR (25.6576 vs. 25.2735) compared to standard convolutions and reduces the number of parameters by approximately 11%, validating the effectiveness of the dual-scale gated design; Removing the residual connections resulted in a significant performance drop of 2.97 dB (25.6576 vs. 22.6871), clearly demonstrating the critical role of residual connections in training deep networks; Enhanced CBAM outperformed traditional CBAM by 0.65 dB in PSNR (25.6576 vs. 25.0054), validating the advantages of variance pooling and multi-scale spatial attention design (Table 2).

Table 2. Result of ablation study on Haze4K dataset. Best results are highlighted in bold.

Replace module	PSNR↑	SSIM↑	Params (M)↓	FLOPs (G)↓
DGFF→3×3Conv	25.2735	0.9578	9.16	27.92
no residual	22.6871	0.9183	**7.17**	**25.19**
ECBAM→CBAM	25.0054	0.9468	8.06	26.41
ours	**25.6576**	**0.9581**	8.15	26.43

5 Conclusion

This paper proposes an image defogging algorithm called DCGNet, which combines an enhanced convolutional block attention module with a dual-scale gated feedforward network. By utilizing an enhanced CBAM module, a dual-scale gated feedforward network, and a residual-enhanced U-Net architecture, DCGNet effectively addresses the issues of insufficient feature extraction and low computational efficiency in existing defogging algorithms when dealing with complex haze scenarios. Experimental results on the Haze4K dataset demonstrate that DCGNet achieves optimal performance in terms of PSNR (25.66 dB) and SSIM (0.9581), while also exhibiting significant computational efficiency advantages. The model has only 8.15 million parameters and achieves an inference speed of 31.38 FPS. Compared to existing mainstream algorithms, it significantly reduces computational complexity while maintaining high-quality fog removal effects, making it more suitable for practical application deployment.

6 Future Work

The core innovations of DCGNet—the Enhanced CBAM for multi-pooling feature refinement and the DGFF for adaptive multi-scale fusion—are not limited to image dehazing. The architecture holds significant potential for generalization to other low-level vision tasks. For instance, in image denoising, the variance pooling in ECBAM could be particularly effective in distinguishing between noise patterns and underlying image textures, while the gating mechanism in DGFF could adaptively blend features from different scales to recover clean details. Similarly, for image deraining, the multi-scale spatial attention is inherently suited to detecting and removing rain streaks of varying sizes and densities. The model's ability to capture both global contextual information (via enhanced attention) and local details (via the convolutional backbone) provides a robust foundation for a wide array of image restoration challenges. Future work will explicitly explore and validate the efficacy of this architecture across these related domains.

Acknowledgements. This work was supported by National Natural Science Foundation Project (72101033), Excellent Science and Technology Innovation Team Project in Tongzhou District (CXTD2023010), Key Project of Science and Technology Plan of Beijing Municipal Education Commission (KZ202210037046).

References

1. He, K., Sun, J., Tang, X.: Single image haze removal using dark channel prior. IEEE Trans. Pattern Anal. Mach. Intell. **33**(12), 2341–2353 (2011)
2. Zhu, Q., Mai, J., Shao, L.: A fast single image haze removal algorithm using color attenuation prior. IEEE Trans. Image Process. **24**(11), 3522–3533 (2015)
3. Cai, B., Xu, X., Jia, K., Qing, C., Tao, D.: Dehazenet: an end-to-end system for single image haze removal. IEEE Trans. Image Process. **25**(11), 5187–5198 (2016)
4. Li, B., Peng, X., Wang, Z., Xu, J., Feng, D.: Aod-net: All-in-one dehazing network. In: Proceedings of the IEEE international conference on computer vision, pp. 4770–4778 (2017)

5. Qin, X., Wang, Z., Bai, Y., Xie, X., Jia, H.: FFA-Net: Feature fusion attention network for single image dehazing. In: Proceedings of the AAAI Conference on Artificial Intelligence, Vol. 34, No. 07, pp. 11908–11915 (2020). https://doi.org/10.1609/aaai.v34i07.6865
6. Chen, D., et al.: Gated context aggregation network for image dehazing and deraining. In: 2019 IEEE Winter Conference on Applications of Computer Vision (WACV), pp. 1375–1383 (2019). https://doi.org/10.1109/WACV.2019.00151
7. Liu, X., Ma, Y., Shi, Z., Chen, J.: Griddehazenet: attention-based multi-scale network for image dehazing. In: Proceedings of the IEEE/CVF International Conference on Computer Vision, pp. 7314–7323 (2019)
8. Dong, H., et al.: Multi-scale boosted dehazing network with dense feature fusion. In: Proceedings of the IEEE/CVF Conference on Computer Vision and Pattern Recognition, pp. 2157–2167 (2020)
9. Sun, S., Ren, W., Gao, M., Wang, R., Cao, X.: Restoring images in adverse weather conditions via histogram transformer. In: European Conference on Computer Vision, pp. 111–129. Springer, Cham (2024). https://doi.org/10.1007/978-3-031-72670-5_7
10. Woo, S., Park, J., Lee, J.Y., Kweon, I.S.: Cbam: Convolutional block attention module. In: Proceedings of the European conference on Computer Vision (ECCV), pp. 3–19 (2018). https://doi.org/10.48550/arXiv.1807.06521
11. Misra, D.: Mish: A self regularized non-monotonic activation function. arxiv preprint arxiv:1908.08681 (2019)
12. Qu, Y., Chen, Y., Huang, J., Xie, Y.: Enhanced pix2pix dehazing network. In: Proceedings of the IEEE/CVF Conference on Computer Vision and Pattern Recognition, pp. 8160–8168 (2019)
13. Liu, X., Suganuma, M., Sun, Z., Okatani, T.: Dual residual networks leveraging the potential of paired operations for image restoration. In: Proceedings of the IEEE/CVF Conference on Computer Vision and Pattern Recognition, pp. 7007–7016 (2019)
14. Dong, Y., Liu, Y., Zhang, H., Chen, S., Qiao, Y.: FD-GAN: Generative adversarial networks with fusion-discriminator for single image dehazing. In: Proceedings of the AAAI Conference on Artificial Intelligence, Vol. 34, No. 07, pp. 10729–10736 (2020). https://doi.org/10.1609/aaai.v34i07.6701
15. Song, Y., He, Z., Qian, H., Du, X.: Vision transformers for single image dehazing. IEEE Trans. Image Process. **32**, 1927–1941 (2023)
16. Ren, W., et al.: Gated fusion network for single image dehazing. In: Proceedings of the IEEE Conference on Computer Vision and Pattern Recognition, pp. 3253–3261 (2018). https://doi.org/10.48550/arXiv.1804.00213
17. Anwar, S., Barnes, N.: Real image denoising with feature attention. In: Proceedings of the IEEE/CVF International Conference on Computer Vision, pp. 3155–3164 (2019)
18. Zamir, S.W., et al.: Multi-stage progressive image restoration. In: Proceedings of the IEEE/CVF Conference on Computer Vision and Pattern Recognition, pp. 14821–14831 (2021). https://doi.org/10.48550/arXiv.2102.02808
19. Zamir, S.W., Arora, A., Khan, S., Hayat, M., Khan, F.S., Yang, M.H.: Restormer: efficient transformer for high-resolution image restoration. In: Proceedings of the IEEE/CVF Conference on Computer Vision and Pattern Recognition, pp. 5728–5739 (2022). https://doi.org/10.48550/arXiv.2111.09881

20. Valanarasu, J.M.J., Patel, V. M.: TransWeather: transformer-based restoration of images degraded by adverse weather conditions. In Proceedings of the IEEE/CVF Conference on Computer Vision and Pattern Recognition, pp. 2353–2363 (2022). https://doi.org/10.48550/arXiv.2111.14813
21. Chen, W.T., Huang, Z.K., Tsai, C.C., Yang, H.H., Ding, J.J., Kuo, S.Y.: Learning multiple adverse weather removal via two-stage knowledge learning and multi-contrastive regularization. In: Proceedings of the IEEE/CVF Conference on Computer Vision and Pattern Recognition, pp. 17653–17662 (2021)
22. Ronneberger, O., Fischer, P., Brox, T.: U-net: convolutional networks for biomedical image segmentation. In: International Conference on Medical Image Computing and Computer-Assisted Intervention, pp. 234–241 (2015). https://doi.org/10.1007/978-3-319-24574-4_28

Tracking-Based Particle Image Velocimetry for Water Surface Velocity Measurement

Yarong Li[1], Wenbin Zhang[1], Xue Cao[1], Yitong He[1], Siyan Zhu[1], Pengyu Liu[2(✉)], and Jiali Chen[2]

[1] North Automatic Control Technology Institute, Shanxi, China
[2] School of Information Science and Technology, Beijing University of Technology, Beijing, China
liupengyu@bjut.edu.cn

Abstract. Water surface velocity measurement plays a pivotal role in watershed flood prevention and water resource allocation, providing essential data for timely flood warnings and scientific water management. Particle Image Velocimetry (PIV) is a conventional approach to estimate flow velocity by tracking small floating particles on water surfaces. However, this method frequently encounters practical limitations due to challenging environmental conditions, including surface glare and sparse particle distribution, which often lead to tracking failures. To overcome these limitations, we develop an advanced particle tracking technique that incorporates deep learning algorithms. The proposed method combines a YOLOv5s-based particle detection system with a DeepSort tracking framework. This integrated approach enables real-time velocity monitoring while demonstrating remarkable robustness against common environmental disturbances such as variable illumination and low particle concentrations. Field validation experiments achieved a mean velocity error (MVE) of 0.015 m/s under realistic operating conditions.

Keywords: Water surface velocity · YOLOv5 · DeepSort · Particle Image Velocimetry

1 Introduction

Water surface velocity constitutes a critical component in hydrological monitoring, flood prevention, and water resource management, playing a pivotal role in flood early warning, hydraulic engineering operations, and aquatic ecosystem studies. Conventional measurement techniques primarily employ acoustic or radar instrumentation. Ordonez et al. [1] deployed an Acoustic Doppler Current Profiler (ADCP) on autonomous gliders, determining flow velocity by analyzing frequency shifts in reflected acoustic signals while compensating for platform motion through vertical velocity gradient integration and bottom tracking. Dong et al. [2] utilized Continuous-Wave Ultrasonic Doppler (CWUD) technology, emitting ultrasonic waves and analyzing frequency shifts via FFT to establish theoretical models correlating Doppler shifts with mean flow velocity. Their subsequent

P. Umapada et al. (Eds.): ICCPR 2025, CCIS 2811, pp. 464–473, 2026.
https://doi.org/10.1007/978-981-95-8315-7_37

work [3] extended this approach by developing flow-pattern-adaptive velocity profile correction models for diverse regimes, significantly enhancing measurement accuracy. Tamari et al. [4] employed handheld radar devices emitting Ku-band microwaves to detect surface velocities rapidly, though measurements exhibited notable susceptibility to wind-wave interference. While instrumental methods provide direct measurements, they incur substantial costs and demonstrate sensitivity to medium properties and environmental variables.

Alternative approaches leverage theoretical or empirical models. Song et al. [5] derived statistical velocity distributions through potential flow theory, modeling water particle kinematics under second-order Stokes waves characterized by two-dimensional wavenumber spectra. Such models, though theoretically rigorous, incorporate idealized assumptions neglecting turbulence and topographic effects. Chiu et al. [6] established empirical wave height-velocity relationships through laboratory experiments, offering computational simplicity but constrained by specific experimental conditions and extensive calibration requirements.

Particle Image Velocimetry (PIV)-based methods [7–9] present cost-effective alternatives with superior adaptability to diverse flow conditions and better generalization than theoretical models. These techniques quantify surface velocities by tracking inter-frame tracer particle displacements. Jodeau et al. [10] implemented Large-Scale PIV (LS-PIV) in mountain streams during reservoir releases, combining elevated mobile video acquisition with artificial seeding while addressing high-flow challenges through dynamic surface correction and intensity thresholding. Liu et al. [11] applied LS-PIV to dam break experiments, reconstructing breach velocity fields complemented by pressure sensor data. Lin et al. [12] developed a real-time hydrological monitoring system integrating water level recognition with PIV-based surface velocimetry through cross-correlation analysis.

Nevertheless, conventional computer vision approaches suffer from limited environmental robustness and computationally intensive parameter tuning. Deep learning-based solutions mitigate these limitations by automating tracer particle detection and matching. Advanced object detection networks enable precise identification of natural tracers under complex conditions, while tracking algorithms facilitate real-time displacement analysis for velocity estimation. This study develops an automated surface velocity measurement framework combining YOLOv5s for particle detection and DeepSort [13] for object tracking. Key contributions include:

(1) Construction of a dedicated dataset featuring natural floating objects as tracer particles.
(2) Implementation of an integrated detection-tracking model (YOLOv5s-DeepSort) for particle displacement analysis and velocity computation.

2 Methods and Principle

This section introduces the water surface detection methods we used and their principles, including the construction of dataset, the overall system structure, and the model of YOLOv5s and DeepSort.

2.1 Construction of Dataset

We collected images of surface tracer particles through web crawling, public datasets, and surveillance acquisition, selecting natural floating objects such as leaves as tracers. The acquisition period covered both normal flow conditions (0.3–1.2 m/s) and flood seasons (1.5–3.8 m/s) to ensure temporal representativeness and comprehensive velocity coverage. During image acquisition, we specifically considered multi-scale tracer particles and challenging conditions including motion blur and occlusion to enhance model robustness. To improve generalization capability, we performed data augmentation on the original images, including spatial transformations (e.g., flipping and cropping) and illumination adjustments (e.g., low-light and highlight conditions). All images were then meticulously annotated using the Labelme tool following standardized protocols.

The final constructed dataset comprises 2,760 accurately labeled images, which were divided into training, validation, and test sets at a ratio of 7:1:2. Notably, the test set contains completely new data not present in training to ensure objective evaluation. Detailed dataset statistics are presented in Fig. 1.

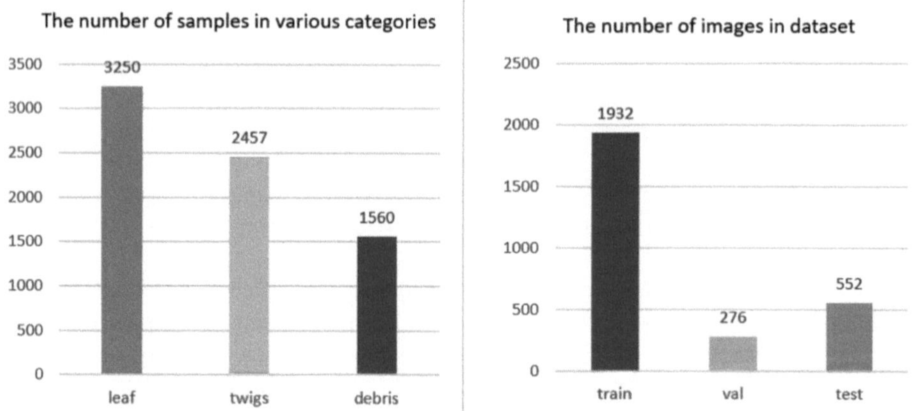

Fig. 1. The Statistics of dataset

2.2 Overall Structure

This study proposes a novel PIV-based surface flow velocity measurement method using object tracking with deep learning. The proposed framework comprises three core components:

(1) An optimized YOLOv5s network for real-time tracer particle detection, which maintains computational efficiency while enhancing multi-scale feature extraction capabilities.
(2) A DeepSort-based tracking module that achieves robust inter-frame association through hybrid motion and appearance feature matching.

By leveraging deep learning advantages in feature representation and target association, the proposed approach overcomes traditional PIV's stringent requirements on particle density and image quality, enabling accurate velocity measurements in complex natural environments.The overall structure of our proposed model is illustrated in Fig. 2.

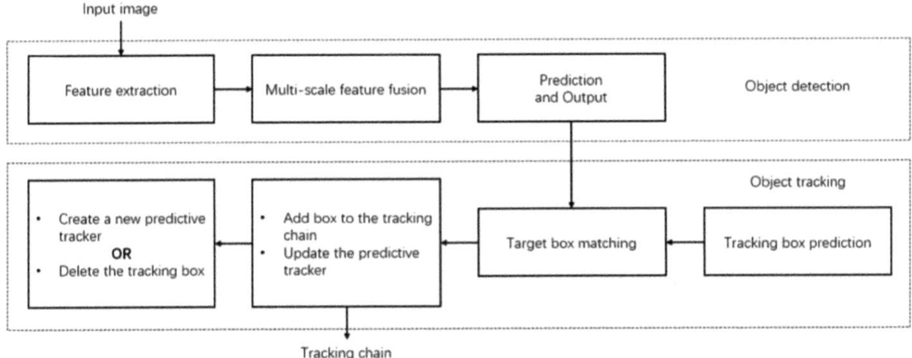

Fig. 2. The Overall Process of Model

2.3 Object Detection Model

We employ the single-stage object detection algorithm YOLOv5s as our base network architecture for constructing the surface tracer particle detection network, achieving real-time high-precision detection. It utilizes a grid-wise prediction approach, where the input image is divided into S × S grids through downsampling, with each grid responsible for both object classification and bounding box regression. Its model architecture consists of four key components: the backbone network, neck for feature fusion, and detection heads. The schematic diagram of this structure is presented in Fig. 3.

Backbone The backbone performs multi-scale feature extraction through stacked convolutions and downsampling. It consists of ConvBNSiLU blocks, C3 modules for efficient gradient propagation, and SPPF modules for expanded receptive fields. Deeper layers capture large objects while shallower layers detect small targets.

Neck The neck combines FPN and PAN structures for multi-scale feature fusion. FPN propagates semantic features top-down, while PAN transmits localization features bottom-up, effectively integrating deep and shallow features for improved detection accuracy.

Head Three detection heads predict at different scales, each outputting bounding boxes, confidence scores, and class probabilities. Scale-specific anchor boxes facilitate regression, with NMS applied during post-processing. CIoU loss optimizes training.

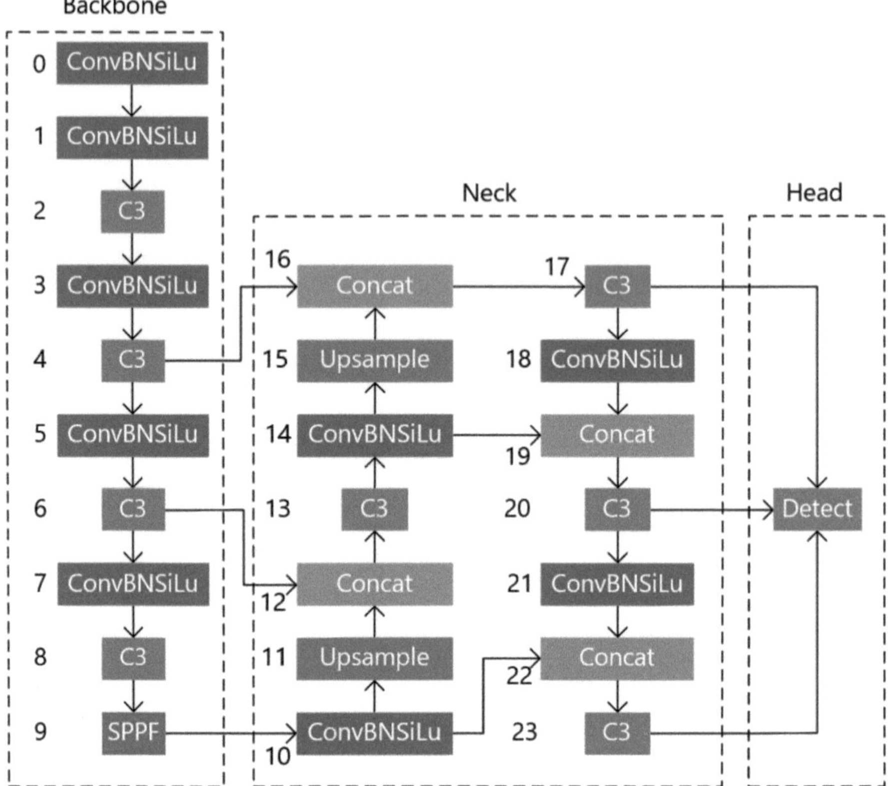

Fig. 3. The Overall Structure of Detection Model

2.4 Object Tracking Model

We employ the DeepSort tracking algorithm to automatically monitor surface tracer particles and estimate flow velocity based on inter-frame displacement of tracked bounding boxes. The algorithm classifies track states into three distinct categories: confirmed, unconfirmed, and deleted. Transition to confirmed state requires consecutive successful matches between unconfirmed tracks and detection boxes, while persistent matching failures trigger state transition to deleted. The complete workflow is illustrated in Fig. 4.

The framework implements a hierarchical matching process: (1) Cascade matching prioritizes association between detector outputs and Kalman filter-predicted confirmed tracks, with successful matches triggering Kalman filter updates and track sequence incorporation. (2) Failed candidates undergo secondary IoU matching, where valid associations similarly update the tracking sequence. (3) Newly detected targets initialize corresponding Kalman filter trackers. The algorithm enforces track termination criteria based on maximum allowable lost frames - unconfirmed tracks are immediately purged upon matching failure, while confirmed tracks persist until exceeding the frame loss threshold. These mechanisms collectively embody DeepSort's core functionalities: state estimation, track handling, and cascade matching.

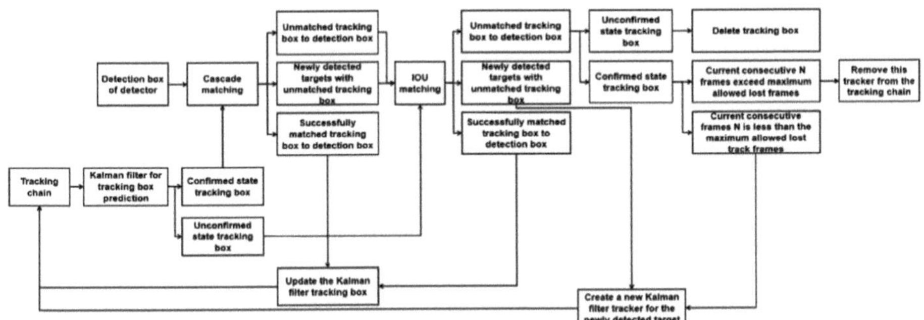

Fig. 4. The Overall Structure of Tracking Model

State Estimation The tracking algorithm employs an 8-dimensional state vector for its estimation model, comprising the bounding box's center coordinates (u, v), aspect ratio γ, height h, and their respective velocities in the image plane ($\dot{u}$, $\dot{v}$, $\dot{\gamma}$, $\dot{h}$). This representation enables trajectory prediction through a constant velocity Kalman filter framework. Each tracked object's trajectory is characterized by three distinct state attributes: confirmed, unconfirmed, and deleted. State transitions between these attributes occur dynamically during the detection-to-track association process.

Track Handling For each tracked trajectory, a frame counter is maintained to record the number of consecutive frames since the last successful measurement association. This counter increments during Kalman filter prediction steps, resets to zero upon successful measurement association, and increments by one for each failed association attempt. When the counter exceeds a predefined threshold, the trajectory is considered to have exited the scene and is consequently removed from the tracking sequence.

Regarding new target initialization: when a detection cannot be associated with any existing trajectories, it is tentatively identified as a potential new target. A provisional trajectory is established only if the target maintains successful associations for a specified number of consecutive frames (typically N = 3 in standard implementations). Failure to maintain consistent associations during this probationary period results in trajectory deletion.

Cascade Matching The cascade matching strategy is employed to address the detection-to-track association problem. When an object experiences prolonged occlusion, the uncertainty in the Kalman filter's predicted state increases significantly, causing the probability mass to disperse in the state space and consequently reducing the observation likelihood. To mitigate this issue, DeepSort implements a hierarchical cascade matching approach that prioritizes tracks based on occlusion duration. Specifically, tracks are stratified into different matching levels according to their occlusion time, with shorter-occluded tracks receiving higher matching priority. This hierarchical organization ensures that during each matching iteration, only tracks with comparable occlusion durations compete for detection associations, thereby eliminating the competition bias between long-term and short-term occluded tracks.

3 Experiments

We implement our model and conduct experiments basedon PyTorch toolbox. The YOLOv5s model was trained, validated and tested on a self-built dataset containing 2760 images of tracer particles on water surfaces. The models were trained with an input image size of 640 × 640 pixels for 100 epochs using a batch size of 16. We use the RTX 4090 GPU for training and evaluating all models in our experiments.

Figure 5 demonstrates the tracking performance of floating leaves on water surfaces. As evidenced, our model achieves accurate identification of leaves with varying densities and shapes while maintaining robustness against background interference.

Fig. 5. Example of The Tracking Result

In all experiments, we selected nine test videos capturing flow velocities ranging from 0.49 m/s to 0.88 m/s. Using handheld radar velocimeter measurement as standard velocity and ground truth, we compared the model's per-frame average velocity output with the standard velocity, with representative comparison results shown in Fig. 6. The results demonstrate excellent agreement between model output and standard velocity, achieving an average velocity error of merely 0.0147 m/s across all test sequences. Figure 6 further illustrates that while initial velocity estimates (first few hundred frames) exhibited relatively higher fluctuations, the results gradually stabilized as the tracking algorithm's matching performance improved.

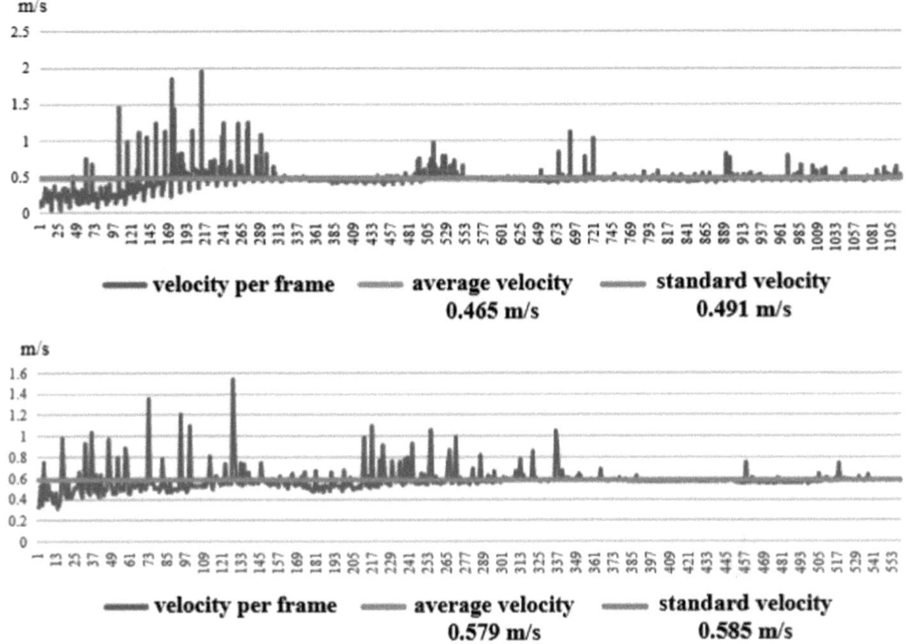

Fig. 6. Comparison between Predicted and Standard Velocities

The quantitative comparison in Table 1 highlights the performance trade-offs between traditional PIV methods and our proposed approach. Cross-Correlation, while achieving a moderate frame rate, exhibits the highest mean velocity error (MVE), likely due to its sensitivity to particle displacement limits and noise. Feature Matching reduces MVE to 0.174 m/s but suffers from significantly lower FPS, making it impractical for real-time applications. Random Forest strikes a balance with 0.196 m/s MVE and 15 FPS, but its reliance on handcrafted features limits both accuracy and speed.

In contrast, our method achieves a remarkably low MVE (0.015 m/s) while maintaining real-time performance (180 FPS), outperforming all baseline techniques. This demonstrates that our approach effectively addresses the limitations of traditional methods by combining high-resolution motion estimation with computational efficiency, making it ideal for dynamic fluid analysis.

Table 1. Quantitative Comparison between PIV Methods.

Methods	MVE(m/s)↓	FPS↑
Cross-Correlation	0.261	50
Feature Matchin	0.174	5
Random Forest	0.196	15
Ours	**0.015**	**180**

Table 2 presents a resolution-wise ablation study of our method, revealing a clear trade-off between accuracy and computational efficiency. At 320 × 320 resolution, the system achieves the highest frame rate but suffers from elevated MVE, indicating insufficient spatial detail for precise motion estimation. Transitioning to 640 × 640 strikes an optimal balance, reducing MVE by 40% while maintaining real-time performance (180 FPS). Further increasing resolution to 960 × 960 and 1280 × 1280 yields diminishing returns in error reduction at the cost of drastically lower FPS, making these configurations suitable only for offline high-precision analysis.

Table 2. Resolution-wise Ablation Analysis of Our Method.

Resolutions	MVE(m/s)↓	FPS↑
320 × 320	0.025	280
640 × 640	0.015	180
960 × 960	0.012	90
1280 × 1280	0.011	50

The 640 × 640 resolution emerges as the "sweet spot", delivering near-minimal error with practical real-time throughput, ideal for field deployments like river flow monitoring.

Disclosure of Interests. The authors have no competing interests to declare that are relevant to the content of this article.

References

1. Ordonez, C.E., Shearman, R.K., Barth, J.A., Welch, P., Erofeev, A., Kurokawa, Z.: Obtaining absolute water velocity profiles from glider-mounted acoustic dopplercurrent profilers. In: 2012 Oceans-Yeosu. pp. 1–7. IEEE (2012)
2. Dong, X., Tan, C., Dong, F.: Water continuous oil-water flow velocity measurement based on continuous waves ultrasonic doppler method. In: 2015 IEEE International Instrumentation and Measurement Technology Conference (I2MTC) Proceedings. pp. 370–375. IEEE (2015)
3. Dong, X., Tan, C., Yuan, Y., Dong, F.: Oil–water two-phase flow velocity measurement with continuous wave ultrasound doppler. Chem. Eng. Sci. **135**, 155–165 (2015)
4. Tamari, S., García, F., Arciniega-Ambrocio, J., Porter, A.: Testing a handheld radar to measure water velocity at the surface of channels. La Houille Blanche **100**(3), 30–36 (2014)
5. Song, J.B., Wu, Y.H.: Statistical distribution of water-particle velocity below the surface layer for finite water depth. Coast. Eng. **40**(1), 1–19 (2000)
6. Chiu, Y., Kuo, Y., et al.: Estimation of water particle velocities by empirical transfer function. China Ocean Eng. **11**(4), 393–410 (1997)
7. Gebremariam, F.T., Tesfay, A.H., Sigtryggsdóttir, F.G., Goitom, H., Lia, L.: Overtopping-induced embankment breaching experiments: state-of-the-art review on measurement and instrumentation. Water **17**(7), 1051 (2025)

8. AlYousif, A., Van Batenburg, T., Memar, S., Melling, G., Hofland, B.: Experimental investigation of overflow on the lee side of river groins due to long-period primary ship-induced waves using particle image velocimetry analysis. J. Waterw. Port Coast. Ocean Eng. **151**(3), 04025004 (2025)
9. Harrison, L.R., Legleiter, C.J., Overstreet, B.T., White, J.S.: Evaluating the potential to quantify salmon habitat via uas-based particle image velocimetry. Water Resour. Res. **61**(3), e2024WR038045 (2025)
10. Jodeau, M., Hauet, A., Paquier, A., Le Coz, J., Dramais, G.: Application and evaluation of ls-piv technique for the monitoring of river surface velocities in high flow conditions. Flow Meas. Instrum. **19**(2), 117–127 (2008)
11. Liu, J., Zhou, X.C., Chen, W., Hong, X.: Breach discharge estimates and surface velocity measurements for an earth dam failure process due to overtopping based on the ls-piv method. Arab. J. Sci. Eng. **44**, 329–339 (2019)
12. Lin, F., Chang, W.Y., Lee, L.C., Hsiao, H.T., Tsai, W.F., Lai, J.S.: Applications of image recognition for real-time water level and surface velocity. In: 2013 IEEE International Symposium on Multimedia. pp. 259–262. IEEE (2013)
13. Wojke, N., Bewley, A., Paulus, D.: Simple online and realtime tracking with a deep association metric. In: 2017 IEEE international conference on image processing (ICIP). pp. 3645–3649. IEEE (2017)

Author Index

P. Umapada et al. (Eds.): ICCPR 2025, CCIS 2811, pp. 475–476, 2026.
https://doi.org/10.1007/978-981-95-8315-7

MIX
Papier aus verantwortungsvollen Quellen
Paper from responsible sources
FSC® C105338

If you have any concerns about our products,
you can contact us on
ProductSafety@springernature.com

In case Publisher is established outside the EU,
the EU authorized representative is:
Springer Nature Customer Service Center GmbH
Europaplatz 3, 69115 Heidelberg, Germany

Printed by Libri Plureos GmbH
in Hamburg, Germany